CONTROL SYSTEM DESIGN

An Introduction to State-Space Methods

BERNARD FRIEDLAND

DOVER PUBLICATIONS, INC.
Mineola, New York

Copyright

Bibliographical Note

This Dover edition, first published in 2005, is an unabridged republication of the work originally published in 1986 by McGraw-Hill, Inc., New York.

Library of Congress Cataloging-in-Publication Data

Friedland, Bernard.
Control system design : an introduction to state-space methods / Bernard Freidland.—Dover ed.
p. cm.
Originally published: New York : McGraw-Hill, c1986.
Includes bibliographical references and index.
ISBN-13: 978-0-486-44278-5 (pbk.)
ISBN-10: 0-486-44278-0 (pbk.)
1. Automatic control. 2. Control theory. 3. System design. 4. State-space methods. I. Title.

TJ213.F865 2005
629.8—dc22

2004063542

Manufactured in the United States by Courier Corporation
44278009 2015
www.doverpublications.com

FOR *Zita, Barbara,*
Irene, and Shelly

CONTENTS

PREFACE

The age of modern control theory was ushered in at the launching of the first sputnik in 1957. This achievement of Soviet technology focused attention of scientists and engineers in general, and the automatic-control community in particular, eastward toward the USSR. By worldwide consensus, Moscow was the appropriate location for the First Congress of the International Federation of Automatic Control in 1960.

In turning their attention to the Soviet Union, control system scientists and engineers discovered a different approach to control theory than the approach with which they were familiar. Differential equations replaced transfer functions for describing the dynamics of processes; stability was approached via the theory of Liapunov instead of the frequency-domain methods of Bode and Nyquist; optimization of system performance was studied by the special form of the calculus of variations developed by Pontryagin instead of by the Wiener-Hopf methods of an earlier era.

In a few years of frenzied effort, Western control theory had absorbed and mastered this new "state-space" approach to control system analysis and design, which has now become the basis of much of modern control theory.

State-space concepts have made an enormous impact on the thinking of those control scientists and engineers who work at the frontiers of technology. These concepts have also been used with notable success in a number of important high-technology projects—the U.S. Apollo project was a highly visible example. Nevertheless, the majority of control systems implemented at the present time are designed by methods of an earlier era.

Many control engineers schooled in the earlier methods have felt that the modern state-space approach is mathematically esoteric and more suited to advanced graduate research than to the design of practical control systems. I can sympathize with the plight of the engineer who has waded through a morass of mathematics with the hope of learning how to solve his practical problem only to return empty-handed; I have been there too. One thesis of this book is that state-space methods can be presented in a style that can be grasped by the engineer who is more interested in using the results than in proving them. Another thesis is that the results *are useful.* I would even go so far as to say that if one had to choose between the frequency-domain methods of the past and the state-space methods of the present, then the latter are the better choice. Fortunately, one does not need to make the choice: both methods are useful and complement each other. Testimony to my continued faith in frequency-domain analysis is a long chapter, Chap. 4, which presents some of the basic methods of that approach, as a review and for those readers who may not be knowledgeable in these methods.

This book is addressed not only to students but also to a general audience of engineers and scientists (e.g., physicists, applied mathematicians) who are interested in becoming familiar with state-space methods either for direct application to control system design or as a background for reading the periodical literature. Since parts of the book may already be familiar to some of these readers, I have tried, at the expense of redundancy, to keep the chapters reasonably independent and to use customary symbols wherever practical. It was impossible, of course, to eliminate all backward references, but I hope the reader will find them tolerable.

Vectors and matrices are the very language of state-space methods; there is no way they can be avoided. Since they are also important in many other branches of technology, most contemporary engineering curricula include them. For the reader's convenience, however, a summary of those facts about vectors and matrices that are used in the book is presented in the Appendix.

Design is an interplay of science and art—the instinct of using exactly the right methods and resources that the application requires. It would be presumptuous to claim that one could learn control system design by reading this book. The most one could claim is to have presented examples of how state-space methods could be used to advantage in several representative applications. I have attempted to do this by selecting fifteen or so examples and weaving them into the fabric of the text and the homework problems. Several of the examples are started in Chap. 2 or 3 and taken up again and again later in the book. (This is one area where backward references are used extensively.) To help the reader follow each example on its course through the book, an applications index is furnished (pages 503 to 505). Many of the examples are drawn from fields I am best acquainted with: aerospace and inertial instrumentation. Many other applications of state-space methods have been studied and implemented: chemical process control, maritime operations, robotics, energy systems, etc. To demonstrate the wide applicability of state-space methods, I have included examples from some of these fields, using dynamic models and data selected from the periodical literature. While not personally familiar with these applications, I have endeavored to emphasize some of their realistic aspects.

The emphasis on application has also motivated the selection of topics. Most of the attention is given to those topics that I believe have the most practical utility. A number of topics of great intrinsic interest do not, in my judgment, have the practical payoff commensurate with the effort needed to learn them. Such topics have received minimal attention. Some important concepts are really quite simple and do not need much explaining. Other concepts, although of lesser importance, require more elaborate exposition. It is easy to fall into the trap of dwelling on subjects in inverse proportion to their significance. I have tried to avoid this by confining the discussion of secondary topics to notes at the end of each chapter, with references to the original sources, or to the homework problems.

Much of practical engineering design is accomplished with the aid of computers. Control systems are no exception. Not only are computers used for on-line, real-time implementation of feedback control laws—in applications as diverse as aircraft autopilots and chemical process controls—but they are also used extensively to perform the design calculations. Indeed, one of the major advantages of state-space design methods over frequency-domain methods is that the former are better suited to implementation by digital computers. Computer-aided design, however, creates a dilemma for the author. On the one hand, he wants to make the concepts understandable to a reader who doesn't have a computer. On the other hand the full power of the method is revealed only through applications that require the use of a computer. My decision has been a compromise. I have tried to keep the examples in the text simple enough to be followed by the reader, at least part of the way, without recourse to a computer for numerical calculation. There are a number of homework problems, however, some of which continue examples from the text, for which a computer is all but essential.

The reader is certainly not expected to write the software needed to perform the numerical calculations. During the past several years a number of organizations have developed software packages for computer-aided control system design (CACSD). Such software is available for mainframes and personal computers at prices to suit almost any budget and with capabilities to match. Several of these packages would be adequate for working the homework problems that require a computer and for other applications. Anyone with more than a casual interest in state-space methods would be well advised to consider acquiring and maintaining such software.

The education of most engineers ends with the bachelor's or master's degree. Hence, if state-space methods are to be widely used by practicing engineers, they must be included in the undergraduate or first-year graduate curriculum—they must not be relegated to advanced graduate courses. In support of my commitment to state-space methods as a useful tool for practicing engineers, I have endeavored to teach them as such. A number of years ago I presented some introductory after-hours lectures on this subject to fellow employees at the Kearfott Division of The Singer Company. These lectures served as the basis of an undergraduate elective I have been teaching at the Polytechnic Institute of New York. For want of a more suitable textbook, I have been distributing hard copies of the overhead transparencies used in the lectures. It occurred to me that the material I had assembled in these overhead transparencies was the nucleus of the book I had needed but had been unable to locate. And so I embarked upon this project.

It is a pleasure to acknowledge the contributions made by a number of individuals to this project. Most of the manuscript was patiently and expertly typed by Win Griessemer. Additional typing and editorial assistance, not to mention moral support, was provided when needed most by my wife and daughters, to whom this book is dedicated. My associates at The Singer Company, Dave Haessig, Appa Madiwale, Jack Richman, and Doug Williams between them read most of the manuscript, found many errors large and small, and offered a number of helpful suggestions. A preliminary version of this book was used as a text for my undergraduate course at the Polytechnic Institute of New York and for a similar course, taught by Professor Nan K. Loh, at Oakland University (Michigan). The students in these courses provided additional feedback used in the preparation of the final manuscript.

The vision of this book has long been in my mind's eye. To all those named above, and others not named but not forgotten, who have helped me realize this vision, my gratitude is boundless.

Bernard Friedland

CHAPTER
ONE

FEEDBACK CONTROL

1.1 THE MECHANISM OF FEEDBACK

No mechanism in nature or technology is more pervasive than the mechanism of feedback.

By the mechanism of feedback a mammal maintains its body temperature constant to within a fraction of a degree even when the ambient temperature fluctuates by a hundred degrees or more.

Through feedback the temperature in an oven or in a building is kept to within a fraction of a degree of a desired setting even though the outside temperature fluctuates by 20 or 30 degrees in one day.

An aircraft can maintain its heading and altitude and can even land, all without human intervention, through feedback.

Feedback is the mechanism that makes it possible for a biped to stand erect on two legs and to walk without falling.

When the Federal Reserve Bank exercises its controls in the interest of stabilizing the national economy, it is attempting to use feedback.

When the Mayor of New York City asks, "How'm I doing?" he is invoking the mechanism of feedback.

Hardly a process occurring in nature or designed by man does not, in one way or another, entail feedback.

Because feedback is ubiquitous, it is taken for granted except when it is not working properly: when the volume control of a public address system in an auditorium is turned up too high and the system whistles; then everyone becomes aware of "feedback." Or when the thermostat in a building is not working properly and all the occupants are freezing, or roasting.

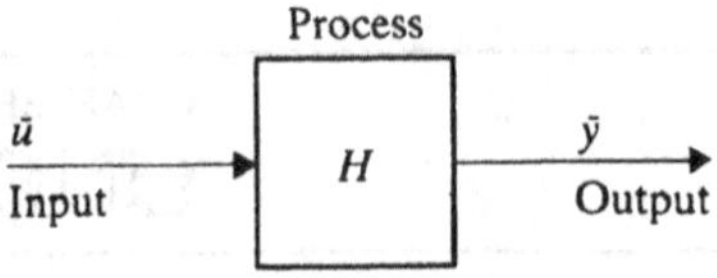

Figure 1.1 Open-loop control. Input $\bar{u}$ is selected to produce desired output $\bar{y}$.

To get an appreciation of the mechanism of feedback, suppose that there is a process H that we wish to control. Call the input to the process u and the output from the process y. Suppose that we have a complete description of the process: we know what the output y will be for any input. Suppose that there is one particular input, say $\bar{u}$, which corresponds to a specified, desired output, say $\bar{y}$. One way of controlling the process so that it produces the desired output $\bar{y}$ is to supply it with the input $\bar{u}$. This is "open-loop control." (Fig. 1.1.) A billiard player uses this kind of control. With an instinctive or theoretical knowledge of the physics of rolling balls that bounce off resilient cushions, an expert player knows exactly how to hit the cue ball to make it follow the planned trajectory. The blow delivered by the cue stick is an open-loop control. In order for the ball to follow the desired trajectory, the player must not only calculate exactly how to impart that blow, but also to execute it faultlessly. Is it any wonder that not everyone is an expert? On the other hand, suppose one wants to cheat at billiards by putting some kind of sensor on the cue ball so that it can always "see" the target—a point on another ball or a cushion—and by some means can control its motion—"steer"—to the target. Finally, put a tiny radio in the ball so that the cheater can communicate the desired target to the cue ball. With such a magic cue ball the cheater cannot but win every game. He has a cue ball that uses the mechanism of feedback.

The magic cue ball has two of the characteristics common to every feedback system: a means of monitoring its own behavior ("How'm I doing") and a means of correcting any sensed deviation therefrom. These elements of a feedback control system are shown in Fig. 1.2. Instead of controlling the output of the process by picking the control signal $\bar{u}$ which produces the desired $\bar{y}$, the control signal u is generated as a function of the "system error," defined as the difference between the desired output $\bar{y}$ and the actual output y

$$e = \bar{y} - y \tag{1.1}$$

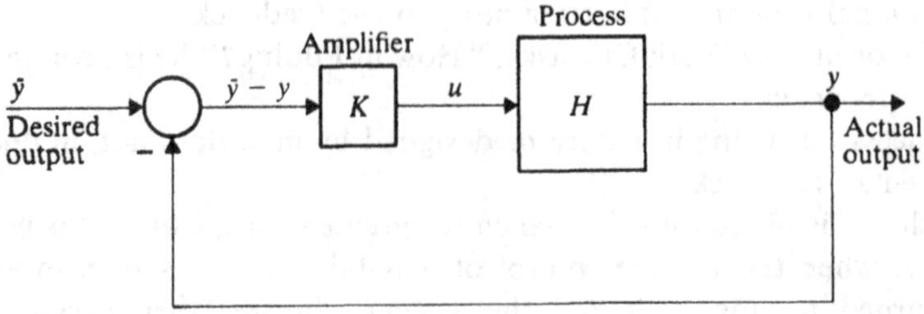

Figure 1.2 Feedback control system. Input u is proportional to difference between desired and actual output.

This error, suitably amplified, as shown by the output of the box labeled "amplifier," is the input to the process.

Suppose that the operation of the process under control can be represented by a simple algebraic relation

$$y = Hu \tag{1.2}$$

and that the amplifier can similarly be described

$$u = Ke \tag{1.3}$$

Combine (1.1), (1.2), and (1.3) into the single relation

$$y = HKe = HK(\bar{y} - y)$$

Solve for y and obtain

$$y = \frac{HK}{1 + HK}\bar{y} \tag{1.4}$$

Although the output y is not exactly equal to the desired output $\bar{y}$, if the amplifier "gain" K is large enough (i.e., $HK \gg 1$) then

$$y \simeq \bar{y} \tag{1.5}$$

We can make the actual output y approach the desired output as closely as we wish simply by making the gain K large enough. Moreover, this result holds, for any desired output! We don't have to know $\bar{y}$ in advance as we did in determining the open-loop control $\bar{u}$. And, even more remarkably, this result holds independent of the process—it doesn't matter what H is. In fact H can even change over the course of time without affecting the basic result. These are among the wonders of feedback and help to explain why it is so useful.

Unfortunately, nature is not as simple as the above analysis would suggest; if it were, there would be no need for this book. The problem is that the process whose input is u and whose output is y cannot be represented by an algebraic equation as simple as (1.2). Because of the process *dynamics*, the relationship between the output and the input is much more complex than (1.2).

The effect of dynamics on the behavior of a feedback system is easily illustrated by a simple example. Suppose that the output of system H is an exact replica of the input, except delayed by a small amount of time, say τ:

$$y(t) = u(t - \tau) \tag{1.6}$$

for any input $u(t)$. (See Fig. 1.3.) We assume that (1.1) and (1.2) continue to hold for every time t. Then

$$u(t - \tau) = Ke(t - \tau) = K[\bar{y}(t - \tau) - y(t - \tau)] \tag{1.7}$$

Substitute (1.7) into (1.6) to obtain

$$y(t) = K[\bar{y}(t - \tau) - y(t - \tau)] \tag{1.8}$$

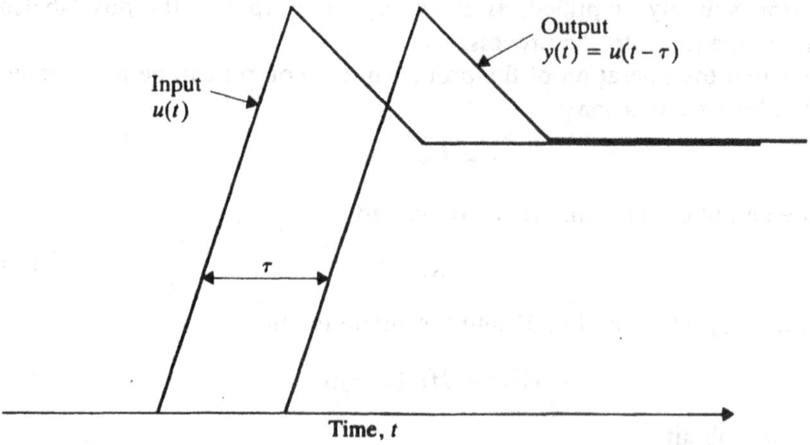

Figure 1.3 Example of input and output of a process in which the output is an exact but delayed replica of the input.

This is an example of a "difference equation" and describes how $y(t)$ evolves as time t increases. Difference equations are a common way of describing the dynamic behavior of discrete-time (sampled-data) systems but they are not studied extensively in this book. This equation, however, is so simple that it can be solved without any theory.

Suppose the desired output is a "unit step" as shown in Fig. 1.4(a):

$$\bar{y}(t) = \begin{cases} 0 & \text{for } t < 0 \\ 1 & \text{for } t > 0 \end{cases} \tag{1.9}$$

and also suppose that $y(t) = 0$ for $t < 0$. Then, by (1.8) and also by looking at Fig. 1.3, we can see that there is no output for the first τ units of time. But the input to the process

$$u(t) = K(1 - 0) = K \qquad \text{for } 0 < t < \tau$$

After an interval of τ units of time the output starts to appear as shown in Fig. 1.4(b)

$$y(t) = K \qquad \text{for } \tau < t < 2\tau$$

For the next τ units of time the input to the process is

$$u(t) = K(1 - K) = K - K^2 \qquad \text{for } \tau < t < 2\tau$$

This is the value of $y(t)$ for the next τ units of time, i.e., for $2\tau < t < 3\tau$. Proceeding in this fashion we see that

$$y(t) = K - K^2 + K^3 + \cdots + (-1)^{n-1}K^{n-1} \qquad \text{for } n\tau < t < (n+1)\tau \tag{1.10}$$

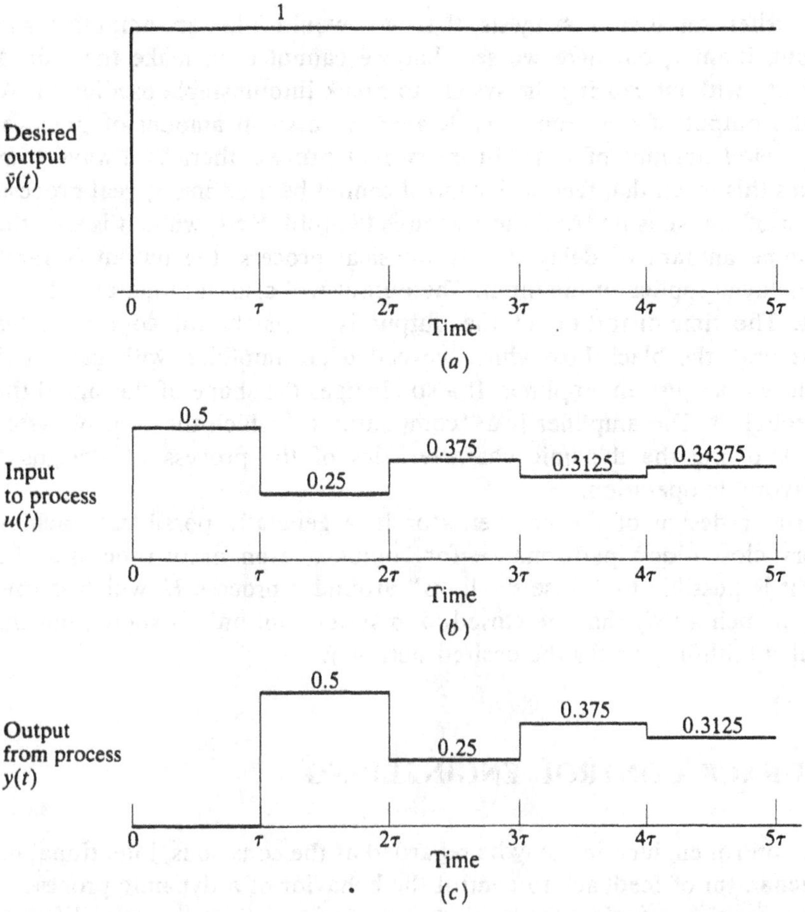

Figure 1.4 Response of feedback control system to $\bar{y} = 1$ when output is a replica of input delayed by τ with gain $K = \frac{1}{2}$. (*a*) Desired output $\bar{y}(t)$; (*b*) Input to process $u(t)$; (*c*) Output from process $y(t)$.

If K is less than 1, then (1.10) implies that $y(t)$ will eventually converge to a limit:

$$\lim_{t\to\infty} y(t) = K - K^2 + K^3 - K^4 + \cdots = \frac{K}{1+K} \tag{1.11}$$

If K is exactly equal to 1, the output $y(t)$ will flip between 0 and 1 indefinitely. And if $K > 1$, the output will flip between positive and negative values, ever increasing in amplitude, and ultimately become infinite.

Thus we see that the amplification factor (or *gain*) K of the amplifier cannot be made as large as 1 if we want the output to stabilize. (Also, as K approaches 1, the output is only half the value of the input. This can be corrected, however, by multiplying the desired output by $(1 + K)$ before comparing it with the actual output.)

Our earlier discussion suggests that we would like an amplifier gain approaching infinity, but here we see that we cannot even make the gain as large as unity without causing the system to break into unstable oscillation. All because the output of the process is delayed by a small amount of time—an *arbitrarily small* amount of time. In every real process there is always some delay. Does this mean that feedback control cannot be used in any real process? The answer of course is no. And the reason is twofold. First, while it is true that there is some amount of delay in any physical process, the output is rarely simply a delayed replica of the input. The output will also not look exactly like the input. The time-distortion of the output is a benefit for control system design. Second, the black box which we called an amplifier, with gain K, is usually more than just an amplifier. It also changes the shape of the signal that passes through it. The amplifier is a "compensator," which the control system engineer, knowing the dynamic characteristics of the process H, designs to achieve favorable operation.

By proper design of the compensator it is generally possible to achieve satisfactory closed-loop performance for complex, even nasty processes. For example, it is possible to "close the loop" around a process H, which is itself unstable, in such a way that the closed-loop system not only is stable, but that the output y faithfully tracks the desired output $\bar{y}$.

1.2 FEEDBACK CONTROL ENGINEERING

Feedback control engineering may be regarded as the conscious, intentional use of the mechanism of feedback to control the behavior of a dynamic process.

The course that a typical feedback system design follows is exemplified by the hypothetical magic cue ball of the previous section. Suppose one has a client who comes prepared to pay the expense of the design and construction of such a technical marvel. How would one proceed?

The performance requirements are easy to imagine. To escape detection, the entire system must fit inside a hollowed-out ball and its weight and inertia must exactly equal those of the material removed. If the cue ball is to be able to home-in on its target, it should be able to sense its position relative to the target. How can this be accomplished? Perhaps a miniature infrared sensor? Will the sensor be able to discriminate between the actual target and another cue ball that resembles the target? Perhaps the billiard table can have a hidden means of generating an electric or magnetic field that is altered by the presence of the balls and this information can be transmitted to the cue ball.

Suppose we have tentatively solved the problem of sensing the motion of the cue ball. Next we need some means to alter its trajectory. Can we use tiny, barely visible gas jets? Perhaps we can use a movable weight inside the cue ball which will displace the center of mass from the geometric center and hence, with the aid of gravity, create moments which, when combined with friction, can change the trajectory. Maybe we can use tiny, almost imperceptible bumps on the surface that can be moved to change the course of the ball.

Conception of the means of measuring the behavior of the process—the cue ball—and affecting or altering its behavior is the first stage of control system design. Without a doubt, this is the stage that requires the greatest degree of inventiveness and understanding of what can be achieved at the current level of technology and at what price.

If the project has not been abandoned at the first stage for want of suitable technological means, the next step is to acquire or design the sensors that have been chosen to measure the motion of the vehicle relative to the target and the actuation means that have been selected to alter the motion of the cue ball.

After the hardware is all selected, the final stage of the design is begun. This is the stage in which it is decided how the feedback loop or loops are to be closed: how the data from the sensor or sensors are to be processed before being sent to the actuator. It is at this stage that the designer decides what the block box labeled "amplifier" must really do in order for the closed loop system to operate properly. This step is the design of the "control law" or "control algorithm."

This last stage of control system design is the entire content of control theory. By the time control theory enters the picture, the system concept has already been established and the control hardware has already been selected. The whole apparatus of control theory, it would appear, deals with only a small, insignificant fraction of the overall problem. In this light, the effort devoted to the development of control theory—the subject matter of this book—hardly seems worth the effort.

The magic cue ball design problem, however, does not represent the typical design problem. Although it is true enough that the control concept must be defined and the hardware must be selected for every control system design, not every design requires such inventiveness. In most cases, the process to be controlled is only slightly different from yesterday's. Today's control hardware is only slightly different from yesterday's, probably better (more accurate, cheaper, and more reliable). Hence the first design steps are taken almost unconsciously. The engineer, not without justification, forgets about the first two steps and believes that the control system design begins at the point that it is almost over.

If today's process and control hardware are not changed much from yesterday's, why can't one simply use yesterday's control law? Oftentimes, one can. Most control laws are probably designed by this very method: Take yesterday's control law and modify its parameters to account for the difference between yesterday's hardware and today's.

But the procedure is not always satisfactory. The new process may not be sufficiently similar to the old one. The new control hardware, although improved (say digital instead of analog) may have different characteristics that cannot be overlooked. And finally, the customer may demand a higher level of performance than yesterday's system was able to deliver.

1.3 CONTROL THEORY BACKGROUND

This book is concerned with the third and final stage of control system engineering—the stage in which the dynamic characteristics of the compensator are designed, after the control concept has been established, after the hardware (sensors and actuators) have been selected, after the performance requirements have been determined.

This aspect of control system engineering is generally called control "*theory*." The term "theory" is appropriate for several reasons. First, it is essentially mathematical in content, and mathematics is often equated to theory. Second, it deals not with the actual devices but with their idealized (theoretical, i.e., mathematical) models. Third, it constitutes a systematic body of knowledge: theorems, design algorithms, graphical methods, and the like which can be applied to control systems independent of the specific technology used in the practical implementation.

The history of control theory can be conveniently divided into three periods. The first, starting in prehistory and ending in the early 1940s, may be termed the *primitive* period. This was followed by a *classical* period, lasting scarcely 20 years, and finally came the *modern* period which includes the content of this book.

The term *primitive* is used here not in a pejorative sense, but rather in the sense that the theory consisted of a collection of analyses of specific processes by mathematical methods appropriate to, and often invented to deal with, the specific processes, rather than an organized body of knowledge that characterizes the classical and the modern period.

Although feedback principles can be recognized in the technology of the Middle Ages and earlier, the intentional use of feedback to improve the performance of dynamic systems was started at around the beginning of the industrial revolution in the late 18th and early 19th centuries. The benchmark development was the ball-governor invented by James Watt to control the speed of his steam engine. Throughout the first half of the 19th century, engineers and "mechanics" were inventing improved governors. The theoretical principles that describe their operation were studied by such luminaries of 18th and 19th century mathematical physics as Huygens,[1] Hooke,[2] Airy,[3] and Maxwell.[4] By the mid 19th century it was understood that the stability of a dynamic system was determined by the location of the roots of the algebraic characteristic equation. Routh[5] in his Adams Prize Essay of 1877 invented the stability algorithm that bears his name.

Mathematical problems that had arisen in the stability of feedback control systems (as well as in other dynamic systems including celestial mechanics) occupied the attention of early 20th century mathematicians Poincaré and Liapunov, both of whom made important contributions that have yet to be superseded.

Development of the gyroscope as a practical navigation instrument during the first quarter of the 20th century led to the development of a variety of autopilots for aircraft (and also for ships). Theoretical problems of stabilizing these systems and improving their performance engaged various mathematicians of the period. Notable among them was N. Minorsky[6] whose mimeographed notes on nonlinear systems was virtually the only text on the subject before 1950.

The *classical* period of control theory begins during World War II in the Radiation Laboratory of the Massachusetts Institute of Technology. (See Note 1.1.) The personnel of the Radiation Laboratory included a number of engineers, physicists, and mathematicians concerned with solving engineering problems that arose in the war effort, including radar and advanced fire control systems. The laboratory that was assigned problems in control systems included individuals knowledgeable in the frequency response methods, developed by people such as Nyquist and Bode for communication systems, as well as by engineers familiar with other techniques. Working together, they evolved a systematic control theory which is not tied to any particular application. Use of frequency-domain (Laplace transform) methods made possible the representation of a process by its transfer function and thus permitted a visualization of the interaction of the various subsystems in a complex system by the interconnection of the transfer functions in the block diagram. The block diagram contributed perhaps as much as any other factor to the development of control theory as a distinct discipline. Now it was possible to study the dynamic behavior of a hypothetical system by manipulating and combining the black boxes in the block diagram without having to know what goes on inside the boxes.

The classical period of control theory, characterized by frequency-domain analysis, is still going strong, and is now in a "neoclassical" phase—with the development of various sophisticated techniques for multivariable systems. But concurrent with it is the *modern* period, which began in the late 1950s and early 1960s.

State-space methods are the cornerstone of *modern control theory.* The essential feature of state-space methods is the characterization of the processes of interest by differential equations instead of transfer functions. This may seem like a throwback to the earlier, primitive, period where differential equations also constituted the means of representing the behavior of dynamic processes. But in the earlier period the processes were simple enough to be characterized by a *single* differential equation of fairly low order. In the modern approach the processes are characterized by systems of coupled, first-order differential equations. In principle there is no limit to the order (i.e., the number of independent first-order differential equations) and in practice the only limit to the order is the availability of computer software capable of performing the required calculations reliably.

Although the roots of modern control theory have their origins in the early 20th century, in actuality they are intertwined with the concurrent development of computers. A digital computer is all but essential for performing the calculations that must be done in a typical application. Only in the simplest examples can the calculations be performed without a digital computer. The fact that calculations for simple applications can be done manually can sometimes be misleading, because the design for such simple applications can usually be achieved more efficiently by classical frequency-domain methods. State-space methods prove their mettle in applications which are intractable by classical methods.

Digital computers of even modest capability can crunch out the numerical calculations of the design for a complicated system in a few seconds or minutes. It is thus very easy to arrive at a design which is correct numerically but not practical. (There is no inherent reason why this can't also happen with a design based on classical methods. But because of the labor entailed in achieving the design, the engineer is more likely to check intermediate results for reasonability rather than to wait for the final design to emerge as a unit.) The realization that there may be practical problems with a computer-aided design ought to make the designer especially cautious: both in making certain that the computer has good data to begin with, i.e., a proper model of the process to be controlled, and in testing the proposed design by all appropriate means including simulation.

1.4 SCOPE AND ORGANIZATION OF THIS BOOK

The vision of the early pioneers of modern control theory was that it would provide a single, unified framework for all feedback control systems: linear and nonlinear, continuous-time, and discrete-time, fixed and time-varying. That vision is a chimera. A few results of broad generality have been achieved, but for the most part the vaunted general theory has been achieved only for linear systems, and furthermore, the required calculations can be performed only for time-invariant, linear systems. This is nevertheless no mean accomplishment, because the theory that does exist is still able to cope with any design problem that the classical theory can cope with, because the frequency-domain approach is entirely predicated on linear, time-invariant models.

Being an introduction to state-space methods, this book does not go beyond systems that can be characterized by linear, time-invariant models. (The sole exception is a missile guidance system which has time-varying dynamics that are so simple that they can easily be handled without the need for any special theory.)

The first few chapters are intended as an introduction to the use of state-space methods for characterizing the behavior of dynamic systems. In particular, in Chap. 2, we learn how linear state-space models can be set up for various kinds of physical processes, and in Chap. 3 we study the basic properties of such models: such things as the state-transition matrix, the resolvent, the characteristic equation. Although the properties of linear, time-invariant systems can be gleaned without use of the Laplace transform, they are more readily obtained through its use. Since most readers of this book are familiar with the basic theory of Laplace transforms, we see no reason for not making use of them. We also see no reason for abandoning classical, frequency-domain methods and the insights they provide. Hence, in Chap. 4, we provide a review of frequency-domain analysis, emphasizing where possible the connection with state-space methods. Notwithstanding the length of Chap. 4, it is still only an overview; the reader is assumed to be already somewhat familiar with the material or prepared to consult one of the standard textbooks in the field to gain a more comprehensive understanding.

Controllability and observability theory, one of the earliest unique achievements of modern control theory, is the subject of Chap. 5.

The first five chapters set the stage for the use of state-space methods for control system design. These are followed by three which show how state-space methods can be used in design. Chapter 6 is concerned with design of controllers that use "full-state" feedback, i.e., design under the assumption that all of the state variables are accessible to measurement, if needed for the control law. This is an unrealistic assumption, and Chap. 7 shows how to design observers which are dynamic systems, the inputs to which are the measured inputs and outputs of the process under control. The state of the observer is an estimate of the state of the process under control. Chapter 8, which concludes the three introductory chapters on design, shows how the full-state feedback control of Chap. 6 can be combined with the observer of Chap. 7, to finally provide the design of a compensator which is typically the goal of the control system designer. Chapter 8 is also concerned with the robustness of compensators designed by the methods of these three chapters: it addresses the question of how well the compensator will work if the mathematical model used in the design is not exactly matched to the actual physical process.

Although the methods of Chaps. 6 through 8 constitute a set of procedures for designing compensators for controllable and observable processes, they do not of themselves arrive at *optimum* designs. Optimization of these designs is the subject of Chaps. 9 through 11.

In Chap. 9, we learn how to optimize the gain matrix of the full-state control law by choosing it to minimize a quadratic integral performance criterion. The weighting matrices in the integral are putatively chosen to correspond, at least approximately, to physical performance requirements. Computing the gain matrix is shown to entail solving for the appropriate (matrix) root of a matrix quadratic equation which has come to be known as the *algebraic Riccati equation.* Numerical solution of this equation is a job for the digital computer.

The selection of the optimum gain for the observer is formulated as a statistical problem: to find the observer gain matrix that minimizes the estimation error variance under the hypothesis that the process is excited by white noise with a known spectral density matrix and that the observations are corrupted with white noise also with known spectral density. The resulting observer gain matrix is also the solution to an algebraic Riccati equation which has a structure quite similar to that of the algebraic Riccati equation for the optimum controller. The theory for the optimum observer, also known as a

Kalman filter, is developed in Chap. 11. A minimal overview of the statistical prerequisites to Chap. 11 is presented in Chap. 10.

Matrices and vectors are the very language of state-space methods. By now they are so commonplace in every branch of technology, that we can hardly imagine a reader unfamiliar with them. Nevertheless we have included an appendix in which the basic facts about matrices are summarized, without an attempt at proofs. If the reader wants proofs, there are innumerable texts available for that purpose.

One of the objectives of this book has been to illustrate the use of state-space methods in various aspects of system analysis and design by means of examples that have some relationship to real-world applications. In line with that objective a number of "running examples" are provided. Each example occurs in several places in the text: to exemplify development of the model of a system an example may appear in Chap. 2. The same example may appear again to illustrate the calculation of open-loop response, and in various aspects of control system design. References to earlier and later appearances of the same example are given each time an example reappears. In addition, an index of the examples is given at the end of the book. By use of this index, the reader should be able to locate all references to an example and thereby trace the course of its development through the book. Thus each example constitutes something of a "case study." Some of the examples are the subject of homework problems: the reader thereby actively participates in the development of the case study.

NOTES

Note 1.1 Historical antecedents

Before World War II, feedback control systems were largely mechanical. The feedback paths, not generally identified as such, were implemented by means of ingenious combinations of springs, dashpots, pneumatic devices, and similar gadgets. The electrical components that were used were magnets and perhaps resistors. Almost every new control system represented a genuine invention and many were in fact patented. Nothwithstanding the ingenuity that these inventions required[7, 8] the variety of functions that could be achieved with these devices was (and still is) extremely limited. Thus a mathematical theory of the function that a feedback compensator must perform would have been of little practical value, since no means of implementing the function was available. Electronic technology of the era was represented by large vacuum tubes enclosed in fragile glass envelopes, massive inductors and capacitors, and similar bulky and unreliable hardware. A few electrical components were used in the 1920s and 1930s,[9] but a control system designer proposing to use "electronics" to implement the feedback loops of a control system would very likely have been the object of ridicule. Before World War II, the only industry with any serious interest in electronics was the communications industry—radio and telephony.

Electronic technology underwent a major transformation during the war. Electronic components (i.e., tubes) became smaller and more reliable, and the functions that electronic systems were able to perform became more sophisticated as a result of concerted efforts by scientists and engineers and mathematicians working together in the war effort. A notable wartime development, among others, was radar.

At the Radiation Laboratory, established at the Massachusetts Institute of Technology in 1940 to aid in the war effort, one of the technical groups was concerned with control—"servomechanism"—problems. Members of the group included physicists, communication engineers, mechanical engineers, mathematicians, and technicians. Each member brought different insights to the problems assigned to the group. The resulting collaboration laid the groundwork for the second, *classical* phase of control technology, in which the frequency-domain methods of communication engineering (transfer function analysis, Bode and Nyquist diagrams, and the like) were applied to the analysis and design of control systems.

After the war, the results of the work at the Radiation Laboratory were published in a multivolume series. The volume by James, Nichols, and Phillips[10] constituted the exposition of the classical frequency-domain methodology developed at that laboratory.

With the war concluded, research in control theory was continued along these lines at a number of universities. One of the centers of research in control theory was at Columbia University. Under the leadership of John R. Ragazzini, much of the classical (i.e., Z transform) theory for sampled-data systems was developed there during the decade of the 1950s. Into the hospitable environment that Ragazzini fostered at Columbia was welcomed an iconoclastic young graduate student who preached against the frequency-domain methods and taught a new doctrine: state-space. That student was Rudolf E. Kalman.

Kalman argued with increasing success that the frequency-domain methods developed for communication systems were not the most appropriate for control systems: the methods were not readily adaptable to time-varying and nonlinear systems and even for linear, time-invariant systems they dealt with the wrong problems. Moreover, Kalman taught, the classical methods of analysis and design obscured the physical nature of the dynamic variables which the state-space methods preserved.

The ranks of adherents to Kalman's state-space approach swelled during the decade of the 1960s and the modern era of control theory thus became firmly established. But not everyone was persuaded that frequency-domain methods had been superseded. Debates, sometimes acrimonious, over the merits of the two approaches, which started then, continue unto the present time.

REFERENCES

1. Huygens, G., *Horologium Oscillatorium (1673) in Oeuvres*, complètes, Nijhoff, Amsterdam, vol. 17-18, 1932.
2. Hooke, R., "Lampas, or Description of Some Mechanical Improvements of Lamps and Waterpones Together with Some Other Physical and Mechanical Discoveries," *Proc. Royal Society (London)*, vol. 8, 1831.
3. Airy, G. B., "On the Regulator of the Clockwork for Effecting Uniform Movement of the Equatoreals," *Memoirs, Royal Astronomical Society*, vol. 11, 1840, pp. 249-267.
4. Maxwell, J. C., "On Governors," *Philosophical Magazine*, vol. 35, 1868, pp. 385-398.
5. Routh, E. J., *A Treatise on the Stability of a Given State of Motion*, Macmillan & Co., London, 1877.
6. Minorsky, N., *Nonlinear Oscillations*, D. Van Nostrand, New York, 1962.
7. Fuller, A. T., "The Early Development of Control Theory," *Trans. ASME (J. Dynamic Systems, Measurement & Control)*, vol. 98G, no. 2, June 1976, pp. 109-118.
8. Fuller, A. T., "The Early Development of Control Theory, II," *Trans. ASME (J. Dynamic Systems, Measurement & Control)*, vol. 98G, no. 3, September 1976, pp. 224-235.
9. Oppelt, W., "A Historical Review of Autopilot Development, Research, and Theory in Germany," *Trans. ASME (J. Dynamic Systems, Measurement & Control)*, vol. 98G, no. 3, September 1976, pp. 215-223.
10. James, H. M., Nichols, N. B., and Phillips, R. S., *Theory of Servomechanisms* (MIT Radiation Laboratory Series, vol. 25), McGraw-Hill Book Co., New York, 1947.

CHAPTER

TWO

STATE-SPACE REPRESENTATION OF DYNAMIC SYSTEMS

2.1 MATHEMATICAL MODELS

The most important task confronting the control system analyst is developing a mathematical model of the process of interest. In many situations the essence of the analytical design problem is in the modeling: once that is done the rest of the analysis falls quickly into place.

The control system engineer is often required to deal with a system having a number of subsystems the physical principles of which depend on entirely different types of physical laws. A chemical process, for example, may comprise a chemical reactor, the dynamics of which are the subject of chemical kinetic theory, a heat exchanger which is governed by thermodynamic principles, and various valves and motors the dynamics of which depend on the physics of mechanical and electrical systems. The control of a typical aircraft entails an understanding of the interaction between the airframe governed by principles of aerodynamics and structural dynamics, the actuators which are frequently hydraulic or electrical, and the sensors (gyroscopes and accelerometers) which operate under laws of rigid body dynamics. And, if the human pilot of the aircraft is to be considered, aspects of physiology and psychology enter into the analysis.

One of the attractions of control system engineering is its interdisciplinary content. The control system engineer sees the "big picture" in the challenge to harmonize the operation of a number of interconnected subsystems, each of which operates under a different set of laws. But at the same time the control system engineer is almost totally dependent on the other disciplines. It is simply impossible to gain a sufficient understanding of the details of each of the subsystems in a typical control process without the assistance of individuals having an intimate understanding of these subsystems. These individuals often have the knowledge that the control system analyst requires, but are not accustomed to expressing it in the form that the analyst would like to have it. The analyst must be able to translate the information he receives from others into the form he needs for his work.

The analyst needs *mathematical models* of the processes in the system under study: equations and formulas that predict how the various devices will behave in response to the inputs to these devices. From the viewpoint of the systems analyst each device is the proverbial "black box," whose operation is governed by appropriate mathematical models. The behavior of the overall process is studied and controlled by studying the interaction of these black boxes.

There are two modeling and analysis approaches in customary use for linear systems: the transfer-function or frequency-domain approach, to be discussed in Chap. 4, and the state-space approach which is the subject of the present chapter.

The feature of the state-space approach that sets it apart from the frequency-domain approach is the representation of the processes under examination by systems of first-order differential equations. This method of representation may appear novel to the engineer who has become accustomed to thinking in terms of transfer functions, but it is not at all a new way of looking at dynamic systems. The state-space is the mode of representation of a dynamic system that would be most natural to the mathematician or the physicist. Were it not that much of classical control theory was developed by electrical engineers, it is arguable that the state-space approach would have been in use much sooner.

State-space methods were introduced to the United States engineering community through the efforts of a small number of mathematically oriented engineers and applied mathematicians during the late 1950s and early 1960s. The spiritual father of much of this activity was Professor Solomon Lefschetz who organized a mathematical systems research group at the Research Institute of Advanced Studies (RIAS) in Baltimore, Md. Lefschetz, already a world-famous mathematician, brought together a number of exceptionally talented engineers and mathematicians committed to the development of mathematical control theory. At Columbia University another group, under the aegis of Professor J. R. Ragazzini, and including R. E. Kalman and J. E. Bertram among others, was also at work developing the foundations of modern control theory.

In the Soviet Union there was less of an emphasis on transfer functions than on differential equations. Accordingly, many of the earliest uses of state-space methods were made by investigators in the Soviet Union. Much of the activity in the United States during the late 1950s entailed translation of the latest Russian papers into English. The Moscow location of the First Congress of the International Federation of Automatic Control (IFAC) in 1960 was entirely appropriate, and provided the first major opportunity for investigators from all over the world to meet and exchange ideas. Although the IFAC congress was concerned with components and applications as well as with control theory, much of the interest of the meeting was on the newest theoretical developments.

2.2 PHYSICAL NOTION OF SYSTEM STATE

The notion of the state of a dynamic system is a fundamental notion in physics. The basic premise of newtonian dynamics is that the future evolution of a dynamic process is entirely determined by its present state. Indeed we might consider this premise as the basis of an abstract definition of the state of a dynamic system:

> The state of a dynamic system is a set of physical quantities, the specification of which (in the absence of external excitation) completely determines the evolution of the system.

The difficulty with this definition, as well as its major advantage, is that the *specific* physical quantities that define the system state are not unique, although their number (called the system *order*) is unique. In many situations there is an obvious choice of the variables (*state variables*) to define the system state, but there are also many cases in which the choice of state variables is by no means obvious.

Newton invented calculus as a means of characterizing the behavior of dynamic systems, and his method continues in use to this very day. In particular, behavior of dynamic systems is represented by systems of ordinary differential equations. The differential equations are said to constitute a *mathematical model* of the physical process. We can predict how the physical process will behave by solving the differential equations that are used to model the process.

In order to obtain a solution to a system of ordinary differential equations, it is necessary to specify a set of *initial conditions*. The number of initial conditions that must be specified defines the *order* of the system. When the differential equations constitute the mathematical model of a physical system, the initial conditions needed to solve the differential equations correspond to physical quantities needed to predict the future behavior of the system. It thus follows that the initial conditions and physical state variables are equal in number.

In analysis of dynamic systems such as mechanical systems, electric networks, etc. the differential equations typically relate the dynamic variables and their time derivatives of various orders. In the state-space approach, all the differential equations in the mathematical model of a system are *first-order* equations: only the dynamic variables and their first derivatives (with respect to time) appear in the differential equations. Since only one initial condition is needed to specify the solution of a first-order equation, it follows that the number of first-order differential equations in the mathematical model is equal to the order of the corresponding system.

The dynamic variables that appear in the system of first-order equations are called the *state variables*. From the foregoing discussion, it should be clear that the *number* of state variables in the model of a physical process is unique, although the identity of these variables may not be unique. A few familiar examples serve to illustrate these points.

Example 2A Mass acted upon by friction and spring forces The mechanical system consisting of a mass which is acted upon by the forces of friction and a spring is a paradigm of a second-order dynamic process which one encounters time and again in control processes.

Consider an object of mass M moving in a line. In accordance with Newton's law of motion, the acceleration of the object is the total force f acting on the object divided by the mass.

$$\frac{d^2x}{dt^2} = \frac{f}{M} \tag{2A.1}$$

where the direction of f is in the direction of x. We assume that the force f is the sum of two forces, namely a friction force f_1 and a spring force f_2. Both of these forces physically tend to resist the motion of the object. The friction force tends to resist the velocity: there is no friction force unless the velocity is nonzero. The spring force, on the other hand, is proportional to the amount that the spring has been compressed, which is equal to the amount that the object has been displaced. Thus

$$f = f_1 + f_2$$

where

$$f_1 = -\beta\left(\frac{dx}{dt}\right)$$

$$f_2 = -\kappa(x)$$

Thus

$$\frac{d^2x}{dt^2} = -\left[\beta\left(\frac{dx}{dt}\right) + \kappa(x)\right]\bigg/ M \tag{2A.2}$$

A more familiar form of (2A.2) is the second-order differential equation

$$M\frac{d^2x}{dt^2} + \beta\left(\frac{dx}{dt}\right) + \kappa(x) = 0 \tag{2A.3}$$

But (2A.2) is a form more appropriate for the state-space representation. Differential equation (2A.2) or its equivalent (2A.3) is a second-order differential equation and its solution requires two initial conditions: x_0, the initial position, and $\dot{x}_0$, the initial velocity.

To obtain a state-space representation, we need two state variables in terms of which the dynamics of (2A.2) can be expressed as two first-order differential equations. The obvious choice of variables in this case are the displacement x and the velocity $v = dx/dt$. The two first-order equations for the process in this case are the equation by which velocity is defined

$$\frac{dx}{dt} = v \tag{2A.4}$$

and (2A.2) expressed in terms of x and v. Since $d^2x/dt^2 = dv/dt$, (2A.2) becomes

$$\frac{dv}{dt} = -[\beta(v) + \kappa(x)]/M \tag{2A.5}$$

Thus (2A.4) and (2A.5) constitute a system of two first-order differential equations in terms of the state variables x and v.

If we wish to control the motion of the object we would include an additional force f_0 external to the system which would be added to the right-hand side of (2A.5)

$$\frac{dv}{dt} = -[\beta(v) + \kappa(x)]/M + f_0/M \tag{2A.6}$$

How such a control force would be produced is a matter of concern to the control system designer. But it is not considered in the present example.

In a practical system both the friction force and the spring force are nonlinear functions of their respective variables and a realistic prediction of the system behavior would entail solution of (2A.4) and (2A.5) in which $\beta(v)$ and $\kappa(x)$ are *nonlinear* functions of their arguments. As an approximation, however, it may be permissible to treat these functions as being linear

$$\beta\left(\frac{dx}{dt}\right) \simeq B\frac{dx}{dt}$$

$$\kappa(x) \simeq Kx$$

where B and K are constants. Often $\beta(\ )$ and $\kappa(\ )$ are treated as linear functions for purposes of control system design, but the accurate nonlinear functions are used in evaluating how the design performs.

A block diagram representation of the differential equations (2A.4) and (2A.6), in accordance with the discussion of Sec. 2.3, is shown in Fig. 2.1.

Example 2B Electric motor with inertia load One of the most common uses of feedback control is to position an inertia load using an electric motor. (See Fig. 2.2.) The inertia load may consist of a very large, massive object such as a radar antenna or a small object such as a precision instrument. An important aspect of the control system design is the selection of a suitable motor, capable of achieving the desired dynamic response and suited to the objective in cost, size, weight, etc. An electric motor is a device that converts electrical energy (input) to mechanical energy (output). The electro-mechanical energy transducer relations are idealizations of Faraday's law of induction and Ampere's law for the force produced on a conductor moving in a magnetic field. In particular, under ideal circumstances the torque developed at the shaft of a motor is proportional to the input current to the motor; the induced emf v ("back emf") is proportional to the speed ω of rotation

$$\tau = K_1 i \tag{2B.1}$$

$$v = K_2\omega \tag{2B.2}$$

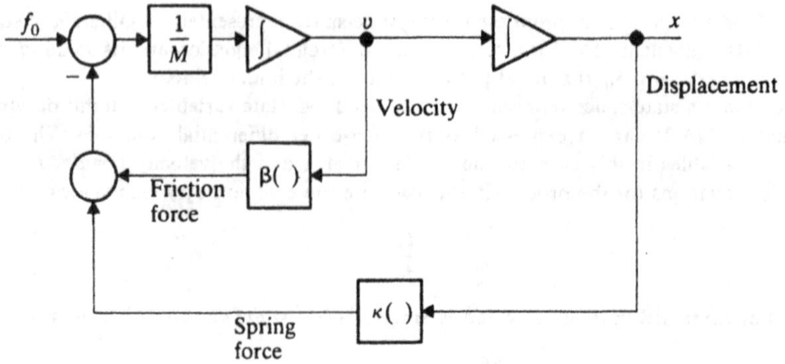

Figure 2.1 Block diagram representing motion of mass with friction and spring reaction forces.

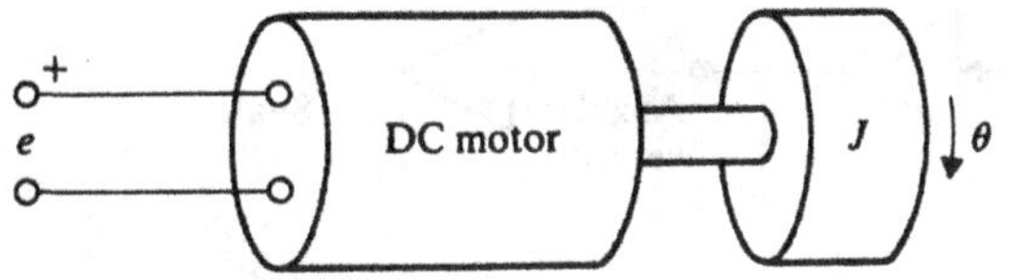

Figure 2.2 DC motor driving inertia load.

The electrical power p_e input to the motor is the product of the current and the induced emf

$$p_e = vi = K_2 \omega \tau / K_1 \tag{2B.3}$$

The mechanical output power is the product of the torque and the angular velocity

$$p_m = \omega \tau$$

Thus, from (2B.3)

$$p_e = \frac{K_2}{K_1} p_m$$

If the energy conversion is 100 percent efficient, then

$$K_1 = K_2 = K$$

If the energy-conversion efficiency is less than 100 percent then $K_2/K_1 > 1$.

To completely specify the behavior of the system we need the relationships between the input voltage e and the induced emf, and between the torque and the angular velocity of the motor. These are given by

$$e - v = Ri \qquad \text{(Ohm's law)} \tag{2B.4}$$

where R is the electrical resistance of the motor armature, and

$$\tau = J \frac{d\omega}{dt} \tag{2B.5}$$

where J is the inertia of the load. From (2B.1), (2B.5), and (2B.4)

$$J \frac{d\omega}{dt} = K_1 i = \frac{K_1}{R}(e - v) \tag{2B.6}$$

On using (2B.2) this becomes

$$J \frac{d\omega}{dt} = \frac{K_1}{R} e - \frac{K_1 K_2}{R} \omega$$

or

$$\frac{d\omega}{dt} = -\frac{K_1 K_2}{JR} \omega + \frac{K_1}{JR} e \tag{2B.7}$$

which is a first-order equation with the angular velocity ω as the state variable and with e serving as the external control input.

The first-order model of (2B.7) is suitable for control of the speed of the shaft rotation. When the position θ of the shaft carrying the inertia J is also of concern, we must add the differential equation

$$\frac{d\theta}{dt} = \omega \tag{2B.8}$$

This and (2B.7) together constitute a second-order system.

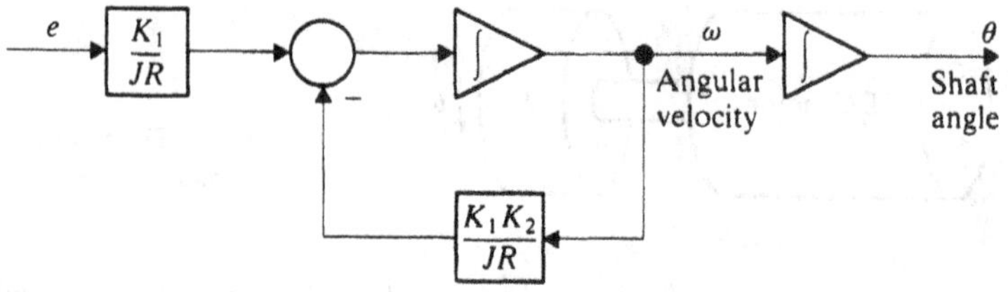

Figure 2.3 Block diagram representing dynamics of dc motor driving inertia load.

Equations (2B.7) and (2B.8) can be arranged in the vector-matrix form

$$\frac{d}{dt}\begin{bmatrix}\theta\\ \omega\end{bmatrix} = \begin{bmatrix}0 & 1\\ 0 & -K_1K_2/JR\end{bmatrix}\begin{bmatrix}\theta\\ \omega\end{bmatrix} + \begin{bmatrix}0\\ K_1/JR\end{bmatrix}e$$

A block-diagram representation of the differential equations that represent this system is given in Fig. 2.3.

Example 2C Electrical network and its thermal analog It is not generally required to design feedback control systems for electrical networks comprising resistors, capacitors, and inductors. But such networks often are mathematically analogous to mechanical systems which one does desire to control, and an engineer experienced in the analysis of electrical networks might be more comfortable with the latter than with the mechanical systems they represent.

One class of mechanical system which is analogous to an electrical network is a thermal conduction system. Electrical voltages are analogous to temperatures and currents are analogous to heat flow rates. The paths of conduction of heat between various points in the system are represented by resistors; the mass storage of heat in various bodies is represented by capacitances; the input of heat by current sources; and fixed temperatures at the boundaries of the system by voltage sources.

Table 2C.1 summarizes the thermal quantities and their electrical analogs.

As an illustration of the use of electrical analogs of thermal systems, consider the system shown in Fig. 2.4 consisting of two masses of temperatures T_1 and T_2 embedded in a thermally

Table 2C.1 Electrical analogs of thermal systems

Thermal system			Electrical system		
Quantity	Symbol	Unit	Quantity	Symbol	Unit
Temperature	T	deg	Voltage	v	volt
Heat flux	q	cal/s	Current	i	ampere
Thermal resistivity	R	deg · s/cal	Resistance	R	ohm
Thermal capacity	C	cal/deg	Capacitance	C	farad
Conduction equation	$q = \frac{1}{R}(T_2 - T_1)$			$i = \frac{1}{R}(v_2 - v_1)$	
Storage equation	$\frac{dT}{dt} = \frac{q}{C}$			$\frac{dv}{dt} = \frac{1}{C}i$	

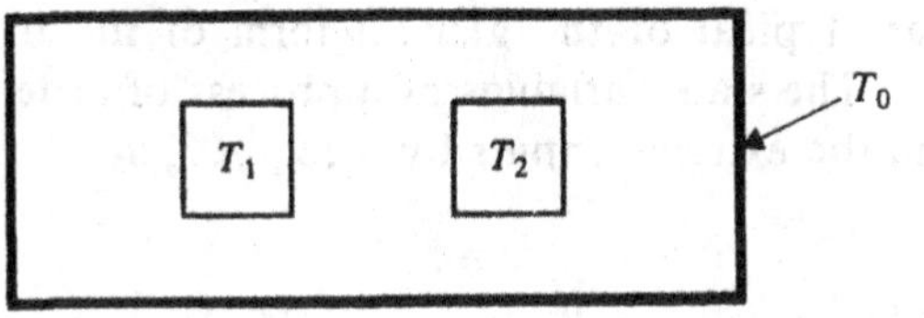

Figure 2.4 Thermal system with two capacitances.

insulating medium contained in a metal container which, because of its high thermal conductivity, may be assumed to have a constant temperature T_0. The temperatures T_1 and T_2 of the masses are to be controlled by controlling the temperature T_0 of the container.

An electrical analog of the system is shown in Fig. 2.5. The capacitors C_1 and C_2 represent the heat capacities of the masses; the resistor R_3 represents the path of heat flow from mass 1 to mass 2; R_1 and R_2 represent the heat flow path from these masses to the metal container.

The differential equations governing the thermal dynamics of the mechanical system are the same as the differential equations of the electrical system, which can be obtained by various standard methods. By use of nodal analysis, for example, it is determined that

$$C_1\frac{dv_1}{dt}+\left(\frac{1}{R_1}+\frac{1}{R_3}\right)v_1-\frac{1}{R_3}v_2-\frac{1}{R_1}e_0=0$$

$$C_2\frac{dv_2}{dt}+\left(\frac{1}{R_2}+\frac{1}{R_3}\right)v_2-\frac{1}{R_3}v_1-\frac{1}{R_2}e_0=0 \tag{2C.1}$$

The appropriate state variables for the process are the capacitor voltages v_1 and v_2. The temperature of the case is represented by a voltage source e_0 which is the input variable to the process. Thus the differential equations of the process are

$$\frac{dv_1}{dt}=-\frac{1}{C_1}\left(\frac{1}{R_1}+\frac{1}{R_3}\right)v_1+\frac{1}{C_1R_3}v_2+\frac{1}{R_1C_1}e_0$$

$$\frac{dv_2}{dt}=\frac{1}{C_2R_3}v_1-\frac{1}{C_2}\left(\frac{1}{R_2}+\frac{1}{R_3}\right)v_2+\frac{1}{R_2C_2}e_0 \tag{2C.2}$$

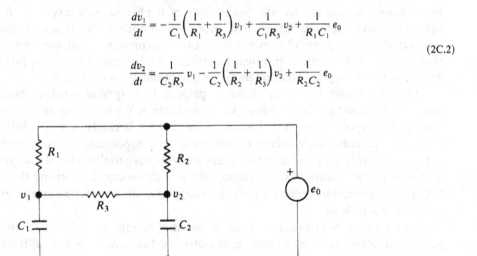

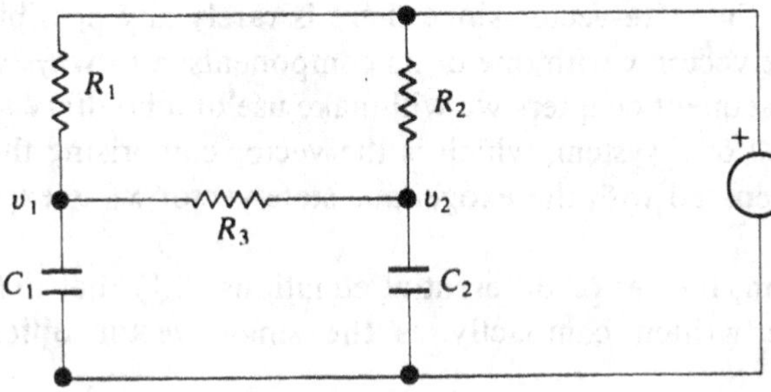

Figure 2.5 Electrical analog of thermal system of Fig. 2.4.

The foregoing examples are typical of the general form of the dynamic equations of a dynamic process. The state variables of a process of order k are designated by $x_1, x_2, \ldots, x_k$ and the external inputs by $u_1, u_2, \ldots, u_l$

$$\begin{aligned}
\dot{x}_1 &= \frac{dx_1}{dt} = f_1(x_1, x_2, \ldots, x_k, u_1, u_2, \ldots, u_l, t) \\
\dot{x}_2 &= \frac{dx_2}{dt} = f_2(x_1, x_2, \ldots, x_k, u_1, u_2, \ldots, u_l, t) \\
&\cdots\cdots\cdots\cdots\cdots\cdots\cdots\cdots\cdots \\
\dot{x}_k &= \frac{dx_k}{dt} = f_k(x_1, x_2, \ldots, x_k, u_1, u_2, \ldots, u_l, t)
\end{aligned} \tag{2.1}$$

These equations express the time-derivatives of each of the state variables as general functions of all the state variables, inputs, and (possibly) time. The dot over a variable is Newton's notation for the derivative with respect to time.

To simplify the notation the state variables $x_1, x_2, \ldots, x_k$ and control variables $u_1, u_2, \ldots, u_l$ are collected in vectors

$$x = \begin{bmatrix} x_1 \\ \vdots \\ x_k \end{bmatrix} \qquad u = \begin{bmatrix} u_1 \\ \vdots \\ u_l \end{bmatrix} \tag{2.2}$$

called the *state vector* and the *input vector*, respectively. These are vectors in the mathematical sense and not necessarily in the physical sense. The components of a *physical* vector are usually projections of a physical quantity (e.g., force, velocity) along a set of reference axes. But the components of the state vector of a dynamic system generally do not have this interpretation and need not even represent the same kind of physical quantities: As our examples show, position and velocity are typical components of a mathematical state vector.

In some books the state vector is printed in a special typeface such as boldface **x**, to distinguish it from a scalar variable x. We have chosen not to use any special typeface for the state vector since there is rarely any possibility of confusing the entire state vector x with one of its components x_i (always written with a subscript). In subsequent chapters we will make use of a boldface symbol **x** to denote the *metastate* of a system, which is the vector comprising the state (or error) vector, concatenated with the exogenous state vector x_0 as explained in Chap. 5 and later.

Using vector notation, the set of differential equations (2.1) that defines a general process can be written compactly as the single vector differential equation

$$\dot{x} = \frac{dx}{dt} = f(x, u, t) \tag{2.3}$$

where $f(x, u, t)$ is understood to be a k-dimensional vector-valued function of $k+l+1$ arguments. When time t does not appear explicitly in any of the functions f_i in (2.1), i.e., in the vector f of (2.3), the system is said to be *time-invariant.* If (2.3) is an accurate model of a physical process, we would expect it to be time-invariant, since we do not have physical laws that change with time. In many situations, however, the differential equations represented by (2.3) are only an approximate model of the physical world, either because a more accurate model is not known, or because it is too complicated to be useful in the intended application. Very often such approximate models are time-varying.

An exact model of a physical process is usually nonlinear. But fortunately many processes can be adequately approximated by linear models over a significant range of operation. In the state-space model of a linear process, the general differential equations of (2.1) take the special form:

$$\begin{aligned}
\dot{x}_1 &= \frac{dx_1}{dt} = a_{11}(t)x_1 + \cdots + a_{1k}(t)x_k + b_{11}(t)u_1 + \cdots + b_{1l}(t)u_l \\
\dot{x}_2 &= \frac{dx_2}{dt} = a_{21}(t)x_1 + \cdots + a_{2k}(t)x_k + b_{21}(t)u_1 + \cdots + b_{2l}(t)u_l \\
&\cdots\cdots\cdots\cdots\cdots\cdots\cdots\cdots\cdots\cdots \\
\dot{x}_k &= \frac{dx_k}{dt} = a_{k1}(t)x_1 + \cdots + a_{kk}(t)x_k + b_{k1}(t)u_1 + \cdots + b_{kl}(t)u_l
\end{aligned} \tag{2.4}$$

In vector notation, using the definitions of the state and control vectors as defined in (2.2), the linear dynamic model of (2.4) is written

$$\dot{x} = \frac{dx}{dt} = A(t)x + B(t)u \tag{2.5}$$

where $A(t)$ and $B(t)$ are *matrices* given by

$$A(t) = \begin{bmatrix} a_{11}(t) \cdots a_{1k}(t) \\ a_{21}(t) \cdots a_{2k}(t) \\ \cdots\cdots\cdots \\ a_{k1}(t) \cdots a_{kk}(t) \end{bmatrix} \qquad B(t) = \begin{bmatrix} b_{11}(t) \cdots b_{1l}(t) \\ b_{21}(t) \cdots b_{2l}(t) \\ \cdots\cdots\cdots \\ b_{k1}(t) \cdots b_{kl}(t) \end{bmatrix} \tag{2.6}$$

It is noted that the matrix $A(t)$ is always a square (k by k) matrix, but that the matrix $B(t)$ need not be square. In most processes of interest the number l of inputs is smaller than the number of state variables: $B(t)$ is a tall, thin matrix. Often there is only one input and the matrix $B(t)$ is only one column wide.

When the system is time-invariant, none of the elements in the matrices A and B depend upon time. Most of this book is concerned with linear, time-invariant processes, having the dynamic equations

$$\dot{x} = Ax + Bu \tag{2.7}$$

where A and B are constant matrices.

Although the concept of the state of a system is fundamental, there are many situations in which one is not interested in the state directly, but only in its effect on the system *output* vector $y(t)$

$$y(t) = \begin{bmatrix} y_1(t) \\ y_2(t) \\ \vdots \\ y_m(t) \end{bmatrix} \tag{2.8}$$

for a system having m outputs. In a linear system the output vector is assumed to be a linear combination of the state and the input

$$y(t) = C(t)x(t) + D(t)u(t) \tag{2.9}$$

where $C(t)$ is an $m \times k$ matrix and $D(t)$ is an $m \times l$ matrix. If the system is time-invariant, $C(t)$ and $D(t)$ are constant matrices.

The outputs of a system are generally those quantities which can be *observed*, i.e., measured by means of suitable sensors. Accordingly, the output vector is called the *observation vector* and (2.9) is called the observation equation.

The presence of the matrix D in (2.9) means that there is a direct connection between the input $u(t)$ and the output $y(t)$, without the intervention of the state $x(t)$. Although there is no general reason for the matrix D to be absent in a practical application, it turns out that it *is* absent in the overwhelming majority of applications. This is fortunate, because the presence of D increases the complexity of much of the theory. Thus most of our development will rest on the assumption that $D = 0$.

The input vector u in (2.7) represents the assemblage of all physical quantities that affect the behavior of the state. From the control system design standpoint, however, the inputs are of two types:

Control inputs, produced intentionally by the operation of the control system, and

"Exogenous" inputs, present in the environment and not subject to control within the system.

It is customary to reserve the symbol u for the control inputs and to use another symbol for the exogenous inputs. (The word "exogenous," widely used in the field of economics and other social sciences, is gaining currency in the field of control theory.) In this book we shall find it convenient to represent the exogenous inputs by the vector x_0. The use of the letter "x" suggests that the exogenous inputs are state variables and so they may be regarded: x_0 may be regarded as the *state of the environment.* (Later in the book we shall concatenate the state x of the system to be controlled with the state x_0 of the environment into a metastate of the overall process.)

Thus, separating the input u of (2.7) into a control input and an exogenous input, (2.7) becomes

$$\dot{x} = Ax + Bu + Ex_0 \tag{2.10}$$

which, together with (2.9) will serve as the general representation of a linear system.

2.3 BLOCK-DIAGRAM REPRESENTATIONS

System engineers often find it helpful to visualize the relationships between dynamic variables and subsystems of a system by means of block diagrams. Each subsystem is represented by a geometric figure (such as a rectangle, a circle, a triangle, etc.) and lines with arrows on them show the inputs and the outputs. For many systems, these block diagrams are more expressive than the mathematical equations to which they correspond.

The relationships between the variables in a linear system (2.4) can be expressed using only three kinds of elementary subsystems:

Integrators, represented by triangles
Summers, represented by circles, and
Gain elements, represented by rectangular or square boxes as shown in Fig. 2.6.

An integrator is a block-diagram element whose output is the integral of the input; put in other words, it is the element whose input is the derivative of the output.

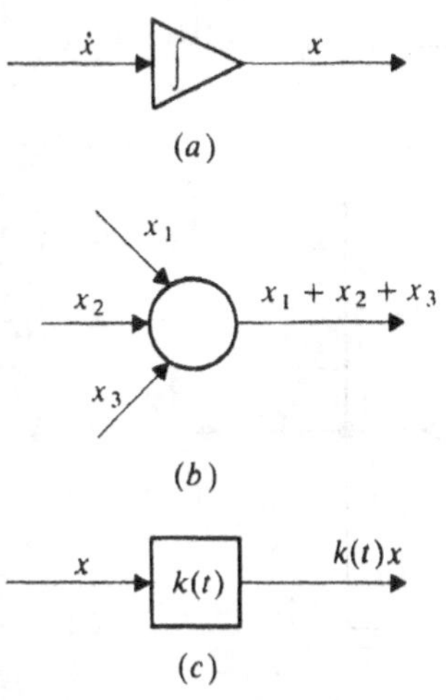

Figure 2.6 Elements used in block-diagram representation of linear systems. (*a*) Integrator; (*b*) Summer; (*c*) Gain element.

A summer is a block-diagram element whose output is the sum of all its inputs.

A gain element is a block-diagram element whose output is proportional to its input. The constant of proportionality, which may be time-varying, is placed inside the box (when space permits) or adjacent to it.

Note that the integrator and the gain element are single-input elements; the summer, on the other hand, always has at least two inputs.

A general block diagram for a second-order system ($k = 2$) with two external inputs u_1 and u_2 is shown in Fig. 2.7. Two integrators are needed, the outputs of which are x_1 and x_2, and the inputs to which are $\dot{x}_1$ and $\dot{x}_2$, respectively. From the general form of the differential equations (2.4) these are given by

$$\dot{x}_1 = a_{11}x_1 + a_{12}x_2 + b_{11}u_1 + b_{12}u_2$$
$$\dot{x}_2 = a_{21}x_1 + a_{22}x_2 + b_{21}u_1 + b_{22}u_2$$

which are the relationships expressed by the outputs of the two summers shown in Fig. 2.7.

The same technique applies in higher-order systems. If the A matrix has many nonzero terms, the diagram can look like a plate of spaghetti and meatballs. In most practical cases, however, the A matrix is fairly sparse, and

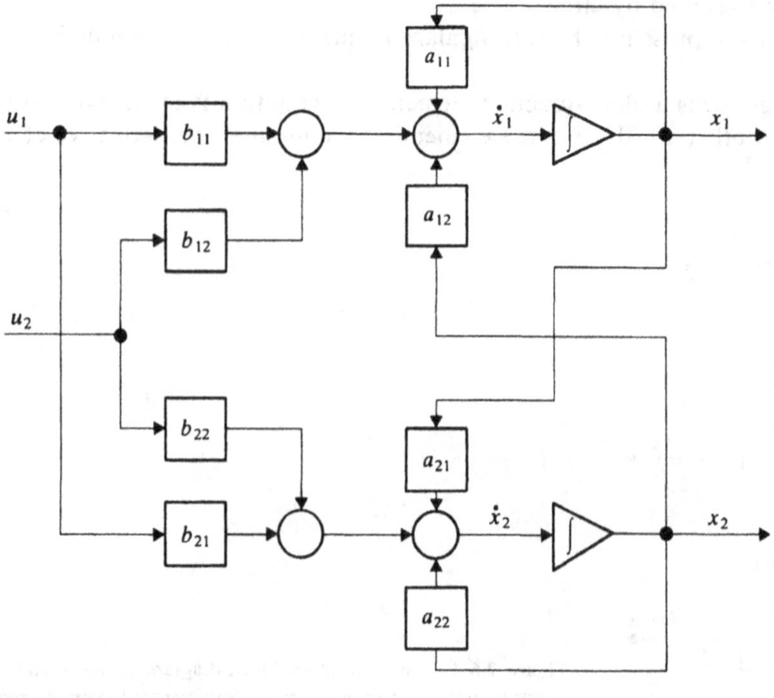

Figure 2.7 Block diagram of general second-order linear system.

with some attention to layout it is possible to draw a block diagram with a minimum of crossed lines.

To simplify the appearance of the block-diagram it is sometimes convenient to use redundant summers. This is shown in Fig. 2.7. Instead of using two summers, one feeding another, in front of each integrator we could have drawn the diagram with only one summer with four inputs in front of each integrator. But the diagram as shown has a neater appearance. Another technique to simplify the appearance of a block diagram is to show a sign reversal by means of a minus sign adjacent to the arrow leading into a summer instead of a gain element with a gain of -1. This usage is illustrated in Figs. 2.1 and 2.3 of the foregoing examples.

Although there are several international standards for block-diagram symbols, these standards are rarely adhered to in technical papers and books. The differences between the symbols used by various authors, however, are not large and are not likely to cause the reader any confusion.

The following examples illustrate the use of matrices and block diagrams to represent the dynamics of various processes.

Often it is convenient to express relationships between vector quantities by means of block diagrams. The block-diagram symbols of Fig. 2.6 can also serve to designate operations on vectors. In particular, when the input to an integrator of Fig. 2.6(a) is a vector quantity, the output is a vector each component of which is the integral of the corresponding input. The summer of Fig. 2.6(b) represents a vector summer, and the gain element box of Fig. 2.6(c) represents a matrix. In the last case, the matrix need not be square and the dimension of the vector of outputs from the box need not equal the dimension of the vector

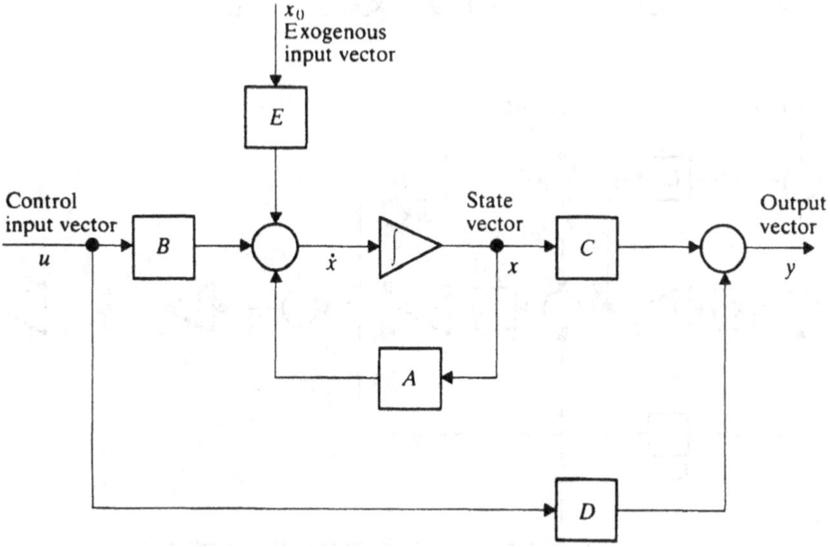

Figure 2.8 Block-diagram representation of general linear system.

of inputs. Using this mode of representation, the block diagram of Fig. 2.8 represents the general system given by (2.9) and (2.10).

Example 2D Hydraulically actuated tank gun turret The control of a hydraulically actuated gun turret in an experimental tank has been studied by Loh, Cheok, and Beck.[1] The linearized dynamic model they used for each axis (elevation, azimuth) is given by

$$\begin{aligned}\dot{\theta} &= \omega \\ \dot{\omega} &= p + d_\tau \\ \dot{p} &= -\Omega_m p + \frac{K_m}{J} q - \frac{K_m}{J}\omega + d_p \\ \dot{q} &= -K_v L_v q - K_v K_{\Delta p} J p + K_v u + d_q\end{aligned} \tag{2D.1}$$

where $x_1 = \theta$ = turret angle
$x_2 = \omega$ = turret angular rate
$x_3 = p$ = angular acceleration produced by hydraulic drive
$x_4 = q$ = hydraulic servo valve displacement
u = control input to servo valve
K_m = servo motor gain
J = turret inertia
Ω_m = motor natural frequency
K_v = servo valve gain
$K_{\Delta p}$ = differential pressure feedback coefficient

The quantities d_τ, d_p, and d_q represent disturbances, including effects of nonlinearities not accounted for by the linearized model (2D.1).

With the state variable definitions given above, the matrices of this process are

$$A = \begin{bmatrix} 0 & 1 & 0 & 0 \\ 0 & 0 & 1 & 0 \\ 0 & K_m/J & -\Omega_m & -K_m/J \\ 0 & 0 & -K_v K_{\Delta p} J & -K_v L_v \end{bmatrix} \qquad B = \begin{bmatrix} 0 \\ 0 \\ 0 \\ K_v \end{bmatrix}$$

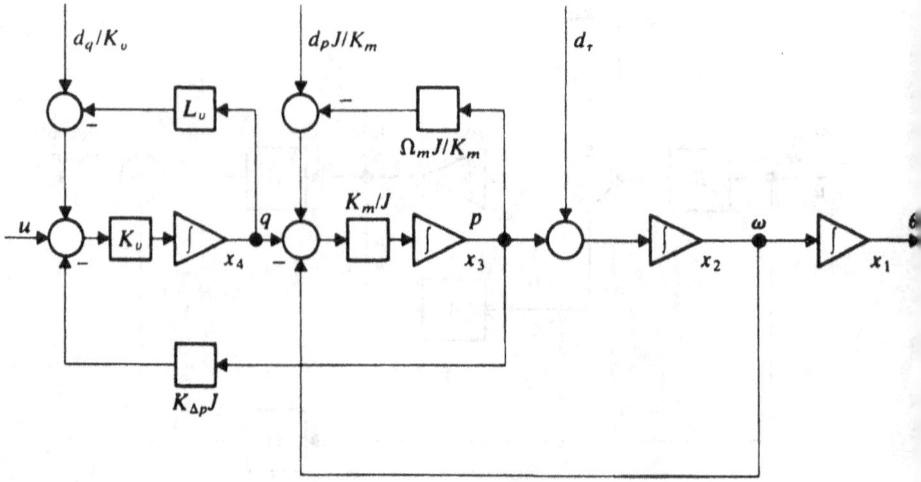

Figure 2.9 Dynamic model of hydraulically actuated tank gun turret.

Table 2D.1 Numerical values of parameters in tank turret control

	Numerical value	
Parameter	Azimuth	Elevation
K_v	94.3	94.3
L_v	1.00	1.07
J (ft-lb · s²)	7900.	2070.
K_m	8.46×10^6	1.96×10^6
ω_m (rad/s)	45.9	17.3
$K_{\Delta p}$	6.33×10^{-6}	3.86×10^{-5}

Numerical data for a specific tank were found by Loh, Cheok, and Beck to be as given in Table 2D.1

A block-diagram representation of the dynamics represented by (2D.1) is shown in Fig. 2.9.

2.4 LAGRANGE'S EQUATIONS

The equations governing the motion of a complicated mechanical system, such as a robot manipulator, can be expressed very efficiently through the use of a method developed by the eighteenth-century French mathematician Lagrange. The differential equations that result from use of this method are known as *Lagrange's equations* and are derived from Newton's laws of motion in most textbooks on advanced dynamics.[2, 3]

Lagrange's equations are particularly advantageous in that they automatically incorporate the constraints that exist by virtue of the different parts of a system being connected to each other, and thereby eliminate the need for substituting one set of equations into another to eliminate forces and torques of constraint. Since they deal with scalar quantities (potential and kinetic energy) rather than with vectors (forces and torques) they also minimize the need for complicated vector diagrams that are usually required to define and resolve the vector quantities in the proper coordinate system. The advantages of Lagrange's equations may also turn out to be disadvantages, because it is necessary to identify the generalized coordinates correctly at the very beginning of the analysis of a specific system. An error made at this point may result in a set of differential equations that look correct but do not constitute the correct model of the physical system under investigation.

The fundamental principle of Lagrange's equations is the representation of the system by a set of generalized coordinates q_i $(i = 1, 2, \ldots, r)$, one for each independent degree of freedom of the system, which completely incorporate the constraints unique to that system, i.e., the interconnections between the parts of the system. After having defined the generalized coordinates, the kinetic energy T is expressed in terms of these coordinates and their derivatives, and the potential energy V is expressed in terms of the generalized coordinates. (The potential energy is a function of only the generalized coordinates and *not* their derivatives.) Next, the *lagrangian* function

$$L = T(q_1, \ldots, q_r, \dot{q}_1, \ldots, \dot{q}_r) - V(q_1, \ldots, q_r)$$

is formed. And finally the desired equations of motion are derived using Lagrange's equations

$$\frac{d}{dt}\left(\frac{\partial L}{\partial \dot{q}_i}\right) - \frac{\partial L}{\partial q_i} = Q_i \qquad i = 1, 2, \ldots, r \tag{2.11}$$

where Q_i denotes generalized forces (i.e., forces and torques) that are external to the system or not derivable from a scalar potential function.

Each of the differential equations in the set (2.11) will be a second-order differential equation, so a dynamic system with r degrees of freedom will be represented by r second-order differential equations. If one *state variable* is assigned to each generalized coordinate and another to the corresponding derivative, we end up with $2r$ equations. Thus a system with r degrees of freedom is of order $2r$.

Example 2E Inverted pendulum on moving cart A typical application of Lagrange's equations is to define the motion of a collection of bodies that are connected together in some manner such as the inverted pendulum on a cart illustrated in Fig. 2.10.

It is observed that the motion of the system is uniquely defined by the displacement of the cart from some reference point, and the angle that the pendulum rod makes with respect to the vertical. Instead of using θ, we could use the horizontal displacement, say y_1, of the bob relative to the pivot point, or the vertical height z_2 of the bob. But, whatever variables are used, it is essential to know that the system has only two degrees of freedom, and that the dynamics must be expressed in terms of the corresponding generalized coordinates.

The kinetic energy of the system is the sum of the kinetic energy of each mass. The cart is confined to move in the horizontal direction so its kinetic energy is

$$T_1 = \tfrac{1}{2}M\dot{y}^2$$

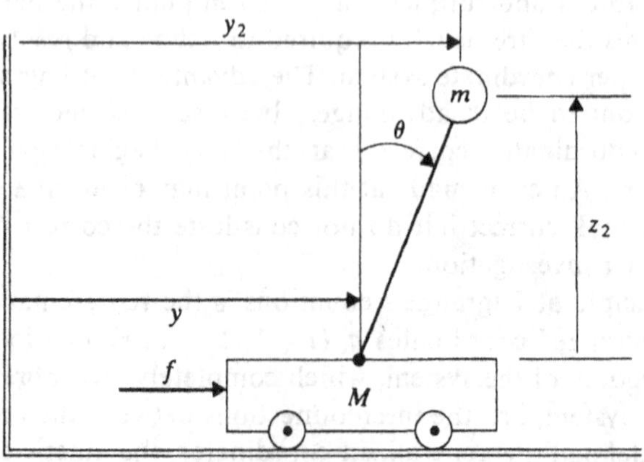

Figure 2.10 Inverted pendulum on moving cart.

The bob can move in the horizontal and in the vertical direction so its kinetic energy is

$$T_2 = \tfrac{1}{2}m(\dot{y}_2^2 + \dot{z}_2^2)$$

But the rigid rod constrains z_2 and y_2

$$y_2 = y + l \sin \theta \qquad \dot{y}_2 = \dot{y} + l\dot{\theta} \cos \theta$$
$$z_2 = l \cos \theta \qquad \dot{z}_2 = -l\dot{\theta} \sin \theta$$

Thus

$$\begin{aligned} T = T_1 + T_2 &= \tfrac{1}{2}M\dot{y}^2 + \tfrac{1}{2}m[(\dot{y} + l\dot{\theta}\cos\theta)^2 + l^2\dot{\theta}^2 \sin^2\theta] \\ &= \tfrac{1}{2}M\dot{y}^2 + \tfrac{1}{2}m[\dot{y}^2 + 2\dot{y}\dot{\theta}l\cos\theta + l^2\dot{\theta}^2] \end{aligned}$$

The only potential energy is stored in the bob

$$V = mgz_2 = mgl \cos \theta$$

Thus the lagrangian is

$$L = T - V = \tfrac{1}{2}(M + m)\dot{y}^2 + ml\cos\theta\dot{y}\dot{\theta} + \tfrac{1}{2}ml^2\dot{\theta}^2 - mgl\cos\theta \tag{2E.1}$$

The generalized coordinates having been selected as (y, θ), Lagrange's equations for this system are

$$\begin{aligned} \frac{d}{dt}\left(\frac{\partial L}{\partial \dot{y}}\right) - \frac{\partial L}{\partial y} &= f \\ \frac{d}{dt}\left(\frac{\partial L}{\partial \dot{\theta}}\right) - \frac{\partial L}{\partial \theta} &= 0 \end{aligned} \tag{2E.2}$$

Now

$$\frac{\partial L}{\partial \dot{y}} = (M + m)\dot{y} + ml\cos\theta\dot{\theta}$$

$$\frac{\partial L}{\partial y} = 0$$

$$\frac{\partial L}{\partial \dot{\theta}} = ml\cos\theta\dot{y} + ml^2\dot{\theta}$$

$$\frac{\partial L}{\partial \theta} = mgl\sin\theta - ml\sin\theta\dot{y}\dot{\theta}$$

Thus (2E.2) become

$$\begin{aligned} (M + m)\ddot{y} + ml\cos\theta\ddot{\theta} - ml\dot{\theta}^2\sin\theta &= f \\ ml\cos\theta\ddot{y} + ml^2\ddot{\theta} - mgl\sin\theta &= 0 \end{aligned} \tag{2E.3}$$

These are the exact equations of motion of the inverted pendulum on a cart shown in Fig. 2.10. They are nonlinear owing to the presence of the trigonometric terms $\sin\theta$ and $\cos\theta$ and the quadratic terms $\dot{\theta}^2$ and $\dot{y}\dot{\theta}$. If the pendulum is stabilized, however, then θ will be kept small. This justifies the approximations

$$\cos\theta \simeq 1 \qquad \sin\theta \simeq \theta$$

We may also assume that $\dot{\theta}$ and $\dot{y}$ will be kept small, so the quadratic terms are negligible. Using these approximations we obtain the linearized dynamic model

$$\begin{aligned} (M + m)\ddot{y} + ml\ddot{\theta} &= f \\ m\ddot{y} + ml\ddot{\theta} - mg\theta &= 0 \end{aligned} \tag{2E.4}$$

A state-variable representation corresponding to (2E.4) is obtained by defining the state vector

$$x = [y, \theta, \dot{y}, \dot{\theta}]'$$

Then

$$\frac{dy}{dt} = \dot{y}$$
$$\frac{d\theta}{dt} = \dot{\theta} \tag{2E.5}$$

constitute the first two dynamic equations and on solving (2E.4) for $\ddot{y}$ and $\ddot{\theta}$, we obtain two more equations

$$\frac{d}{dt}(\dot{y}) = \ddot{y} = \frac{f}{M} - \frac{mg}{M}\theta$$
$$\frac{d}{dt}(\dot{\theta}) = \ddot{\theta} = -\frac{f}{Ml} + \left(\frac{M+m}{Ml}\right)g\theta \tag{2E.6}$$

The four equations can be put into the standard matrix form

$$\dot{x} = Ax + Bu$$

with

$$A = \begin{bmatrix} 0 & 0 & 1 & 0 \\ 0 & 0 & 0 & 1 \\ 0 & -mg/M & 0 & 0 \\ 0 & (M+m)g/Ml & 0 & 0 \end{bmatrix} \qquad B = \begin{bmatrix} 0 \\ 0 \\ 1/M \\ -1/Ml \end{bmatrix}$$

and

$$u = f = \text{external force}$$

A block-diagram representation of the dynamics (2E.5) and (2E.6) is shown in Fig. 2.11.

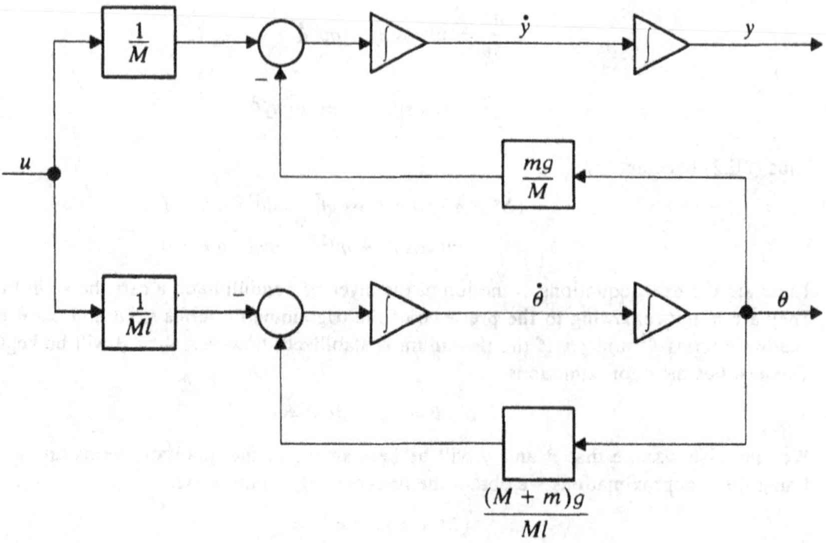

Figure 2.11 Block diagram of dynamics of inverted pendulum on moving cart.

2.5 RIGID BODY DYNAMICS

The motion of a single rigid body has six dynamic degrees of freedom: three of these define the location of a reference point (usually the center of mass) in the body, and three define the orientation (attitude) of the body. Since each of the six degrees of freedom takes two state variables (one position and one velocity) a total of 12 first-order differential equations are required to completely describe the motion of the body. In most applications, however, not all of these 12 state variables are of interest and not all the differential equations are needed. In a gyroscope, for example, only the orientation is of interest.

The motion of a rigid body is, of course, governed by the familiar newtonian laws of motion

$$\frac{d\vec{p}}{dt} = \vec{f} \tag{2.12}$$

$$\frac{d\vec{h}}{dt} = \vec{\tau} \tag{2.13}$$

where $\vec{p} = [p_x, p_y, p_z]'$ is the linear momentum of the body
$\vec{h} = [h_x, h_y, h_z]'$ is the angular momentum of the body
$\vec{f} = [f_x, f_y, f_z]'$ is force acting on the body
$\vec{\tau} = [\tau_x, \tau_y, \tau_z]'$ is torque acting on the body

It is important to understand that (2.12) and (2.13) are valid only when the axes along which the motion is resolved are an inertial frame of reference, i.e., they are neither accelerating nor rotating. If the axes are accelerating linearly or rotating, then (2.12) and (2.13) must be modified to account for the motion of the reference axes.

The rotational dynamics of a rigid body are more complicated than the translational dynamics for several reasons: the mass M of a rigid body is a scalar, but the moment of inertia J is a 3×3 matrix. If the body axes are chosen to coincide with the "principal axes," the moment of inertia matrix is diagonal; otherwise the matrix J has off-diagonal terms. This is not the only complication, however, or even the main one. The main complication is in the description of the attitude or orientation of the body in space. To define the orientation of the body in space, we can define three axes (x_B, y_B, z_B) fixed in the body, as shown in Fig. 2.12. One way of defining the attitude of the body is to define the angles between the body axes and the inertial reference axes (x_I, y_I, z_I). These angles are not shown in the diagram. Not only are they difficult to depict in a two-dimensional picture, but they are not always defined the same way. In texts on classical mechanics, the orientation of the body is defined by a set of three angles, called *Euler angles,* which describe the orientation of a set of non-orthogonal axes fixed in the body with respect to the inertial reference axes. In aircraft and space mechanics it is now customary to define the orientation of a set of *orthogonal* axes in the body (*body axes*) with respect to the inertial reference.

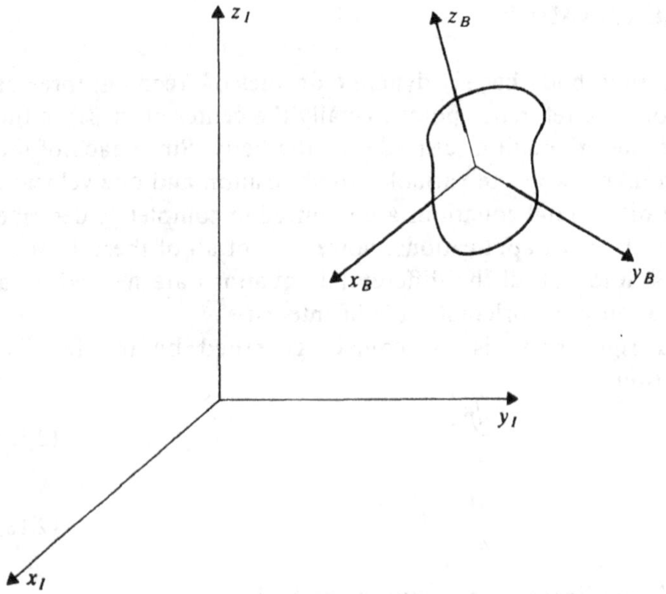

Figure 2.12 Inertial and body-fixed axes.

Suppose the body axes are initially aligned with the inertial reference axes. Then, the following sequence of rotations are made to bring the body axes into *general position*:

First, a rotation ψ (yaw) about the z axis
Second, a rotation θ (pitch) about the resulting y axis
Third, a rotation ϕ (roll) about the resulting x axis

By inspection of the diagrams of Fig. 2.13 we see that

$$\begin{bmatrix} x_{B1} \\ y_{B1} \\ z_{B1} \end{bmatrix} = \begin{bmatrix} \cos\psi & \sin\psi & 0 \\ -\sin\psi & \cos\psi & 0 \\ 0 & 0 & 1 \end{bmatrix} \begin{bmatrix} x_I \\ y_I \\ z_I \end{bmatrix} \tag{2.14}$$

$$\begin{bmatrix} x_{B2} \\ y_{B2} \\ z_{B2} \end{bmatrix} = \begin{bmatrix} \cos\theta & 0 & -\sin\theta \\ 0 & 1 & 0 \\ \sin\theta & 0 & \cos\theta \end{bmatrix} \begin{bmatrix} x_{B1} \\ y_{B1} \\ z_{B1} \end{bmatrix} \tag{2.15}$$

$$\begin{bmatrix} x_B \\ y_B \\ z_B \end{bmatrix} = \begin{bmatrix} x_{B3} \\ y_{B3} \\ z_{B3} \end{bmatrix} = \begin{bmatrix} 1 & 0 & 0 \\ 0 & \cos\phi & \sin\phi \\ 0 & -\sin\phi & \cos\phi \end{bmatrix} \begin{bmatrix} x_{B2} \\ y_{B2} \\ z_{B2} \end{bmatrix} \tag{2.16}$$

Thus we see that

$$\begin{bmatrix} x_B \\ y_B \\ z_B \end{bmatrix} = T_{BI} \begin{bmatrix} x_I \\ y_I \\ z_I \end{bmatrix}$$

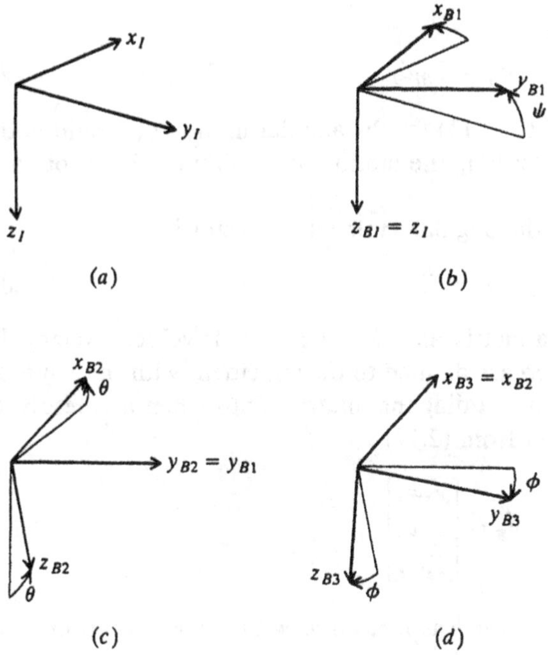

Figure 2.13 Sequence of rotations of body axes from reference to "general" orientation (z axis down in aircraft convention). (a) Axes in reference position; (b) First rotation—about z axis—yaw (ψ); (c) Second rotation—about y axis—pitch (θ); (d) Third and final rotation—about x axis—roll (ϕ).

where T_{BI} is the matrix that rotates the body axes from reference position, and is the product of the three matrices in (2.14)-(2.16).

$$T_{BI} = \begin{bmatrix} 1 & 0 & 0 \\ 0 & \cos\phi & \sin\phi \\ 0 & -\sin\phi & \cos\phi \end{bmatrix} \begin{bmatrix} \cos\theta & 0 & -\sin\theta \\ 0 & 1 & 0 \\ \sin\theta & 0 & \cos\theta \end{bmatrix} \begin{bmatrix} \cos\psi & \sin\psi & 0 \\ -\sin\psi & \cos\psi & 0 \\ 0 & 0 & 1 \end{bmatrix} \tag{2.17}$$

Each factor of T_{BI} is an orthogonal matrix and hence T_{BI} is orthogonal, i.e.,

$$T_{IB} = T_{BI}^{-1} = T'_{BI} \tag{2.18}$$

Note that $T_{BI}^{-1} = T_{IB}$ is the matrix that returns the body axes from the general position to the reference position.

Note that the *order of rotations* implicit in T_{BI} *is important*: the three matrices in (2.17) *do not commute.*

Since any vector in space can be resolved into its components in body axes or in inertial axes, we can use the transformation (2.17) to obtain the components of a vector in one set of axes, given its components in the other. In particular suppose $\vec{a}$ is any vector in space. When it is resolved into components along an inertial reference we attach the subscript I; when it is resolved in body axes, we attach the subscript B

$$\vec{a}_I = \begin{bmatrix} a_{xI} \\ a_{yI} \\ a_{zI} \end{bmatrix} \qquad \vec{a}_B = \begin{bmatrix} a_{xB} \\ a_{yB} \\ a_{zB} \end{bmatrix}$$

Using (2.17) we obtain

$$\vec{a}_B = T_{BI}\vec{a}_I \tag{2.19}$$

This relationship can be applied to (2.13) for the angular motion of a rigid body and, as we shall see later, for describing the motion of an aircraft along rotating body axes.

In the case of a rigid body, the angular momentum vector is

$$\vec{h} = J\vec{\omega} \tag{2.20}$$

where J is the moment of inertia matrix and $\vec{\omega}$ is the angular velocity vector. If the axes along which $\vec{h}$ is resolved are defined to be coincident with the *physical* principal axes of the body, then J is a diagonal matrix. Thus when $\vec{h}$ is resolved along principal body axes, we get from (2.17)

$$\vec{h}_B = \begin{bmatrix} J_x\omega_x \\ J_y\omega_y \\ J_z\omega_z \end{bmatrix} \tag{2.21}$$

But (2.13) holds only when the vector $\vec{h}$ is measured with respect to an inertial reference: In the notation established above

$$\frac{d\vec{h}_I}{dt} = \frac{d}{dt}(T_{IB}\vec{h}_B) = \vec{\tau}_I \tag{2.22}$$

The transformation T_{IB}, however, is not constant. Hence (2.22) must be written

$$T_{IB}\dot{\vec{h}}_B + \dot{T}_{IB}\vec{h}_B = \vec{\tau}_I$$

or, multiplying both sides by $T_{BI} = T_{IB}^{-1}$:

$$\dot{\vec{h}}_B + T_{BI}\dot{T}_{IB}\vec{h}_B = T_{BI}\vec{\tau}_I = \vec{\tau}_B \tag{2.23}$$

which, in component form can be written

$$\begin{bmatrix} J_x\dot{\omega}_{xB} \\ J_y\dot{\omega}_{yB} \\ J_z\dot{\omega}_{zB} \end{bmatrix} + T_{BI}\dot{T}_{IB}\begin{bmatrix} J_x\omega_{xB} \\ J_y\omega_{yB} \\ J_z\omega_{zB} \end{bmatrix} = \begin{bmatrix} \tau_{xB} \\ \tau_{yB} \\ \tau_{zB} \end{bmatrix} \tag{2.24}$$

These differential equations relate the components of the angular velocity vector, $\vec{\omega}$ projected onto rotating body axes

$$\vec{\omega}_B = [\omega_{xB}, \omega_{yB}, \omega_{zB}]'$$

to the torque vector also projected along body axes. To complete (2.24) we need the matrix $T_{BI}\dot{T}_{IB}$. It can be shown that

$$\dot{T}_{IB} = T_{IB}\begin{bmatrix} 0 & -\omega_{zB} & \omega_{yB} \\ \omega_{zB} & 0 & -\omega_{xB} \\ -\omega_{yB} & \omega_{xB} & 0 \end{bmatrix} \tag{2.25}$$

So that

$$T_{BI}\dot{T}_{IB} = \begin{bmatrix} 0 & -\omega_{zB} & \omega_{yB} \\ \omega_{zB} & 0 & -\omega_{xB} \\ -\omega_{yB} & \omega_{xB} & 0 \end{bmatrix} \tag{2.26}$$

(See Note 2.1.)

Hence (2.24) becomes

$$\begin{aligned} J_x\dot{\omega}_{xB} + (J_z - J_y)\omega_{yB}\omega_{zB} &= \tau_{xB} \\ J_y\dot{\omega}_{yB} + (J_x - J_z)\omega_{xB}\omega_{zB} &= \tau_{yB} \\ J_z\dot{\omega}_{zB} + (J_y - J_x)\omega_{xB}\omega_{yB} &= \tau_{zB} \end{aligned} \tag{2.27}$$

These are the famous *Euler equations* that describe how the body-axis components of the angular velocity vector evolve in time, in response to torque components in body axes.

In order to completely define the attitude (orientation), we need to relate the rotation angles ϕ, θ, and ψ to the angular velocity components. One way—not the easiest, however—of obtaining the required relations is via (2.17) and (2.25). It can be shown that

$$\begin{aligned} \dot{\phi} &= \omega_x + (\omega_y \sin\phi + \omega_z \cos\phi)\tan\theta \\ \dot{\theta} &= \omega_y \cos\phi - \omega_z \sin\phi \\ \dot{\psi} &= (\omega_x \sin\phi + \omega_y \cos\phi)/\cos\theta \end{aligned} \tag{2.28}$$

These relations, also nonlinear, complete the description of the rigid body dynamics.

Example 2F The gyroscope One of the most interesting applications of Euler's equations is to the study of the gyroscope. This device (also the spinning top) has fascinated mathematicians and physicists for over a century. (See Note 2.2.) And the gyroscope is an extremely useful sensor of aircraft and spacecraft motion. Its design and control has been an important technological problem for half a century.

In an ideal gyroscope the rotor, or "wheel," is kept spinning at a constant angular velocity. (A motor is provided to overcome the inevitable friction torques present even in the best of instruments. The precise control of wheel speed is another important control problem.) Suppose that the axis through the wheel is the body z axis. We assume that τ_{zB} is such that $\dot{\omega}_{zB} = 0$, i.e., that

$$H_z = J_z\omega_z = \text{const} \tag{2F.1}$$

(J_z is called the "polar" moment of inertia in gyro parlance.) We can also assume that the gyroscope wheel is a "true" wheel: that the z axis is an axis of symmetry, and hence that

$$J_x = J_y = J_d \qquad \text{(the "diametrical" moment of inertia)}$$

The first two equations of (2.27) then become

$$\begin{aligned} \dot{\omega}_{xB} + \frac{H}{J_d}\omega_{yB} &= \frac{\tau_x}{J_d} \\ \dot{\omega}_{yB} - \frac{H}{J_d}\omega_{xB} &= \frac{\tau_y}{J_d} \end{aligned} \tag{2F.2}$$

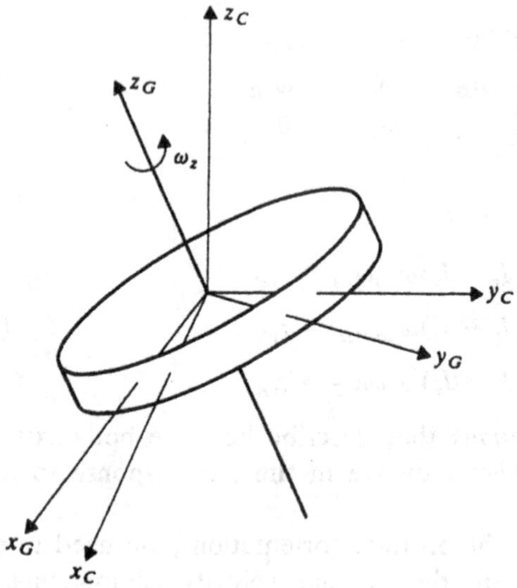

Figure 2.14 Two-degrees-of-freedom gyro wheel.

where

$$H = H_z\left(1 - \frac{J_d}{J_z}\right)$$

To use a gyro as a sensor, the wheel is mounted in an appropriate system of gimbals which permit it to move with respect to the outer case of the gyro. In a two-axis gyro, the wheel is permitted two degrees of freedom with respect to the case, as depicted in Fig. 2.14. The case of the gyro is rigidly attached to the body whose motion is to be measured.

The range of motion of the wheel about its x and y body axes relative to the gyro case is very small (usually a fraction of a degree). Hence the gyro must be "torqued" about the axes in the plane normal to the spin axis to make the wheel keep up with its case, and as we shall see shortly, the torque required to do this is a measure of the angular velocity of the case.

Since the motion of the wheel relative to the case is very small, we do not need equations like (2.27) to relate the angular displacements of the gyro wheel from its null positions in the case. We can write

$$\begin{aligned} \dot{\delta}_x &= \omega_{xB} - \omega_{xE} \\ \dot{\delta}_y &= \omega_{yB} - \omega_{yE} \end{aligned} \tag{2F.3}$$

where ω_{xE} and ω_{yE} are the external angular velocities that the gyro is to measure.

These equations, together with (2F.2), constitute the basic equations of an ideal gyro. A block-diagram representation of (2F.2) and (2F.3), and a closed-loop feedback system for controlling the gyro is shown in Fig. 2.15. The feedback system shows the control torques generated as functions of the displacements δ_x and δ_y. These displacements can be measured by means of "pick-offs"—small magnetic sensors located on the case and capable of measuring small tilts of the wheel. The control torque needed to drive the "pick-off angles" δ_x and δ_y to zero can also be generated magnetically. In some designs the pick-off and torquer functions can be combined in a single device. The control system is designed to drive the angular displacements δ_x and δ_y to zero. If this is accomplished

$$\omega_{xB} = \omega_{xE} \qquad \omega_{yB} = \omega_{yE} \tag{2F.4}$$

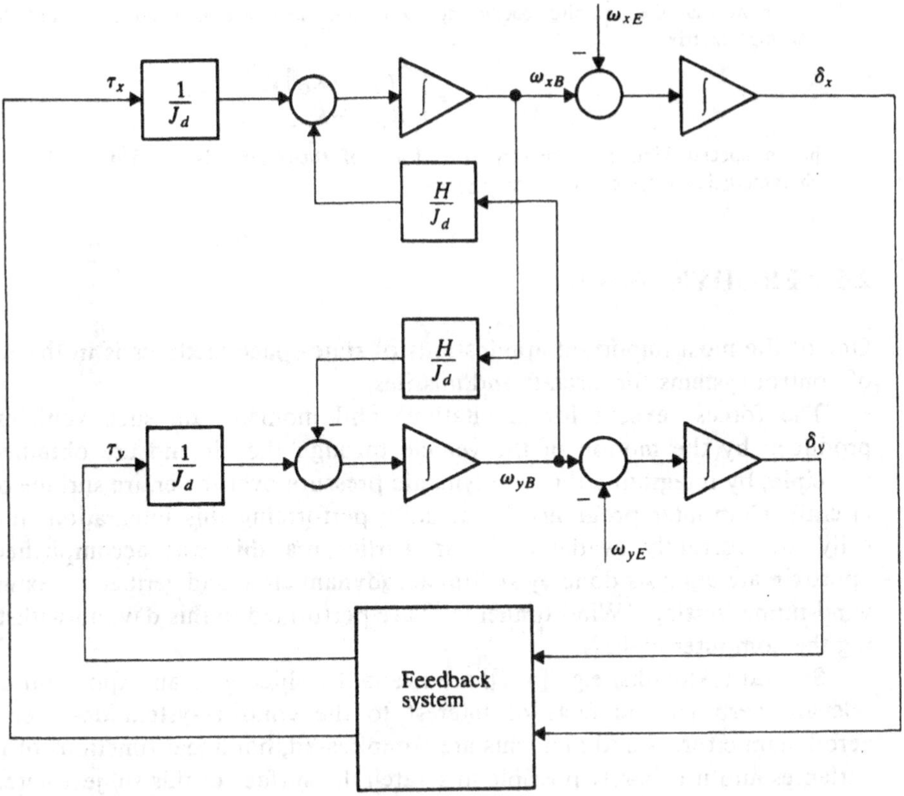

Figure 2.15 Block diagram of two-axis gyro dynamics showing "capture" control system.

If the angular velocity components ω_{xB} and ω_{yB} are constant

$$\begin{aligned} \tau_x &= H\omega_{yB} = H\omega_{yE} \\ \tau_y &= -H\omega_{xB} = -H\omega_{xE} \end{aligned} \tag{2F.5}$$

where H is a constant of the gyro. If this constant is accurately calibrated, and if the input torque to the gyro is accurately metered, then the steady state torques about the respective axes that keep the wheel from tilting relative to its case (i.e., "capture" the wheel) are proportional to the measured external angular velocity components.

The control system that keeps the wheel captured is an important part of every practical gyro. Some of the issues in the design of such a control system will be the subject of problems in later chapters.

The differential equations of (2F.2) are idealized to the point of being all but unrealistic. In addition to the control torques acting on the gyro, other torques, generated internal to the gyro, are also inevitably present. These include damping torques (possibly aerodynamic). And in a so-called tuned-rotor gyro, the gimbals are implemented by a special flexure hinge which produces small but not insignificant spring torques. When these torques are included, (2F.2) becomes

$$\dot{\omega}_{xB} = \frac{H}{J_d}\omega_{yB} - \frac{B}{J_d}(\omega_{xB} - \omega_{xE}) - \frac{K_D}{J_d}\delta_x - \frac{K_Q}{J_d}\delta_y + \frac{\tau_x}{J_d}$$

$$\dot{\omega}_{yB} = -\frac{H}{J_d}\omega_{xB} - \frac{B}{J_d}(\omega_{yB} - \omega_{yE}) - \frac{K_D}{J_d}\delta_y + \frac{K_Q}{J_d}\delta_x + \frac{\tau_y}{J_d}$$

Note that the damping coefficients D in both axes are assumed equal and that the "spring" matrix

$$K = \begin{bmatrix} -K_D & -K_Q \\ K_Q & -K_D \end{bmatrix}$$

has a special kind of symmetry. This form of the matrix is justified by the physical characteristics of typical tuned-rotor gyros.

2.6 AERODYNAMICS

One of the most important applications of state-space methods is in the design of control systems for aircraft and missiles.

The forces (except for gravitation) and moments on such vehicles are produced by the motion of the vehicle through the air and are obtained, in principle, by integrating the aerodynamic pressure over the entire surface of the aircraft. Computer programs for actually performing this integration numerically are currently available. In an earlier era this was accomplished by approximate analysis done by skillful aerodynamicists, and verified by extensive wind-tunnel testing. (Wind-tunnel tests are performed to this day, notwithstanding the computer codes.)

Several textbooks, e.g., [4, 5], are available which give an exposition of the relevant aerodynamic facts of interest to the control system designer. The aerodynamic forces and moments are complicated, nonlinear functions of many variables and it is barely possible to scratch the surface of this subject here. The purpose of this section is to provide only enough of the principles as are needed to motivate the design examples to be found later on in the book.

The aerodynamic forces and moments depend on the velocity of the aircraft relative to the air mass. In still air (no winds) they depend on the velocity of the aircraft along its own body axes: the orientation of the aircraft is not relevant in determining the aerodynamic forces and moments. But, since the natural axes for resolving the aerodynamic forces and moments are moving (rotating and accelerating), it is necessary to formulate the equations of motion in the moving coordinate system.

The rotation motion of a general rigid body has been given in (2.24). In aircraft terminology the projections of the angular velocity vector on the body x, y, and z axes have standard symbols:

$$\begin{aligned} \omega_x &= p \quad \text{(roll rate)} \\ \omega_y &= q \quad \text{(pitch rate)} \qquad (2.29) \\ \omega_z &= r \quad \text{(yaw rate)} \end{aligned}$$

(The logic of using three consecutive letters of the alphabet (p, q, r) to denote the projections of the angular velocity vector on the three consecutive body axes is unassailable. But the result is "amnemonic" (hard to remember): p does *not* represent pitch rate and r does *not* represent roll rate.)

Thus, assuming that the body axes are the principal axes of the aircraft, the rotational dynamics are expressed as

$$\begin{aligned}
\dot{p} &= \frac{L}{J_x} - \frac{J_z - J_y}{J_x} qr \\
\dot{q} &= \frac{M}{J_y} - \frac{J_x - J_z}{J_y} pr \\
\dot{r} &= \frac{N}{J_z} - \frac{J_y - J_x}{J_z} pq
\end{aligned} \tag{2.30}$$

where L, M, and N are the aerodynamic moments about the body x, y, and z axes respectively. Thus L is the rolling moment, M is the pitching moment, and N is the yawing moment. These are functions of various dynamic variables, as explained later.

To define the translational motion of an aircraft it is customary to project the velocity vector onto body fixed axes

$$\vec{v}_B = \begin{bmatrix} u \\ v \\ w \end{bmatrix} \tag{2.31}$$

where u, v, and w are the projections of the vehicle velocity vector onto the body x, y, and z axes. The linear momentum of the body, in an inertial frame, is

$$\vec{p} = m\vec{v}_I = mT_{IB}\vec{v}_B$$

Hence, the dynamic equations for translation are

$$\frac{d}{dt}(mT_{IB}\vec{v}_B) = m\left(T_{IB}\frac{d\vec{v}_B}{dt} + \dot{T}_{IB}\vec{v}_B \right) = \vec{f}_I \tag{2.32}$$

where $\vec{f}_I$ are the external forces acting on the aircraft referred to an inertial frame. Proceeding as we did in developing (2.24) we find that

$$\frac{d\vec{v}_B}{dt} = -T_{BI}\dot{T}_{IB}\vec{v}_B + \frac{1}{m}\vec{f}_B \tag{2.33}$$

where $\vec{f}_B = T_{BI}\vec{f}_I$ is the force acting on the aircraft resolved along the body-fixed axes and

$$T_{BI}\dot{T}_{IB} = \begin{bmatrix} 0 & -r & q \\ r & 0 & -p \\ -q & p & 0 \end{bmatrix} \tag{2.34}$$

as given by (2.26) but using the p, q, r notation defined in (2.29).

In component form (2.33) becomes

$$\dot{u} = rv - qw + \frac{1}{m} f_{xB}$$

$$\dot{v} = -ru + pw + \frac{1}{m} f_{yB} \tag{2.35}$$

$$\dot{w} = qu - pv + \frac{1}{m} f_{zB}$$

where f_{xB}, f_{yB}, and f_{zB} are the total forces (engine, aerodynamic, *and* gravitational) acting on the body. Since the aircraft axes are not in general in the direction of the gravity vector, each component f_{xB}, f_{yB}, and f_{zB} will have a term due to gravity. In addition to the force of gravity, there is the thrust force produced by the aircraft engine—generally assumed to act along the vehicle x axis—and the aerodynamic forces—the lift and drag forces. The acceleration terms rv, qw, etc., are Coriolis accelerations due to the rotation of the body axes.

Complete dynamic equations of the vehicle consist of (2.30) which give the angular accelerations, (2.35) which give the linear accelerations, (2.28) which give the angular orientation, and finally the equations for the vehicle position:

$$\begin{bmatrix} \dot{x} \\ \dot{y} \\ \dot{z} \end{bmatrix} = \dot{\vec{r}} = \vec{v}_I = T_{IB} \begin{bmatrix} u \\ v \\ w \end{bmatrix} \tag{2.36}$$

This system of 12 first-order differential equations, with the moments and forces evaluated as functions of whatever they depend upon constitute the complete six-degrees-of-freedom description of the aircraft behavior.

The aerodynamic forces and moments all depend on the *dynamic pressure*

$$Q = \tfrac{1}{2}\rho V^2 \tag{2.37}$$

where ρ is the air density and

$$V = (u^2 + v^2 + w^2)^{1/2}$$

is the speed of the aircraft. (Dynamic pressure has the dimension of force per unit area.) Thus the aerodynamic forces and moments can be expressed in the form

$$\begin{aligned} f_{xA} &= QAC_x \\ f_{yA} &= QAC_y \\ f_{zA} &= QAC_z \\ L &= lQAC_L \\ M &= lQAC_M \\ N &= lQAC_N \end{aligned} \tag{2.38}$$

where C_x, C_y, C_z, C_L, C_M, C_N, are *dimensionless* aerodynamic "coefficients," A

is a reference area (usually the frontal area of the vehicle), and l is a reference length. (In some treatments different reference lengths are used for roll, pitch, and yaw.)

The aerodynamic coefficients in turn are functions of the vehicle velocity (linear and angular) components, and, for movable control surfaces, also functions of the deflections of the surfaces from their positions of reference. The variables of greatest influence on the coefficients are the vehicle speed (or, more precisely, the Mach number), the *angle-of-attack* α and the *side-slip angle* β. These, respectively, define the direction of the velocity vector relative to the vehicle body axes; α is the angle that the velocity vector makes with respect to the longitudinal axis in the pitch direction and β is the angle it makes with respect to the longitudinal axis in the yaw direction. (See Fig. 2.16.) From the figure

$$
\begin{aligned}
\alpha &= \tan^{-1}\left(\frac{w}{(u^2 + v^2)^{1/2}}\right) \simeq \frac{w}{u} \\
\beta &= \tan^{-1}\left(\frac{v}{u}\right) \simeq \frac{v}{u}
\end{aligned}
\tag{2.39}
$$

with the approximate expressions being valid for small angles.

For purposes of control system design, the aircraft dynamics are frequently linearized about some operating condition or "flight regime," in which it is assumed that the aircraft velocity and attitude are constant. The control surfaces and engine thrust are set, or "trimmed," to these conditions and the control system is designed to maintain them, i.e., to force any perturbations from these conditions to zero.

If the forward speed is approximately constant, then the angle of attack and angle of side slip can be used as state variables instead of w and v, respectively.

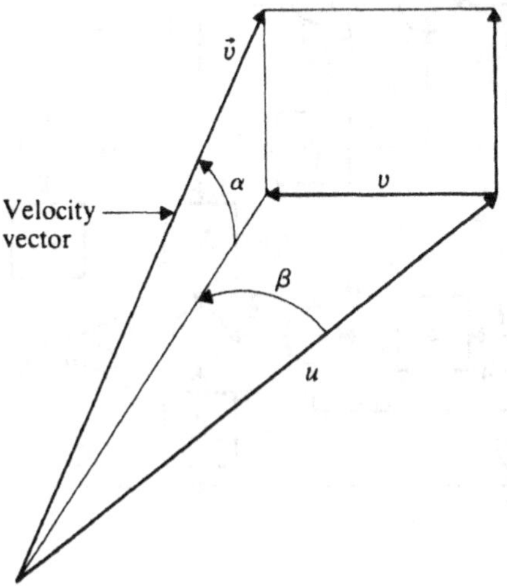

Figure 2.16 Definitions of angle-of-attack α and side-slip angle β.

Table 2.1 Aerodynamic variables

	Longitudinal	Lateral
Rates	α: angle of attack q: pitch rate Δu: change in speed	β: side slip angle p: roll rate r: yaw rate
Positions	θ: pitch z: altitude	ϕ: roll angle ψ: yaw angle x: forward displacement y: cross-track displacement
Controls	δ_E: elevator deflection	δ_A: aileron deflection δ_R: rudder deflection

Also in studying small perturbations from trim conditions it is customary to separate the longitudinal motion from the lateral motion. In many cases the lateral and longitudinal dynamics are only lightly coupled, and the control system can be designed for each channel without regard to the other. The variables are grouped as shown in Table 2.1.

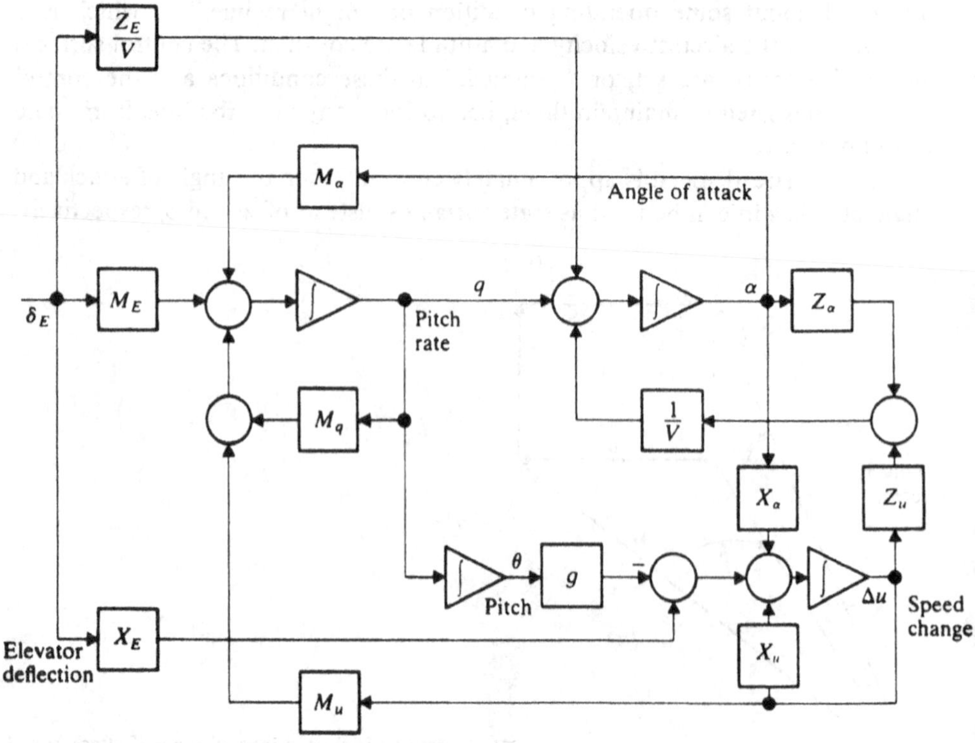

Figure 2.17 Aircraft longitudinal dynamics.

The aircraft pitch motion is typically controlled by a control surface called the *elevator* (or by canards in the front of the vehicle). The roll is controlled by a pair of ailerons, and the yaw is controlled by a rudder. These are also shown in Table 2.1.

The function of most control system designs is to regulate *small* motions rather than to control absolute position (x, y, and z). Thus the inertial position is frequently not included in the state equations. This leaves nine equations, four in the longitudinal channel and five in the lateral channel. These can be written in the following form:

Longitudinal dynamics (See Fig. 2.17)

$$
\begin{aligned}
\Delta \dot{u} &= X_u \Delta u + X_\alpha \alpha - g\theta + X_E \delta_E \\
\dot{\alpha} &= \frac{Z_u}{V} \Delta u + \frac{Z_\alpha}{V} \alpha + q + \frac{Z_E}{V} \delta_E \\
\dot{q} &= M_u \Delta u + M_\alpha \alpha + M_q q + M_E \delta_E \\
\dot{\theta} &= q
\end{aligned}
\tag{2.40}
$$

Lateral dynamics (See Fig. 2.18)

$$
\begin{aligned}
\dot{\beta} &= \frac{Y_\beta}{V} \beta + \frac{Y_p}{V} p + \left(\frac{Y_r}{V} - 1\right) r + \frac{g}{V} \phi + \frac{Y_A}{V} \delta_A + \frac{Y_R}{V} \delta_R \\
\dot{p} &= L_\beta \beta + L_p p + L_r r + L_A \delta_A + L_R \delta_R \\
\dot{r} &= N_\beta \beta + N_p p + N_r r + N_A \delta_A + N_R \delta_R \\
\dot{\phi} &= p \\
\dot{\psi} &= r
\end{aligned}
\tag{2.41}
$$

The symbols X, Y, Z, L, M, and N, with subscripts have become fairly standardized in the field of aircraft and missile control, although the sign conventions often differ from one user to another, which can often cause consternation. The symbols with the capital-letter subscripts, E, A, and R (for elevator, ailerons, and rudder), however, are not standard. It is customary to use cumbersome double subscript notation for these quantities.

2.7 CHEMICAL AND ENERGY PROCESSES

It is often necessary to control large industrial processes which involve heat exchangers, chemical reactors, evaporators, furnaces, boilers, driers, and the like.

Because of their large physical size, such processes have very slow dynamic behavior—measured on a scale of minutes or hours rather than seconds as in the case for aircraft and instrument controls. Such processes are often slow

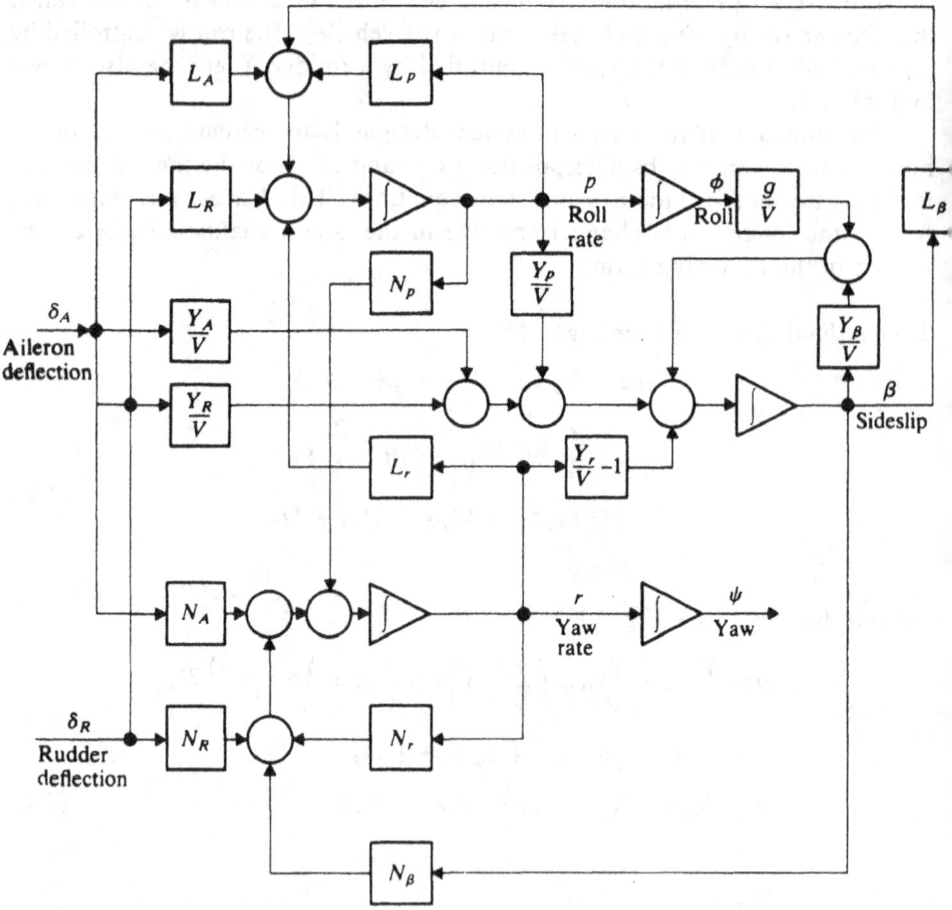

Figure 2.18 Aircraft lateral dynamics.

enough to be controlled manually: an experienced plant operator can monitor the instruments in the control room and (remotely) open and close the valves to maintain a satisfactory equilibrium condition. But slow as such processes are, they are not necessarily stable. If the operator is not constantly monitoring and actively intervening in its operation the process may run away with itself. The "Three-Mile Island" nuclear plant accident (See Note 2.3), is perhaps the most dramatic episode of this kind in recent years, but other episodes, less dramatic than Three-Mile Island, occur with regrettable frequency.

Gross failures of the type of Three-Mile Island are probably not traceable to inadequate dynamic performance of the control system, but rather to failures in hardware that inadequate procedures and training permit to go without prompt repair. The considerations that apply to design of proper procedures and personnel practices are outside the scope of this book. Here we are concerned with the design of systems for normal, closed-loop operation, i.e., under the assumption that the sensors and actuators are maintained in good working order. Often so much of the engineering effort is spent in selecting suitable hardware—and this effort is totally justified—that little time is left to consider efficient operation under normal conditions. Large industrial processes are costly to operate, however, and even small percentage improvements (such as in reduced energy consumption) can be worth a considerable sum. There is a celebrated design (See Note 2.4) in the paper industry in which a small improvement in the product quality ("base weight" of the paper) returned many times the cost of installing a computer control system.

One of the difficulties in working with large industrial processes is that they involve subsystems the behavior of which are not readily characterized by simple mathematical models. The physics and chemistry of devices like evaporators, heat exchangers, driers, and the like, are not as amenable to mathematical representation as are the physics of simple electromechanical systems, or even of aircraft. It is often necessary to work with empirical models obtained by fitting curves to test data. And test data is often hard to come by because the processes are slow and there is considerable reluctance to shut them down long enough to amass a sufficient quantity of data with which to construct an empirical model.

Still another difficulty in dealing with industrial processes is the large number of dynamic variables that must be considered. Unless suitable simplifications are made, the number of variables can run into the hundreds. Although the methods described in this book can be used for designing control systems for very high order processes, the insights that the engineer often can develop using low-order models will be lacking.

To show how state-space methods can be applied to industrial processes we have selected several examples that have been described in the literature and are actually in operation. These examples show that it is possible to deal effectively with processes of considerable complexity using models of reasonably low order.

Example 2G Distillation column A distillation column is a complex process. A large number of variables (upward of 100) are needed to accurately model its dynamic behavior.

In the interest of applying modern control techniques to the design of a control system for a distillation column, Gilles and Retzbach[6, 7, 8] manage ingeniously to reduce the number of state variables to only 4. Their study deals with an extractive column intended for separating isopropanol from a mixture with water, using glycol as an extractant. A schematic diagram of the column is shown in Fig. 2.19. The mixture of water and isopropanol is introduced at the feed stage F_A and the glycol extractant is introduced near the top of the column. A controlled amount of heating steam is introduced near the bottom of the column where the *bottom product*—the extractant, glycol, is drawn off. In addition, the *vapor side stream* flow rate can be controlled by another valve. The objective of the process is to produce nearly pure isopropanol at the top of the column.

The key to the simplified model of the distillation column developed by Gilles and Retzbach are the profiles of concentration and temperature in the column, sketched in Fig. 2.19. There are two vertical locations in the tower at which the principal physical changes occur: z_1 at which there is an interphase change between water and isopropanol, and a second location z_2 where there is an interphase change between the water and the glycol extractant. At each of these locations there is a sharp temperature gradient.

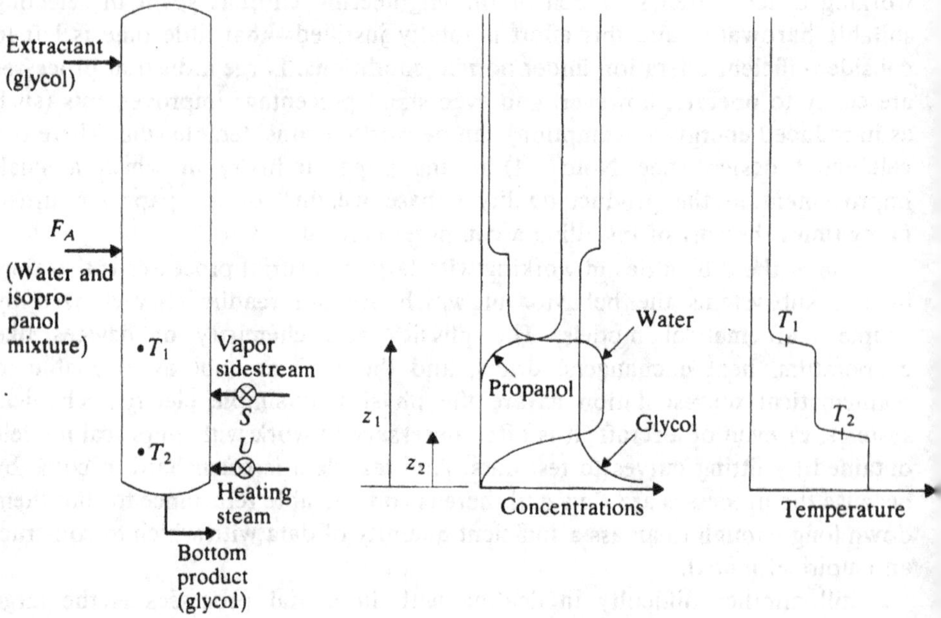

Figure 2.19 Distillation column with profiles of concentration and temperature. *(After Gilles and Retzbach.)*

By varying the flow rates of the water-isopropanol mixture, the heating steam, and the vapor side stream, the positions of these loci or "fronts" can be moved up and down, but the shapes of the distributions are otherwise hardly changed. Thus by controlling the positions of these fronts, the distribution of temperature and concentration can be controlled throughout the column. This property of the distributions motivated Gilles and Retzbach to use the positions z_1 and z_2 as state variables that can adequately represent the behavior of this complex process.

In addition to these state variables, other state variables needed to represent the steam boiler are included in the overall model.

The boiler dynamics are represented by

$$\begin{aligned} \Delta \dot{Q}_l &= a_{11}\Delta Q_l + b_{11}\Delta u_1 \\ \Delta \dot{V} &= a_{21}\Delta Q_l + a_{22}\Delta V_l \end{aligned} \tag{2G.1}$$

where ΔQ_l = heat flow to reboiler "holdup"
ΔV_l = vapor flow rate
Δu_1 = steam flow rate
} changes from equilibrium

Gilles and Retzbach in [6] show that rates of change $\Delta \dot{z}_1$ and $\Delta \dot{z}_2$ in the position of the interphase loci (fronts) are linearly related to the various flow rates:

$$\begin{aligned} \Delta \dot{z}_1 &= b_{32}\Delta S + f_{31}\Delta x_{FA1} + f_{32}\Delta F_A \\ \Delta \dot{z}_2 &= b_{42}\Delta S + f_{42}\Delta F_A \end{aligned} \tag{2G.2}$$

where ΔS = flow rate of vapor side stream
Δx_{FA1} = feed composition
ΔF_A = feed flow rate
} changes from equilibrium

As noted, the steam flow rate and the flow rate of the vapor side-stream are control variables. Changes in the feed composition and flow rate are disturbances that the control system is to be designed to counteract.

The positions of the fronts are determined in this process by measuring the temperatures with thermomocouples located near the desired positions of the fronts. It has been found that the temperature changes are approximately proportional to the front position changes:

$$\begin{aligned} \Delta T_1 &= c_{13}\Delta z_1 \\ \Delta T_2 &= c_{24}\Delta z_2 \end{aligned} \tag{2G.3}$$

The state, input, disturbance, and observation vectors are defined respectively by

$$x = \begin{bmatrix} \Delta Q_l \\ \Delta V_l \\ \Delta z_1 \\ \Delta z_2 \end{bmatrix} \qquad u = \begin{bmatrix} \Delta u_1 \\ \Delta S \end{bmatrix} \qquad x_0 = \begin{bmatrix} \Delta x_{FA1} \\ \Delta F_A \end{bmatrix} \qquad y = \begin{bmatrix} \Delta T_1 \\ \Delta T_2 \end{bmatrix}$$

In terms of these variables, the process has the standard state-space representation

$$\begin{aligned} \dot{x} &= Ax + Bu + Ex_0 \\ y &= Cx \end{aligned} \tag{2G.4}$$

with

$$A = \begin{bmatrix} a_{11} & 0 & 0 & 0 \\ a_{21} & a_{22} & 0 & 0 \\ 0 & a_{32} & 0 & 0 \\ 0 & a_{42} & 0 & 0 \end{bmatrix} \qquad B = \begin{bmatrix} b_{11} & 0 \\ 0 & 0 \\ 0 & b_{32} \\ 0 & b_{42} \end{bmatrix} \qquad E = \begin{bmatrix} 0 & 0 \\ 0 & 0 \\ f_{31} & f_{32} \\ 0 & f_{42} \end{bmatrix} \tag{2G.5}$$

$$C = \begin{bmatrix} 0 & 0 & c_{13} & 0 \\ 0 & 0 & 0 & c_{24} \end{bmatrix}$$

Numerical data for a specific process considered by Gilles and Retzbach are as follows:

$$\begin{aligned} a_{11} &= -30.3 & b_{11} &= 6.15 \times 10^5 & f_{31} &= 62.2 \\ a_{21} &= 0.120 \times 10^{-3} & b_{32} &= 3.04 & f_{32} &= 5.76 \\ a_{22} &= -6.02 & b_{42} &= 0.052 & f_{42} &= 5.12 \\ a_{32} &= -3.77 \\ a_{42} &= -2.80 & c_{13} &= -7.3 \\ & & c_{24} &= -25.0 \end{aligned}$$

Time is measured in hours, and temperature in degrees Celsius.

Example 2H Double effect evaporator Over a period of several years in the mid 1970s a group of chemical engineering faculty members and students at the University of Alberta developed a laboratory pilot plant which could be used to test various concepts and control system design techniques. The results of some of these studies have been published in a number of technical journals and reprinted as a case study [9].

The pilot plant is a double-effect evaporator shown schematically in Fig. 2.20. According to Professors D. G. Fisher and D. E. Seborg, leaders of the project and authors of the case study: "The first effect is a short-tube vertical calandria-type unit with natural circulation. The 9-in diameter unit has an operating holdup of 2 to 4 gallons, and its 32 stainless steel tubes, $\frac{3}{4}$-in o.d. by 18 in. long, provide approximately 10 square feet of heat transfer surface altogether.

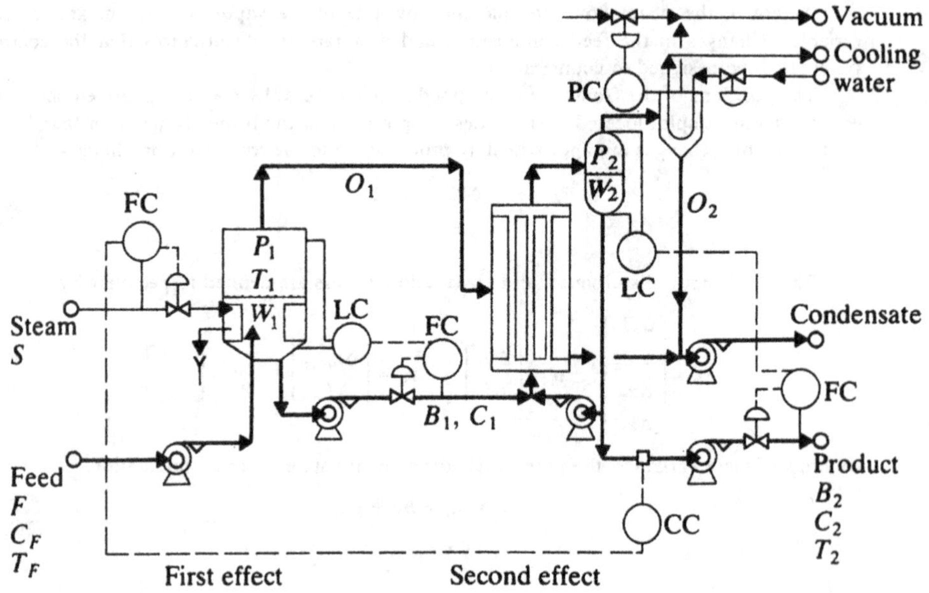

Figure 2.20 Schematic diagram of pilot-scale evaporator and a conventional multiloop control system. (*Reproduced with permission of "Industrial Engineering Chemistry, Process Design and Development" Copyright 1972, American Chemical Society.*)

"The second stage is a long-tube vertical effect setup for either natural or forced circulation. It has a heat transfer area of 5 square feet and is made up of three 6-ft long 1-in o.d. tubes. Capacity of the circulating system is about 3 gallons."[10]

The inputs to the plant are steam and a concentrated solution of triethylene glycol. The outputs are glycol, the concentration and flow rate of which is to be controlled, and the condensate.

The system is a relatively complicated dynamic process requiring many state variables for its accurate description. A number of studies, reported in [11], were undertaken aimed at developing a model that represents a reasonable compromise between fidelity to the real process and amenability to control system designs. On the basis of such considerations the investigators found that a fifth-order model is in close agreement with a tenth-order model, the latter fitting the pilot plant test data very well.

The fifth-order model uses the state variables

$$\begin{aligned} x_1 &= W_1 = \text{first-effect "holdup"} \\ x_2 &= C_1 = \text{first-effect concentration} \\ x_3 &= H_1 = \text{first-effect enthalpy} \\ x_4 &= W_2 = \text{second-effect "holdup"} \\ x_5 &= C_2 = \text{second-effect concentration} \end{aligned} \tag{2H.1}$$

and control variables

$$\begin{aligned} u_1 &= S_1 = \text{first-effect steam flow rate} \\ u_2 &= B_1 = \text{first-effect "bottoms" flow rate} \\ u_3 &= B_2 = \text{second-effect "bottoms" flow rate} \end{aligned} \tag{2H.2}$$

In addition to the state and control variables there are also disturbance inputs to the process

$$\begin{aligned} d_1 &= F_1 = \text{variations in feed flow rate} \\ d_2 &= C_{F1} = \text{variations in feed concentration} \\ d_3 &= H_{F1} = \text{variations in feed enthalpy} \end{aligned} \tag{2H.3}$$

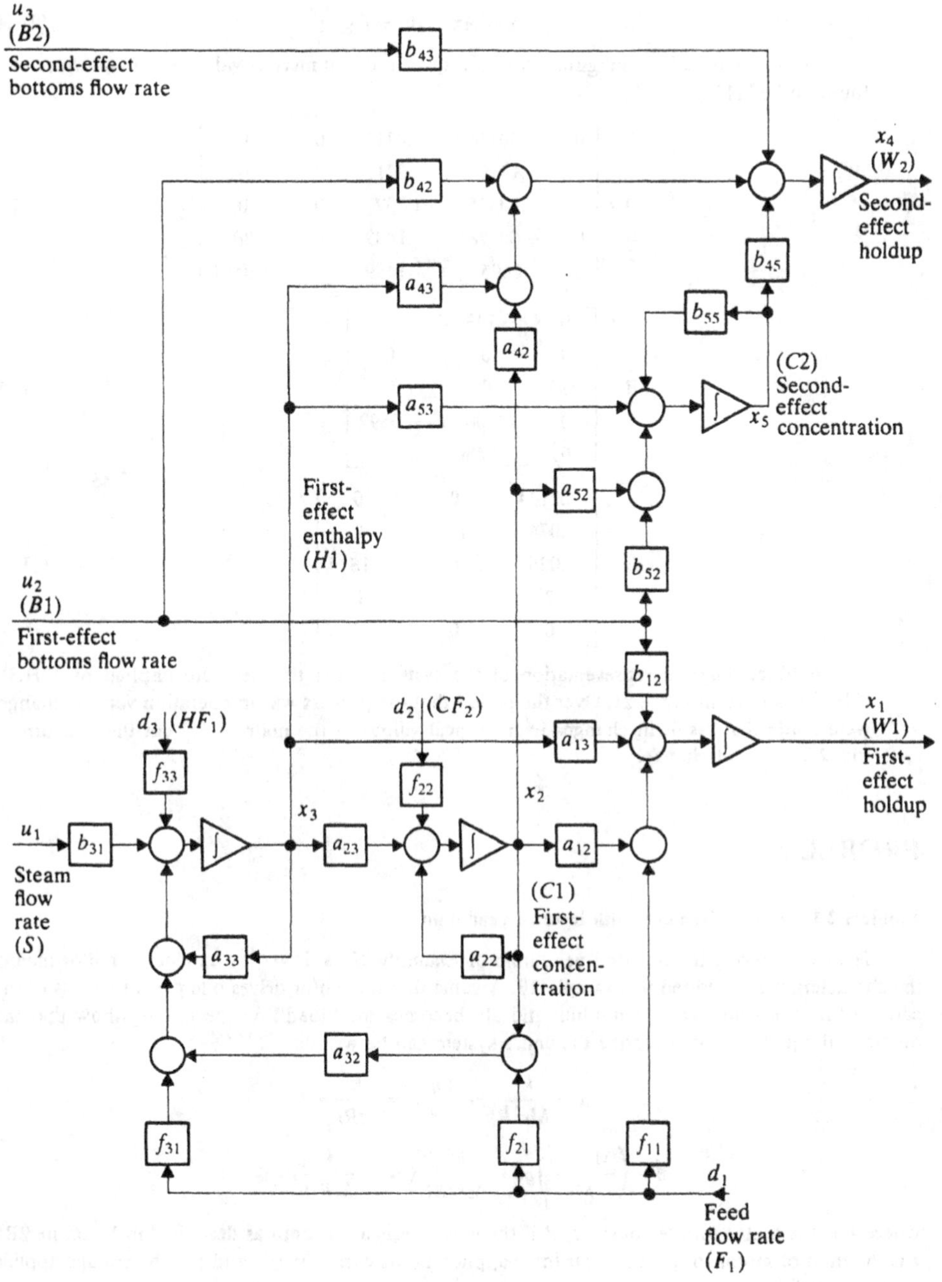

Figure 2.21 Fifth-order evaporator dynamic model.

The linearized differential equations for this process have been developed by Newell and Fisher[9] and are in the standard state space form

$$\dot{x} = Ax + Bu + Ex_0 \tag{2H.4}$$

For one particular configuration of the system, the numerical values of the matrices were found to be [11]

$$A = \begin{bmatrix} 0 & -.00156 & -.0711 & 0 & 0 \\ 0 & -.1419 & .0711 & 0 & 0 \\ 0 & -.00875 & -1.102 & 0 & 0 \\ 0 & -.00128 & -.1489 & 0 & -.0013 \\ 0 & .0605 & .1489 & 0 & -.0591 \end{bmatrix} \tag{2H.5}$$

$$B = \begin{bmatrix} 0 & -.143 & 0 \\ 0 & 0 & 0 \\ .392 & 0 & 0 \\ 0 & .108 & -.0592 \\ 0 & -.0486 & 0 \end{bmatrix} \tag{2H.6}$$

$$E = \begin{bmatrix} .2174 & 0 & 0 \\ -.074 & .1434 & 0 \\ -.036 & 0 & .1814 \\ 0 & 0 & 0 \\ 0 & 0 & 0 \end{bmatrix} \tag{2H.7}$$

A block-diagram representation of the system, using the structure implied by (2H.5)-(2H.7) is shown in Fig. 2.21. Over the period that the process was in operation various changes were made that result in changes in numerical values in the matrices,[9] but the structure of Fig. 2.21 did not change.

PROBLEMS

Problem 2.1 Motor-driven cart with inverted pendulum

The cart carrying the inverted pendulum of Example 2E is driven by an electric motor having the characteristics described in Example 2B. Assume that the motor drives one pair of wheels of the cart, so that the whole cart, pendulum and all, becomes the "load" on the motor. Show that the differential equations that describe the entire system can be written

$$\ddot{x} + \frac{k^2}{Mr^2R}\dot{x} + \frac{mg}{M}\theta = \frac{k}{MRr}e$$

$$\ddot{\theta} - \left(\frac{M+m}{Ml}\right)g\theta - \frac{k^2}{Mr^2Rl}\dot{x} = -\frac{k}{MRrl}e$$

where k is the motor torque constant, R is the motor resistance (both as described in Example 2B), r is the ratio of motor torque to linear force applied to the cart ($\tau = rf$), and e is the voltage applied to the motor.

Problem 2.2 Motor-driven inverted pendulum

Derive the dynamic model for an inverted pendulum pivoted at its lower end and driven by an electric motor, as shown in Fig. 6.3.

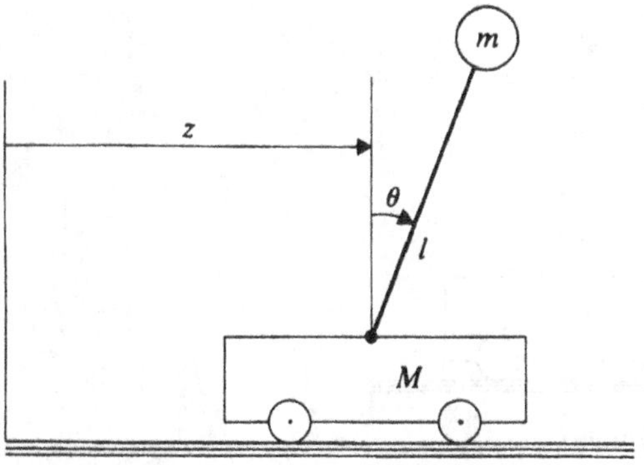

Figure P2.1 Inverted pendulum on cart.

Show that the dynamic equations of the inverted pendulum on a cart of Prob. 2.1 reduce to that of a pendulum fixed at its lower end as the mass of the cart becomes infinite.

Problem 2.3 Three-capacitance thermal system

A conducting bar (Fig. P2.3(*a*)) is insulated along its length but exposed to the ambient temperature at one end, and heated at the other end. An approximate electrical equivalent, based on "lumping" the bar into three finite lengths, is shown in Fig. P2.3(*b*).

Write the differential equations for the system using as state variables v_1, v_2, and v_3, the capacitor voltages. The input u is the temperature e_0 at the heated end, and the output y is the temperature v_3 at point 3 on the rod, as would be determined by a thermocouple, for example.

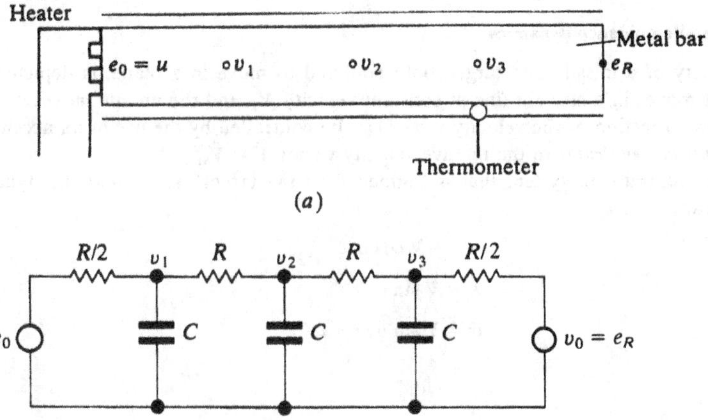

(*a*)

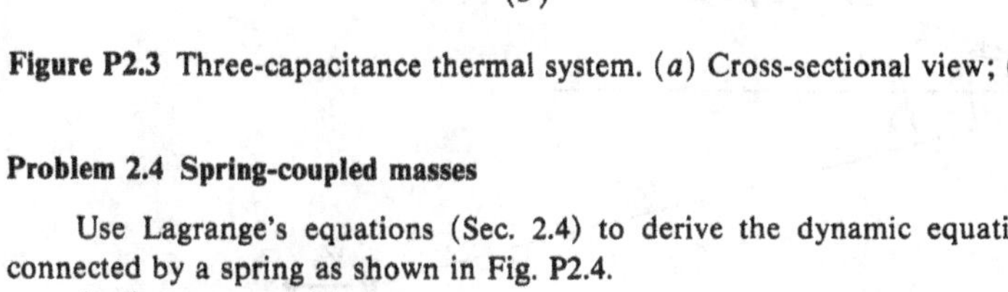

(*b*)

Figure P2.3 Three-capacitance thermal system. (*a*) Cross-sectional view; (*b*) Electrical analog.

Problem 2.4 Spring-coupled masses

Use Lagrange's equations (Sec. 2.4) to derive the dynamic equations of a pair of masses connected by a spring as shown in Fig. P2.4.

As the state variables use

$$x_1 = z_1 \qquad x_2 = z_2 \qquad x_3 = \dot{z}_1 \qquad x_4 = \dot{z}_2$$

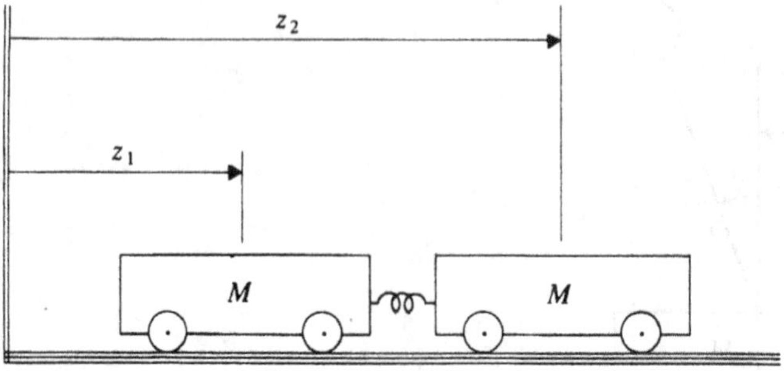

Figure P2.4 Spring-coupled masses (two-car train).

Problem 2.5 Two-car train

An idealized two-car train consists of a pair of masses coupled by a spring, as shown in Fig. P2.4. The wheels of each car are independently driven by an electric motor such as described in Example 2B. (Also see Prob 2.1.)

(*a*) Express the differential equations of the system in state-space form. (Find the A and B matrices.) Assume R is the motor resistance, K is the spring constant, k is the motor torque constant, and $r = \tau/f$ is the ratio of the motor torque to the linear force applied to the car. Use the following state and control variables

$$x_1 = z_1 \qquad x_2 = z_2 \qquad x_3 = \dot{z}_1 \qquad x_4 = \dot{z}_2 \qquad u_1 = e_1 \qquad u_2 = e_2$$

where e_1 and e_2 are the voltages on the drive motors.

(*b*) Draw the block diagram of the system.

Problem 2.6 Missile guidance dynamics

The geometry of a missile and target, both confined to move in a plane, is depicted in Fig. P2.6. The target moves in a straight line at constant velocity V_T and the missile moves at constant speed V_M but the direction of the velocity vector can be controlled by the use of an acceleration a which is assumed *perpendicular* to the relative velocity vector $\vec{V} = \vec{V}_M - \vec{V}_T$.

(*a*) Using a coordinate system that is "attached to the target" show that the dynamics of relative motion are

$$\begin{aligned} r &= -V \cos \sigma \\ \dot{\lambda} &= V \sin \sigma / r \\ \dot{\sigma} &= V \sin \sigma / r + a / V \end{aligned} \qquad \text{(P2.6}(a))$$

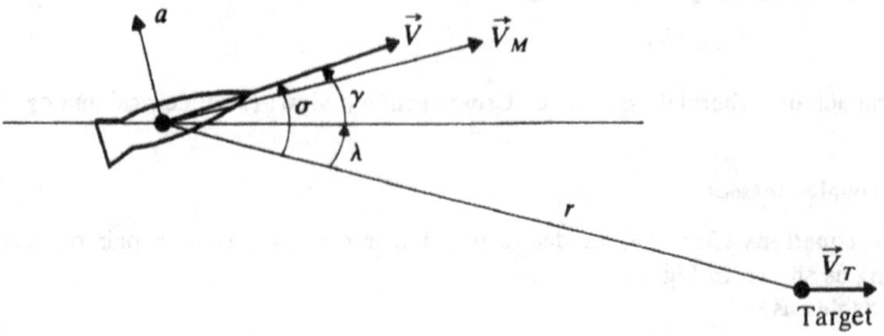

Figure P2.6 Missile dynamics.

where r is the range to the target, λ is the inertial line-of-sight angle, σ is the angle subtended at the missile by the velocity vector and the line of sight, and a is the applied acceleration.

(*b*) Let z be the "distance of closest approach" of the missile to the target, under the assumption that the missile continues in a straight line without any further acceleration. (Sometimes z is called the *projected miss distance.*) Show that

$$z = r \sin \sigma$$

and, using P2.6(*a*)

$$\dot{z} = (r \cos \sigma / V) a \tag{P2.6(b)}$$

(*c*) Assume that σ is a small angle. Then $\dot{r} \approx -V$. Thus $r(t) = r_0 - Vt$, then $r/V \approx T_0 - t = \bar{T}$ where $T_0 = r_0/V$: $\bar{T}$ is often called "time-to-go." Show that the following equations represent the approximate dynamics

$$\dot{\lambda} = \frac{1}{V\bar{T}^2} z$$
$$\dot{z} = \bar{T} a \tag{P2.6(c)}$$

These equations are studied further in Chap. 3, Example 3B.

NOTES

Note 2.1 Rigid body dynamics

The representation of the motion of a rigid body in a noninertial coordinate system (i.e., a coordinate system in which the reference axes rotate and accelerate linearly) is treated in most standard textbooks in classical dynamics, such as Goldstein.[2] The basic relationship with regard to axes fixed in a rotating body are expressed by

$$\left.\frac{d\vec{\mu}}{dt}\right|_{\text{inertial}} = \left.\frac{d\vec{\mu}}{dt}\right|_{\text{body}} + \vec{\omega} \times \vec{\mu}$$

where $\vec{\mu}$ is any (three-component) vector. The symbol $\times$ denotes the *vector cross product* and "body" means that the derivatives are taken as if the body axes were inertially fixed; $\vec{\omega}$ is the angular velocity of the body axes. Thus (2.27) can be expressed as

$$\vec{\tau} = \left.\frac{d\vec{h}}{dt}\right|_{\text{inertial}} = \left.\frac{d\vec{h}}{dt}\right|_{\text{body}} + \vec{\omega} \times \vec{h}$$

Since $\vec{h} = J\vec{\omega}$, this becomes

$$\vec{\tau} = J\frac{d\vec{\omega}}{dt} + \vec{\omega} \times J\vec{\omega}$$

which is the form in which the "Euler equation" appears in many books.

Note 2.2 The gyroscope

The gyroscope is one of the two basic components of all inertial navigation systems. (The other is an accelerometer.) Since the 1920s, gyroscopes (or "gyros" as they are now known) have been used in navigation systems, first in gyro compasses and more recently (i.e., since about 1950) in complete inertial navigation systems. In addition to being used in navigation systems, gyros are also used as motion sensors for stabilizing the motion of ships, aircraft, and other mechanical systems. The inner ear of a human has a vestibular system that includes three gyroscopic sensors known as "semicircular canals" which are important in the biological feedback system that maintains the human body upright.

The remarkable properties of spinning bodies have always been a source of fascination, not only to children, but also to mathematicians and physicists. The renowned mathematician Felix Klein, one of the founders of the field of topology, also wrote a famous treatise on the theory of tops.[12]

Note 2.3 Three-Mile Island

The near disaster caused by the sequence of failures at the Three-Mile Island (Pennsylvania) nuclear plant in 1979, has a number of valuable, if costly, lessons. The failure was not due to use of novel, untested design concepts nor to new state-of-the-art hardware having been insufficiently tested. Neither the design nor the hardware were flawed in principle. The combination of misfortune, lack of training, and deficient critical judgment were in part responsible for the accident that may well have spelled doom for the nuclear industry in the United States. An outstanding account of the Three-Mile Island incident and its implications were presented in the November 1979 issue of the IEEE Spectrum.[13]

Note 2.4 Swedish papermaking industry

The benefits of using modern control concepts in the field of process control were vividly demonstrated by Karl J. Åström, now a professor at the Lund Institute of Technology in Sweden. During the late 1950s and early 1960s, Åström, under sponsorship of IBM, in association with a group of investigators (including R. E. Kalman, J. E. Bertram) at Columbia University, initiated the investigation of the use of state-space methods for improved process control design, particularly in papermaking. After his return to Sweden he succeeded in persuading the management of a paper company that the improved performance using modern methods, and implemented by means of a digital computer, would more than justify the cost of the new installation. With the cooperation of the plant management, he performed the tests needed to get the required dynamic characteristics of the plant and then installed the new computer control. The results were outstandingly successful; within a few years much of the Swedish paper industry adopted the new control system design approach. A technical account of Åström's work is found in Chap. 6 of [14].

REFERENCES

1. Loh, N. K., Cheok, K. C., and Beck, R. R., "Modern Control Design for Gun-Turret Control System," Southcon/85 Convention Record, Atlanta GA, Paper no. 1/5, 1983.
2. Goldstein, H., *Classical Mechanics*, Addison-Wesley Publishing Co., Reading MA, 1953.
3. Synge, J. L., and Griffith, B. A., *Principles of Mechanics*, McGraw-Hill Book Co., New York, 1949.
4. Etkin, B., *Dynamics of Flight*, John Wiley & Sons, New York, 1959.
5. Seckel, E., *Stability and Control of Airplanes and Helicopters*, Academic Press, New York, 1964.
6. Gilles, E. D., and Retzbach, B., "Reduced Models and Control of Distillation Columns with Sharp Temperature Profiles," *IEEE Trans. on Automatic Control*, vol. AC-28, no. 5, May 1983, pp. 628–630.
7. Gilles, E. D., Retzbach, B., and Silberberger, F., "Modeling, Simulation, and Control of an Extractive Distillation Column," *Computer Applications to Chem. Eng.* (ACS Symp. ser. no. 124), 1980, pp. 481–492.
8. Retzbach, B., "Einsatz von systemtechnische Methoden am Beispeil einer Mehrstoffdestillation," *Chem.-Ing.-Techn.*, vol. 55, no. 3, 1983, p. 235.
9. Fisher, D. G., and Seborg, D. E. (eds.), *Multivariable Computer Control: A Case Study*, North-Holland Publishing Co., Amsterdam, 1976.
10. Fisher, D. G., and Seborg, D. E., "Advanced Computer Control Improves Process Performance," *Instrumentation Technology*, vol. 20, no. 9, September 1973, pp. 71–77.

11. Fisher, D. G., and Seborg, D. E., "Model Development, Reduction, and Experimental Evaluation for an Evaporator," *Ind. Eng. Chem. Process Design and Development*, vol. 11, no. 2, February 1972, pp. 213-221.
12. Klein, F., and Sommerfeld, A., *Über die Theorie des Kreisels* (2 vol.), Teubner, Leipzig, 1897.
13. Special Issue: "Three-Mile Island and the Future of Nuclear Power," *IEEE Spectrum*, vol. 16, no. 11, November 1979.
14. Åström, K. J., *Introduction to Stochastic Control Theory*, Academic Press, New York, 1970.

CHAPTER
THREE

DYNAMICS OF LINEAR SYSTEMS

3.1 DIFFERENTIAL EQUATIONS REVISITED

In the last chapter we saw that the dynamic behavior of many dynamic systems is quite naturally characterized by systems of first-order differential equations. For a general system these equations in state space notation take the form

$$\dot{x} = f(x, u, t)$$

and in a linear system they take the special form

$$\dot{x} = A(t)x + B(t)u \tag{3.1}$$

where $x = [x_1, x_2, \ldots, x_k]'$ is the system state vector and $u = [u_1, u_2, \ldots, u_m]'$ is the input vector.

If the matrices A and B in (3.1) are constant matrices, i.e., not functions of time, the system is said to be "time-invariant." Time-varying systems are conceptually and computationally more difficult to handle than time-invariant systems. For this reason our attention will be devoted primarily to time-invariant systems. Fortunately many processes of interest can be approximated by linear, time-invariant models.

In using the conventional, frequency-domain approach the differential equations are converted to transfer functions as soon as possible, and the dynamics of a system comprising several subsystems is obtained by combining the transfer functions of the subsystems using well-known techniques (reviewed in Chap. 4). With the state-space methods, on the other hand, the description of the system dynamics in the form of differential equations is retained throughout the analysis and design. In fact, if a subsystem is characterized by a transfer function it is often necessary to convert the transfer function to differential equations in order to proceed by state-space methods.

In this chapter we shall develop the general formula for the solution of a vector-matrix differential equation in the form of (3.1) in terms of a very important matrix known as the *state-transition matrix* which describes how the state $x(t)$ of the system at some time t evolves into (or from) the state $x(\tau)$ at some other time τ. For time-invariant systems, the state-transition matrix is the matrix exponential function, which is easily calculated. For most time-varying systems, however, the state-transition matrix, although known to exist, cannot be expressed in terms of simple functions (such as real or complex exponentials) or even not-so-simple functions (such as Bessel functions, hypergeometric functions). Thus, while many of the results developed for time-invariant systems apply to time-varying systems, it is very difficult as a practical matter to carry out the required calculations. This is one reason why our attention is confined mainly (but not exclusively) to time-invariant systems. The world of real applications contains enough of the latter to keep a design engineer occupied.

3.2 SOLUTION OF LINEAR DIFFERENTIAL EQUATIONS IN STATE-SPACE FORM

Time-invariant dynamics The simplest form of the general differential equation of the form (3.1) is the "homogeneous," i.e., unforced equation

$$\dot{x} = Ax \tag{3.2}$$

where A is a constant k by k matrix. The solution to (3.2) can be expressed as

$$x(t) = e^{At}c \tag{3.3}$$

where e^{At} is the matrix exponential function

$$e^{At} = I + At + A^2\frac{t^2}{2} + A^3\frac{t^3}{3!} + \cdots \tag{3.4}$$

and c is a suitably chosen constant vector. To verify (3.3) calculate the derivative of $x(t)$

$$\frac{dx(t)}{dt} = \frac{d}{dt}(e^{At})c \tag{3.5}$$

and, from the defining series (3.4),

$$\frac{d}{dt}(e^{At}) = A + A^2t + A^3\frac{t^2}{2!} + \cdots = A\left(I + At + A^2\frac{t^2}{2!} + \cdots\right) = A\,e^{At}$$

Thus (3.5) becomes

$$\frac{dx(t)}{dt} = Ae^{At}c = Ax(t)$$

which was to be shown. To evaluate the constant c suppose that at some time τ the state $x(\tau)$ is given. Then, from (3.3),

$$x(\tau) = e^{A\tau}c \tag{3.6}$$

Multiplying both sides of (3.6) by the inverse of $e^{A\tau}$ we find that

$$c = (e^{A\tau})^{-1}x(\tau)$$

Thus the general solution to (3.2) for the state $x(t)$ at time t, given the state $x(\tau)$ at time τ, is

$$x(t) = e^{At}(e^{A\tau})^{-1}x(\tau) \tag{3.7}$$

The following property of the matrix exponential can readily be established by a variety of methods—the easiest perhaps being the use of the series definition (3.4)—

$$e^{A(t_1+t_2)} = e^{At_1}e^{At_2} \tag{3.8}$$

for any t_1 and t_2. From this property it follows that

$$(e^{A\tau})^{-1} = e^{-A\tau} \tag{3.9}$$

and hence that (3.7) can be written

$$x(t) = e^{A(t-\tau)}x(\tau) \tag{3.10}$$

The matrix $e^{A(t-\tau)}$ is a special form of the *state-transition matrix* to be discussed subsequently.

We now turn to the problem of finding a "particular" solution to the nonhomogeneous, or "forced," differential equation (3.1) with A and B being constant matrices. Using the "method of the variation of the constant,"[1] we seek a solution to (3.1) of the form

$$x(t) = e^{At}c(t) \tag{3.11}$$

where $c(t)$ is a function of time to be determined. Take the time derivative of $x(t)$ given by (3.11) and substitute it into (3.1) to obtain:

$$Ae^{At}c(t) + e^{At}\dot{c}(t) = Ae^{At}c(t) + Bu(t)$$

or, upon cancelling the terms $A\,e^{At}c(t)$ and premultiplying the remainder by e^{-At},

$$\dot{c}(t) = e^{-At}Bu(t) \tag{3.12}$$

Thus the desired function $c(t)$ can be obtained by simple integration (the mathematician would say "by a quadrature")

$$c(t) = \int_T^t e^{-A\lambda}Bu(\lambda)\,d\lambda$$

The lower limit T on this integral cannot as yet be specified, because we will need to put the particular solution together with the solution to the homogeneous equation to obtain the complete (general) solution. For the present, let T be undefined. Then the particular solution, by (3.11), is

$$x(t) = e^{At} \int_T^t e^{-A\lambda} Bu(\lambda)\, d\lambda = \int_T^t e^{A(t-\lambda)} Bu(\lambda)\, d\lambda \tag{3.13}$$

In obtaining the second integral in (3.13), the exponential e^{At}, which does not depend on the variable of integration λ, was moved under the integral, and property (3.8) was invoked to write $e^{At}e^{-A\lambda} = e^{A(t-\lambda)}$.

The complete solution to (3.1) is obtained by adding the "complementary solution" (3.10) to the particular solution (3.13). The result is

$$x(t) = e^{A(t-\tau)}x(\tau) + \int_T^t e^{A(t-\lambda)} Bu(\lambda)\, d\lambda \tag{3.14}$$

We can now determine the proper value for lower limit T on the integral. At $t = \tau$ (3.14) becomes

$$x(\tau) = x(\tau) + \int_T^\tau e^{A(t-\lambda)} Bu(\lambda)\, d\lambda \tag{3.15}$$

Thus, the integral in (3.15) must be zero for any $u(t)$, and this is possible only if $T = \tau$. Thus, finally we have the complete solution to (3.1) when A and B are constant matrices

$$x(t) = e^{A(t-\tau)}x(\tau) + \int_\tau^t e^{A(t-\lambda)} Bu(\lambda)\, d\lambda \tag{3.16}$$

This important relation will be used many times in the remainder of the book. It is worthwhile dwelling upon it. We note, first of all, that the solution is the sum of two terms: the first is due to the "initial" state $x(\tau)$ and the second—the integral—is due to the input $u(\tau)$ in the time interval $\tau \leqq \lambda \leqq t$ between the "initial" time τ and the "present" time t. The terms initial and present are enclosed in quotes to denote the fact that these are simply convenient definitions. There is no requirement that $t \geqq \tau$. The relationship is perfectly valid even when $t \leqq \tau$.

Another fact worth noting is that the integral term, due to the input, is a "convolution integral": the contribution to the state $x(t)$ due to the input u is the convolution of u with $e^{At}B$. Thus the function $e^{At}B$ has the role of the impulse response[1] of the system whose output is $x(t)$ and whose input is $u(t)$.

If the output y of the system is not the state x itself but is defined by the observation equation

$$y = Cx$$

then this output is expressed by

$$y(t) = Ce^{A(t-\tau)}x(t) + \int_\tau^t Ce^{A(t-\lambda)} Bu(\lambda)\, d\lambda \tag{3.17}$$

and the impulse response of the system with y regarded as the output is $Ce^{A(t-\lambda)}B$.

The development leading to (3.16) and (3.17) did not really require that B and C be constant matrices. By retracing the steps in the development it is readily seen that when B and C are time-varying, (3.16) and (3.17) generalize to

$$x(t) = e^{A(t-\tau)}x(\tau) + \int_\tau^t e^{A(t-\lambda)}B(\lambda)u(\lambda)\,d\lambda \tag{3.18}$$

and

$$y(t) = C(t)\,e^{A(t-\tau)}x(\tau) + \int_\tau^t C(t)\,e^{A(t-\lambda)}B(\lambda)u(\lambda)\,d\lambda \tag{3.19}$$

Time-varying dynamics Unfortunately, however, the results expressed by (3.18) and (3.19) do *not* hold when A is time-varying.

In any unforced (homogeneous) system the state at time t depends only on the state at time τ. In a linear system, this dependence is linear; thus we can always write the solution to $\dot{x} = A(t)x$ as

$$x(t) = \Phi(t, \tau)x(\tau) \tag{3.20}$$

The matrix $\Phi(t, \tau)$ that relates the state at time t to the state at time τ is generally known as the *state-transition matrix* because it defines how the state $x(\tau)$ evolves (or "transitions") into (or from) the state $x(t)$. In a time-invariant system $\Phi(t, \tau) = e^{A(t-\tau)}$, but there is no simple expression for the state-transition matrix in a time-varying system. The absence of such an expression is rarely a serious problem, however. It is usually possible to obtain a control system design from only a knowledge of the dynamics matrix $A(t)$, without having an expression for the transition matrix.

The complete solution to (3.1) can be expressed in the form of (3.18), with the general transition matrix $\Phi(t, \tau)$ replacing the matrix exponential of a time-invariant system. The general solution is thus given by

$$x(t) = \Phi(t, \tau)x(\tau) + \int_\tau^t \Phi(t, \lambda)B(\lambda)u(\lambda)\,d\lambda \tag{3.21}$$

$$y(t) = C(t)\Phi(t, \tau)x(\tau) + \int_\tau^t C(t)\Phi(t, \lambda)B(\lambda)u(\lambda)\,d\lambda \tag{3.22}$$

The derivation of (3.21) follows the same pattern as was used to obtain (3.18). The reader might wish to check his comprehension of the development by deriving (3.21). The development can also be found in a number of textbooks on linear systems, [1] for example.

The state-transition matrix The state-transition matrix for a time-invariant system can be calculated by various methods. One of these is to use the series definition (3.4) as will be illustrated in Example 3A. This is generally not a convenient method for pencil-and-paper calculations. It sometimes may be appropriate for numerical calculations, although there are better methods. (See Note 3.1.) For pencil-and-paper calculations, the Laplace transform method, to be developed in Sec. 3.4, is about as good a method as any.

It should be noted that the state-transition matrix for a time-invariant system is a function only of the *difference* $t - \tau$ between the initial time τ and the present time t as would be expected for a time-invariant system. (See Note 3.2.) Thus, in a time-invariant system, there is no loss in generality in taking the initial time τ to be zero and in computing $\Phi(t) = e^{At}$. If, for a subsequent calculation the initial time is not zero, and $\Phi(t, \tau)$ is needed, it is obtained from $\Phi(t)$ by replacing t by $t - \tau$.

In a time-varying system this procedure is of course not valid; both the initial time and the present time must be treated as general variables. A knowledge of $\Phi(t, 0)$ is not adequate information for the determination of $\Phi(t, \tau)$.

Although the state transition matrix cannot be calculated analytically in general, it is *sometimes* possible to do so because of the very simple structure of the dynamics matrix $A(t)$, as will be illustrated in the missile-guidance example below. Thus, if an application arises in which an expression is necessary for the transition matrix of a time-varying system, the engineer should consider "having a go at it," using whatever ad hoc measures appear appropriate.

Example 3A Motion of mass without friction The differential equation for the position of a mass to which an external force f is applied is

$$\ddot{x} = f/m = u \tag{3A.1}$$

(The control variable $u = f/m$ in this case is the total acceleration.)

Defining the state variables by

$$x_1 = x \qquad x_2 = \dot{x}$$

results in the state-space form

$$\begin{aligned} \dot{x}_1 &= x_2 \\ \dot{x}_2 &= u \end{aligned} \tag{3A.2}$$

Thus, for this example,

$$A = \begin{bmatrix} 0 & 1 \\ 0 & 0 \end{bmatrix} \qquad B = \begin{bmatrix} 0 \\ 1 \end{bmatrix}$$

Using the series definition (3.4) we obtain the state transition matrix

$$\Phi(t) = e^{At} = \begin{bmatrix} 1 & 0 \\ 0 & 1 \end{bmatrix} + \begin{bmatrix} 0 & 1 \\ 0 & 0 \end{bmatrix} t = \begin{bmatrix} 1 & t \\ 0 & 1 \end{bmatrix}$$

The series terminates after only two terms.

The integral in (3.18) with $\tau = 0$ is given by

$$\int_0^t \begin{bmatrix} 1 & \lambda \\ 0 & 1 \end{bmatrix} \begin{bmatrix} 0 \\ 1 \end{bmatrix} u(\lambda)\, d\lambda = \begin{bmatrix} \int_0^t \lambda u(\lambda)\, d\lambda \\ \int_0^t u(\lambda)\, d\lambda \end{bmatrix}$$

Thus, the solution to (3A.2), using the general formula (3.18) is given by

$$x_1(t) = x_1(0) + tx_2(0) + \int_0^t \lambda u(\lambda)\, d\lambda$$

$$x_2(t) = x_2(0) + \int_0^t u(\lambda)\, d\lambda$$

Obviously these answers could have been obtained directly from (3A.1) without using all the state-space apparatus being developed. This apparatus has its greatest utility when simple methods fail.

Example 3B Missile guidance The equations of motion (assumed to be confined to a plane) of a missile moving at constant speed, relative to a target also moving at constant speed, can be approximated by

$$\dot{\lambda} = \frac{1}{V\bar{T}^2} z \qquad \dot{z} = \bar{T} u \tag{3B.1}$$

where λ is the line-of-sight angle to the target
z is the projected miss distance
V is the velocity of the missile relative to the target
$\bar{T} = T - t$ is the "time-to-go"
u is the acceleration normal to the missile relative velocity vector

It is assumed that the terminal time T is a known quantity. (The reader should review the discussion in Prob. 2.6 for the significance of these variables and the derivation of (3B.1).)

Using the state-variable definitions

$$x_1 = \lambda \qquad x_2 = z$$

results in the matrices

$$A(t) = \begin{bmatrix} 0 & \frac{1}{V\bar{T}^2} \\ 0 & 0 \end{bmatrix} \qquad B(t) = \begin{bmatrix} 0 \\ \bar{T} \end{bmatrix} \tag{3B.2}$$

Since $A(t)$ is time-varying (through $\bar{T}$), the transition matrix is *not* the matrix exponential and cannot be found using the series (3.4). In this case, however, we can find the transition matrix by an ad hoc method. First we note that the transition matrix $\Phi(t, \tau)$ expresses the solution to the unforced system

$$\dot{\lambda} = \frac{1}{V\bar{T}^2} z \tag{3B.3}$$

$$\dot{z} = 0 \tag{3B.4}$$

The general form of this solution is

$$\begin{aligned} \lambda(t) &= \phi_{11}(t, \tau)\lambda(\tau) + \phi_{12}(t, \tau) z(\tau) \\ z(t) &= \phi_{21}(t, \tau)\lambda(\tau) + \phi_{22}(t, \tau) z(\tau) \end{aligned} \tag{3B.5}$$

The terms $\phi_{ij}(t, \tau)$ $(i, j) = 1, 2$, which we will now calculate, are the elements of the required transition matrix.

From (3B.4) we have immediately

$$z(t) = z(\tau) = \text{const} \tag{3B.6}$$

Hence

$$\phi_{21}(t, \tau) = 0 \qquad \phi_{22}(t, \tau) = 1 \tag{3B.7}$$

The easiest way to get the first row (ϕ_{11} and ϕ_{12}) of the transition matrix is to use (3B.3) which can be written

$$V(T - \xi)^2\dot{\lambda}(\xi) = z(\xi) \qquad \text{for all } \xi$$

Thus

$$V(T - \tau)^2\dot{\lambda}(\tau) = z(\tau)$$

But, from (3B.6), $z(\xi) = z(\tau)$. Hence

$$\dot{\lambda}(\xi) = \frac{1}{(T - \xi)^2} z(\tau) \tag{3B.8}$$

Integrate both sides of (3B.8) from τ to t

$$\int_\tau^t \dot{\lambda}(\xi)\, d\xi = \int_\tau^t \frac{d\xi}{(T - \xi)^2} z(\tau)\, d\xi$$

or

$$\lambda(t) - \lambda(\tau) = \left(\frac{1}{T - t} - \frac{1}{T - \tau}\right) z(\tau) \tag{3B.9}$$

Thus, from (3B.9), we obtain

$$\phi_{11}(t, \tau) = 1 \qquad \phi_{12}(t, \tau) = \frac{1}{T - t} - \frac{1}{T - \tau} \tag{3B.10}$$

Combining (3B.10) with (3B.7) gives the state transition matrix

$$\Phi(t, \tau) = \begin{bmatrix} 1 & \dfrac{1}{T - t} - \dfrac{1}{T - \tau} \\ 0 & 1 \end{bmatrix} \tag{3B.11}$$

3.3 INTERPRETATION AND PROPERTIES OF THE STATE-TRANSITION MATRIX

The state-transition matrix, which is fundamental to the theory of linear dynamic systems, has a number of important properties which are the subject of this section.

We note, first of all, that the state-transition matrix is an expression of the solution to the homogeneous equation

$$\frac{dx(t)}{dt} = A(t)x(t) \tag{3.23}$$

where $x(t)$ is given by (3.20). The time derivative of $x(t)$ in (3.20) must of course satisfy (3.23) for any t and $x(t)$. In (3.20) $x(\tau)$ represents initial data and is not a time function. Thus

$$\frac{dx(t)}{dt} = \frac{\partial\Phi(t, \tau)}{\partial t} x(\tau) \tag{3.24}$$

(Since the transition matrix is a function of two arguments t and τ, it is necessary to write its time derivative as a *partial* derivative. The transition matrix also has a derivative with respect to the "initial" time τ which is investigated in Prob. 3.4.) Substitution of (3.24) and (3.20) into (3.23) gives

$$\frac{\partial \Phi(t, \tau)}{\partial t} x(\tau) = A(t)\Phi(t, \tau)x(\tau)$$

Since this must hold for any $x(\tau)$, we may cancel $x(\tau)$ on both sides to finally obtain

$$\frac{\partial \Phi(t, \tau)}{\partial t} = A(t)\Phi(t, \tau) \tag{3.25}$$

In other words, the transition matrix Φ satisfies the same differential equation as the state x. This can be emphasized by writing (3.25) simply as

$$\dot{\Phi} = A\Phi \tag{3.26}$$

which does not explicitly exhibit the time dependence of A and Φ. The dot on top of Φ must be interpreted to designate differentiation with respect to the first argument. (Because of the possibility of confusion of arguments use of the full expression (3.25) is recommended in analytical studies.)

We note that (3.20) holds for any t and τ, including $t = \tau$. Thus

$$x(t) = \Phi(t, t)x(t)$$

for any $x(t)$. Thus we conclude that

$$\Phi(t, t) = I \qquad \text{for any } t \tag{3.27}$$

This becomes the initial condition for (3.25) or (3.26).

Other properties of the transition matrix follow from the fact that the differential equation (3.23) not only possesses a solution for any initial state $x(\tau)$ and any time interval $[\tau, t]$ but that this solution is unique. This is a basic theorem in the theory of ordinary differential equations and is proved in standard textbooks on the subject, e.g., [2, 3]. There are certain restrictions on the nature of permissible time variations of $A(t)$ but these are always satisfied in real-world systems. When A is a constant matrix, of course, not only do we know that Φ exists but we have an expression for it, namely $\Phi(t) = e^{At}$.

Assuming the existence and uniqueness of solutions, we can write

$$x(t_3) = \Phi(t_3, t_1)x(t_1) \qquad \text{for any } t_3, t_1 \tag{3.28}$$

and also

$$x(t_3) = \Phi(t_3, t_2)x(t_2) \qquad \text{for any } t_3, t_2 \tag{3.29}$$

$$x(t_2) = \Phi(t_2, t_1)x(t_1) \qquad \text{for any } t_2, t_1 \tag{3.30}$$

Thus, substituting (3.30) into (3.29)

$$x(t_3) = \Phi(t_3, t_2)\Phi(t_2, t_1)x(t_1) \tag{3.31}$$

Comparing (3.31) with (3.28) we see that

$$\Phi(t_3, t_1) = \Phi(t_3, t_2)\Phi(t_2, t_1) \qquad \text{for any } t_3, t_2, t_1 \tag{3.32}$$

This very important property—known as the *semigroup* property—of the state-transition matrix is a direct consequence of the fact that whether we go from state $x(t_1)$ to $x(t_3)$ directly or via an "intermediate" state $x(t_2)$, we must end at the same point. Note, however, that the time t_2 of the intermediate state need not be between t_1 and t_3.

The semigroup properties (3.32) and (3.27) gives

$$I = \Phi(t, \tau)\Phi(\tau, t)$$

or

$$\Phi(\tau, t) = [\Phi(t, \tau)]^{-1} \qquad \text{for any } t, \tau \tag{3.33}$$

This of course means that the state-transition matrix is never singular even if the dynamics matrix A is singular, as it often is.

In a time-invariant system, the transition matrix is characterized by a single argument, as already discussed:

$$\Phi(t_1, t_2) = \Phi(t_1 - t_2)$$

Thus, for time-invariant systems, the properties (3.27), (3.32), and (3.33) become

$$\Phi(0) = I \tag{3.34}$$

$$\Phi(t)\Phi(\tau) = \Phi(t + \tau) \tag{3.35}$$

$$\Phi^{-1}(t) = \Phi(-t) \tag{3.36}$$

It is readily verified that $\Phi(t) = e^{At}$ possesses these properties:

$$e^{A0} = I \tag{3.37}$$

$$e^{At}e^{A\tau} = e^{A(t+\tau)} \tag{3.38}$$

$$(e^{At})^{-1} = e^{-At} \tag{3.39}$$

The first relation (3.37) is apparent from the series definition (3.4) and the second relation (3.38) can be verified by multiplying the series for e^{At} by the series for $e^{A\tau}$. (The calculations are a bit tedious, but the skeptical reader is invited to perform them.) The third relation (3.39) follows from the first two.

By analogy with (3.38) the reader might be tempted to conclude that $e^{At}e^{Bt} = e^{(A+B)t}$. This is generally *not* true, however. In order for it to be true A and B must commute (i.e., $AB = BA$) and this condition is rarely met in practice.

3.4 SOLUTION BY THE LAPLACE TRANSFORM: THE RESOLVENT

As the reader is no doubt aware, Laplace transforms are very useful for solving time-invariant differential equations. Indeed Laplace transforms are the basis of the entire frequency-domain methodology, to which the next chapter is devoted.

The Laplace transform of a signal $f(t)$ which may be an input variable or a state variable is defined by

$$\mathscr{L}[f(t)] = \mathsf{f}(s) = \int_0^\infty f(t)e^{-st}\,dt \tag{3.40}$$

where s is a complex variable generally called *complex frequency.* A discussion of the region of convergence of $\mathsf{f}(s)$ in the complex s plane, and many other details about the Laplace transform are to be found in many standard textbooks such as [1] and [4].

The sans-serif letter f used to designate the Laplace transform of $f(t)$ was chosen advisedly. In texts in which the signals are all scalars, capital letters are used to denote Laplace transforms (viz., $X(s) = \mathscr{L}[x(t)]$, $Y(s) = \mathscr{L}[y(t)]$, etc.). But in this book capital letters have been preempted for designating matrices. The use of sans-serif letters for Laplace transforms avoids the risk of confusion.

The lower limit on the integral has been written as 0. In accordance with engineering usage, this is understood to be 0^-, that is, the instant just prior to the occurrence of discontinuities, impulses, etc., in the signals under examination. The reader who is unfamiliar with this usage should consult a standard text such as [1] or [4].

The Laplace transform is useful for solving (3.1) only when A and B are constant matrices, which we will henceforth assume. In order to use the Laplace transform, we need an expression for the Laplace transform of the time derivative of $f(t)$

$$\mathscr{L}[\dot{f}(t)] = \int_0^\infty e^{-st}\frac{df}{dt}\,dt = e^{-st}f(t)\Big|_0^\infty - \int_0^\infty -s\,e^{-st}f(t)\,dt \tag{3.41}$$

upon integration by parts. Assuming

$$\lim_{t\to\infty} e^{-st}f(t) \to 0$$

(3.41) becomes

$$\mathscr{L}[\dot{f}(t)] = s\int_0^\infty e^{-st}f(t)\,dt - f(0) = s\mathsf{f}(s) - f(0) \tag{3.42}$$

We also note that (3.42) applies when $f(t)$ is a vector:

$$\mathscr{L}[f(t)] = \mathscr{L}\begin{bmatrix} f_1(t) \\ \vdots \\ f_n(t) \end{bmatrix} = \begin{bmatrix} \mathscr{L}[f_1(t)] \\ \vdots \\ \mathscr{L}[f_n(t)] \end{bmatrix} = \begin{bmatrix} \mathsf{f}_1(s) \\ \vdots \\ \mathsf{f}_n(s) \end{bmatrix} = \mathsf{f}(s) \tag{3.43}$$

and also that

$$\mathscr{L}[Ax(t)] = A\mathsf{x}(s) \tag{3.44}$$

Applying all of these to (3.1) with A and B *constant* gives

$$s\mathsf{x}(s) - x(0) = A\mathsf{x}(s) + B\mathsf{u}(s)$$

or

$$(sI - A)\mathsf{x}(s) = x(0) + B\mathsf{u}(s)$$

Solve for $\mathsf{x}(s)$ to obtain

$$\mathsf{x}(s) = (sI - A)^{-1}x(0) + (sI - A)^{-1}B\mathsf{u}(s) \tag{3.45}$$

On taking the inverse Laplace transform of $\mathsf{x}(s)$ as given by (3.45) we obtain the desired solution for $x(t)$. We note that $\mathsf{x}(s)$ is the sum of two terms, the first due to the initial condition $x(0)$ multiplied by the matrix $(sI - A)^{-1}$ and the second being the product of this matrix and the term due to the input $B\mathsf{u}(s)$. Knowing the inverse Laplace transform of $(sI - A)^{-1}$ would permit us to find the inverse Laplace transform of (3.45) and hence obtain $x(t)$. In the scalar case we recall that

$$\mathscr{L}[e^{at}] = \frac{1}{s - a} = (s - a)^{-1} \tag{3.46}$$

We have not yet discussed calculating the Laplace transform of a matrix function of time. But we should not be very much surprised to learn that

$$\mathscr{L}[e^{At}] = (sI - A)^{-1} \tag{3.47}$$

which is simply the matrix version of (3.46). It can be shown by direct calculation (see Note 3.3) that (3.47) is in fact true. And if this be the case then the inverse Laplace transform of (3.45) is

$$x(t) = e^{At}x(0) + \int_0^t e^{A(t-\lambda)}Bu(\lambda)\, d\lambda \tag{3.48}$$

which is the desired solution. The integral term in (3.48) is given by the well-known *convolution theorem* for the Laplace transform [1]

$$\mathscr{L}\left[\int_0^t f(t - \lambda)g(\lambda)\, d\lambda\right] = \mathsf{f}(s)\mathsf{g}(s)$$

which is readily extended from scalar functions to matrices.

The solution for $x(t)$ given by (3.48) is a special case (namely $\tau = 0$) of the general solution (3.16) obtained by another method of analysis. This confirms, if confirmation is necessary, the validity of (3.47).

The exponential matrix e^{At} is known as the state transition matrix (for a time invariant system) and its Laplace transform

$$\Phi(s) = (sI - A)^{-1} \tag{3.49}$$

is known in mathematical literature as the *resolvent* of A. In engineering literature this matrix has been called the *characteristic frequency* matrix[1] or simply the *characteristic matrix*.[4] Regrettably there doesn't appear to be a standard symbol for the resolvent, which we have designated as $\Phi(s)$ in this book.

The fact that the state transition matrix is the inverse Laplace transform of the resolvent matrix facilitates the calculation of the former. It also characterizes the dynamic behavior of the system, the subject of the next chapter. The steps one takes in calculating the state-transition matrix using the resolvent are:

(*a*) Calculate $sI - A$.
(*b*) Obtain the resolvent by inverting $(sI - A)$.
(*c*) Obtain the state-transition matrix by taking the inverse Laplace transform of the resolvent, element by element.

The following examples illustrate the process.

Example 3C DC motor with inertial load In Chap. 2 (Example 2B) we found that the dynamics of a dc motor driving an inertial load are

$$\dot{\theta} = \omega$$

$$\dot{\omega} = -\alpha\omega + \beta u$$

The matrices of the state-space characterization are

$$A = \begin{bmatrix} 0 & 1 \\ 0 & -\alpha \end{bmatrix} \qquad B = \begin{bmatrix} 0 \\ \beta \end{bmatrix}$$

Thus the resolvent is

$$\Phi(s) = (sI - A)^{-1} = \begin{bmatrix} s & -1 \\ 0 & s+\alpha \end{bmatrix}^{-1} = \frac{1}{s(s+\alpha)}\begin{bmatrix} s+\alpha & 1 \\ 0 & s \end{bmatrix} = \begin{bmatrix} \dfrac{1}{s} & \dfrac{1}{s(s+\alpha)} \\ 0 & \dfrac{1}{s+\alpha} \end{bmatrix}$$

Finally, taking the inverse Laplace transforms of each term in $\Phi(s)$ we obtain

$$e^{At} = \Phi(t) = \begin{bmatrix} 1 & (1 - e^{-\alpha t})/\alpha \\ 0 & e^{-\alpha t} \end{bmatrix}$$

Example 3D Inverted pendulum The equations of motion of an inverted pendulum were determined to be (approximately)

$$\dot{\theta} = \omega$$

$$\dot{\omega} = \Omega^2\theta + u$$

Hence the matrices of the state-space characterization are

$$A = \begin{bmatrix} 0 & 1 \\ \Omega^2 & 0 \end{bmatrix} \qquad B = \begin{bmatrix} 0 \\ 1 \end{bmatrix}$$

The resolvent is

$$\Phi(s) = (sI - A)^{-1} = \begin{bmatrix} s & -1 \\ -\Omega^2 & s \end{bmatrix}^{-1} = \frac{1}{s^2 - \Omega^2}\begin{bmatrix} s & 1 \\ \Omega^2 & s \end{bmatrix}$$

and the state-transition matrix is

$$\Phi(t) = e^{At} = \begin{bmatrix} \cosh \Omega t & \sinh \Omega t/\Omega \\ \Omega \sinh \Omega t & \cosh \Omega t \end{bmatrix}$$

For a general kth-order system the matrix $sI - A$ has the following appearance

$$sI - A = \begin{bmatrix} s - a_{11} & -a_{12} & \cdots & -a_{1k} \\ -a_{21} & s - a_{22} & \cdots & -a_{2k} \\ \cdots & \cdots & \cdots & \cdots \\ -a_{k1} & -a_{k2} & \cdots & s - a_{kk} \end{bmatrix} \tag{3.50}$$

We recall (see Appendix) that the inverse of any matrix M can be written as the adjoint matrix, adj M, divided by the determinant $|M|$. Thus

$$(sI - A)^{-1} = \frac{\operatorname{adj}(sI - A)}{|sI - A|}$$

If we imagine calculating the determinant $|sI - A|$ we see that one of the terms will be the product of the diagonal elements of $sI - A$:

$$(s - a_{11})(s - a_{22}) \cdots (s - a_{kk}) = s^k + c_1 s^{k-1} + \cdots + c_k$$

a polynomial of degree k with the leading coefficient of unity. There will also be other terms coming from the off-diagonal elements of $sI - A$ but none will have a degree as high as k. Thus we conclude that

$$|sI - A| = s^k + a_1 s^{k-1} + \cdots + a_k \tag{3.51}$$

This is known as the *characteristic polynomial* of the matrix A. It plays a vital role in the dynamic behavior of the system. The roots of this polynomial are called the *characteristic roots*, or the *eigenvalues*, or the *poles*, of the system and determine the essential features of the unforced dynamic behavior of the system, since they determine the inverse Laplace transform of the resolvent, which is the transition matrix. See Chap. 4.

The adjoint of a k by k matrix is itself a k by k matrix whose elements are the cofactors of the original matrix. Each cofactor is obtained by computing the determinant of the matrix that remains when a row and a column of the original matrix are deleted. It thus follows that each element in adj $(sI - A)$ is a polynomial in s of maximum degree $k - 1$. (The polynomial cannot have degree k when any row and column of $sI - A$ is deleted.) Thus it is seen that the adjoint of $sI - A$ can be written

$$\operatorname{adj}(sI - A) = E_1 s^{k-1} + E_2 s^{k-2} + \cdots + E_k$$

Thus we can express the resolvent in the following form

$$(sI - A)^{-1} = \frac{E_1 s^{k-1} + \cdots + E_k}{s^k + a_1 s^{k-1} + \cdots + a_k} \tag{3.52}$$

An interesting and useful relationship for the coefficient matrices E_i of the adjoint matrix can be obtained by multiplying both sides of (3.52) by $|sI - A|(sI - A)$. The result is

$$|sI - A|I = (sI - A)(E_1 s^{k-1} + E_2 s^{k-2} + \cdots + E_k) \tag{3.53}$$

or

$$s^k I + a_1 s^{k-1} I + \cdots + a_k I = s^k E_1 + s^{k-1}(E_2 - AE_1) + \cdots + s(E_k - AE_{k-1}) - AE_k$$

Equating the coefficients of s^i on both sides of (3.53) gives

$$\begin{aligned} E_1 &= I \\ E_2 - AE_1 &= a_1 I \\ E_3 - AE_2 &= a_2 I \\ &\cdots\cdots \\ E_k - AE_{k-1} &= a_{k-1} I \\ -AE_k &= a_k I \end{aligned} \tag{3.54}$$

We have thus determined that the leading coefficient matrix of adj $(sI - A)$ is the identity matrix, and that the subsequent coefficients can be obtained recursively:

$$\begin{aligned} E_2 &= AE_1 + a_1 I \\ E_3 &= AE_2 + a_2 I \\ &\cdots\cdots \\ E_k &= AE_{k-1} + a_{k-1} I \end{aligned} \tag{3.55}$$

The last equation in (3.54) is redundant, but can be used as a check, when the recursion equations (3.55) are used as the basis of a numerical algorithm. In this case the "check equation" can be written

$$E_{k+1} = AE_k + a_k I = 0 \tag{3.56}$$

An algorithm based on (3.55) requires the coefficients a_i $(i = 1, \ldots, k)$ of the characteristic polynomial. Fortunately, the determination of these coefficients can be included in the algorithm, for it can be shown that

$$\begin{aligned} a_1 &= -\mathrm{tr}\,(AE_1) \\ a_2 &= -\tfrac{1}{2}\,\mathrm{tr}\,(AE_2) \end{aligned}$$

More generally

$$a_i = -\frac{1}{i}\,\mathrm{tr}\,(AE_i) \qquad i = 1, 2, \ldots, k \tag{3.57}$$

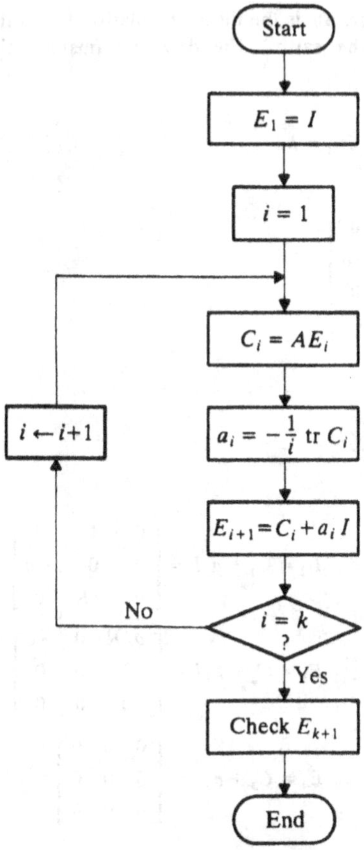

Figure 3.1 Algorithm for computing

$$(sI - A)^{-1} = \frac{E_1 s^{k-1} + \cdots + E_k}{s^k + a_1 s^{k-1} + \cdots + a_k}$$

An algorithm for computing the numerator matrices E_i and the coefficients a_i, starting with $E_1 = I$, is illustrated in the form of a flow chart in Fig. 3.1.

A proof of (3.57) is found in many textbooks such as [5, 6]. The algorithm based on (3.56) and (3.57) appears to have been discovered several times in various parts of the world. The names of Leverrier, Souriau, Faddeeva, and Frame are often associated with it.

This algorithm is convenient for hand calculation and easy to implement on a digital computer. Unfortunately, however, it is not a very good algorithm when the order k of the system is large (higher than about 10). The check matrix E_{k+1}, which is supposed to be zero, usually turns out to be embarrassingly large, and hence the resulting coefficients a_i and E_i are often suspect.

Example 3E Inertial navigation The equations for errors in an inertial navigation system are approximated by

$$\begin{aligned} \Delta\dot{x} &= \Delta v \\ \Delta\dot{v} &= -g\Delta\psi + E_A \\ \Delta\dot{\psi} &= \frac{1}{R}\Delta v + E_G \end{aligned} \tag{3E.1}$$

where Δx is the position error, Δv is the velocity error, $\Delta\psi$ is the tilt of the platform, g is the acceleration of gravity, and R is the radius of the earth. (The driving terms are the accelerometer error E_A and the gyro error E_G.)

For the state variables defined by

$$x_1 = \Delta x \qquad x_2 = \Delta v \qquad x_3 = \Delta\psi$$

the A matrix is given by

$$A = \begin{bmatrix} 0 & 1 & 0 \\ 0 & 0 & -g \\ 0 & 1/R & 0 \end{bmatrix}$$

and, regarding E_A and E_G as inputs, the B matrix is

$$B = \begin{bmatrix} 0 & 0 \\ 1 & 0 \\ 0 & 1 \end{bmatrix}$$

The matrices appearing in the recursive algorithm are

$$C_1 = AE_1 = \begin{bmatrix} 0 & 1 & 0 \\ 0 & 0 & -g \\ 0 & 1/R & 0 \end{bmatrix} \qquad a_1 = -\operatorname{tr} C_1 = 0 \qquad E_2 = C_1 + a_1 I = \begin{bmatrix} 0 & 1 & 0 \\ 0 & 0 & -g \\ 0 & 1/R & 0 \end{bmatrix}$$

$$C_2 = AE_2 = \begin{bmatrix} 0 & 0 & -g \\ 0 & -g/R & 0 \\ 0 & 0 & -g/R \end{bmatrix} \qquad \begin{aligned} a_2 &= -\tfrac{1}{2}(-2g/R) \\ &= g/R \end{aligned} \qquad E_3 = C_2 + a_2 I = \begin{bmatrix} g/R & 0 & -g \\ 0 & 0 & 0 \\ 0 & 0 & 0 \end{bmatrix}$$

$$C_3 = AE_3 = \begin{bmatrix} 0 & 0 & 0 \\ 0 & 0 & 0 \\ 0 & 0 & 0 \end{bmatrix} \qquad a_3 = 0 \qquad E_4 = C_3 + a_3 I = \begin{bmatrix} 0 & 0 & 0 \\ 0 & 0 & 0 \\ 0 & 0 & 0 \end{bmatrix}$$

Thus

$$(sI - A)^{-1} = \begin{bmatrix} s^2 + g/R & s & -g \\ 0 & s^2 & -gs \\ 0 & s/R & s^2 \end{bmatrix} \frac{1}{s^3 + (g/R)s}$$

$$= \begin{bmatrix} \dfrac{1}{s} & \dfrac{1}{s^2 + g/R} & \dfrac{-g}{s(s^2 + g/R)} \\ 0 & \dfrac{s}{s^2 + g/R} & \dfrac{-g}{s^2 + g/R} \\ 0 & \dfrac{1/R}{s^2 + g/R} & \dfrac{s}{s^2 + g/R} \end{bmatrix} \tag{3E.2}$$

The state transition matrix corresponding to the resolvent (3E.2) is obtained by taking its inverse Laplace transform.

$$\Phi(t) = \begin{bmatrix} 1 & \dfrac{\sin \Omega t}{\Omega} & \dfrac{g}{\Omega^2}(\cos \Omega t - 1) \\ 0 & \cos \Omega t & -\dfrac{g}{\Omega}\sin \Omega t \\ 0 & \dfrac{\sin \Omega t}{\Omega R} & \cos \Omega t \end{bmatrix} \qquad \Omega = \sqrt{g/R} \tag{3E.3}$$

The elements of the state transition matrix, with the exception of ϕ_{11} are all oscillatory with a frequency $\Omega = \sqrt{g/R}$ which is the natural frequency of a pendulum of length equal to the earth's radius; $\Omega = 0.001\,235$ rad/s corresponding to a period $T = 2\pi/\Omega = 84.4$ min., which is known as the "Schuler period." (See Note 3.4.)

Because the error equations are undamped, the effects of even small instrument biases can result in substantial navigation errors. Consider, for example, a constant gyro bias

$$E_G = \frac{c}{s}$$

The Laplace transform of the position error is given by

$$\Delta \mathsf{x}(s) = \phi_{13}(s)\frac{c}{s} = -\frac{g}{s^2(s^2+\Omega^2)}c \tag{3E.4}$$

and the corresponding position error, as a function of time, is the inverse Laplace transform of (3E.4)

$$\Delta x(t) = -\frac{g}{\Omega^2}\left(t - \frac{1}{\Omega}\sin\Omega t\right)c \tag{3E.5}$$

The position error consists of two terms: a periodic term at the Schuler period and a term which grows with time (also called a *secular term* at a rate of $-(g/\Omega^2)c = -Rc$. The position error thus grows at a rate proportional to the earth's radius. The position error will grow at a rate of about 70 m/h for each degree-per-hour "drift" ($E_G = c$) of the gyro.

3.5 INPUT-OUTPUT RELATIONS: TRANSFER FUNCTIONS

In conventional (frequency-domain) analysis of system dynamics attention is focused on the relationship between the output y and the input u. The focus shifts to the state vector when state space analysis is used, but there is still an interest in the input-output relation. Usually when an input-output analysis is made, the initial state $x(0)$ is assumed to be zero. In this case the Laplace transform of the state is given by

$$\mathsf{x}(s) = (sI - A)^{-1}B\mathsf{u}(s) \tag{3.58}$$

If the output is defined by

$$y(t) = Cx(t) \tag{3.59}$$

Then its Laplace transform is

$$\mathsf{y}(s) = C\mathsf{x}(s) \tag{3.60}$$

and, by (3.58)

$$\mathsf{y}(s) = C(sI - A)^{-1}B\mathsf{u}(s) \tag{3.61}$$

The matrix

$$\mathsf{H}(s) = C(sI - A)^{-1}B \tag{3.62}$$

that relates the Laplace transform of the output to the Laplace transform of the input is known as the *transfer-function* matrix.

The inverse Laplace transform of the transfer-function matrix

$$H(t) = \mathscr{L}^{-1}[\mathsf{H}(s)] = Ce^{At}B \tag{3.63}$$

is known as the *impulse-response* matrix. In the time domain $y(t)$ can be expressed by the convolution of the impulse-response matrix with the input

$$y(t) = \int_0^t H(t-\lambda)u(\lambda)\,d\lambda = \int_0^t Ce^{A(t-\lambda)}Bu(\lambda)\,d\lambda \tag{3.64}$$

This relationship is equivalent to (3.48) in which the initial state $x(0)$ is assumed to be zero, with (3.59) relating $y(t)$ to $x(t)$.

If there is a direct path from the input to the output owing to the presence of a matrix D

$$y(t) = Cx(t) + Du(t)$$

Then

$$\mathsf{y}(s) = C\mathsf{x}(s) + D\mathsf{u}(s)$$

and the transfer-function matrix

$$\mathsf{H}(s) = C(sI - A)^{-1}B + D \tag{3.65}$$

with the corresponding impulse-response matrix

$$H(t) = Ce^{At}B + D\delta(t) \tag{3.66}$$

The delta function (unit impulse) appears in (3.66) because of the direct connection, through D, from the input to the output. Since the impulse response of a system is defined as the output $y(t)$ when the input $u(t) = \delta(t)$, it is clear that the output must contain $D\delta(t)$. If the direct connection from the input to the output is absent, the impulse response does not contain an impulse term. This implies that the degree of the numerator in $\mathsf{H}(s)$ must be lower than the degree of the denominator. Since the adjoint matrix of $sI - A$ is of the degree $k-1$ (see (3.52)) then the degree of $\mathsf{H}(s)$ is no higher than $k-1$. Specifically, with $D = 0$

$$\begin{aligned}\mathsf{H}(s) &= \frac{C[E_1 s^{k-1} + E_2 s^{k-2} + \cdots + E_k]B}{|sI - A|} \\ &= \frac{CBs^{k-1} + CE_2Bs^{k-2} + \cdots + CE_kB}{s^k + a_1 s^{k-1} + \cdots + a_k}\end{aligned} \tag{3.67}$$

Thus the transfer-function matrix is a rational function of s with the numerator of degree $k-1$ (or less) and the denominator of degree k.

Example 3F Missile dynamics Except for difference in size, weight, and speed a missile is simply a pilotless aircraft. Hence the aerodynamic equations of a missile are the same as those of an aircraft, namely (2.40) and (2.41).

In many cases the coupling of the change of velocity Δu normal to the longitudinal axis into the equations for angle of attack α and pitch rate q is negligible: Z_u, M_α, X_u are

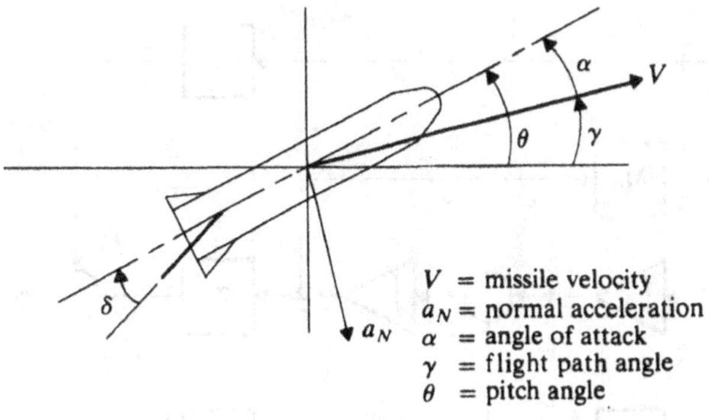

Figure 3.2 Missile dynamic variables.

insignificant. In this case (2.40) gives the following pitch dynamics:

$$\dot{\alpha} = \frac{Z_\alpha}{V}\alpha + q + \frac{Z_\delta}{V}\delta$$
$$\dot{q} = M_\alpha \alpha + M_q q + M_\delta \delta \tag{3F.1}$$

where δ is the control surface deflection. (The control surface may be located in front of the missile—in which case it is called a *canard*—or in the more familiar aft position. Its location with respect to the center of mass of the missile will determine the signs of the Z_δ and M_δ, used here instead of Z_E and M_E which were introduced in Chap. 2.)

The pitch angle θ is usually not of interest, hence the differential equation $\dot{\theta} = q$ can be omitted.

Missile guidance laws are generally expressed in terms of the component of acceleration normal to the velocity vector of the missile; in proportional navigation, for example, it is desired that this acceleration be proportional to the inertial line-of-sight rate. (See Example 9G.) Thus the output of interest in a typical missile is the "normal" component of acceleration a_N. In the planar case (see Fig. 3.2)

$$a_N \approx -V\dot{\gamma} \tag{3F.2}$$

where γ is the flight path angle. But

$$\gamma = \theta - \alpha$$

or

$$\dot{\gamma} = q - \dot{\alpha} \tag{3F.3}$$

Thus, using (3F.2) and (3F.1),

$$a_N \approx Z_\alpha \alpha + Z_\delta \delta \tag{3F.4}$$

With the state, input, and output of the missile defined respectively by

$$x = \begin{bmatrix} \alpha \\ q \end{bmatrix} \qquad u = \delta \qquad y = a_N$$

the matrices of the standard representation $\dot{x} = Ax + Bu$, $y = Cx + Du$ are

$$A = \begin{bmatrix} Z_\alpha/V & 1 \\ M_\alpha & M_q \end{bmatrix} \qquad B = \begin{bmatrix} Z_\delta/V \\ M_\delta \end{bmatrix}$$

$$C = [Z_\alpha \quad 0] \qquad\qquad D = [Z_\delta]$$

A block-diagram representation of the system is shown in Fig. 3.3.

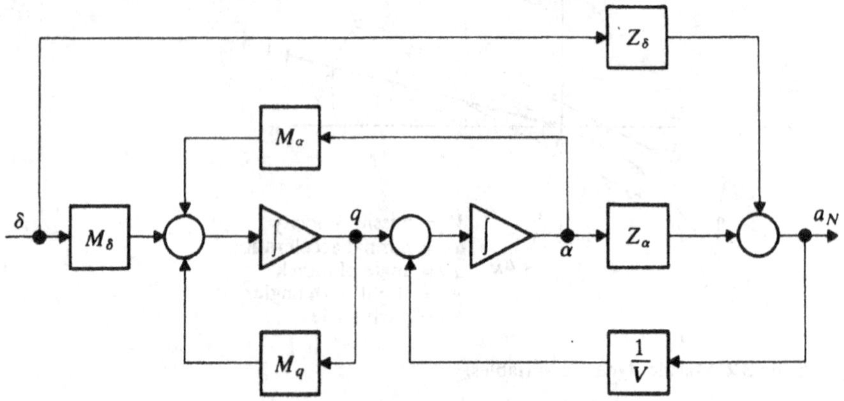

Figure 3.3 Block-diagram of missile dynamics showing normal acceleration as output.

The transfer function from the input $u = \delta$ to the output $y = a_N$ is given by

$$
\begin{aligned}
\mathsf{H}(s) &= C(sI - A)^{-1}B + D \\
&= [Z_\alpha \quad 0]\begin{bmatrix} s - Z_\alpha/V & -1 \\ -M_\alpha & s - M_q \end{bmatrix}^{-1}\begin{bmatrix} Z_\delta/V \\ M_\delta \end{bmatrix} + Z_\delta \\
&= \frac{Z_\delta(s^2 - M_q s - M_\alpha) + Z_\alpha M_\delta}{s^2 - \left(M_q + \dfrac{Z_\alpha}{V}\right)s + \dfrac{Z_\alpha}{V}M_q - M_\alpha}
\end{aligned}
\tag{3F.5}
$$

In a typical missile Z_α, M_α, Z_δ, and M_δ are all negative. Thus the coefficient of s^2 in the numerator of $\mathsf{H}(s)$ in (3F.5) is negative. The constant term $Z_\alpha M_\delta - M_\alpha Z_\delta$, on the other hand, is typically positive. This implies that the numerator of $\mathsf{H}(s)$ has a zero in the right half of the s plane. A transfer function having a right-half plane zero is said to be "nonminimum-phase" and can be the source of considerable difficulty in design of a well-behaved closed-loop control system. One can imagine the problem that might arise by observing that the dc gain $-(Z_\alpha M_\delta - M_\alpha Z_\delta)/M_\alpha$ is (typically) positive but the high-frequency gain $-Z_\delta/M_\alpha$ is (typically) negative. So if a control law is designed to provide negative feedback at dc, unless great care is exercised in the design, it is liable to produce positive feedback at high frequencies. Another peculiarity of the transfer function of (3F.5) is that its step response starts out

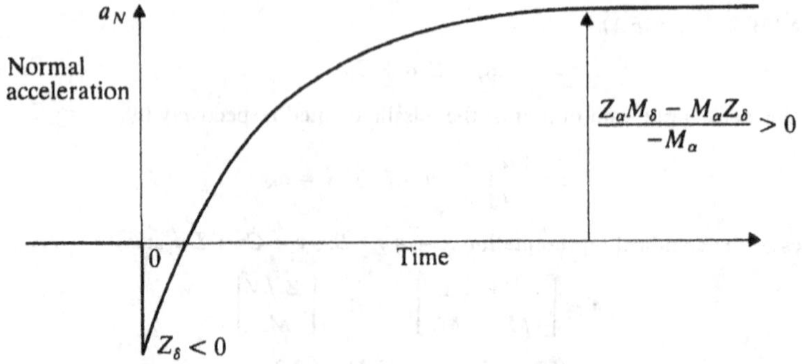

Figure 3.4 Normal acceleration step response (open-loop) of tactical missile showing reversal in sign.

negative and then turns positive, as shown in Fig. 3.4. The initial value of the step response is

$$\lim_{s\to\infty} s\left[\frac{1}{s}\,\mathrm{H}(s)\right] = Z_\delta < 0 \quad \text{(typically)}$$

but the final value of the step response is

$$\lim_{s\to 0} s\left[\frac{1}{s}\,\mathrm{H}(s)\right] = \frac{Z_\alpha M_\delta - M_\alpha Z_\delta}{-M_\alpha} > 0 \qquad \text{(typically)}$$

Example 3G Dynamics of two-axis gyroscope In Example 2F we used the general theory of rigid-body dynamics, and made small angle approximations to develop the equations of motion for a two-axis gyroscope ("gyro"):

$$\begin{aligned}
\dot{\delta}_x &= \omega_{xB} - \omega_{xE} \\
\dot{\delta}_y &= \omega_{yB} - \omega_{yE} \\
\dot{\omega}_{xB} &= -\frac{H}{J_d}\omega_{yB} - \frac{B}{J_d}(\omega_{xB} - \omega_{xE}) - \frac{K_D}{J_d}\delta_x - \frac{K_Q}{J_d}\delta_y + \frac{\tau_x}{J_d} \\
\dot{\omega}_{yB} &= \frac{H}{J_d}\omega_{xB} - \frac{B}{J_d}(\omega_{yB} - \omega_{yE}) + \frac{K_Q}{J_d}\delta_x - \frac{K_D}{J_d}\delta_y + \frac{\tau_y}{J_d}
\end{aligned} \tag{3G.1}$$

where δ_x and δ_y are the angular displacements of the gyro rotor about x and y axes with respect to the case; ω_{xB} and ω_{yB} are the components of the inertial velocity of the rotor projected onto the x and y axes of the gyro; ω_{xE}, ω_{yE} are the angular velocity components of the gyro case projected onto the same axes; τ_x and τ_y are the externally supplied control torques. The parameters H, J_D, K_D, K_Q are physical parameters of the gyro, as explained in Example 2F.

With respect to the dynamic model of (3G.1), there are two kinds of inputs: *control* inputs, represented by the control torques τ_x and τ_y, and *exogenous* inputs, represented by the case angular-velocity components ω_x and ω_y. These exogenous inputs are not "disturbances" in the sense of being unwanted; their presence is the raison d'être for the gyro.

The standard vector matrix form of (3G.1) is thus

$$\dot{x} = Ax + Bu + Ex_0$$

where

$$x = \begin{bmatrix} \delta_x \\ \delta_y \\ \omega_{xB} \\ \omega_{yB} \end{bmatrix} \qquad u = \begin{bmatrix} \tau_x \\ \tau_y \end{bmatrix} \qquad x_0 = \begin{bmatrix} \omega_{xE} \\ \omega_{yE} \end{bmatrix}$$

$$A = \left[\begin{array}{cc|cc} 0 & 0 & 1 & 0 \\ 0 & 0 & 0 & 1 \\ \hline -K_D/J_d & -K_Q/J_d & -B/J_d & -H/J_d \\ K_Q/J_d & -K_D/J_d & H/J_d & -B/J_d \end{array}\right] \tag{3G.2}$$

$$B = \begin{bmatrix} 0 & 0 \\ 0 & 0 \\ 1/J_d & 0 \\ 0 & 1/J_d \end{bmatrix} \qquad E = \begin{bmatrix} -1 & 0 \\ 0 & -1 \\ B/J_d & 0 \\ 0 & B/J_d \end{bmatrix} \tag{3G.3}$$

The special structure of the lower half of the A matrix is noteworthy: The 2×2 submatrix in the lower right-hand corner is

$$\begin{bmatrix} -B/J_d & -H/J_d \\ H/J_d & -B/J_d \end{bmatrix} = -\frac{B}{J_d}\begin{bmatrix} 1 & 0 \\ 0 & 1 \end{bmatrix} + \frac{H}{J_d}\begin{bmatrix} 0 & -1 \\ 1 & 0 \end{bmatrix} \tag{3G.4}$$

The B/J_d terms are conventional damping terms (torque proportional to angular velocity) which tend to dissipate the initial energy of the gyro. The H/J_d terms (which appear in a skew symmetric matrix) have an entirely different effect: They do not cause the energy of the gyro to dissipate but rather produce a high-frequency oscillation called "nutation," a phenomenon present in all gyros, to be discussed at greater length later.

The 2×2 submatrix in the lower left-hand corner of the A matrix is also of interest. This matrix is

$$\begin{bmatrix} -K_D/J_d & -K_Q/J_d \\ K_Q/J_d & -K_D/J_d \end{bmatrix} = -\frac{K_D}{J_d}\begin{bmatrix} 1 & 0 \\ 0 & 1 \end{bmatrix} + \frac{K_Q}{J_d}\begin{bmatrix} 0 & -1 \\ 1 & 0 \end{bmatrix} \tag{3G.5}$$

The K_D/J_d terms are conventional spring terms. In a gyro they give rise to a low-frequency oscillatory motion known as "precession."

We can evince these phenomena by studying the characteristic equation of the gyro:

$$|sI - A| = \left|\begin{array}{cc|cc} s & 0 & -1 & 0 \\ 0 & s & 0 & -1 \\ \hline c_1 & c_2 & s + b_1 & b_2 \\ -c_2 & c_1 & -b_2 & s + b_1 \end{array}\right| = 0 \tag{3G.6}$$

where

$$b_1 = B/J_d \qquad b_2 = H/J_d$$

$$c_1 = K_D/J_d \qquad c_2 = K_Q/J_d$$

The determinant appearing in (3G.6) can be evaluated in a variety of ways—some simpler than others. The result is

$$|sI - A| = (s^2 + b_1 s + c_1)^2 + (b_2 s + c_2)^2 = 0 \tag{3G.7}$$

or

$$(s^2 + b_1 s + c_1)^2 = -(b_2 s + c_2)^2$$

Thus the eigenvalues are the roots of

$$s^2 + b_1 s + c_1 = \pm j(b_2 s + c_2)$$

or

$$s^2 + (b_1 \mp jb_2)s + c_1 \mp jc_2 = 0 \tag{3G.8}$$

The eigenvalues of the system are thus the four roots of (3G.8)

$$s = \frac{-(b_1 \mp jb_2) \pm \sqrt{(b_1 \mp jb_2)^2 - 4(c_1 \mp jc_2)}}{2} \tag{3G.9}$$

In an ideal gyro the "spring" coefficients c_1 and c_2 are zero; they are not zero in some types of real gyros, but in any case they are very small; i.e.,

$$|c_1 + jc_2| \ll |b_1 + jb_2|^2 \tag{3G.10}$$

Taking note of this, we write the radical in (3G.9) as

$$\sqrt{(b_1 \mp jb_2)^2 - 4(c_1 \mp jc_2)} = (b_1 \mp jb_2)\sqrt{1 - \frac{4(c_1 \mp jc_2)}{(b_1 \mp jb_2)^2}} \tag{3G.11}$$

Using the approximation:

$$(1 + \varepsilon)^{1/2} \approx 1 + \tfrac{1}{2}\varepsilon \qquad \text{for } \varepsilon \ll 1$$

we obtain for (3G.11)

$$\sqrt{(b_1 \mp jb_2)^2 - 4(c_1 \mp jc_2)} \approx b_1 \mp jb_2 - \frac{2(c_1 \mp jc_2)}{b_1 \mp jb_2}$$

Now

$$\frac{c_1 \mp jc_2}{b_1 \mp jb_2} = \frac{c_1 \mp jc_2}{b_1 \mp jb_2}\,\frac{b_1 \pm jb_2}{b_1 \pm jb_2} = \frac{(b_1c_1 + b_2c_2) \mp j(b_1c_2 - b_2c_1)}{b_1^2 + b_2^2}$$

Hence, by (3G.9), the approximate poles are given by

$$s = -b_1 + \frac{b_1c_1 + b_2c_2}{b_1^2 + b_2^2} \pm j\left(b_2 + \frac{b_2c_1 - b_1c_2}{b_1^2 + b_2^2}\right) \qquad \text{(3G.12)}$$

and

$$s = -\frac{b_1c_1 + b_2c_2}{b_1^2 + b_2^2} \pm j\frac{b_2c_1 - b_1c_2}{b_1^2 + b_2^2} \qquad \text{(3G.13)}$$

On the complex plane, the four eigenvalues are positioned as shown in Fig. 3.5. Two eigenvalues are located relatively close to the origin at a natural frequency

$$\omega_p = \frac{b_2c_1 - b_1c_2}{b_1^2 + b_2^2} \qquad \text{(3G.14)}$$

which is known as the *precession* frequency. The pole is stable with a (negative) real part

$$\alpha_p = -\frac{b_1c_1 + b_2c_2}{b_1^2 + b_2^2} \qquad \text{(3G.15)}$$

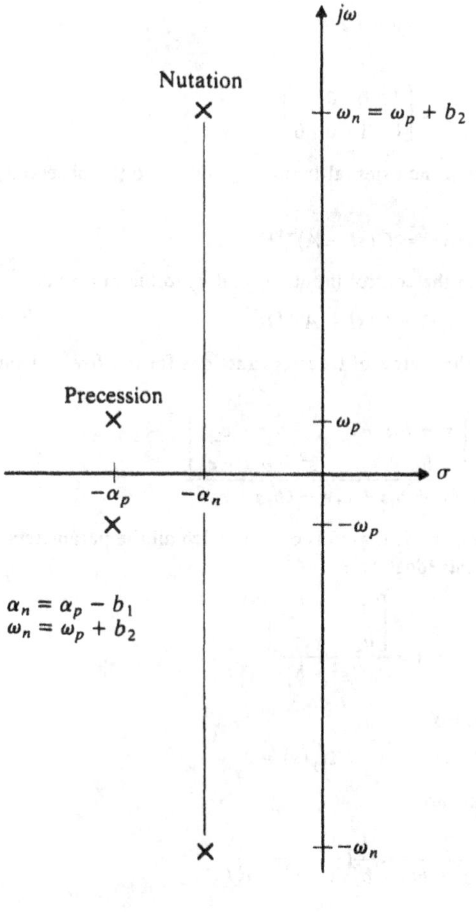

Figure 3.5 Poles of two-axis gyroscope.

The other two poles are located much farther from the origin, at a natural frequency

$$\omega_n = b_2 + \omega_p$$

which is known as the "nutation" frequency. This pole is also stable with a (negative) real part of

$$\alpha_n = -b_1 + \alpha_p \tag{3G.16}$$

The precession poles are due to the presence of the spring terms c_1 and c_2. In an ideal gyro in which these terms are absent, the precession poles move to the origin and the nutation terms become

$$\begin{aligned} \omega_n &= b_2 = H/J_d \\ \alpha_n &= -b_1 = B/J_d \end{aligned} \tag{3G.17}$$

With the precession terms present, the nutation frequency changes from H/J_d by the amount of the precession frequency, and the damping is decreased.

The outputs of the gyro are the signal measured at the pick-off angles. Thus the output equations are

$$y_1 = \delta_x$$

$$y_2 = \delta_y$$

or, in vector-matrix notation

$$y = Cx$$

with

$$C = \begin{bmatrix} 1 & 0 & 0 & 0 \\ 0 & 1 & 0 & 0 \end{bmatrix}$$

The transfer-function matrix from the external inputs ω_x and ω_y to the observed outputs δ_x and δ_y is

$$\mathsf{H}_I(s) = C(sI - A)^{-1}E \tag{3G.18}$$

and the transfer-function matrix from the control inputs τ_x and τ_y to the output is

$$\mathsf{H}_u(s) = C(sI - A)^{-1}B \tag{3G.19}$$

On evaluating (3G.18) we find the matrix of transfer functions for the *free* (uncontrolled) gyro

$$\mathsf{H}_I(s) = \frac{\begin{bmatrix} s^2 + b_1 s + c_1 & -b_2 s - c_2 \\ b_2 s + c_2 & s^2 + b_1 s + c_1 \end{bmatrix}}{(s^2 + b_1 s + c_1)^2 + (b_2 s + c_2)^2}$$

For inertial navigation purposes, an ideal gyro is one in which all the parameters are zero with the exception of $b_2 = H/J$. In this ideal case

$$\mathsf{H}_I(s) = \frac{\begin{bmatrix} s^2 & -b_2 s \\ b_2 s & s^2 \end{bmatrix}}{s^2(s^2 + b_2^2)}$$

For a step input of angular velocity, say

$$\Omega_x(s) = 1/s \qquad \Omega_y(s) = 0$$

the Laplace transforms of the outputs are

$$\Delta_x(s) = \frac{1}{s(s^2 + b_2^2)} = \frac{1}{b_2^2}\left(\frac{1}{s} - \frac{s}{s^2 + b_2^2}\right)$$

$$\Delta_y(s) = -\frac{1}{s^2(s^2+b_2^2)} = -\frac{1}{b_2}\left(\frac{1}{s^2} - \frac{1}{s^2+b_2^2}\right)$$

and the corresponding time functions are

$$\delta_x(t) = \frac{1}{b_2^2}(1 - \cos b_2 t)$$

$$\delta_y(t) = -\frac{1}{b_2}\left(t - \frac{1}{b_2}\sin b_2 t\right)$$

as shown in Fig. 3.6. The output angle δ_x for an angular velocity input about the x axis is a sinusoid of amplitude $1/b_2^2$ with a dc value of $1/b_2^2$. The cross-axis output, however, oscillates about a line having a slope of $1/b_2$. Thus, a constant angular velocity input produces an output in the cross axis with a constantly increasing mean value. Because of this output, an ideal gyro is also called a *rate-integrating* gyro, since its long-term outputs (the pick-off angles δ_x, δ_y) are proportional to the integrals of the angular velocity components about the corresponding cross axes. (Note that the constant of proportionality for one input-output pair has the same numerical magnitude as that of the other input-output pair, but is of opposite sign.)

Since the pick-off angles (i.e., the angular displacements of the wheel plane) cannot be large in a typical gyro, a rate-integrating gyro is not suitable for applications in which the integrals of the body rates (i.e., the displacement of the gyro case relative to the rotor, whose axis tends to remain stationary in space) are appreciable. Since the motion of the craft (air, sea, or space) which carries the gyro cannot be confined to such small angles, the gyros are

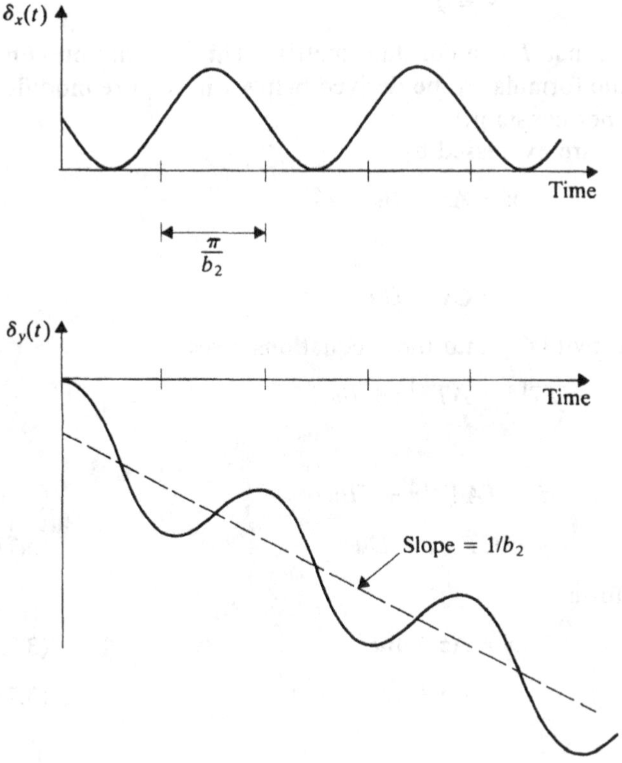

Figure 3.6 Outputs on two axes of gyro for constant angular velocity on x-axis.

typically mounted on a *stable platform* which is connected to the carrying craft by means of a set of gimbals that permit the stable platform to maintain a fixed orientation in space while the carrying craft undergoes arbitrary motion. Any tendency of the stable platform to rotate in space is immediately sensed by the gyro pick-offs and the output signals are used to generate feedback signals that drive gimbal torquers which move the gimbals to maintain the pick-off angles very close to null.

3.6 TRANSFORMATION OF STATE VARIABLES

It frequently happens that the state variables used in the original formulation of the dynamics of a system are not as convenient as another set of state variables. Instead of having to reformulate the system dynamics, it is possible to transform the matrices A, B, C, and D of the original formulation to a new set of matrices $\bar{A}$, $\bar{B}$, $\bar{C}$, and $\bar{D}$. The change of variables is represented by a linear transformation

$$z = Tx \tag{3.68}$$

where z is the state vector in the new formulation and x is the state vector in the original formulation. It is assumed that the transformation matrix T is a nonsingular k by k matrix, so that we can always write

$$x = T^{-1}z \tag{3.69}$$

We assume, moreover, that T is a constant matrix. (This assumption is not necessary, however, but the formulas to be derived below will require modification to include $\dot{T}$, if T is not constant.)

The original dynamics are expressed by

$$\dot{x} = Ax + Bu$$

and the output by

$$y = Cx + Du$$

Substitution of x as given by (3.69) into these equations gives

$$T^{-1}\dot{z} = AT^{-1}z + Bu$$

or

$$\dot{z} = TAT^{-1}z + TBu \tag{3.70}$$

$$y = CT^{-1}z + Du \tag{3.71}$$

These are in the normal form

$$\dot{z} = \bar{A}z + \bar{B}u \tag{3.72}$$

$$y = \bar{C}z + \bar{D}u \tag{3.73}$$

with

$$\bar{A} = TAT^{-1} \qquad \bar{B} = TB \qquad \bar{C} = CT^{-1} \qquad \bar{D} = D \tag{3.74}$$

In the language of matrix algebra, the dynamics matrix of the transformed system $\bar{A} = TAT^{-1}$ is said to be *similar* to the dynamics matrix A of the original system. A well-known fact of matrix algebra is that similar matrices have the same characteristic polynomial. If we didn't already know this we could show it using the argument that the input-output relations for the system, i.e., the transfer function from the input to the output, should not depend on how the state variables are defined. Using the original state variables, we found in the previous section that the transfer function is given by

$$\mathsf{H}(s) = \frac{CBs^{k-1} + CE_2Bs^{k-2} + \cdots + CE_kB}{s^k + a_1s^{k-1} + \cdots + a_k} + D \tag{3.75}$$

Using the new state variables, the transfer function is given by

$$\mathsf{H}(s) = \frac{\bar{C}\bar{B}s^{k-1} + \bar{C}\bar{E}_2\bar{B}s^{k-2} + \cdots + \bar{C}\bar{E}_k\bar{B}}{s^k + \bar{a}_1s^{k-1} + \cdots + \bar{a}_k} + \bar{D} \tag{3.76}$$

where

$$s^k + \bar{a}_1s^{k-1} + \cdots + \bar{a}_k = |sI - \bar{A}|$$

and

$$\operatorname{adj}(sI - \bar{A}) = Is^{k-1} + \bar{E}_2s^{k-2} + \cdots + \bar{E}_k$$

In order for the two transfer functions given by (3.75) and (3.76) to be equal, we need $\bar{D} = D$, which we have already determined, and we also must have

$$CB = \bar{C}\bar{B} \tag{3.77}$$

$$CE_iB = \bar{C}\bar{E}_i\bar{B} \qquad i = 1, 2, \ldots, k \tag{3.78}$$

$$a_i = \bar{a}_i \qquad i = 1, 2, \ldots, k \tag{3.79}$$

Using (3.74) $\bar{C}\bar{B} = CT^{-1}TB = CB$, so (3.77) is satisfied. The condition that $a_i = \bar{a}_i$ is a verification of the condition that the characteristic polynomials of similar matrices are equal. Finally, we must verify that (3.78) is satisfied. This is done with the aid of (3.56). For the original system, (3.56) gives

$$CE_{i+1}B = CAE_iB + a_iCB \tag{3.80}$$

and, from (3.74) $C = \bar{C}T$ and $B = T^{-1}\bar{B}$. Thus (3.80) becomes

$$\bar{C}TE_{i+1}T^{-1}\bar{B} = \bar{C}TA(T^{-1}T)E_iT^{-1}\bar{B} + \bar{a}_i\bar{C}\bar{B} \tag{3.81}$$

Note that $T^{-1}T = I$ has been inserted and that (3.79) and (3.77) have been used. It is thus seen that (3.81) reduces to

$$\bar{C}\bar{E}_{i+1}\bar{B} = \bar{C}\bar{A}\bar{E}_i\bar{B} + \bar{a}_i\bar{C}\bar{B}$$

which will satisfy (3.78) provided that

$$\bar{E}_i = TE_iT^{-1}$$

which means that each coefficient matrix E_i of the adjoint matrix of $\bar{A}$

transforms from the corresponding coefficient matrix E_i of the original matrix A, in the same way as $\bar{A}$ transforms from A, i.e.,

$$\bar{A} = TAT^{-1}$$

as given by (3.74). This is another fact of matrix algebra, which has been verified by the requirement that transfer functions between the input and the output must not depend on the definition of the state vector.

Example 3H Spring-coupled masses The equations of motion of a pair of masses M_1 and M_2 coupled by a spring, and sliding in one dimension in the absence of friction (see Fig. 3.7(*a*)) are

$$\ddot{x}_1 + \frac{K}{M_1}(x_1 - x_2) = \frac{u_1}{M_1}$$
$$\ddot{x}_2 + \frac{K}{M_2}(x_2 - x_1) = \frac{u_2}{M_2} \tag{3H.1}$$

where u_1 and u_2 are the externally applied forces and K is the spring constant. Defining the state

$$x = [x_1 \quad x_2 \quad \dot{x}_1 \quad \dot{x}_2]'$$

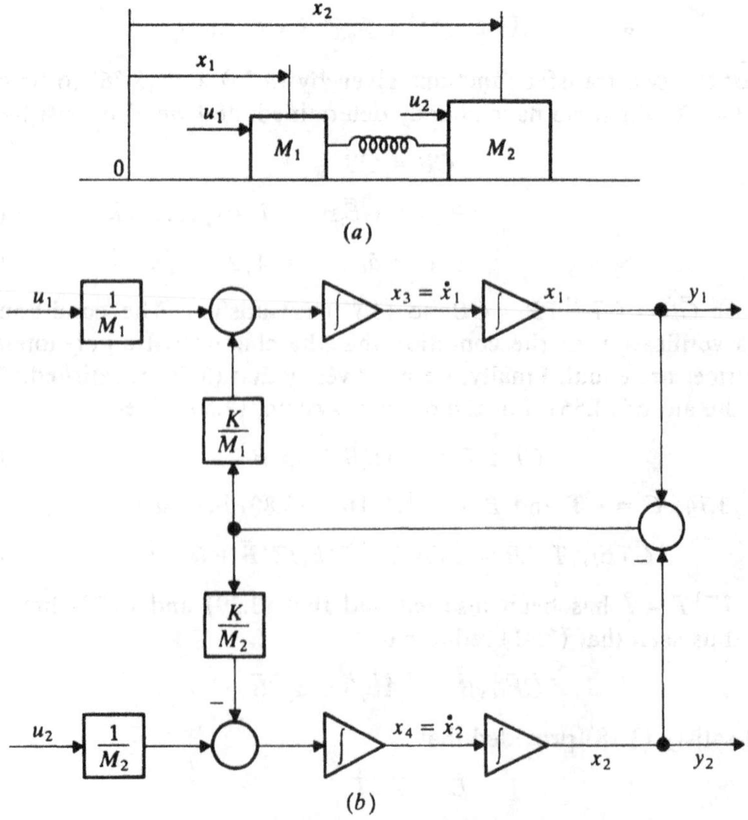

Figure 3.7 Dynamics of spring-coupled masses. (*a*) System configuration; (*b*) Block diagram.

results in the following matrices

$$A = \begin{bmatrix} 0 & 0 & 1 & 0 \\ 0 & 0 & 0 & 1 \\ -K/M_1 & K/M_1 & 0 & 0 \\ K/M_2 & -K/M_2 & 0 & 0 \end{bmatrix} \qquad B = \begin{bmatrix} 0 & 0 \\ 0 & 0 \\ 1/M_1 & 0 \\ 0 & 1/M_2 \end{bmatrix} \tag{3H.2}$$

It might be more convenient, however, to define the motion of the system by the motion of the center-of-mass

$$\bar{x} = \frac{M_1}{M} x_1 + \frac{M_2}{M} x_2 \qquad (M = M_1 + M_2) \tag{3H.3}$$

and the difference

$$\delta = x_1 - x_2 \tag{3H.4}$$

between the positions of the two masses. We let

$$z = [\bar{x}, \delta, \dot{\bar{x}}, \dot{\delta}]$$

From (3H.3) and (3H.4)

$$\dot{\bar{x}} = \frac{M_1}{M} \dot{x}_1 + \frac{M_2}{M} \dot{x}_2$$

$$\dot{\delta} = \dot{x}_1 - \dot{x}_2$$

Thus we have

$$\begin{bmatrix} \bar{x} \\ \delta \\ \dot{\bar{x}} \\ \dot{\delta} \end{bmatrix} = \underbrace{\begin{bmatrix} M_1/M & M_2/M & 0 & 0 \\ 1 & -1 & 0 & 0 \\ 0 & 0 & M_1/M & M_2/M \\ 0 & 0 & 1 & -1 \end{bmatrix}}_{T} \begin{bmatrix} x_1 \\ x_2 \\ \dot{x}_1 \\ \dot{x}_2 \end{bmatrix} \tag{3H.5}$$

The 4 by 4 matrix in (3H.5) is the transformation matrix T, the inverse of which is easily found to be

$$T^{-1} = \begin{bmatrix} 1 & M_2/M & 0 & 0 \\ 1 & -M_1/M & 0 & 0 \\ 0 & 0 & 1 & M_2/M \\ 0 & 0 & 1 & -M_1/M \end{bmatrix} \tag{3H.6}$$

Thus we find

$$\bar{A} = TAT^{-1} = \begin{bmatrix} 0 & 0 & 1 & 0 \\ 0 & 0 & 0 & 1 \\ 0 & 0 & 0 & 0 \\ 0 & -KM/M_1M_2 & 0 & 0 \end{bmatrix} \qquad \bar{B} = TB = \begin{bmatrix} 0 & 0 \\ 0 & 0 \\ 1/M & 1/M \\ 1/M_1 & -1/M_2 \end{bmatrix} \tag{3H.7}$$

The differential equations corresponding to $\bar{A}$ and $\bar{B}$ are

$$\begin{aligned} \ddot{\bar{x}} &= \frac{u_1 + u_2}{M} \\ \ddot{\delta} &= -\frac{KM}{M_1 M_2}\delta + \frac{u_1}{M_1} - \frac{u_2}{M_2} \end{aligned} \tag{3H.8}$$

In this case, these equations could readily have been obtained directly from the original equations (3H.1).

3.7 STATE-SPACE REPRESENTATION OF TRANSFER FUNCTIONS: CANONICAL FORMS

In Sec. 3.5 we learned how to determine the transfer function of a linear, time-invariant system, given the state-space representation. Sometimes it is necessary to go in the other direction: from the transfer-function to the state-space representation. This need may arise because the only available description of a subsystem within a larger system is the transfer function of that subsystem. In order to use state-space methods, the transfer function must be turned into a set of first-order differential equations. Another reason for converting a transfer-function representation into a state-space representation is for the purpose of transient response simulation. Many algorithms and numerical integration computer programs designed for solution of systems of first-order equations are available, but there is not much software for numerical inversion of Laplace transforms. Thus, if a reliable method is needed for calculating the transient response of a system, one may be better off converting the transfer function of the system to state-space form and numerically integrating the resulting differential equations rather than attempting to compute the inverse Laplace transform by numerical methods.

In the last section we saw that there are innumerable systems that have the same transfer function. Hence the representation of a transfer function in state-space form is obviously not unique. In this section we shall develop several standard, or "canonical" representations of transfer functions that can always be used for single-input, multiple-output or multiple-input, single-output systems. One canonical representation has no general advantage over any other, and, moreover, there is no reason why a canonical representation is to be preferred over a noncanonical representation.

First companion form The development starts with a transfer function of a single-input, single-output system of the form

$$\mathsf{H}(s) = \frac{\mathsf{y}(s)}{\mathsf{u}(s)} = \frac{1}{s^k + a_1 s^{k-1} + \cdots + a_k} \tag{3.82}$$

which can be written

$$(s^k + a_1 s^{k-1} + \cdots + a_k)\mathsf{y}(s) = \mathsf{u}(s) \tag{3.83}$$

The differential equation corresponding to (3.83) is

$$D^k y + a_1 D^{k-1} y + \cdots + a_k y = u \tag{3.84}$$

where $D^k y$ stands for $d^k y/dt^k$. Solve for the highest derivative in (3.84)

$$D^k y = -a_1 D^{k-1} y - a_2 D^{k-2} y - \cdots - a_k y + u \tag{3.85}$$

Now consider a chain of k integrators as shown in Fig. 3.8(a), and suppose that the output of the last integrator is y. Then the output of the next-to-last integrator is $Dy = dy/dt$, and so forth. The output from the first integrator is

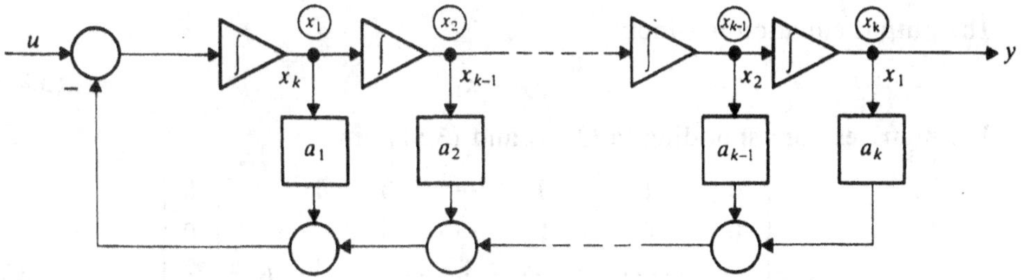

(a)

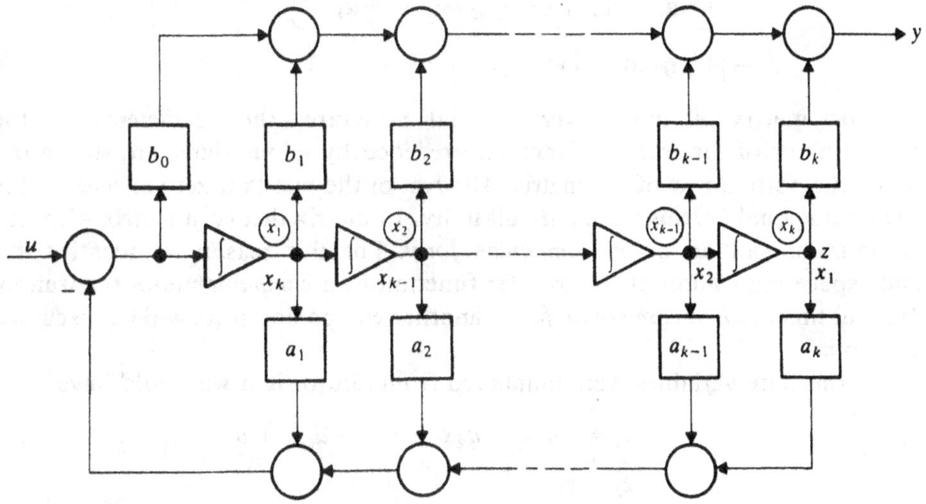

(b)

Figure 3.8 State-space realization of transfer functions in first companion form

$$(a)\ \mathrm{H}(s) = \frac{1}{s^k + a_1 s^{k-1} + \cdots + a_k} \qquad (b)\ \mathrm{H}(s) = \frac{b_0 s^k + b_1 s^{k-1} + \cdots + b_k}{s^k + a_1 s^{k-1} + \cdots + a_k}$$

$D^{k-1}y$ and the input to this integrator is thus $D^k y$. From (3.85) it follows that Fig. 3.8(a) represents the given transfer function (3.82) provided that the feedback gains are chosen as shown in the figure. To get one state-space representation of the system, we identify the output of each integrator with a state variable, starting at the right and proceeding to the left. The corresponding differential equations using this identification of state variables are

$$\begin{aligned} \dot{x}_1 &= x_2 \\ \dot{x}_2 &= x_3 \\ &\cdots\cdots\cdots\cdots\cdots\cdots \\ \dot{x}_{k-1} &= x_k \\ \dot{x}_k &= -a_k x_1 - a_{k-1} x_2 - \cdots - a_1 x_k + u \end{aligned} \tag{3.86}$$

The output equation is simply

$$y = x_1 \tag{3.87}$$

The matrices corresponding to (3.86) and (3.87) are

$$A = \begin{bmatrix} 0 & 1 & 0 & \cdots & 0 \\ 0 & 0 & 1 & \cdots & 0 \\ \cdots & \cdots & \cdots & \cdots & \cdots \\ 0 & 0 & 0 & \cdots & 1 \\ -a_k & -a_{k-1} & -a_{k-2} & \cdots & -a_1 \end{bmatrix} \qquad B = \begin{bmatrix} 0 \\ 0 \\ \vdots \\ 0 \\ 1 \end{bmatrix} \tag{3.88}$$

$$C = [1 \quad 0 \quad 0 \quad \cdots \quad 0]$$

The matrix A has a very special structure: the coefficients of the denominator of the transfer function, preceded by minus signs, are strung out along the bottom row of the matrix. The rest of the matrix is zero except for the "superdiagonal" terms which are all unity. In matrix theory, a matrix with this structure is said to be *in companion form.* For this reason we identify this state-space realization of the transfer function as a *companion-form* realization. We call this the *first companion form*; another companion form will be discussed later on.

If the state variables were numbered from left to right we would have

$$\begin{aligned} \dot{x}_1 &= -a_1 x_1 - a_2 x_2 - \cdots - a_k x_k + u \\ \dot{x}_2 &= x_1 \\ &\cdots\cdots\cdots\cdots\cdots \\ \dot{x}_{k-1} &= x_{k-2} \\ \dot{x}_k &= x_{k-1} \end{aligned} \tag{3.89}$$

and

$$y = x_k$$

The corresponding matrices would be

$$A = \begin{bmatrix} -a_1 & -a_2 & \cdots & -a_{k-1} & -a_k \\ 1 & 0 & \cdots & 0 & 0 \\ 0 & 1 & \cdots & 0 & 0 \\ \cdots & \cdots & \cdots & \cdots & \cdots \\ 0 & 0 & \cdots & 1 & 0 \end{bmatrix} \qquad B = \begin{bmatrix} 1 \\ 0 \\ 0 \\ \vdots \\ 0 \end{bmatrix} \tag{3.90}$$

$$C = [0 \quad 0 \quad \cdots \quad 0 \quad 1]$$

This representation is also called a companion form, but is less frequently used than the form (3.88). There is nothing sacred about numbering the integrators systematically from right to left or from left to right. A perfectly valid, if perverse, representation would result if the integrators were numbered at random.

Having developed a state-space representation of the simple transfer function (3.82), we are now in a position to consider the more general transfer function

$$\mathsf{H}(s) = \frac{\mathsf{y}(s)}{\mathsf{u}(s)} = \frac{b_0 s^k + b_1 s^{k-1} + \cdots + b_k}{s^k + a_1 s^{k-1} + \cdots + a_k} \tag{3.91}$$

The development is aided by the introduction of an intermediate variable $\mathsf{z}(s)$

$$\frac{\mathsf{y}(s)}{\mathsf{u}(s)} = \frac{\mathsf{y}(s)}{\mathsf{z}(s)} \frac{\mathsf{z}(s)}{\mathsf{u}(s)} = \frac{b_0 s^k + b_1 s^{k-1} + \cdots + b_k}{s^k + a_1 s^{k-1} + \cdots + a_k}$$

We identify the first factor with the numerator and the second factor with the denominator:

$$\frac{\mathsf{y}(s)}{\mathsf{z}(s)} = b_0 s^k + b_1 s^{k-1} + \cdots + b_k \tag{3.92}$$

and

$$\frac{\mathsf{z}(s)}{\mathsf{u}(s)} = \frac{1}{s^k + a_1 s^{k-1} + \cdots + a_k} \tag{3.93}$$

The realization of the transfer function from u to z has already been developed. And, from (3.92)

$$\mathsf{y}(s) = (b_0 s^k + b_1 s^{k-1} + \cdots + b_k)\mathsf{z}(s)$$

i.e.,

$$y = b_0 D^k z + b_1 D^{k-1} z + \cdots + b_k z$$

The inputs to the integrators in the chain are the k successive derivatives of z as shown in Fig. 3.8(b), hence we have the required state-space representation. All that remains to be done is to write the corresponding differential equations. The state equations are the same as (3.86) or (3.89) and hence the A and B matrices are the same. The output equation is found by careful examination of the block diagram of Fig. 3.8(b). Note that there are *two* paths from the output of each integrator to the system output: one path upward through the box labeled b_i and a second path down through the box labeled a_i and thence through the box labeled b_0. As a consequence, when the right-to-left state variable numbering is used

$$y = (b_k - a_k b_0)x_1 + (b_{k-1} - a_{k-1} b_0)x_2 + \cdots + (b_1 - a_1 b_0)x_k + b_0 u$$

Hence

$$C = [b_k - a_k b_0,\ b_{k-1} - a_{k-1} b_0, \ldots, b_1 - a_1 b_0], \qquad D = [b_0] \tag{3.94}$$

If the direct path through b_0 is absent, then the D matrix is zero and the C matrix contains only the b_i coefficients.

If left-to-right numbering is used, then

$$C = [b_1 - a_1 b_0, b_2 - a_2 b_0, \ldots, b_k - a_k b_0], \qquad D = [b_0] \tag{3.95}$$

The structure of the first canonical form is very easy to remember ("automnemonic"). The string of integrators can be visualized as the fraction bar of the transfer function (3.91) that is realized. The numerator coefficients appear above the chain of integrators in the same order as they appear above the fraction bar in (3.91) and the denominator coefficients appear below the chain of integrators in the same order as they appear below the fraction bar in (3.91). Not too much imagination is needed to "see" the transfer function (3.91) in Fig. 3.8.

A generalized version of the first companion form can be used to realize a single input, multiple output system represented by l transfer functions, one from the single input to each of the l different outputs

$$\frac{\mathrm{y}_1(s)}{\mathrm{u}(s)} = \frac{b_{01}s^k + b_{11}s^{k-1} + \cdots + b_{k1}}{s^k + a_1 s^{k-1} + \cdots + a_k}$$

$$\cdots\cdots\cdots\cdots\cdots\cdots$$

$$\frac{\mathrm{y}_l(s)}{\mathrm{u}(s)} = \frac{b_{0l}s^k + b_{1l}s^{k-1} + \cdots + b_{kl}}{s^k + a_1 s^{k-1} + \cdots + a_k}$$

The same set of state variables serves for each transfer function. Each numerator, however, is realized by a different set of gains, as shown in Fig. 3.9. Thus the A and B matrices are exactly as given earlier. From Fig. 3.9 it is also seen that the C and D matrices are

$$C = \begin{bmatrix} b_{k1} - a_k b_{01} & b_{k-1,1} - a_{k-1} b_{01} & \cdots & b_{11} - a_1 b_{01} \\ \cdots & \cdots & \cdots & \cdots \\ b_{kl} - a_k b_{0l} & b_{k-1,l} - a_{k-1} b_{0l} & \cdots & b_{1l} - a_1 b_{0l} \end{bmatrix} \qquad D = \begin{bmatrix} b_{01} \\ \vdots \\ b_{0l} \end{bmatrix} \tag{3.96}$$

for the right-to-left numbering, or

$$C = \begin{bmatrix} b_{11} - a_1 b_{01} & b_{21} - a_2 b_{01} & \cdots & b_{k1} - a_k b_{01} \\ \cdots & \cdots & \cdots & \cdots \\ b_{1l} - a_1 b_{0l} & b_{2l} - a_2 b_{0l} & \cdots & b_{kl} - a_k b_{0l} \end{bmatrix} \qquad D = \begin{bmatrix} b_{01} \\ \vdots \\ b_{0l} \end{bmatrix} \tag{3.97}$$

for the left-to-right numbering.

In the first canonical form realizations of Figs. 3.1 through 3.9 the input is connected directly to the first integrator in the chain and the output is a linear combination of the outputs of the integrators (and the input, when the D matrix is nonzero). This form is useful not only for single-input, single-output systems, but also, as we have seen, for single-input, multiple-output systems. A variant of the structure of Fig. 3.8, in which the output is taken directly from the last integrator but the input is connected to all the integrators, is shown in Fig. 3.10. A realization of a multiple-input, single-output system based on the structure of Fig. 3.10 is shown in Fig. 3.11.

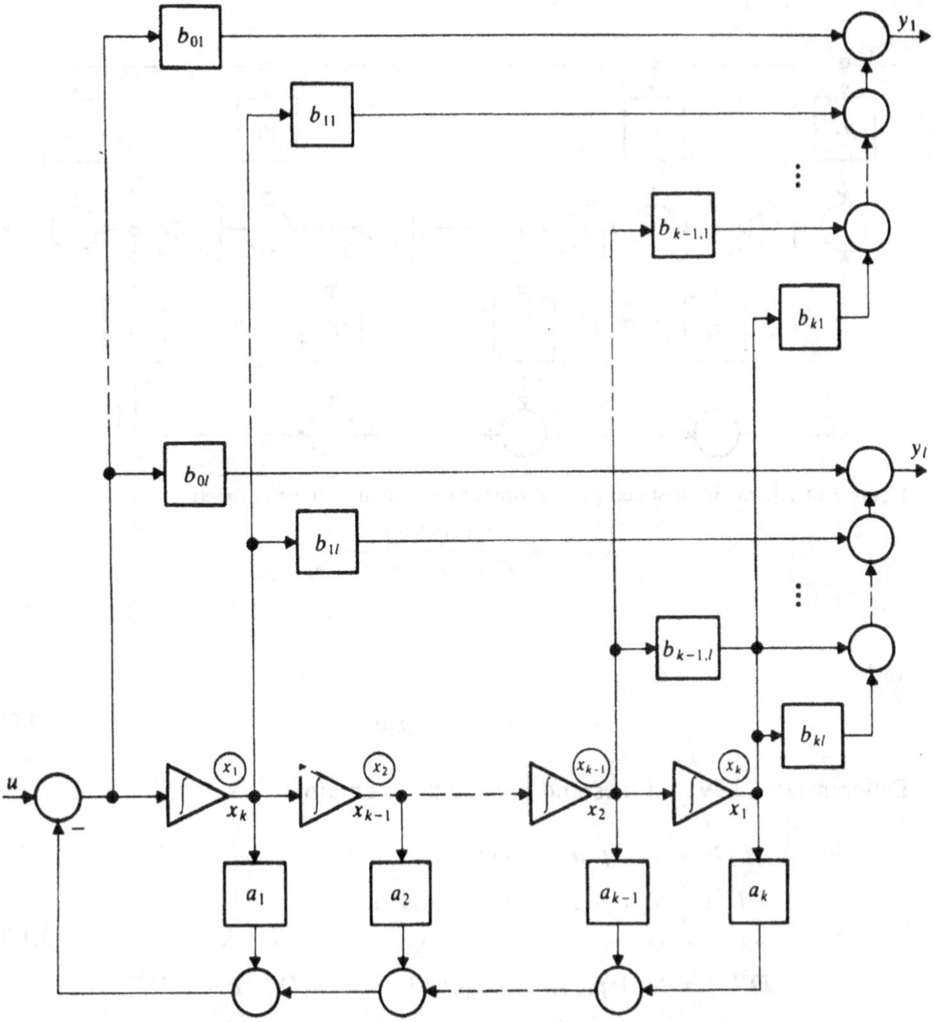

Figure 3.9 Realization of single-input, multiple-output system in first companion form.

The "feedforward" gains $p_1, p_2, \ldots, p_k$ in Fig. 3.10 are in general not equal to the coefficients $b_1, b_2, \ldots, b_k$ of the transfer function but must be obtained by solution of a set of linear algebraic equations which may be derived as follows. From Fig. 3.10 it is easy to see that

$$
\begin{aligned}
\dot{x}_1 &= x_2 + p_1 u \\
\dot{x}_2 &= x_3 + p_2 u \\
&\cdots\cdots\cdots\cdots\cdots\cdots \\
\dot{x}_{k-1} &= x_k + p_{k-1} u \\
\dot{x}_k &= -a_1 x_k - \cdots - a_k x_1 + p_k u
\end{aligned}
\tag{3.98}
$$

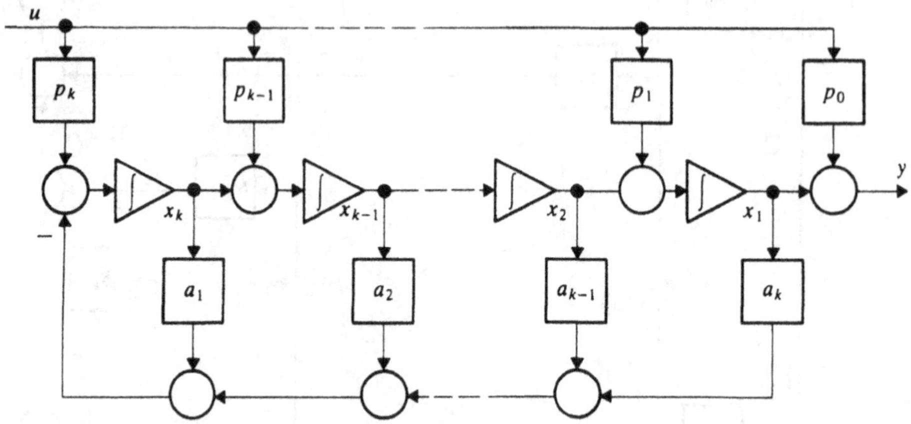

Figure 3.10 Alternative first companion form of realization of transfer function

$$H(s) = \frac{b_0 s^k + b_1 s^{k-1} + \cdots + b_k}{s^k + a_1 s^{k-1} + \cdots + a_k}$$

and

$$y = x_1 + p_0 u \tag{3.99}$$

Differentiate (3.99) k times and use (3.98) to obtain

$$\begin{aligned}
Dy &= x_2 + p_1 u + p_0 Du \\
D^2 y &= x_3 + p_2 u + p_1 Du + p_2 D^2 u \\
&\cdots\cdots\cdots\cdots\cdots\cdots\cdots\cdots\cdots\cdots \\
D^{k-1} y &= x_k + p_{k-1} u + p_{k-2} Du + \cdots + p_1 D^{k-2} u + p_0 D^{k-1} u \\
D^k y &= -a_1 x_k - a_2 x_{k-1} - \cdots - a_k x_1 + p_k u + p_{k-1} Du + \cdots \\
&\quad + p_1 D^{k-1} u + p_0 D^k u
\end{aligned} \tag{3.100}$$

From (3.100) and (3.99) we thus get

$$\begin{aligned}
D^k y + a_1 D^{k-1} y + \cdots + a_{k-1} Dy + a_k y \\
= (p_k + a_1 p_{k-1} + \cdots + a_{k-1} p_1 + \quad a_k p_0) u \\
+ \quad (p_{k-1} + \cdots + a_{k-2} p_1 + a_{k-1} p_0) Du \\
+ \cdots + (p_1 + a_1 p_0) D^{k-1} u \\
p_0 D^k u
\end{aligned} \tag{3.101}$$

In order for (3.101) to represent the differential equation corresponding to the

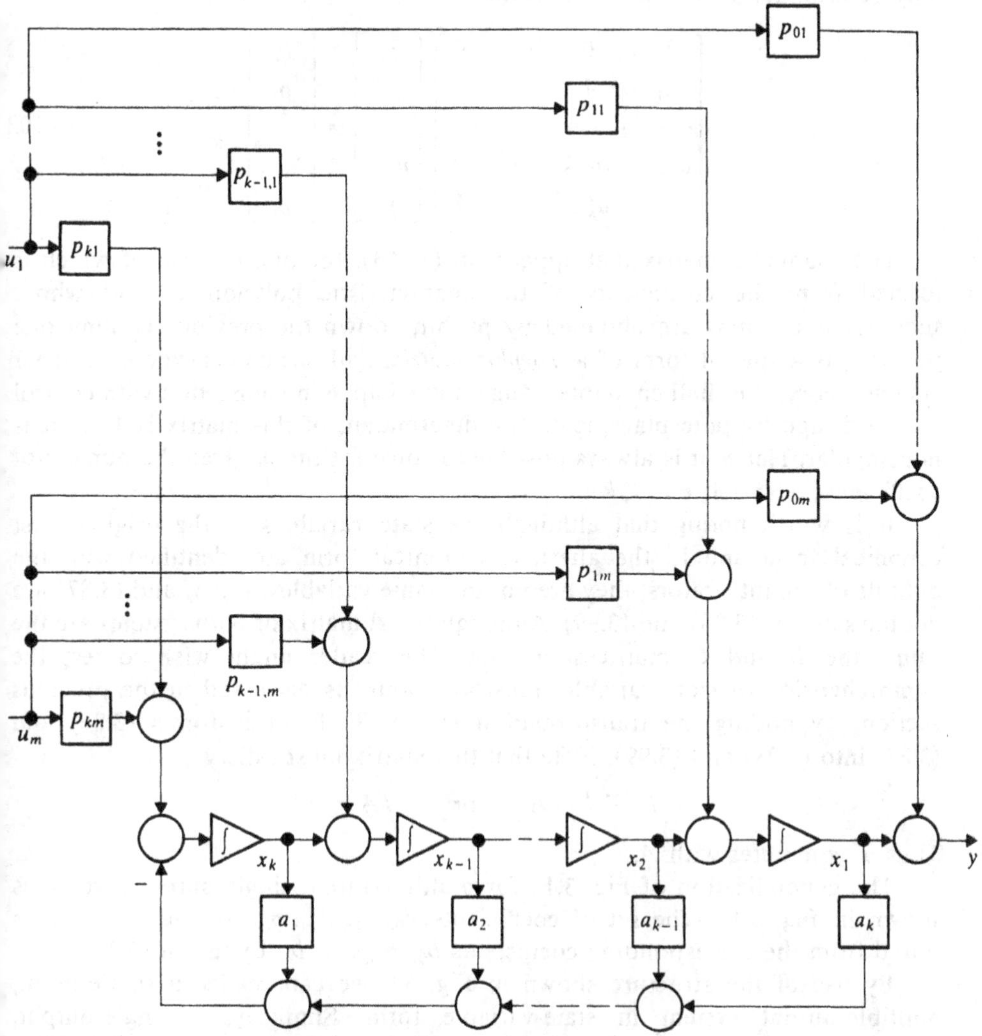

Figure 3.11 Use of alternative first companion form for realizing multiple-input single-output transfer function.

transfer function (3.91) it is necessary that

$$
\begin{aligned}
p_0 &= b_0 \\
p_1 + a_1 p_0 &= b_1 \\
&\cdots\cdots \\
p_{k-1} + \cdots + a_{k-2} p_1 + a_{k-1} p_0 &= b_{k-1} \\
p_k + \cdots + a_{k-1} p_1 + a_k p_0 &= b_k
\end{aligned}
\tag{3.102}
$$

which constitute a set of $k + 1$ simultaneous equations for $p_0, p_1, \ldots, p_k$. These may be arranged in vector-matrix form

$$\begin{bmatrix} 1 & 0 & \cdots & 0 \\ a_1 & 1 & \cdots & 0 \\ \cdots & \cdots & \cdots & \cdots \\ a_{k-1} & a_{k-2} & \cdots & 0 \\ a_k & a_{k-1} & \cdots & 1 \end{bmatrix} \begin{bmatrix} p_0 \\ p_1 \\ \vdots \\ p_{k-1} \\ p_k \end{bmatrix} = \begin{bmatrix} b_0 \\ b_1 \\ \vdots \\ b_{k-1} \\ b_k \end{bmatrix} \tag{3.103}$$

The triangular matrix that appears in (3.103), the first column of which is formed from the coefficients of the characteristic polynomial, and whose subsequent columns are obtained by pushing down the previous column one position, is a special form of a *Toeplitz matrix*, and occurs elsewhere in linear system theory. We shall encounter it again in Chap. 6 in connection with control system design by pole placement. The determinant of this matrix is 1, so it is nonsingular. Hence it is always possible to solve for the p_i given the numerator coefficients b_i $(i = 1, 2, \ldots, k)$.

It is worth noting that although the state variables in the original first canonical form and in the alternate canonical form are identified with the outputs of the integrators, they are not the same variables: (3.86) and (3.87) are not the same as (3.98) and (3.99). Although the A matrix of both systems are the same, the B and C matrices are not. The reader might wish to test the comprehension of state-variable transformations, as discussed in the previous section, by finding the transformation matrix T that transforms (3.86) and (3.87) into (3.98) and (3.99). Note that this matrix must satisfy

$$TAT^{-1} = A \qquad \text{or} \qquad TA = AT$$

Thus T commutes with A.

The generalization of Fig. 3.10 for multiple-input, single-output systems is shown in Fig. 3.11. The set of coefficients $p_{0i}, p_{1i}, \ldots, p_{ki}$ for the ith input is found from the corresponding coefficients $b_{0i}, b_{1i}, \ldots, b_{ki}$ by use of (3.103).

By use of the structure shown in Fig. 3.9 we can realize a single-input, multiple-output system in state-variable form. Similarly, a single-output, multiple-input system can be realized with the structure of Fig. 3.11. One might think that a multiple-input, multiple-output system can be realized with only k integrators using a combination of Figs. 3.9 and 3.11. A bit of reflection, however, will soon convince one that in general this is not possible. It is obvious, however, that one way of realizing a multiple-input, multiple-output system is by using a number of structures of the form of Fig. 3.9 or Fig. 3.11 in parallel. If the number l of outputs is smaller than the number m of inputs, then l structures of Fig. 3.11 are used in parallel; if the number of outputs is greater than the number of inputs then m structures of the form of Fig. 3.9 are used. Hence it is always possible to realize an m-input, l-output system with no more than $k \cdot \min(l, m)$ integrators. But there is no assurance that there is not a realization that requires still fewer integrators. The determination of a "minimum" realization was the subject of considerable research during the

1970s. There are now several algorithms for finding a minimum realization and the matrices A, B, C, and D that result. (See Note 3.5 for a more complete discussion of this subject.)

Second companion form In the first companion form, the coefficients of the denominator of the transfer function appear in one of the rows of the A matrix. There is another set of companion forms in which the coefficients appear in a column of the A matrix. For a single-input, single-output system, this form can be obtained by writing (3.91) as

$$(s^k + a_1 s^{k-1} + \cdots + a_k)\mathrm{y}(s) = (b_0 s^k + b_1 s^{k-1} + \cdots + b_k)\mathrm{u}(s)$$

or

$$s^k[\mathrm{y}(s) - b_0\mathrm{u}(s)] + s^{k-1}[a_1\mathrm{y}(s) - b_1\mathrm{u}(s)] + \cdots + [a_k\mathrm{y}(s) - b_k\mathrm{u}(s)] = 0$$

On dividing by s^k and solving for $\mathrm{y}(s)$, we obtain

$$\mathrm{y}(s) = b_0\mathrm{u}(s) + \frac{1}{s}[b_1\mathrm{u}_1(s) - a_1\mathrm{y}(s)] + \cdots + \frac{1}{s^k}[b_k\mathrm{u}(s) - a_k\mathrm{y}(s)] \quad (3.104)$$

Noting that the multiplier $1/s^j$ is the transfer function of a chain of j integrators, immediately leads to the structure shown in Fig. 3.12. The signal y is fed back to each of the integrators in the chain and the signal u is fed forward. Thus the signal $b_k u - a_k y$ passes through k integrators, as required by (3.104), the signal $b_{k-1}u - a_{k-1}y$ passes through $k - 1$ integrators, and so forth to complete the realization of (3.104). The structure retains the ladder-like shape of the first companion form, but the feedback paths are in different directions.

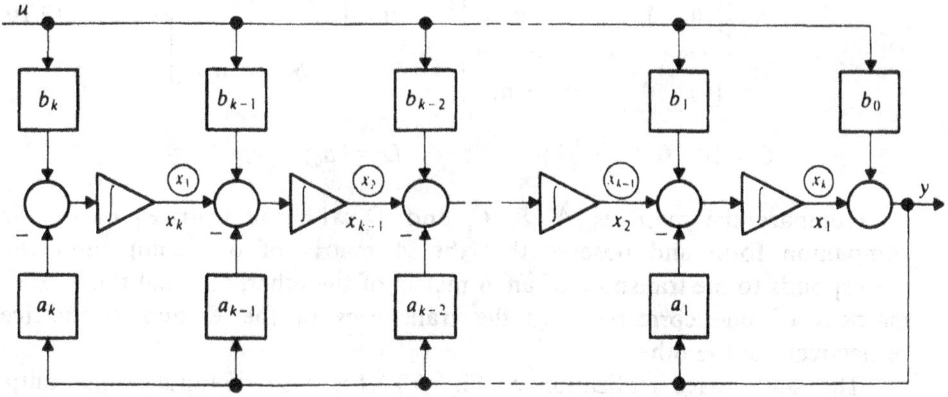

Figure 3.12 State-space realization of transfer function

$$\mathrm{H}(s) = \frac{b_0 s^k + b_1 s^{k-1} + \cdots + b_k}{s^k + a_1 s^{k-1} + \cdots + a_k}$$

in second companion form.

Using the right-to-left numbering of state variables, the differential equations corresponding to Fig. 3.12 are

$$\begin{aligned}
\dot{x}_1 &= x_2 - a_1(x_1 + b_0 u) + b_1 u \\
\dot{x}_2 &= x_3 - a_2(x_1 + b_0 u) + b_2 u \\
\dot{x}_{k-1} &= x_k - a_{k-1}(x_1 + b_0 u) + b_{k-1} u \\
\dot{x}_k &= -a_k(x_1 + b_0 u) + b_k u
\end{aligned} \tag{3.105}$$

and the output equation is

$$y = x_1 + b_0 u$$

Thus the matrices that describe the state-space realization are given by

$$A = \begin{bmatrix} -a_1 & 1 & 0 & \cdots & 0 \\ -a_2 & 0 & 1 & \cdots & 0 \\ \cdots & \cdots & \cdots & \cdots & \cdots \\ -a_{k-1} & 0 & 0 & \cdots & 0 \\ -a_k & 0 & 0 & \cdots & 1 \end{bmatrix} \qquad B = \begin{bmatrix} b_1 - a_1 b_0 \\ b_2 - a_2 b_0 \\ \vdots \\ b_{k-1} - a_{k-1} b_0 \\ b_k - a_k b_0 \end{bmatrix} \tag{3.106}$$

$$C = [1 \quad 0 \quad 0 \quad \cdots \quad 0] \qquad D = [b_0]$$

If the left-to-right numbering convention is employed, then instead of (3.106) we obtain

$$A = \begin{bmatrix} 0 & 0 & \cdots & -a_k \\ 1 & 0 & \cdots & -a_{k-1} \\ 0 & 1 & \cdots & -a_{k-2} \\ \cdots & \cdots & \cdots & \cdots \\ 0 & 0 & \cdots & -a_1 \end{bmatrix} \qquad B = \begin{bmatrix} b_k - a_k b_0 \\ b_{k-1} - a_{k-1} b_0 \\ \vdots \\ b_1 - a_1 b_0 \end{bmatrix} \tag{3.107}$$

$$C = [0 \quad 0 \quad \cdots \quad 1] \qquad D = [b_0]$$

Compare the matrices A, B, C, and D with the matrices of the first companion form and observe that the A matrix of one companion form corresponds to the transpose of an A matrix of the other, and that the B and C matrices of one correspond to the transposes of the C and B matrices, respectively of the other.

The state space realization of Fig. 3.12 for a single-input, single-output system can readily be generalized to a multiple-input, single-output system; the upper part of the block diagram representing the realization would have the same general form as the upper part of Fig. 3.11, with one path from every input to the summer in front of each integrator. The gains are obtained from the elements of the B matrix.

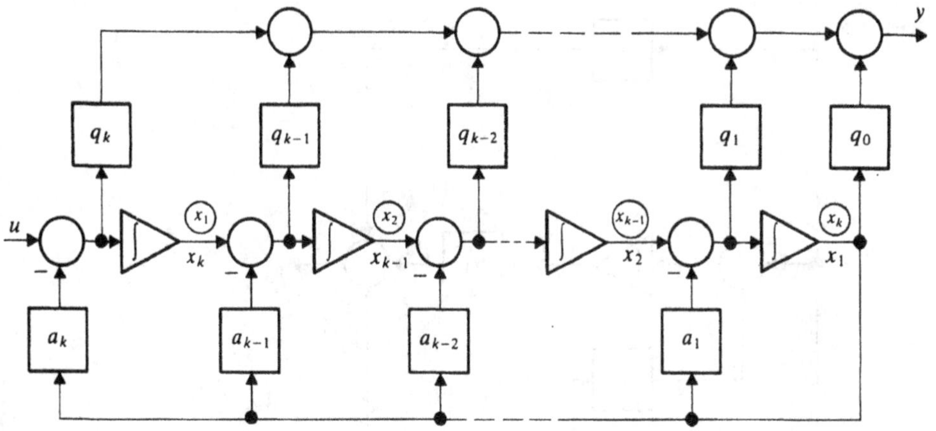

Figure 3.13 Alternate second companion form realization of transfer function.

Just as there are two versions of the first companion form, there are two versions of the second companion form. The second version has the structure shown in Fig. 3.13. The reader by now can probably guess the relationships between the gains $q_1, \ldots, q_k$ and the coefficients of the numerator of the transfer function. (See Problem 3.5.) It is also noted that the structure of Fig. 3.13 can be generalized to the realization of a single-input, multiple-output system.

Jordan Form: Partial Fraction Expansion

Another of the canonical forms of the realization of a transfer function is the *Jordan form*, so named because of the nature of the A matrix that results. This canonical form follows directly from the partial fraction expansion of the transfer functions.

The results are simplest when the poles of the transfer function are all different—no repeated poles. The partial fraction expansion of the transfer function then has the form

$$\mathrm{H}(s) = b_0 + \frac{r_1}{s - s_1} + \frac{r_2}{s - s_2} + \cdots + \frac{r_k}{s - s_k} \tag{3.108}$$

The coefficients r_i $(i = 1, 2, \ldots, k)$ are the residues of the reduced transfer function $\mathrm{H}(s) - b_0$ at the corresponding poles. In the form of (3.108) the transfer function consists of a direct path with gain b_0, and k first-order transfer functions in parallel. A block diagram representation of (3.108) is shown in Fig. 3.14. The gains corresponding to the residues have been placed at the outputs of the integrators. This is quite arbitrary. They could have been located on the input sides, or indeed split between the input and the output.

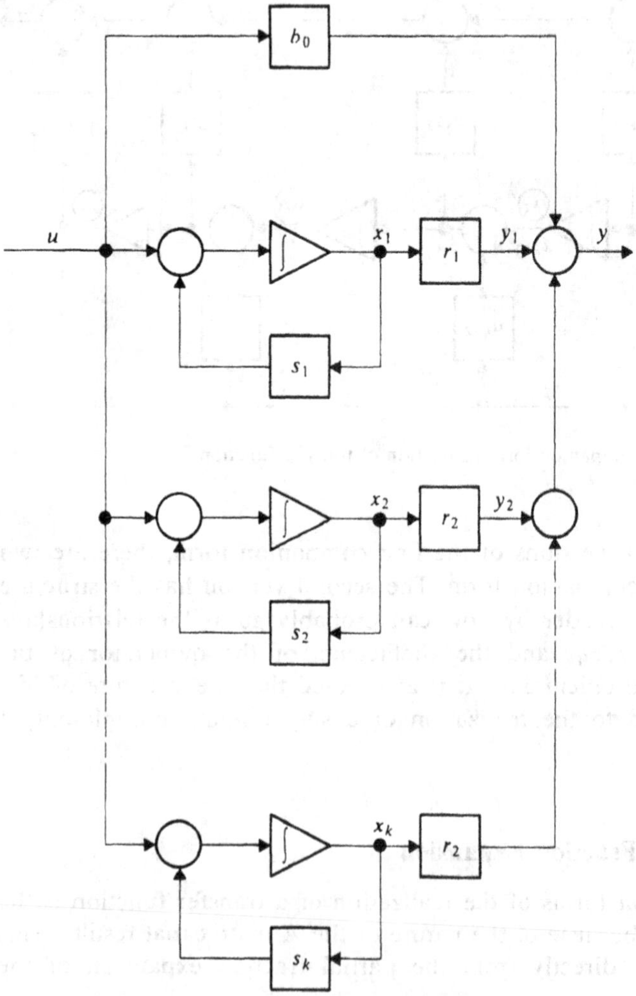

Figure 3.14 Complex Jordan form of transfer function with distinct roots.

Identifying the outputs of the integrators with the state variables results in the following differential equations:

$$
\begin{aligned}
\dot{x}_1 &= s_1 x_1 + u \\
\dot{x}_2 &= s_2 x_2 + u \\
&\cdots\cdots\cdots \\
\dot{x}_k &= s_k x_k + u
\end{aligned}
\tag{3.109}
$$

and an observation equation

$$y = r_1 x_1 + r_2 x_2 + \cdots + r_k x_k + b_0 u \tag{3.110}$$

Hence the matrices corresponding to this realization are

$$A = \begin{bmatrix} s_1 & 0 & \cdots & 0 \\ 0 & s_2 & \cdots & 0 \\ \cdots & \cdots & \cdots & \cdots \\ 0 & 0 & \cdots & s_k \end{bmatrix} \qquad B = \begin{bmatrix} 1 \\ 1 \\ \vdots \\ 1 \end{bmatrix}$$

$$C = [r_1 \quad r_2 \quad \cdots \quad r_k] \qquad D = [b_0]$$

Note that A is a diagonal matrix, which in matrix theory is the Jordan form of a matrix having nonrepeated eigenvalues.

The block-diagram representation of Fig. 3.14 can be turned into hardware only if all the poles $s_1, s_2, \ldots, s_k$ are real. If they are complex, the feedback gains and the gains corresponding to the residues are complex. In this case the representation must be considered as being purely conceptual: valid for theoretical studies, but not physically realizable. If a physically realizable representation is desired, it is possible to combine a pair of complex poles and residues into a single second-order transfer function with real coefficients. The resulting second-order transfer function of the subsystem is then realized in one of the companion forms. Suppose, for example, that s_1 and s_2 are a complex conjugate pair. For a transfer function having real coefficients (as it must in a real system), the residues at a pair of complex conjugate poles must be themselves complex conjugates. Thus a pair of complex conjugate poles, say $s_1 = -\sigma + j\omega$ and $s_2 = -\sigma - j\omega$ with corresponding residues $r = \lambda + j\gamma$ and $r_2 = \lambda - j\gamma$ give rise to the sum

$$\mathsf{H}_{1,2} = \frac{\lambda + j\gamma}{s + \sigma - j\omega} + \frac{\lambda - j\gamma}{s + \sigma + j\omega} = \frac{2[\lambda s + (\lambda\sigma - \omega\gamma)]}{s^2 + 2\sigma s + \sigma^2 + \omega^2}$$

This is a second-order transfer function having the companion-form realization shown in Fig. 3.15. This will give rise to a second-order system in state-space form

$$\begin{aligned} \begin{bmatrix} \dot{\bar{x}}_1 \\ \dot{\bar{x}}_2 \end{bmatrix} & \begin{bmatrix} 0 & 1 \\ -(\sigma^2 + \omega^2) & -2\sigma \end{bmatrix} \begin{bmatrix} \bar{x}_1 \\ \bar{x}_2 \end{bmatrix} + \begin{bmatrix} 0 \\ 1 \end{bmatrix} u \\ y_{1,2} &= [2(\lambda\sigma - \omega\gamma) \quad 2\lambda] \begin{bmatrix} \bar{x}_1 \\ \bar{x}_2 \end{bmatrix} \end{aligned} \tag{3.111}$$

A second-order subsystem such as (3.111) can be used to represent every complex conjugate pair of terms in the partial fraction expansion.

When the system has repeated roots, the partial fraction expansion of the transfer function $\mathsf{H}(s)$ will not be as simple as (3.108). Instead it will be of the form

$$\mathsf{H}(s) = b_0 + \mathsf{H}_1(s) + \cdots + \mathsf{H}_{\bar{k}}(s) \tag{3.112}$$

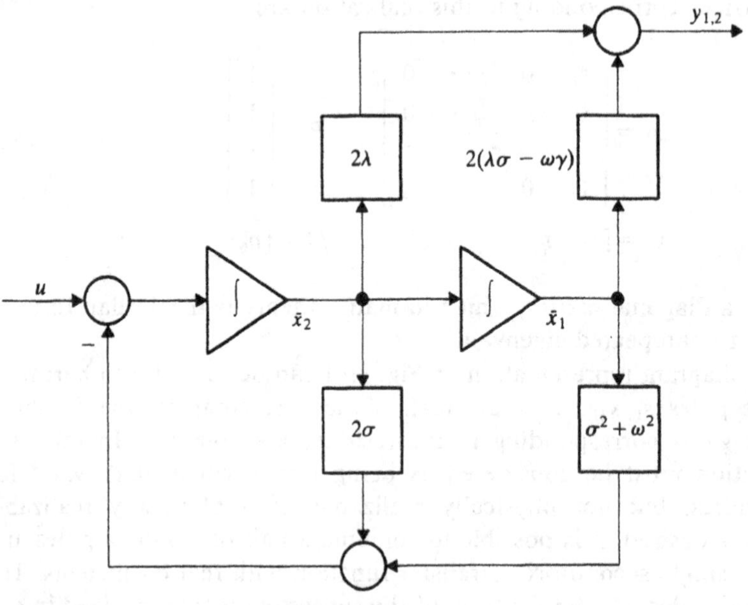

Figure 3.15 Companion-form realization of pair of complex conjugate terms as a real second-order subsystem.

where $\bar{k} < k$ is the number of *distinct* poles of $\mathsf{H}(s)$ and where

$$\mathsf{H}_i(s) = \frac{r_{1i}}{s - s_i} + \frac{r_{2i}}{(s - s_i)^2} + \cdots + \frac{r_{\nu_i i}}{(s - s_i)^{\nu_i}}$$

where ν_i is the multiplicity of the ith pole ($i = 1, 2, \ldots, \bar{k}$). The last term in $\mathsf{H}_i(s)$ can be synthesized as a chain of ν_i identical, first-order systems, each having transfer function $1/(s - s_i)$. The preceding terms in the chain of fewer than ν_i of such transfer functions. Thus the entire transfer function $\mathsf{H}_i(s)$ can be synthesized by the system having the block diagram in Fig. 3.16.

Using the left-to-right numbering convention gives the differential equations

$$\begin{aligned} \dot{x}_{1i} &= s_i x_{1i} + u \\ \dot{x}_{2i} &= x_{1i} + s_i x_{2i} \\ &\cdots\cdots\cdots\cdots\cdots \\ \dot{x}_{\nu_i i} &= x_{(\nu_i - 1)i} + s_i x_{\nu_i i} \end{aligned} \tag{3.113}$$

and the output is given by

$$y_i = r_{1i} x_{1i} + r_{2i} x_{2i} + \cdots + r_{\nu_i i} x_{\nu_i i} \tag{3.114}$$

If the state vector for the subsystem is defined by

$$x^i = [x_{1i} \quad x_{2i} \quad \cdots \quad x_{\nu_i i}]'$$

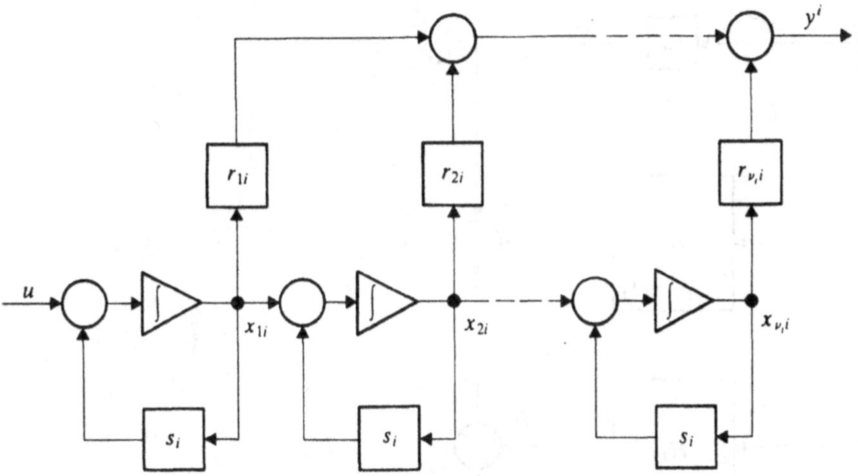

Figure 3.16 Jordan-block realization of part of transfer function having repeated pole

$$H_i(s) = \frac{r_{1i}}{s - s_i} + \cdots + \frac{r_{\nu_i i}}{(s - s_i)^{\nu_i}}$$

then (3.113) and (3.114) can be written in the standard form

$$\begin{aligned} \dot{x}^i &= A_i x^i + b_i u \\ y_i &= C_i x^i \end{aligned} \tag{3.115}$$

where

$$A_i = \begin{bmatrix} s_i & 0 & 0 & \cdots & 0 \\ 1 & s_i & 0 & \cdots & 0 \\ 0 & 1 & s_i & \cdots & 0 \\ \cdots & \cdots & \cdots & \cdots & \cdots \\ 0 & 0 & 0 & \cdots & s_i \end{bmatrix} \qquad B_i = \begin{bmatrix} 1 \\ 0 \\ 0 \\ \vdots \\ 0 \end{bmatrix} \tag{3.116}$$

$$C_i = [r_{1i} \quad r_{2i} \quad \cdots \quad r_{\nu_i i}]$$

Note that the A matrix of the subsystem consists of two diagonals: the principal diagonal has the corresponding characteristic root (pole) and the subdiagonal has all 1's. In matrix theory a matrix having this structure is said to be *in Jordan form*, which is the name used for this realization of the transfer function.

If the right-to-left numbering convention were employed it is easy to see that the A matrix would have 1's on the superdiagonal instead of on the subdiagonal. This is an alternate Jordan form.

According to (3.112) the overall transfer function consists of a direct path with gain b_0 and $\bar{k}$ subsystems, each of which is in the Jordan canonical form, as shown in Fig. 3.17. The state vector of the overall system consists of the

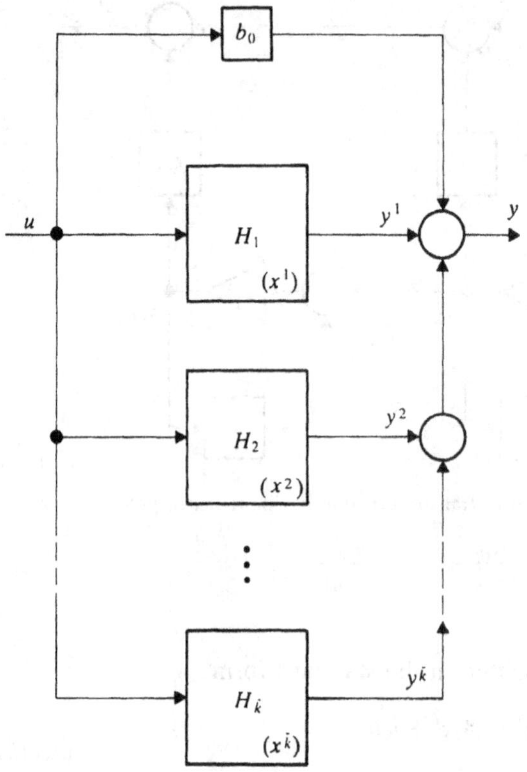

Figure 3.17 Subsystems in Jordan canonical form combined into overall system.

concatenation of the state vectors of each of the Jordan blocks

$$x = \begin{bmatrix} x^1 \\ x^2 \\ \vdots \\ x^{\bar{k}} \end{bmatrix} \tag{3.117}$$

Since there is no feedback from any of the subsystems to the others, the A matrix of the overall system is "block diagonal":

$$A = \begin{bmatrix} A_1 & 0 & \cdots & 0 \\ 0 & A_2 & \cdots & 0 \\ \cdots & \cdots & \cdots & \cdots \\ 0 & 0 & \cdots & A_{\bar{k}} \end{bmatrix} \tag{3.118}$$

where each of the submatrices is in the Jordan canonical form shown in (3.116). The B and C matrices of the overall system are the concatenations of the B_i and C_i matrices of each of the subsystems:

$$B = \begin{bmatrix} B_1 \\ \vdots \\ B_{\bar{k}} \end{bmatrix} \qquad C = [C_1, \ldots, C_{\bar{k}}] \tag{3.119}$$

It is noted that the Jordan blocks are only conceptual if the poles are complex. Pairs of Jordan blocks can be combined to give a real Jordan block of order $2\nu_i$. The details are easy to work out, but the general notation and calculations are quite messy. If the need ever arises (which is highly unlikely) for such a real Jordan block, the procedures used in this section can be followed to obtain the required result.

To conclude this discussion it is noted that the Jordan normal form can be extended directly to either a multiple-input, single-output system, or a multiple-output, single-input system. In the former case, each input has a path to each of the integrators; in the latter, each integrator has a path to each of the outputs.

Example 3I. Spring-coupled masses (continued) It is readily established, either by use of the general relationship (3.65) applied to (3H.2), or by simpler means, that the input-output relationship for the spring-coupled mass system is given by

$$\begin{bmatrix} \mathsf{y}_1(s) \\ \mathsf{y}_2(s) \end{bmatrix} = \begin{bmatrix} \dfrac{s^2 + K/M_2}{s^2(s^2 + K/\bar{M})} & \dfrac{K/M_1}{s^2(s^2 + K/\bar{M})} \\ \dfrac{K/M_2}{s^2(s^2 + K/\bar{M})} & \dfrac{s^2 + K/M_1}{s^2(s^2 + K/\bar{M})} \end{bmatrix} \begin{bmatrix} \mathsf{u}_1(s) \\ \mathsf{u}_2(s) \end{bmatrix} \tag{3I.1}$$

where

$$\frac{1}{\bar{M}} = \frac{1}{M_1} + \frac{1}{M_2}$$

The block diagram of Fig. 3.7 already gives a state-variable realization of the system. For illustrative purposes, however, we assume that $u_2 = 0$, and hence we have a single-input, two-output system. The transfer functions of interest are

$$\begin{aligned} \mathsf{H}_1(s) &= \frac{\mathsf{y}_1(s)}{\mathsf{u}_1(s)} = \frac{s^2 + K/M_2}{s^2(s^2 + K/\bar{M})} \\ \mathsf{H}_2(s) &= \frac{\mathsf{y}_2(s)}{\mathsf{u}_1(s)} = \frac{K/M_2}{s^2(s^2 + K/\bar{M})} \end{aligned} \tag{3I.2}$$

The first companion form, using the structure of Fig. 3.9 for a single-input, multiple-output, system is obtained directly from (3I.2) and is shown in Fig. 3.18(a). The corresponding matrices are

$$A = \begin{bmatrix} 0 & 1 & 0 & 0 \\ 0 & 0 & 1 & 0 \\ 0 & 0 & 0 & 1 \\ 0 & 0 & -K/\bar{M} & 0 \end{bmatrix} \qquad B = \begin{bmatrix} 0 \\ 0 \\ 0 \\ 1 \end{bmatrix} \tag{3I.3}$$

$$C = \begin{bmatrix} K/M_2 & 0 & 1 & 0 \\ K/M_2 & 0 & 0 & 0 \end{bmatrix} \qquad D = 0$$

Although the structure and gains for the single-input, multiple-output version of the second companion form were not given explicitly, it is readily established that the block diagram of Fig. 3.18(b) correctly represents the transfer functions from u_1 to y_1 and y_2. The relevant matrices are

$$A = \begin{bmatrix} 0 & 1 & 0 & 0 \\ -K/\bar{M} & 0 & 1 & 0 \\ 0 & 0 & 0 & 1 \\ 0 & 0 & 0 & 0 \end{bmatrix} \qquad B = \begin{bmatrix} 0 \\ 0 \\ 0 \\ 1 \end{bmatrix} \tag{3I.4}$$

$$C = \begin{bmatrix} -K/M_1 & 0 & 1 & 0 \\ 0 & K/M_2 & 0 & 0 \end{bmatrix} \qquad D = 0$$

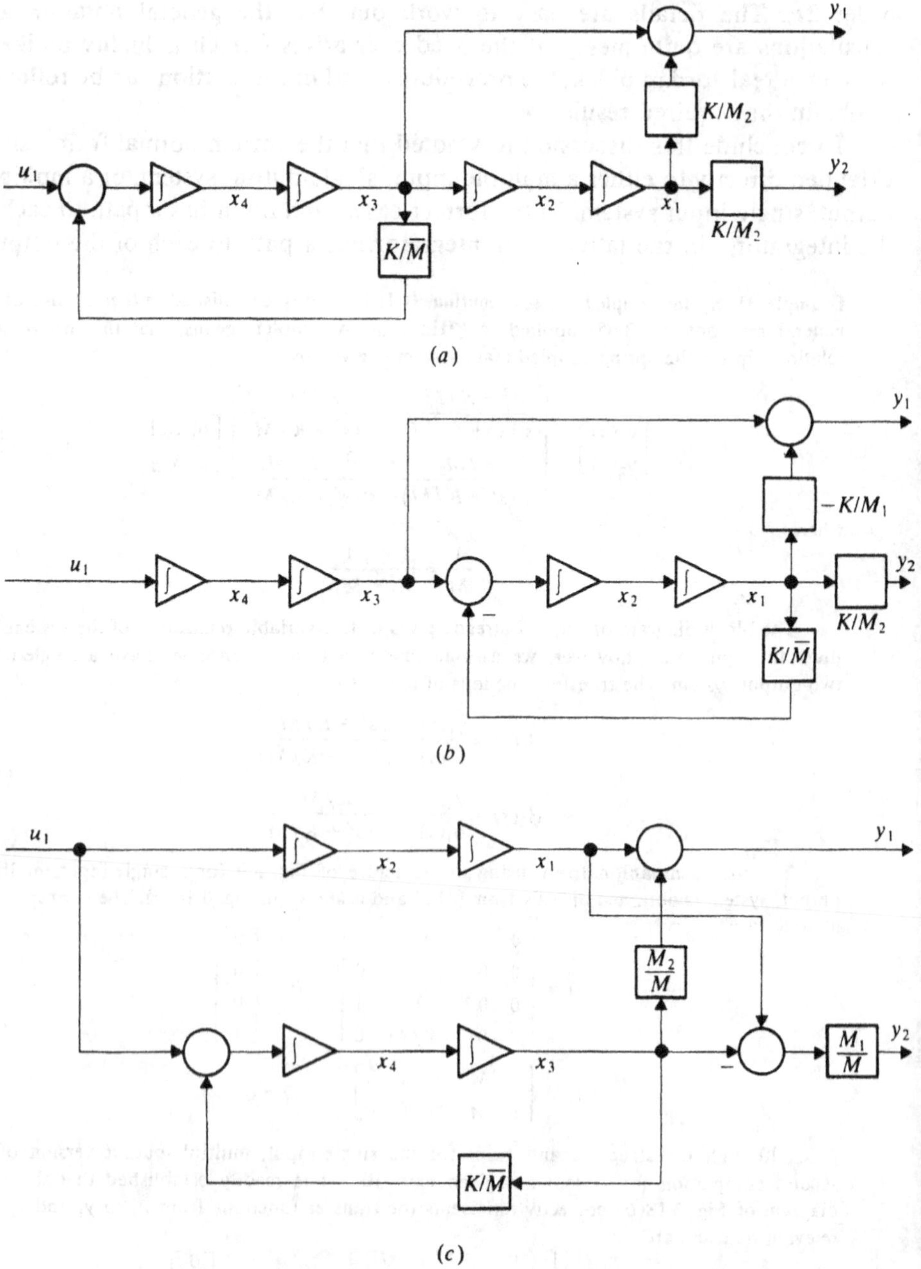

Figure 3.18 Canonical realizations of transfer functions of spring-coupled mass system. (*a*) First companion form, (*b*) second companion form, (*c*) Jordan canonical form.

To obtain the Jordan canonical form we expand the transfer functions in partial fractions

$$\begin{aligned} \mathsf{H}_1(s) &= \frac{1}{s^2} + \frac{M_2 M}{s^2 + K/\bar{M}} \\ \mathsf{H}_2(s) &= \frac{M_1/M}{s^2} + \frac{M_1/M}{s^2 + K/\bar{M}} \end{aligned} \qquad (M = M_1 + M_2) \tag{3I.5}$$

The system has a double pole at the origin and a pair of imaginary poles at $s = \pm j\sqrt{K/\bar{M}}$. To the real form, the two terms with the imaginary poles are already combined in (3I.5). The block diagram representation of (3I.5) in the form appropriate for a single-input, two-output system is shown in Fig. 3.18(c). The system matrices corresponding to this realization are

$$A = \left[\begin{array}{cc|cc} 0 & 1 & 0 & 0 \\ 0 & 0 & 0 & 0 \\ \hline 0 & 0 & 0 & 1 \\ 0 & 0 & -K/\bar{M} & 0 \end{array}\right] \qquad B = \begin{bmatrix} 0 \\ 1 \\ 0 \\ 1 \end{bmatrix}$$

$$C = \begin{bmatrix} 1 & 0 & M_2/M & 0 \\ M_1/M & 0 & -M_1/M & 0 \end{bmatrix} \qquad D = 0$$

The A matrix has been partitioned to show the block-diagonal form. The upper left-hand matrix is in the (superdiagonal) Jordan form for a repeated pole at the origin; the lower right-hand matrix is in the companion form for a second-order system.

PROBLEMS

Problem 3.1 Exercises in resolvents and transition matrices

Find the resolvents and transition matrices for each of the following:

(a) $A_1 = \begin{bmatrix} -1 & 0 & 0 \\ 1 & -2 & 0 \\ 1 & 2 & -3 \end{bmatrix}$

(b) $A_2 = \begin{bmatrix} -1 & 0 & 0 \\ 1 & -1 & 0 \\ 0 & 1 & -1 \end{bmatrix}$

(c) $A_3 = \begin{bmatrix} -2 & 1 & 1 \\ 1 & -2 & 1 \\ 1 & 1 & -2 \end{bmatrix}$

Problem 3.2 Exercises on canonical forms

Determine the canonical forms (companion and Jordan) for each of the following transfer functions:

(a) $\mathsf{H}(s) = \dfrac{(s+2)(s+4)}{(s+1)(s+3)(s+5)}$

(b) $\mathsf{H}(s) = \dfrac{s+2}{s[(s+1)^2+4]}$

(c) $\mathsf{H}(s) = \dfrac{s+3}{(s+1)^2(s+2)}$

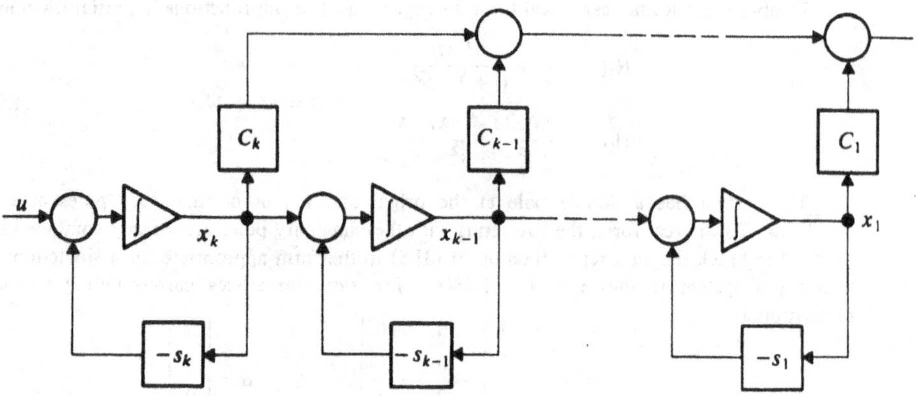

Figure P3.3 Tandem canonical form.

Problem 3.3 Another canonical form

An alternative to the Jordan canonical form for single-input, single-output systems is the "tandem form" shown in Fig. P3.3.

(*a*) Write the A, B, and C matrices for this form.

(*b*) Given the system in Jordan form $\dot{x} = \Lambda x + Bu$ where $\Lambda = \text{diag}\,[-s_1, -s_2, \ldots, -s_k]$, find the transformation matrix T that transforms it to the tandem form.

Problem 3.4 Adjoint equation

Show that the state transition matrix satisfies the following differential equation

$$\frac{\partial \Phi(t, \tau)}{\partial \tau} = -\Phi(t, \tau)A(\tau) \qquad \text{(P3.4)}$$

Hint: Use $dX^{-1}(t)/dt = -X^{-1}(t)(dX(t)/dt)X^{-1}(t)$.

Equation (P3.4) is sometimes called the "adjoint" equation, or the "backward-evolution" equation.

Problem 3.5 Coefficients in second companion form

Find the relationship between the coefficients $q_1, \ldots, q_k$ of the second companion form, Fig. 3.13, to the coefficients of the numerator and denominator of the transfer function $H(s)$.

Problem 3.6 Motor-driven cart with pendulum

Consider the inverted pendulum on a cart driven by an electric motor that was studied in Prob. 2.1. Let the state vector, control, and outputs be defined by

$$x = [x, \dot{x}, \theta, \dot{\theta}]' \qquad u = e \qquad y = [x, \theta]'$$

(*a*) Find the matrices A, B, C, and D of the state-space characterization of the system.

(*b*) Draw the block-diagram representation of the system.

(*c*) Find the resolvent and the state-transition matrix.

(*d*) Find the transfer functions from the input u to the two outputs.

The following numerical data may be used if you would rather use numbers than letters:

$$m = 0.1\text{ kg} \qquad M = 1.0\text{ kg} \qquad l = 1.0\text{ m} \qquad g = 9.8\text{ m}\cdot\text{s}^{-2}$$

$$k = 1\text{ V}\cdot\text{s} \qquad R = 100\,\Omega \qquad r = 0.02\text{ m}$$

Problem 3.7 Three-capacitance thermal system

For the insulated conducting bar of Prob. 2.3, using as the state, vector, control, and exogenous variables

$$x = [v_1, v_2, v_3]'$$

$$u = e_0$$

$$x_0 = v_0$$

(*a*) Find the matrices *A*, *B*, and *E* of the state-space characterization of the system.
(*b*) Find the resolvent and the state-transition matrix.
(*c*) Find the transfer function from the input $u = e_0$ to the output $y = v_3$.
Use $R = 1$, $C = 2$.

Problem 3.8. Eigenvalues of *R*-*C* network

Consider a *passive* electrical network (consisting of only capacitors, resistors, plus voltage, and current sources). Show that all the eigenvalues lie on the *negative real* axis.

Problem 3.9 Two-car train

Consider the two-car train of Prob. 2.5 with the following numerical data:

Trains: $M_1 = M_2 = 1.0$ kg, $K = 40$ N/m.
Motors: $k = 2$ V · s, $R = 100\,\Omega$, $r = 2$ cm.

(*a*) Find the transfer functions from the input voltages to the motor positions.
(*b*) Find the open-loop poles of the system.

NOTES

Note 3.1 Numerical calculation of the transition matrix

It might seem that the numerical determination of the state-transition matrix

$$\Phi(T) = e^{AT}$$

with *T* fixed is a fairly routine numerical task. Algorithms can be based on the series definition

$$\Phi(T) = e^{AT} = I + AT + A^2T^2/2! + \cdots$$

or on the basic definition of an exponential

$$e^{AT} = \lim_{n\to\infty}(I + AT/n)^n$$

The transition matrix can also be computed by numerical integration of the matrix differential equation $\dot{\Phi} = A\Phi$ with the initial condition $\Phi(0) = I$. A variety of numerical integration algorithms (e.g., Runge-Kutta, predictor-corrector, implicit) and implemented computer codes are available.

It is also possible to transform *A* to Jordan canonical form (diagonal form for nonrepeated eigenvalues)

$$A = V\bar{A}V^{-1}$$

where $\bar{A}$ is in the Jordan form as given by (3.118). Then

$$e^{AT} = Ve^{\bar{A}T}V^{-1}$$

and $e^{\bar{A}T}$ has a particularly simple form. (When $\bar{A} = \Lambda = \text{diag}[s_1, s_2, \ldots, s_k]$ then $e^{\bar{A}T} = \text{diag}[e^{s_1T}, e^{s_2T}, \ldots, e^{s_kT}]$.) A number of algorithms are available for finding the eigenvalues of *A* (i.e., $s_1, \ldots, s_k$) and the corresponding transformation matrix *V*.

Notwithstanding the abundance of potentially suitable algorithms, when the dimension of A is large and when the eigenvalues have a range of several orders of magnitude, an accurate efficient algorithm for computing e^{AT} is not trivial.

Note 3.2 Time-varying systems

If we assume that the laws of nature do not change with time, we should not expect to encounter time-varying differential equations in the description of physical processes. Nonlinear, yes; but time-varying, no. Even if we accept this hypothesis, however, it is often necessary to deal with time-varying systems as an *approximate* representation of the physical world. Consider, for example, the motion of an aircraft, for which a set of time-invariant, but highly nonlinear equations can be written using established methods. These differential equations would be appropriate for use in an accurate simulation of the aircraft behavior. But for purposes of design it may be necessary to use a simplified, linear model. When the dynamics are linearized, the resulting differential equations (i.e., the A and B matrices) will have coefficients that depend on such variables as dynamic pressure $Q = \rho v^2/2$ which depend on time.

Example 3B is another example of how a nonlinear time-invariant system is approximated by a linear, time-varying system.

Note 3.3 Laplace transform of exponential

To show that the Laplace transform of e^{At} is $(sI - A)^{-1}$ consider the special case in which A is similar to a diagonal matrix $A = V\Lambda V^{-1}$ where $\Lambda = \text{diag}[s_1, s_2, \ldots, s_k]$. Then $e^{At} = V[e^{s_1 t}, \ldots, e^{s_k t}]V^{-1}$. Then the Laplace transform of e^{At} is $V[(s - s_1)^{-1}, \ldots, (s - s_k)^{-1}]V^{-1} = V(sI - \Lambda)^{-1}V^{-1} = (sI - A)^{-1}$. There are many other ways of showing this.

Note 3.4 Schuler period; inertial navigation

The period of a pendulum is $T = 2\pi\sqrt{l/g}$ (independent of the mass of the bob, which is why a pendulum clock can be extremely accurate). A pendulum having a length l equal to the earth's radius has a period of 84.4 minutes which is commonly called the Schuler period in honor of the German applied physicist Max Schuler,[7] who showed in 1923 that any pendulum having this length would remain vertical even if the pivot moves. This principle is the basis of inertial navigation systems. The orientation of the accelerometers in the system is kept constant by locating them on a "synthetic Schuler pendulum" in which the effect of the long pendulum arm is achieved by use of precise gyros.

Having become extremely sophisticated after World War II, inertial navigation technology is critical in strategic missiles and most military aircraft. It is also used extensively for navigation of modern transoceanic aircraft. Some of the analytical methods of inertial navigation may be found in [8].

Note 3.5 Minimal realizations

Several methods are displayed in Sec. 3.7 for realizing the transfer functions of a system with one input and l outputs, or with m inputs and one output, by a system of order k, where k is the degree of the characteristic polynomial of the system, i.e., the lowest common denominator of all the scalar transfer functions. By using several realizations in parallel it is possible to realize a system with m inputs and l outputs by a system of order $r = k \cdot \min(l, m)$. But it may be possible to realize the matrix of transfer functions by a system of order lower than r. For example, the system of transfer functions may have been obtained from a known system of differential equations of kth order as in Examples 2G or 2H. No matter how many inputs or outputs such a system may have, we know how to realize the transfer functions from all the inputs to all the outputs with a system of kth order.

If the transfer functions alone from the inputs to the outputs are given, however, the minimum number of differential equations (or integrators, in the block diagram representation) is not obvious, and the determination of this "minimum realization" is a significant and nontrivial problem. The problem is important not out of a desire to economize on hardware—a few integrators more or less is hardly of consequence—but because a nonminimum realization is either uncontrollable or unobservable (or both) in the sense defined and explained in Chap. 5, and may cause theoretical or computational difficulties.

The theory of minimum realizations is fundamental to the algebraic treatment of linear systems, as presented by Kailath,[4] for example. Unfortunately, this theory falls far outside the scope of the present text.

REFERENCES

1. Schwarz, R. J., and Friedland, B., *Linear Systems*, McGraw-Hill Book Co., New York, 1965.
2. Coddington, E. A., and Levinson, N., *Theory of Ordinary Differential Equations*, McGraw-Hill Book Co., New York, 1955.
3. Brockett, R. W., *Finite Dimensional Linear Systems*, John Wiley & Sons, New York, 1965.
4. Kailath, T., *Linear Systems*, Prentice-Hall, Inc., Englewood Cliffs, NJ, 1980.
5. Householder, A. S., *The Theory of Matrices in Numerical Analysis*, Blaisdell Publishing Co., Waltham, MA, 1964.
6. Gantmacher, F. R., *The Theory of Matrices*, Chelsea Publishing Company, New York, 1959.
7. Schuler, M., "Die Störung von Pendel- und Kreiselapparaten durch die Beschleunigung des Fahrzeuges," *Physicalische Zeitschrift*, vol. 24, 1923, pp. 344–350.
8. Farrell, J., *Integrated Aircraft Navigation*, Academic Press, New York, 1976.

CHAPTER
FOUR
FREQUENCY-DOMAIN ANALYSIS

4.1 STATUS OF FREQUENCY-DOMAIN METHODS

For a period of about twenty years—from the early 1940s through the early 1960s—frequency-domain methods were the only systematic tools for the analysis and design of control systems. These methods were developed by physicists and electrical engineers in response to the World War II need for improved servomechanisms to be used in various weapons systems, and were based upon the frequency response/operational calculus methods then in use for designing electrical networks for communication systems. It is no coincidence that the pioneering work of Nyquist[1] and Bode[2] in the early part of the century, and even the very invention of the feedback amplifier by Black,[3] all products of the Bell Telephone Laboratories, were done in the interest of improved communication systems.

(The connection between frequency-domain methods and communication systems is a possible explanation of why the development of control theory took place and still continues mostly in academic departments of electrical engineering, even though the electrical hardware in many control systems is all but negligible.)

Through the interdisciplinary activities of individuals such as the late Rufus Oldenburger, a mechanical engineer who understood and appreciated the significance of frequency-domain methods, these techniques were introduced to other branches of engineering and became widely used throughout the entire field of automatic control.

Just at the time that frequency-domain methods had reached their peak of development, in the late 1950s and early 1960s, the alternative state-space methods began to make their appearance. But while the new state-space methods developed rapidly in the decades following and found new adherents and apostles, the vigor of frequency-domain methods hardly diminished. Notwithstanding the level to which state-space methods have been developed, most control systems continue to be analyzed and designed by frequency-domain methods. Concepts such as "bandwidth," "phase and gain margins," and "corner frequencies" are entrenched in control system technology and are not likely to be displaced. They continue to be useful.

Starting in the mid 1970s, new impetus was imparted to frequency-domain methods for multivariable systems through the efforts of a number of investigators centered in Great Britain around Rosenbrock and MacFarlane. (See Note 4.1.) Among the fruits of this effort was a growing recognition that frequency-domain methods and state-space methods enhance and complement each other. The burgeoning theory of robust control systems, which was started only in the past few years, is further evidence of the symbiosis of frequency-domain and state-space methods.

4.2 FREQUENCY-DOMAIN CHARACTERIZATION OF DYNAMIC BEHAVIOR

The fundamental concept of frequency-domain analysis is the "transfer function" which expresses the relationship between the Laplace transform $\mathsf{y}(s)$ of the system output $y(t)$ and the Laplace transform $\mathsf{u}(s)$ of the input $u(t)$

$$\mathsf{y}(s) = \mathsf{H}(s)\mathsf{u}(s) \tag{4.1}$$

where $\mathsf{H}(s)$ is the transfer function of the system. This relationship is valid for any time-invariant linear system, even when the system cannot be represented by sets of ordinary differential equations of finite order. The representation (4.1) is valid, for example, for systems whose physical properties are described by partial differential equations, or by pure "transport" delays.

The validity of (4.1) is a consequence of the linearity and time invariance of the system under examination. In the time domain such a system can be represented by the convolution integral

$$y(t) = \int_0^t H(t-\tau)u(\tau)\,d\tau \tag{4.2}$$

where $H(t)$ is the "impulse-response" (matrix) of the system.

The basic frequency-domain relation (4.1) follows from (4.2) as a result of the well-known "convolution theorem" proved in many texts (see [4], for example) which asserts that the Laplace transform of a convolution of two functions is the product of the respective Laplace transforms of these functions. Thus, the transfer function $\mathsf{H}(s)$ is the Laplace transform of the impulse response:

$$\mathsf{H}(s) = \mathscr{L}[H(t)] = \int_0^\infty e^{-st} H(t)\, dt \tag{4.3}$$

When the number of inputs and/or outputs is greater than 1, then $\mathsf{H}(s)$ is a matrix of appropriate dimension: if there are m inputs and l outputs, then $\mathsf{H}(s)$ is an l-by-m matrix, the elements of which are the transfer functions from the individual components of the input vector to the individual components of the output vector.

When the system of interest has the standard state-space representation

$$\dot{x} = Ax + Bu$$

$$y = Cx + Du$$

then, as shown in Chap. 3, the transfer function (matrix) is given explicitly by

$$\begin{aligned} \mathsf{H}(s) &= C(sI - A)^{-1}B + D \\ &= \frac{C(E_1 s^{k-1} + E_2 s^{k-2} + \cdots + E_k)B}{s^k + a_1 s^{k-1} + \cdots + a_k} + D \end{aligned} \tag{4.4}$$

where the denominator of $\mathsf{H}(s)$ is the characteristic polynomial

$$\mathsf{D}(s) = |sI - A| = s^k + a_1 s^{k-1} + \cdots + a_k \tag{4.5}$$

and $E_1 = I, E_2, \ldots, E_k$ are the coefficient matrices of the adjoint matrix for the resolvent $(sI - A)^{-1}$, as discussed in Chap. 3. The roots of the *characteristic equation* $|sI - A| = 0$ are called the *characteristic roots* or *eigenvalues* of the system.

If the D matrix is nonzero, there is a direct path from some input to some output. The transfer functions from those inputs that are directly connected to the output will be polynomials of degree k in s. All the other transfer functions are *proper* rational functions, that is, ratios of polynomials in s in which the degree of the numerator is strictly less than the degree of the denominator.

The variable s of the Laplace transform is a complex variable

$$s = \sigma + j\omega \qquad j = \sqrt{-1}$$

called *complex frequency. Frequency-domain* analysis owes its name to this identification of s with complex frequency.

A transfer function $\mathsf{H}(s)$ which is a proper rational function of s can be expanded in partial fractions

$$\begin{aligned} \mathsf{H}(s) &= \frac{N_1 s^{k-1} + \cdots + N_k}{s^k + d_1 s^{k-1} + \cdots + d_k} \\ &= \mathsf{H}_1(s) + \mathsf{H}_2(s) + \cdots + \mathsf{H}_{\bar{k}}(s) \end{aligned} \tag{4.6}$$

where

$$\mathsf{H}_i(s) = \frac{R_{1i}}{s - s_i} + \frac{R_{2i}}{(s - s_i)^2} + \cdots + \frac{R_{\nu_i i}}{(s - s_i)^{\nu_i}} \tag{4.7}$$

The complex frequencies $s_i (i = 1, 2, \ldots, \bar{k} < k)$ are the *distinct* roots of the denominator of (4.6) and the ν_i are corresponding multiplicities of these roots. These roots of the denominator are called the *poles* of the transfer function because $\mathsf{H}(s)$ becomes infinite at these complex frequencies and a contour map of the complex plane appears as if it has poles sticking up from these points.

If $\mathsf{H}(s)$ is a matrix, then the coefficients N_i of the numerator polynomial of (4.6) are matrices and so are the coefficient matrices R_{ij} of the partial fraction expansion.

The impulse response $H(t)$ of the system is given by the inverse Laplace transform of (4.6):

$$H(t) = H_1(t) + H_2(t) + \cdots + H_{\bar{k}}(t) \tag{4.8}$$

where

$$H_i(t) = (R_{1i} + R_{2i}t + \cdots + R_{\nu_i}t^{\nu_i - 1}/(\nu_i - 1)!)\, e^{s_i t} \tag{4.9}$$

Thus the impulse response of a time-invariant linear system having a proper rational function of s as its transfer function is a sum of time-weighted exponentials of the form of (4.9). The exponents of the exponentials are the poles of the transfer function, and the time-weighting functions are polynomials in t of one degree less than the multiplicity of the corresponding poles.

If the numerator of the transfer function (4.6) is the same degree as the denominator, the constant term can be removed and the remainder written as a proper rational function, i.e.,

$$\mathsf{H}(s) = \frac{N_0 s^k + N_1 s^{k-1} + \cdots + N_k}{s^k + d_1 s^{k-1} + \cdots + d_k} = N_0 + \frac{\bar{N}_1 s^{k-1} + \cdots + \bar{N}_k}{s^k + d_1 s^{k-1} + \cdots + d_k} \tag{4.10}$$

$$\bar{N}_i = N_i - N_0 d_i \qquad (i = 1, 2, \ldots, k) \tag{4.11}$$

The corresponding impulse response has the form

$$H(t) = N_0 \delta(t) + \sum_{i=1}^{\bar{k}} (R_{1i} + \cdots + R_{\nu_i} t^{\nu_i - 1}/(\nu_i - 1)!)\, e^{s_i t} \tag{4.12}$$

where $\delta(t)$ is the unit impulse function (*Dirac delta* function).

It is certainly possible to conceive of systems having transfer functions in which the degree of the numerator is higher than the denominator. For example, an electrical inductor has the transfer function (complex impedance)

$$\frac{\mathsf{v}(s)}{\mathsf{i}(s)} = \mathsf{z}(s) = Ls$$

when the voltage $v(t)$ is regarded as the output and the current is regarded as the input. The impulse response of such systems, in general, contains not only impulses, but various derivatives (doublets, etc.) of impulses. These are bothersome and can generally be avoided by suitable reformulation of the problem. If the voltage, in the case of the inductor, is regarded as the input and the current is regarded as the output, then the transfer function is the admittance

$$\frac{\mathsf{i}(s)}{\mathsf{v}(s)} = \mathsf{y}(s) = \frac{1}{Ls}$$

which is a perfectly acceptable, proper rational function.

The general form of the transfer function (4.10) is consistent with the transfer function of the state-space representation given by (4.4). In particular

$$N_0 = D$$

and

$$\begin{aligned} \bar{N}_i &= CE_iB \\ d_i &= a_i \end{aligned} \qquad i = 1, 2, \ldots, k$$

Thus the impulse response of a system in the standard state-space representation is a sum of time-weighted exponentials $e^{s_i t}$ with the exponents s_i being the roots of the characteristic polynomial, i.e.,

$$|sI - A| = s^k + a_1 s^{k-1} + \cdots + a_k = (s - s_i)^{\nu_i} \cdots (s - s_{\bar{k}})^{\nu_{\bar{k}}} \qquad (4.13)$$

Multiple poles (i.e., repeated characteristic roots) occur quite frequently at the origin ($s = 0$). For example a pure mass with the transfer function $\mathsf{H}(s) = 1/ms^2$ has a double pole at $s = 0$. But multiple poles at other complex frequencies rarely occur in practical problems. To simplify a derivation it is often convenient to assume that multiple poles of a system occur only at the origin.

4.3 BLOCK-DIAGRAM ALGEBRA

One reason for the popularity of frequency-domain analysis is that the dynamic behavior of a system can be studied using only algebraic operations. The transfer functions of subsystems can be combined algebraically to yield the transfer function of the overall system, and its response to various inputs can be obtained by multiplying the Laplace transform of the input by the transfer function, as prescribed by (4.1), to obtain the Laplace transform of the output. The actual output time function, if needed, is calculated by finding the inverse Laplace transform of $\mathsf{y}(s)$, using algebraic techniques (partial fractions) in conjunction with a table of Laplace transforms. Nowhere in this analysis, except possibly in deriving the transfer functions of the subsystems, is it necessary to have any dealings with differential equations.

The basic techniques of manipulating block diagrams consist of combining transfer functions in parallel and in tandem and eliminating feedback loops. The three operations are illustrated in Fig 4.1.

Figure 4.1(a) shows a system comprising two subsystems with transfer functions (matrices) $\mathsf{H}_1(s)$ and $\mathsf{H}_2(s)$. The summing junction, represented by the

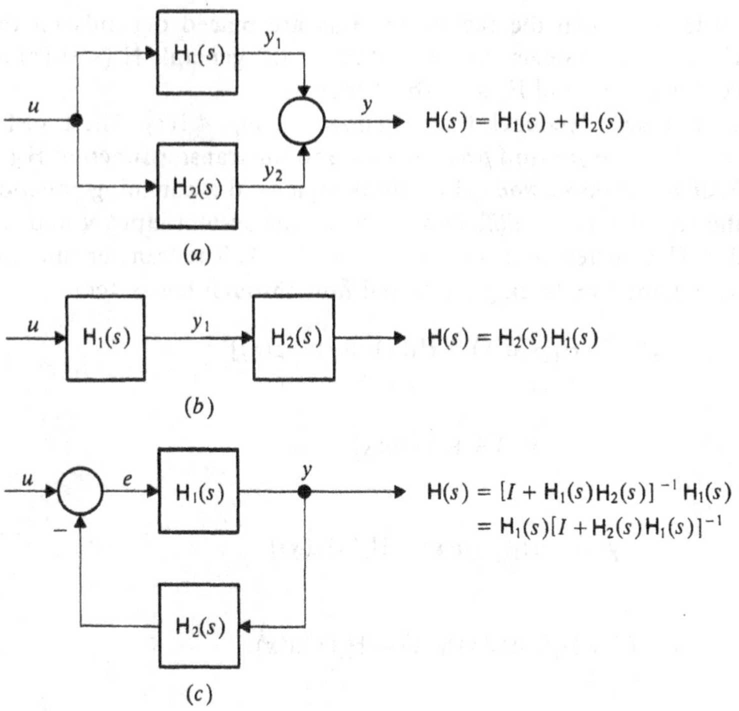

Figure 4.1 Subsystems in combination. (*a*) Subsystems in parallel; (*b*) Subsystems in tandem; (*c*) Single-loop feedback system.

circle, makes sense only when each subsystem has the same number of outputs, i.e., dimensions of y_1 and y_2 are equal. Then

$$\begin{aligned} \mathsf{y}(s) &= \mathsf{y}_1(s) + \mathsf{y}_2(s) = \mathsf{H}_1(s)\mathsf{u}(s) + \mathsf{H}_2(s)\mathsf{u}(s) \\ &= [\mathsf{H}_1(s) + \mathsf{H}_2(s)]\mathsf{u}(s) \end{aligned} \tag{4.14}$$

Thus the transfer function of a parallel combination of subsystems is the sum of the transfer functions.

The *tandem* (or *series*) combination of two subsystems is shown in Fig. 4.1(*b*). For this combination

$$\mathsf{y}_1(s) = \mathsf{H}_1(s)\mathsf{u}(s)$$

and

$$\mathsf{y}(s) = \mathsf{H}_2(s)\mathsf{y}_1(s)$$

Thus

$$\mathsf{y}(s) = \mathsf{H}_2(s)\mathsf{H}_1(s)\mathsf{u}(s)$$

and the transfer function of the tandem combination is the *product* of the transfer functions:

$$\mathsf{H}(s) = \mathsf{H}_2(s)\mathsf{H}_1(s) \tag{4.15}$$

Note that the order in which the factors of $\mathsf{H}(s)$ are placed depends on the order in which the subsystems are connected. In general $\mathsf{H}_2(s)\mathsf{H}_1(s) \neq \mathsf{H}_1(s)\mathsf{H}_2(s)$, except when H_1 and H_2 are 1-by-1 matrices.

A system containing a feedback loop is shown in Fig. 4.1(*c*). The transfer function $\mathsf{H}_1(s)$ is called the *forward transmission* and the transfer function $\mathsf{H}_2(s)$ is called the *feedback transmission.* The minus sign at the summing junction indicates that the signal e is the *difference* between the system input u and the feedback signal z. This corresponds to *negative* feedback. The transfer function for Fig. 4.1(*c*) is obtained by tracing the signal flow through the system:

$$\mathsf{y}(s) = \mathsf{H}_1(s)\mathsf{e}(s) = \mathsf{H}_1(s)[\mathsf{u}(s) - \mathsf{z}(s)]$$

But

$$\mathsf{z}(s) = \mathsf{H}_2(s)\mathsf{y}(s)$$

Thus

$$\mathsf{y}(s) = \mathsf{H}_1(s)[\mathsf{u}(s) - \mathsf{H}_2(s)\mathsf{y}(s)]$$

or

$$[I + \mathsf{H}_1(s)\mathsf{H}_2(s)]\mathsf{y}(s) = \mathsf{H}_1(s)\mathsf{u}(s)$$

and, finally,

$$\mathsf{y}(s) = [I + \mathsf{H}_1(s)\mathsf{H}_2(s)]^{-1}\mathsf{H}_1(s)\mathsf{u}(s)$$

Thus, the transfer function (matrix) of the system containing a feedback loop is

$$\mathsf{H}(s) = [I + \mathsf{H}_1(s)\mathsf{H}_2(s)]^{-1}\mathsf{H}_1(s) \tag{4.16}$$

The matrix

$$\mathsf{F}(s) = I + \mathsf{H}_1(s)\mathsf{H}_2(s)$$

which may be called the *return-difference* (matrix)—a generalization of the terminology introduced by Bode[2]—has an inverse except at isolated values of s at which the transfer matrix becomes infinite. These values of s are the poles of the system. Since

$$[I + \mathsf{H}_1(s)\mathsf{H}_2(s)]^{-1} = \frac{\operatorname{adj}[I + \mathsf{H}_1(s)\mathsf{H}_2(s)]}{|I + \mathsf{H}_1(s)\mathsf{H}_2(s)|}$$

it follows that the characteristic equation of a single-loop feedback system is

$$|I + \mathsf{H}_1(s)\mathsf{H}_2(s)| = 0 \tag{4.17}$$

In words: the zeros of the determinant of the return difference are the poles of the system.

Alternative expressions for the transfer function are obtained by following different sequences of steps. In particular,

$$\mathsf{y}(s) = \mathsf{H}_1(s)\mathsf{e}(s)$$

and

$$\mathsf{e}(s) = \mathsf{u}(s) - \mathsf{z}(s) = \mathsf{u}(s) - \mathsf{H}_2(s)\mathsf{H}_1(s)\mathsf{e}(s)$$

Thus

$$[I + \mathsf{H}_2(s)\mathsf{H}_1(s)]\mathsf{e}(s) = \mathsf{u}(s)$$

or

$$\mathsf{e}(s) = [I + \mathsf{H}_2(s)\mathsf{H}_1(s)]^{-1}\mathsf{u}(s) \tag{4.18}$$

Finally

$$\mathsf{y}(s) = \mathsf{H}_1(s)[I + \mathsf{H}_2(s)\mathsf{H}_1(s)]^{-1}\mathsf{u}(s)$$

Thus

$$\mathsf{H}(s) = \mathsf{H}_1(s)[I + \mathsf{H}_2(s)\mathsf{H}_1(s)]^{-1} \tag{4.19}$$

From (4.19) it is seen that another form of the characteristic equation of the system is

$$|I + \mathsf{H}_2(s)\mathsf{H}_1(s)| = 0 \tag{4.20}$$

One should not make the mistake of assuming that $\mathsf{H}_1(s)$ and $\mathsf{H}_2(s)$ commute just because the order in which they are multiplied does not matter in setting up the characteristic equation. It does follow, however, that $\mathsf{H}_1(s)$ and $\mathsf{H}_2(s)$ are *conformable*, in whatever order they are multiplied. Since $\mathsf{H}_1(s)\mathsf{H}_2(s)$ may be a higher-dimension (or lower-dimension) matrix than $\mathsf{H}_2(s)\mathsf{H}_1(s)$, calculations can be simplified by working with the product having the smaller dimension.

When $\mathsf{H}_1(s)$ and $\mathsf{H}_2(s)$ are transfer functions of single-input, single-output systems, then both (4.16) and (4.19) reduce to the well-known formula

$$\mathsf{H}(s) = \frac{\mathsf{H}_1(s)}{1 + \mathsf{H}_1(s)\mathsf{H}_2(s)} \tag{4.21}$$

and the return difference is

$$\mathsf{F}(s) = 1 + \mathsf{H}_1(s)\mathsf{H}_2(s)$$

By repeated combination of subsystems in parallel, in tandem, and with feedback loops it is often possible to obtain the transfer function of a fairly complex system without performing a great deal of matrix algebra. Instead of by the repeated combination of elements, the block diagram of a single-input, single-output system can be reduced in a single operation by the use of the general-gain formula developed by S. J. Mason.[5] Mason's rule is fraught with possibility of error, however, unless the user is very careful with bookkeeping.

Example 4A Distillation column The fourth-order dynamic model of a distillation column, as developed by Gilles and Retzbach, was given in Chap. 2 (Example 2G on p. 47). The transfer functions from the inputs to the state variables and outputs can be obtained using the matrix calculations described in Chap. 3. But in this example the transfer functions are more readily calculated by block-diagram manipulations.

The block diagram corresponding to the differential equations is shown in Fig. 4.2. The overall system has been subdivided into two subsystems as shown in Fig. 4.3, each corresponding to a different physical aspect of the process. The first subsystem, having a single input Δu_1 and a single output x_2, represents the boiler. The second subsystem then represents the inner operation of the distillation column. The integrators have been represented by their transfer functions, $1/s$. Subsystem 1 itself comprises two single-loop feedback systems, separated by a gain element. Thus, by (4.15) and (4.21)

$$\mathsf{H}_1(s) = \frac{\mathsf{x}_2(s)}{\Delta\mathsf{u}_1(s)} = \frac{1/s}{1-(1/s)a_{22}}\, a_{21}\, \frac{1/s}{1-(1/s)a_{11}} \cdot b_{11} = \frac{a_{21}b_{11}}{(s-a_{22})(s-a_{11})} \tag{4A.1}$$

The second subsystem has two inputs, x_2 and Δu_2, and two outputs, Δz_1 and Δz_2. The input-output relation can be expressed as

$$\begin{bmatrix} \Delta\mathsf{z}_1(s) \\ \Delta\mathsf{z}_2(s) \end{bmatrix} = \mathsf{H}_{21}(s)\Delta\mathsf{s}(s) + \mathsf{H}_{22}(s)\mathsf{x}_2(s) \tag{4A.2}$$

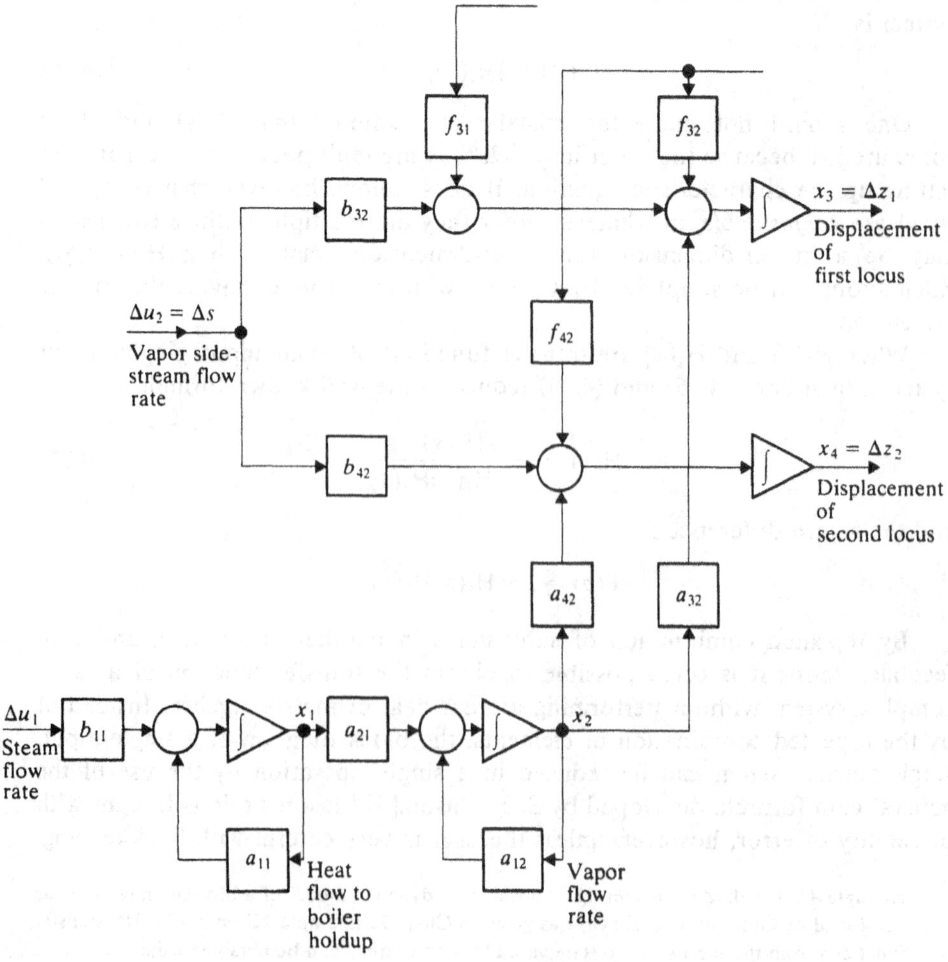

Figure 4.2 Dynamic model of distillation column.

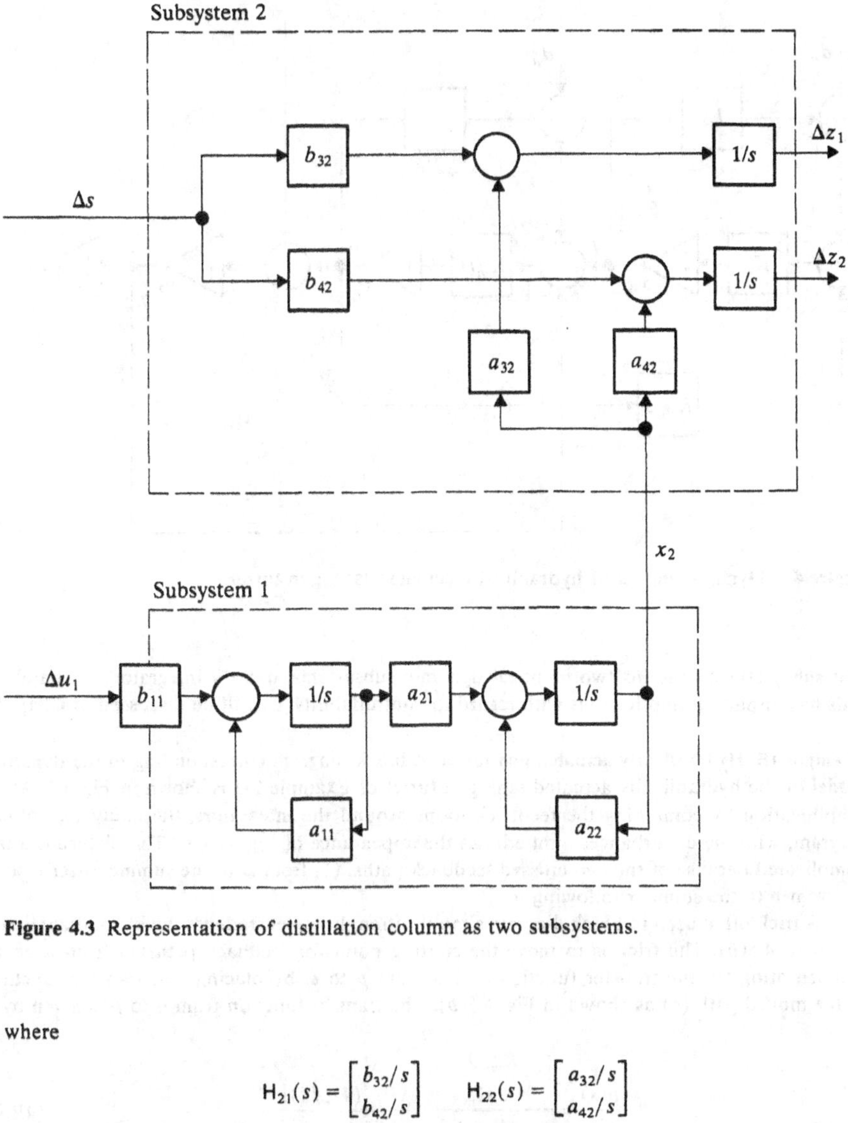

Figure 4.3 Representation of distillation column as two subsystems.

where

$$\mathsf{H}_{21}(s) = \begin{bmatrix} b_{32}/s \\ b_{42}/s \end{bmatrix} \qquad \mathsf{H}_{22}(s) = \begin{bmatrix} a_{32}/s \\ a_{42}/s \end{bmatrix}$$

Substituting $\mathsf{x}_2(s) = \mathsf{H}_1(s)\Delta\mathsf{u}_1(s)$, as given by (4A.1) into (4A.2) gives

$$\begin{bmatrix} \Delta z_1(s) \\ \Delta z_2(s) \end{bmatrix} = H_{21}(s)\Delta \mathsf{s}(s) + \begin{bmatrix} b_{11}a_{32}a_{21}/s(s-a_{22})(s-a_{11}) \\ b_{11}a_{42}a_{21}/s(s-a_{22})(s-a_{11}) \end{bmatrix} \Delta \mathsf{u}_1(s)$$

$$= \left[\begin{array}{c|c} \dfrac{b_{32}}{s} & \dfrac{b_{11}a_{32}a_{21}}{s(s-a_{22})(s-a_{11})} \\ \hline \dfrac{b_{42}}{s} & \dfrac{b_{11}a_{42}a_{21}}{s(s-a_{22})(s-a_{11})} \end{array}\right] \begin{bmatrix} \Delta \mathsf{s}(s) \\ \Delta \mathsf{u}_1(s) \end{bmatrix}$$

Note that the poles of the system are located at

$$s = 0 \qquad s = a_{11} \qquad s = a_{22}$$

There are only three different poles, although the system is fourth-order. The reason for this is

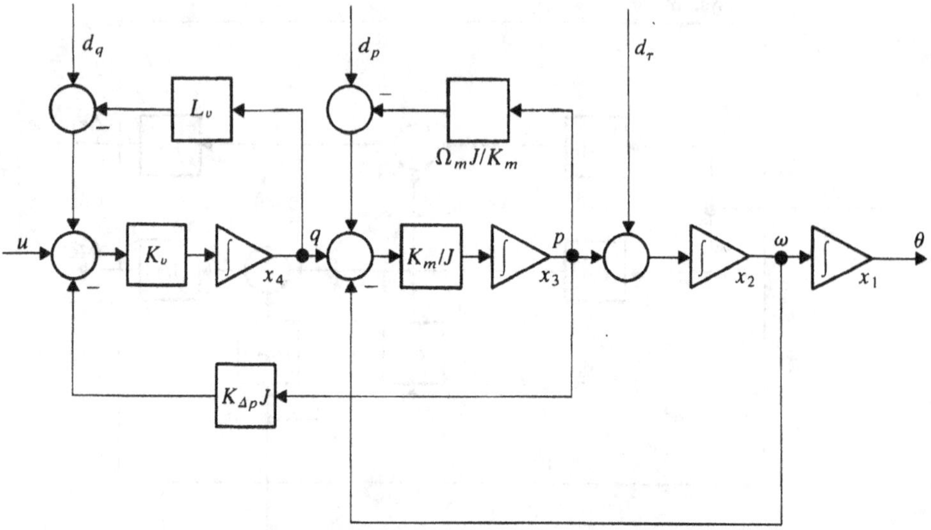

Figure 4.4 Dynamic model of hydraulically actuated tank gun turret.

that subsystem 2 contains two identical dynamic subsystems, namely integrators, in parallel. This has important implications with regard to controllability, as will be discussed in Chap. 5.

Example 4B Hydraulically actuated gun turret A block diagram corresponding to the dynamic model of the hydraulically actuated tank gun turret of Example 2D is shown in Fig 4.4. After simplification by combining the feedback loops around the integrators, the equivalent block diagram, with the disturbances omitted, has the appearance of Fig. 4.5(a). The picture is a bit complicated because of the two crossed feedback paths, (1) from ω to the summer after q, and (2) from p to the summer following u.

A trick often used in block-diagram simplification, however, reduces the block diagram of 4.5(a) to 4.5(b). The trick is to move the starting point for feedback path (1) from ω to p, compensating for the transfer function of $1/s$ from p to ω by placing that transfer function in the moved path (1) as shown in Fig. 4.5(b). The transfer function from q to p is given by

$$\frac{\mathrm{p}(s)}{\mathrm{q}(s)} = \frac{\dfrac{K_m/J}{s+\Omega_m}}{1+\dfrac{K_m/J}{s+\Omega_m}\dfrac{1}{s}} = \frac{(K_m/J)s}{s(s+\Omega_m)+K_m/J} \tag{4B.1}$$

The block-diagram resulting in this simplification is shown in Fig. 4.5(c). From this figure it is seen that the transfer function from u to p is

$$\frac{\mathrm{p}(s)}{\mathrm{u}(s)} = \frac{\dfrac{K_v}{s+K_vL_v}\,\dfrac{(K_m/J)s}{s(s+\Omega_m)+K_m/J}}{1+K_{\Delta p}J\dfrac{K_v}{s+K_vL_v}\,\dfrac{(K_m/J)s}{s(s+\Omega_m)+K_m/J}}$$

$$= \frac{(K_vK_m/J)s}{(s+K_vL_v)[s(s+\Omega_m)+K_m/J]+K_{\Delta p}K_vK_ms} \tag{4B.2}$$

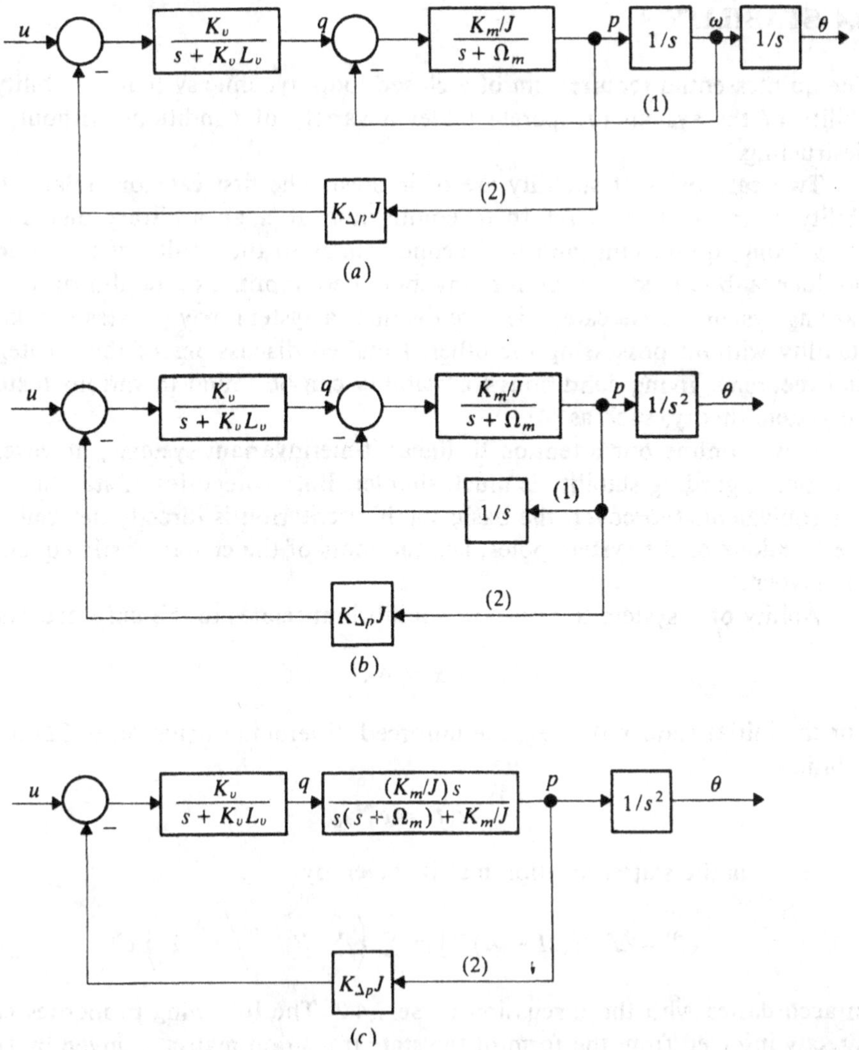

Figure 4.5 Block-diagram simplification of model of hydraulically actuated tank gun turret. (*a*) Figure 4.4 after reduction of loops around integrators; (*b*) Path from ω to q moved to p and integrator added; (*c*) Final simplification.

And the transfer function from the system input u to the angle θ is $1/s^2$ times $\mathsf{p}(s)/\mathsf{u}(s)$. Thus

$$\mathsf{H}(s) = \frac{\theta(s)}{\mathsf{u}(s)} = \frac{K_v K_m / J}{s\{(s + K_v L_v)[s(s + \Omega_m) + K_m/J] + K_{\Delta p} K_v K_m s\}}$$

The denominator of $\mathsf{H}(s)$ is the characteristic polynomial $\mathsf{D}(s)$ of the open-loop system. On expansion it is found that

$$\mathsf{D}(s) = s\left[s^3 + (\Omega_m + K_v L_v)s^2 + \left(\frac{K_m}{J} + \Omega_m K_v L_v + K_{\Delta p} K_v K_m\right)s + \frac{K_m K_v L_v}{J}\right]$$

4.4 STABILITY

The quintessential requirement of a closed-loop dynamic system is stability: the ability of the system to operate under a variety of conditions without "self-destructing."

Two categories of stability are of interest. The first category relates to the ability of the system to return to equilibrium after an arbitrary displacement away from equilibrium, and the second relates to the ability of the system to produce a bounded output for any bounded input. For nonlinear or time-varying systems these categories are distinct: a system may possess one kind of stability without possessing the other. Detailed discussions of these categories and theorems giving conditions for stability can be found in various textbooks on system theory, such as [4].

If we confine our attention to linear, time-invariant systems, however, the situation regarding stability is much simpler. Both categories of stability are all but equivalent. Moreover, the basic stability criterion is directly determined by the locations of the system poles, i.e., the roots of the characteristic equation of the system.

Ability of a system to return to equilibrium relates to the unforced system

$$\dot{x} = Ax \tag{4.22}$$

For the initial state $x(0) = x_0$, the unforced differential equation (4.22) has the solution

$$x(t) = e^{At}x_0 \tag{4.23}$$

where e^{At} is the state-transition matrix, given by

$$e^{At} = \mathscr{L}^{-1}[(sI - A)^{-1}] = \sum_{i=1}^{k} \left(\sum_{j=1}^{\nu_i} R_{ji} t^{j-1} \Big/ j - 1! \right) e^{s_i t} \tag{4.24}$$

in accordance with the discussion of Sec. 4.2. The following properties can be directly inferred from the form of the state transition matrix as given by (4.24):

1. If the real parts of *all* the characteristic roots are *strictly* negative (i.e., not zero or positive), then e^{At} tends asymptotically to zero. Hence, no matter how large the initial state x_0 is, $x(t) \to 0$ as $t \to \infty$. The system is said to be *asymptotically stable.*
2. If *any* characteristic root has a strictly positive real part, the state-transition matrix given by (4.24) will have at least one term which will tend to infinity as $t \to \infty$. In this case it is always possible to find some initial state which will cause $x(t)$ to become infinite. The system is said to be *unstable.*
3. If *all* the characteristic roots have nonpositive real parts, but one or more of the characteristic roots has a *zero* real part, the situation is somewhat more complicated: if all the characteristic roots having zero real parts are simple roots, then the corresponding terms in the state-transition matrix are of the

form

$$R_i e^{j\omega_i t} \qquad j = \sqrt{-1}$$

Since $|e^{j\omega_i t}| = 1$, it is clear that these terms in the state transition matrix are bounded. Hence the state $x(t)$ that evolves from any initial state x_0 will also remain bounded. But there will be some initial states from which the subsequent solution will *not* approach zero asymptotically. Systems of this type are said to be *stable*, but not asymptotically stable. If, on the other hand, any of the characteristic roots that has a zero real part is a *repeated* root, then, owing to the polynomial in t that multiplies $e^{j\omega_i t}$, there will be at least one term in e^{At} which will tend to infinity as $t \to \infty$. Hence there will be some initial state for which $x(t) \to \infty$, and the system is unstable. (In the strict sense, the multiplicity of the roots of the *minimum* equation, i.e., the equation of lowest degree satisfied by the matrix A, as discussed in the Appendix, rather than the multiplicity of the roots of the characteristic equation, must be examined to test for the stability of systems with such roots on the imaginary axis.)

The above conclusions are summarized in Table 4.1.

Stability of the second category: bounded-input bounded-output (BIBO) stability is determined using the convolution integral (4.2). Consider only a single-input, single-output system, having a scalar impulse response $h(t)$. For this system

$$y(t) = \int_0^t h(t-\tau)u(\tau)\, d\tau \tag{4.25}$$

It is easy to show that

$$|y(t)| \leqq \int_0^t |h(t-\tau)||u(\tau)|\, d\tau \tag{4.26}$$

The meaning of the input $u(t)$ being bounded is that there is a constant c such that

$$|u(t)| \leqq c \qquad \text{for all } t \tag{4.27}$$

Table 4.1 Stability conditions for linear systems

Condition		Implication
1. $\text{Re}(s_i) < 0$	for all i	System is asymptotically stable
2. $\text{Re}(s_i) > 0$	for some i	System is unstable
3. $\text{Re}(s_i) = 0$	for some $i = j$, and	
(*a*) s_j is simple root for all such j		System is stable, but not asymptotically stable
(*b*) s_j is multiple root for some such j		System is unstable

In this case, by (4.26),

$$|y(t)| \leqq c \int_0^t |h(t-\tau)|\, d\tau \tag{4.28}$$

In accordance with (4.8) and (4.9), the impulse response of a time-invariant system is a sum of time-weighted exponentials. If the system is asymptotically stable, then the exponentials all tend asymptotically to zero; no matter how large the time-weighting on the exponentials, the integral in (4.28) will be finite for all t (including $t \to \infty$), and hence $|y(t)|$ will be finite. Thus we see that an asymptotically stable time-invariant system produces a bounded output for every bounded input. On the other hand, suppose the system produces an unbounded output for some bounded inputs. This output must result from some term in the impulse response that does not tend asymptotically to zero, which implies that the system is not asymptotically stable. Thus a linear time-invariant system in which a bounded input produces an unbounded output cannot be asymptotically stable. (Although the system is not asymptotically stable, it may still be stable. The simplest example is an integrator for which $h(t) = 1$. For a bounded input, say $u(t) = 1$, $y(t) = t$ which tends to infinity with t, so for this example a bounded input does not produce a bounded output. There are many similar examples.)

The foregoing discussion may be summarized as follows:

Asymptotically stable system ⇒ every bounded input produces a bounded output

Unstable system ⇒ some bounded input produces an unbounded output

Note that the implications go only in one direction. We may *not* conclude that a system for which every bounded input produces a bounded output is asymptotically stable. As we shall see in the next chapter, it is possible that some unstable state variables are not excited by the input. It is also not permissible to conclude that if some bounded input produces an unbounded output, the system is unstable. An ideal integrator, already cited, is an example of a stable system for which a bounded input (say a step function) produces an unbounded output (a ramp).

Example 4C Aircraft longitudinal motion The linear dynamic equations for the longitudinal motion of an aircraft were given in (2.40). Using the state and control definitions

$$x = \begin{bmatrix} \Delta u \\ \alpha \\ q \\ \theta \end{bmatrix} \qquad u = \delta_E$$

we obtain the dynamics and control matrices

$$A = \begin{bmatrix} X_u & X_\alpha & 0 & -g \\ Z_u/V & Z_\alpha/V & 1 & 0 \\ M_u & M_\alpha & M_q & 0 \\ 0 & 0 & 1 & 0 \end{bmatrix} \qquad B = \begin{bmatrix} X_E \\ Z_E/V \\ M_E \\ 0 \end{bmatrix} \tag{4C.1}$$

The resolvent for this system is

$$\Phi(s) = (sI - A)^{-1} = \begin{bmatrix} s - X_u & -X_\alpha & 0 & g \\ -Z_u/V & s - Z_\alpha/V & -1 & 0 \\ -M_u & -M_\alpha & s - M_q & 0 \\ 0 & 0 & -1 & s \end{bmatrix}^{-1} \tag{4C.2}$$

From the resolvent we obtain the characteristic polynomial

$$|sI - A| = s^4 + a_1 s^3 + a_2 s^2 + a_3 s + a_4 \tag{4C.3}$$

where $a_1 = -\dfrac{Z_\alpha}{V} - M_q - X_u$

$$a_2 = \frac{Z_\alpha}{V} M_q - M_\alpha + X_u\left(\frac{Z_\alpha}{V} + M_q\right) - \frac{Z_u}{V} X_\alpha$$

$$a_3 = -X_u\left(\frac{Z_\alpha}{V} M_q - M_\alpha\right) + X_\alpha\left(\frac{Z_u}{V} M_q - M_u\right) + gM_u \tag{4C.4}$$

$$a_4 = g\frac{Z_u M_\alpha - Z_\alpha M_u}{V}$$

Contribution to the characteristic equation of the terms due to the change in speed (those with the subscript u) are usually quite small relative to the other terms. Thus, as an approximation, the characteristic polynomial is

$$\begin{aligned} |sI - A| &\approx s^2\left(s^2 - \left(\frac{Z_\alpha}{V} + M_q\right)s + \frac{Z_\alpha}{V} M_q - M_\alpha\right) \\ &= s^2[s^2 + 2\zeta_s \Omega_s s + \Omega_s^2] \end{aligned} \tag{4C.5}$$

The double pole at the origin is due to the translation of the aircraft as an ideal mass, and the quadratic factor is due to the rotation of the aircraft about the center of mass. This motion is seen to be that of a mass-spring-damper system with a damping factor ζ_s and a natural frequency Ω_s, and is called the *short-period* motion of the aircraft.

If Ω_s^2 and ζ_s are both positive, the poles of the short-period motion lie in the left half-plane and the short-period motion is stable. The aircraft in this case is said to be *aerodynamically stable.* Until very recently, it was the responsibility of the aerodynamicist to design the aircraft to ensure aerodynamic stability for all operating regimes of the aircraft. It is of course possible to stabilize an unstable aircraft by means of a properly designed control system, but the hardware (i.e., sensors and actuators) used to implement the control system must be extremely reliable—as reliable as the airframe itself. With the advent of multiply-redundant hardware, it is possible to achieve a very high degree of reliability, and it is now considered safe to operate aerodynamically unstable aircraft having suitable multiply redundant stability augmentation systems.

As an example we consider the numerical data for an actual aircraft, the AFTI-16 (a modified version of the F-16 fighter) in the landing approach configuration, as given in Table 4C.1 [6].

Using the data in Table 4C.1 we find that

$$\frac{Z_\alpha}{V} + M_q = 2\zeta_s \Omega_s = 1.01$$

$$\frac{Z_\alpha}{V} M_q - M_\alpha = \Omega_s^2 = -1.1621 \tag{4C.6}$$

Since $\Omega_s^2 < 0$, the aircraft is aerodynamically unstable in this regime, having poles at

$$s_1 = -1.695 \qquad \text{and} \qquad s_2 = 0.685 \tag{4C.7}$$

Table 4C.1 Aerodynamic parameters for AFTI-16 on landing approach

	$V = 139$ Kt		
$X_u = -0.0507$	$X_\alpha = -3.861$		$X_E = 0$
$Z_u/V = -0.00117$	$Z_\alpha/V = -0.5164$		$Z_E/V = -0.0717$
$M_u = -0.000129$	$M_\alpha = 1.4168$	$M_q = -0.4932$	$M_E = -1.645$

The short-period poles as given by (4C.7) are only approximate, since the effects of the speed changes have not been accounted for. To take these effects into account we must calculate the coefficients of the characteristic polynomial using (4C.4). For the data of Table 4C.1 the coefficients are found to be

$$a_1 = 1.0603$$

$$a_2 = -1.1154$$

$$a_3 = -0.0565$$

$$a_4 = -0.0512$$

Numerical solution of the characteristic equation yields the pole locations

$$s_1 = -1.705 \qquad s_2 = 0.724 \qquad s_{3,4} = -0.0394 \pm j0.200$$

We observe that the short-period poles s_1 and s_2, when speed changes are accounted for, are located very close to the approximate locations given in (4C.7). Another pair of poles (which are at the origin in the approximate analysis), with a natural frequency of $[(0.200)^2 + (0.0394)^2]^{1/2} = 0.204$ and a damping factor of 0.19, also appears due to speed changes. The motion due to these poles is known as *phugoid* motion and is manifest as a slight oscillation in altitude. (See Note 4.2.)

4.5 ROUTH-HURWITZ STABILITY ALGORITHMS

In the previous section we saw that the imaginary axes of the complex frequency plane (*the s plane*) separates the region of stability from the region of instability. If all poles lie in the left half-plane the system is asymptotically stable; otherwise the system is not asymptotically stable.

It is now a routine exercise for a digital computer to find the roots of a polynomial of very high degree. Before the advent of digital computers, however, testing the stability of a system by calculating the zeros of the characteristic equation was not practical. Methods were needed that did not require actual calculation of these roots. The earliest contribution to this problem was the Adams Prize Essay (1874–1877) of E. J. Routh[7] who developed a simple tabular algorithm by which it is possible to determine whether a given polynomial has all its roots in the left half-plane without finding the roots. A different algorithm was developed by A. Hurwitz[8] in 1895. And in 1962, P. C. Parks[9] showed that these algorithms could be derived by use of a stability theorem that M. A. Liapunov developed[10] in 1892–1907.

The algebraic criteria are derived in textbooks such as Schwarz and Friedland[4] on linear systems, and will not be repeated here. For convenience of the reader, the resulting algorithms are presented here without proof.

Table 4.2 Routh table

	1	a_2	a_4	$a_6 \cdots$
	a_1	a_3	a_5	$a_7 \cdots$
$\alpha_1 = \dfrac{1}{a_1}$	$b_1 = a_2 - \alpha_1 a_3$	$b_2 = a_4 - \alpha_1 a_5$	$b_3 = a_6 - \alpha_1 a_7$	$\cdots\cdots$
$\alpha_2 = \dfrac{a_1}{b_1}$	$c_1 = a_3 - \alpha_2 b_2$	$c_2 = a_5 - \alpha_2 b_3$	$\cdots\cdots\cdots\cdots$	
$\alpha_3 = \dfrac{b_1}{c_1}$	$d_1 = b_2 - \alpha_3 c_2$	$\cdots\cdots\cdots\cdots$		
$\alpha_4 = \dfrac{c_1}{d_1}$	$\cdots\cdots\cdots\cdots$			
$\vdots$				

The characteristic polynomial of the system to be tested for stability is assumed to be of the form

$$\mathsf{D}(s) = s^k + a_1 s^{k-1} + \cdots + a_{k-1}s + a_k$$

The Routh table corresponding to $\mathsf{D}(s)$ is constructed as shown in Table 4.2. The first two rows are obtained by transcribing the coefficients of $\mathsf{D}(s)$ in alternate rows as shown. Each succeeding row of the table is completed using entries in the two preceding rows, until there are no more terms to be computed. In the left margin are found a column of exactly k numbers $\alpha_1, \alpha_2, \ldots, \alpha_k$ for a kth-order system. The theorem of the Routh algorithm is that the roots of $\mathsf{D}(s) = 0$ lie in the left half-plane, excluding the imaginary axis, if and only if all the α's are strictly positive.

The Hurwitz criterion, which is equivalent to the Routh algorithm, is based on the construction of a $k \times k$ Hurwitz matrix

$$H = \begin{bmatrix} a_1 & a_3 & \cdots & \cdots \\ 1 & a_2 & \cdots & \cdots \\ \hline 0 & a_1 & a_3 & \cdots \\ 0 & 1 & a_2 & \cdots \\ \hline 0 & 0 & a_1 & \cdots \\ 0 & 0 & 1 & \cdots \end{bmatrix} \quad k \text{ rows}$$

$$\longleftarrow k \text{ columns} \longrightarrow$$

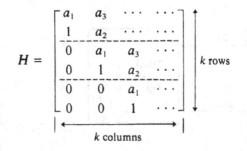

The first two rows of H are formed from the coefficients of $\mathsf{D}(s)$, with zeros used for a_{k+1} through a_{2k-1}. Each row following is obtained by shifting one step to the right the entries of the row two positions above, and padding the empty positions with zeros. This process is continued until the $k \times k$ matrix is completed. The stability theory based on the Hurwitz matrix is that the zeros of $\mathsf{D}(s)$ are in the left half-plane, excluding the imaginary axis, if and only if the determinants

$$D_1 = a_1$$

$$D_2 = \begin{vmatrix} a_1 & a_3 \\ 1 & a_2 \end{vmatrix}$$

$$D_3 = \begin{vmatrix} a_1 & a_3 & a_5 \\ 1 & a_2 & a_4 \\ 0 & a_1 & a_3 \end{vmatrix}$$

. . . .

$$D_k = |H|$$

are all strictly positive.

Example 4D Distillation column—continued A closed-loop control system for the distillation column of the previous example is proposed by making the change in the steam flow rate Δu_1 proportional to the error between Δz_1 and some desired set point value, say $\overline{\Delta z}$. Thus, as shown in Fig 4.6,

$$\Delta u_1(s) = u(s) = K[\overline{\Delta z} - \Delta z_1(s)] \tag{4D.1}$$

and from the analysis of Example 4A

$$u(s) = \frac{s(s - a_{11})(s - a_{22})}{b_{11}a_{32}a_{21}} \Delta z_1(s) \tag{4D.2}$$

The closed-loop transfer function $H_c(s)$ is obtained by substituting (4D.2) into (4D.1) and finding the ratio of Δz_1 to $\overline{\Delta z}$

$$H_c(s) = \frac{\bar{K}}{s(s - a_{11})(s - a_{22}) + \bar{K}} \tag{4D.3}$$

where $\bar{K} = b_{11}a_{32}a_{21}K$.

The characteristic polynomial of the closed-loop system is

$$s^3 - (a_{11} + a_{22})s^2 + a_{11}a_{22}s + \bar{K} \tag{4D.4}$$

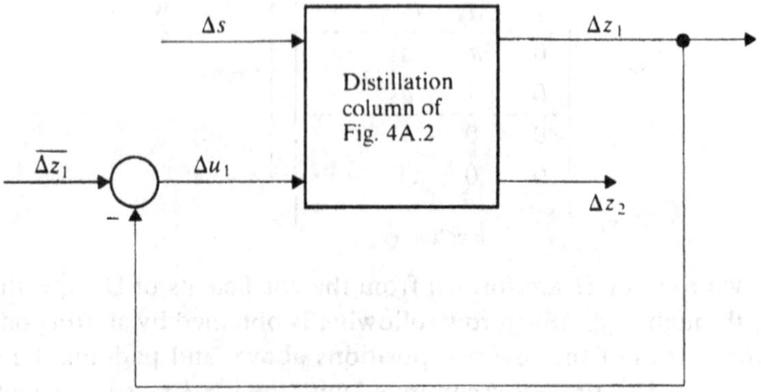

Figure 4.6 Single-loop control of distillation column.

Thus,

$$a_1 = -a_{11} - a_{22}$$

$$a_2 = a_{11}a_{22}$$

$$a_3 = \bar{K}$$

The Routh table for this example is

$$
\begin{array}{c|cc}
 & 1 & a_2 \\
 & a_1 & \bar{K} \\
\hline
\alpha_1 = \dfrac{1}{a_1} & a_2 - \dfrac{\bar{K}}{a_1} & \\
\hdashline
\alpha_2 = \dfrac{a_1}{a_2 - \dfrac{\bar{K}}{a_1}} & \bar{K} & \\
\hdashline
\alpha_3 = \dfrac{a_2 - \dfrac{\bar{K}}{a_1}}{\bar{K}} & &
\end{array}
$$

Thus, for stability of the closed-loop system we must have

$$a_1 > 0 \tag{4D.5}$$

$$a_2 - \frac{\bar{K}}{a_1} > 0 \quad \text{or} \quad \bar{K} < a_1 a_2 \tag{4D.6}$$

$$\bar{K} > 0 \tag{4D.7}$$

The first condition is a requirement on the open-loop dynamics. From the data given about the process a_{11} and a_{22} are both negative, so (4D.5) is automatically satisfied. The second and third conditions are combined to give

$$0 < \bar{K} < a_1 a_2 \tag{4D.8}$$

which means that the gain $\bar{K}$ (which is a negative feedback gain) must be positive—i.e., only negative feedback is permissible, and that $\bar{K}$ must be smaller than a fixed positive number.

The Hurwitz matrix for this example is

$$H = \begin{bmatrix} a_1 & \bar{K} & 0 \\ 1 & a_2 & 0 \\ 0 & a_1 & \bar{K} \end{bmatrix}$$

and the stability requirements are

$$D_1 = a_1 > 0$$

$$D_2 = a_1 a_2 - \bar{K} > 0$$

$$D_3 = \bar{K} D_2 > 0$$

which are the same conditions as obtained using the Routh algorithm.

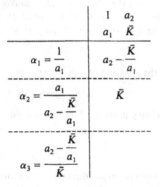

In the root-locus method to be studied in the next section we will be concerned with the variation of the closed-loop poles with the loop gain $\bar{K}$. By the methods to be explained more fully in that section, we find that the roots move from the open-loop poles to infinity. The open-loop poles occur at

$$s = 0$$

$$s = a_{11} \tag{4D.9}$$

$$s = a_{22}$$

and the loci of the closed-loop poles have the appearance shown in Fig 4.7. One locus moves along the negative real axis, and the other two, after moving together, separate from the real axis and move to asymptotes at angles of ±60 degrees from the positive real axis. The gain $\bar{K}$ at which the loci cross the imaginary axis is the gain at which (4D.6) is an equality:

$$\bar{K} = a_1 a_2 \tag{4D.10}$$

The frequency ω at which the crossing occurs is obtained by substituting $s = j\omega$ into (4D.4)

$$-j\omega^3 - \omega^2 a_1 + j\omega a_2 + \bar{K} = 0 \tag{4D.11}$$

The real and the imaginary parts of (4D.11) must simultaneously be zero:

$$\omega^3 - \omega a_2 = 0 \tag{4D.12}$$

$$-\omega^2 a_1 + \bar{K} = 0 \tag{4D.13}$$

From (4D.12) we obtain $\omega = 0$ (corresponding to the open loop pole at the origin) and $\omega^2 = a_2$. From (4D.13) we obtain $\omega^2 = \bar{K}/a_1$. Since by (4D.10) the critical gain $\bar{K} = a_1 a_2$, the second expression for ω^2 is consistent with the first.

To obtain the "breakaway frequency" $s = -c_2$ at which the root loci join before leaving the real axis, we note that at that point, there is a double pole, so the characteristic equation must be

$$(s + c_1)(s + c_2)^2 = s^3 + (c_1 + 2c_2)s^2 + c_2(2c_1 + c_2)s + c_1 c_2^2$$

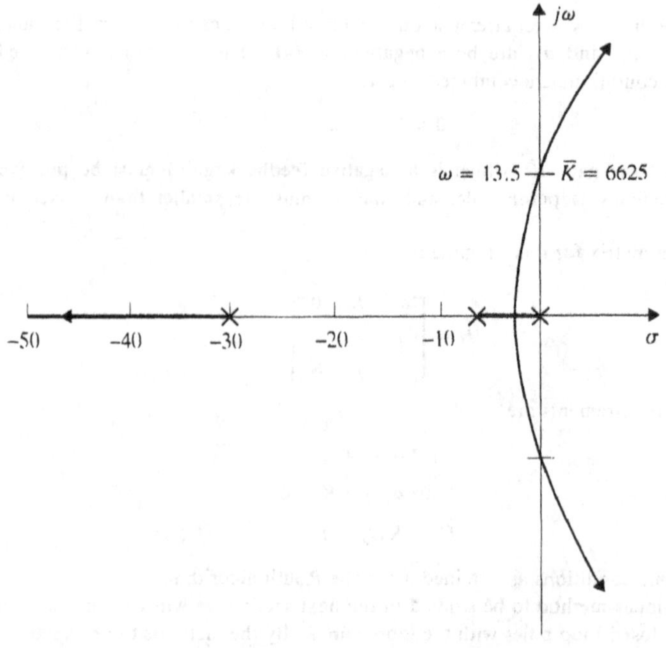

Figure 4.7 Root-locus for feedback of steam flow rate.

Thus we must have

$$a_1 = c_1 + 2c_2$$
$$a_2 = c_2(2c_1 + c_2)$$
$$\bar{K} = c_1 c_2^2$$

which can be solved simultaneously to give c_1, c_2, and $\bar{K}$. Another method of finding the breakaway frequency is given in discussion of the root locus method of the next section. Using the numerical data for the parameters of this process as given in Example 2G, namely

$$a_{11} = -30.3 \qquad a_{22} = -6.02$$

gives

$$a_1 = 36.32 \qquad a_2 = 182.4$$

Thus the gain at which the roots cross into the right half-plane is

$$\bar{K} = 6625$$

and the frequency at the crossing of the axis is

$$\omega = 13.5$$

The root loci separate from the real axis at

$$s = -c_2 = -2.84$$

and this occurs for a gain $\bar{K} = 247$.

4.6 GRAPHICAL METHODS

The algebraic tests of Routh and Hurwitz give the precise range(s) of parameter(s) for which a system is stable, and do not require the calculation of the closed-loop poles. They are most useful for testing whether a design is satisfactory but are not as convenient as some of the graphical methods (root-loci, Bode and Nyquist plots) for design purposes. Since frequency-domain *design* methods are not considered in this book, we will not dwell at length on these graphical methods, but refer the reader instead to one of the standard textbooks on the subject.[4, 11, 12] On the other hand, graphical representations can often serve as an aid to interpreting the design results that are obtained by state-space methods. For this reason, it is worth considering them at least briefly.

Except for the recent extensions to multivariable systems (as typified by the work of Rosenbrock and MacFarlane) the graphical methods are addressed to a single-loop system having a return-difference function

$$\mathrm{T}(s) = 1 + K\mathrm{G}(s) \tag{4.29}$$

where K is a scalar gain (the "loop gain") and $\mathrm{G}(s)$ is a rational function known as the "open-loop" transfer function. A return difference of the form of (4.29) arises directly in the systems shown in Fig. 4.8, but it is always possible to manipulate the block diagram of a system so that the characteristic equation of the system appears in this form for any system parameter represented by K. The graphical methods are devices for elucidating the dynamic characteristics of

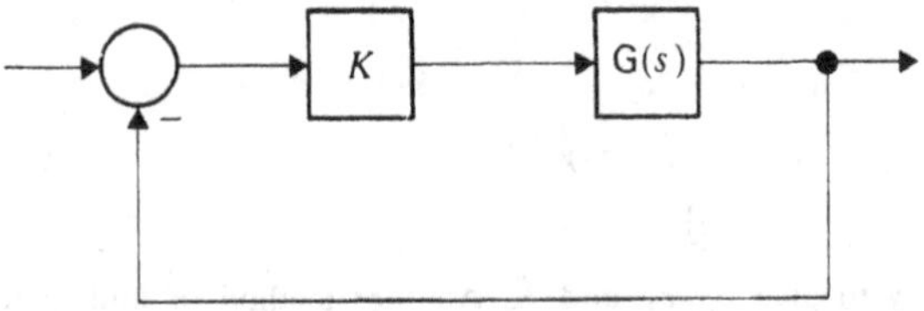

Figure 4.8 Single-loop feedback system return difference $T(s) = 1 + KG(s)$.

a system having an open-loop transfer function $G(s)$ as the loop gain K is varied.

Root-locus method The root-locus method, developed by Evans[13] in 1948 is simply a plot of the locations in the complex plane of the roots of $T(s) = 0$, (i.e., the poles of the closed-loop system) as the loop gain is varied. The open-loop transfer function is assumed to be a rational function of s, i.e.,

$$G(s) = C\frac{\prod_{i=1}^{n_z}(s - z_i)}{\prod_{i=1}^{n_p}(s - p_i)} \tag{4.30}$$

where C is a real constant, z_i $(i = 1, \ldots, n_z)$ are the open-loop zeros and p_i $(i = 1, \ldots, n_p)$ are open-loop poles. If desired the constant C can be absorbed in the loop gain, by defining $\bar{K} = KC$. It is seen that as $K \to 0$, the closed-loop poles, which are the roots of (4.29), tend to the open-loop poles p_i. On the other hand, as the gain K tends to infinity, the closed-loop poles tend to the open-loop zeros. If $G(s)$ is a proper rational function, however, there are fewer open-loop zeros than open-loop poles. Since the number of closed-loop poles does not change as K is varied, where do the closed-loop poles go that do not go to the open-loop zeros? They go to infinity. The manner in which they go to infinity depends on the excess of poles over zeros. Imagine viewing the complex plane from a great distance. From this vantage point all the poles and zeros appear to be at the origin and $G(s)$ looks like $1/s^{(n_p - n_z)}$. Thus, from this vantage point the root-locus equation looks like

$$1 + K\frac{1}{s^e} = 0 \qquad e = n_p - n_z$$

or

$$s^e + K = 0 \tag{4.31}$$

where e is the excess of poles over zeros in the open-loop transfer function. Thus, as K becomes very large, the root loci that do not terminate at the open-loop zeros tend to infinity in the same way as the solutions of (4.31) tend to infinity, namely as the eth roots of $-K$. Since there are exactly e such roots at equal angles around a circle, these lines are the asymptotes of the root loci that tend to infinity. Figure 4.9 illustrates the asymptotic behavior for large values of loop gain K of those branches of the root loci that tend to infinity.

The asymptotic behavior of the root loci can be rationalized another way: we can say that the number of poles and zeros are always equal and that the

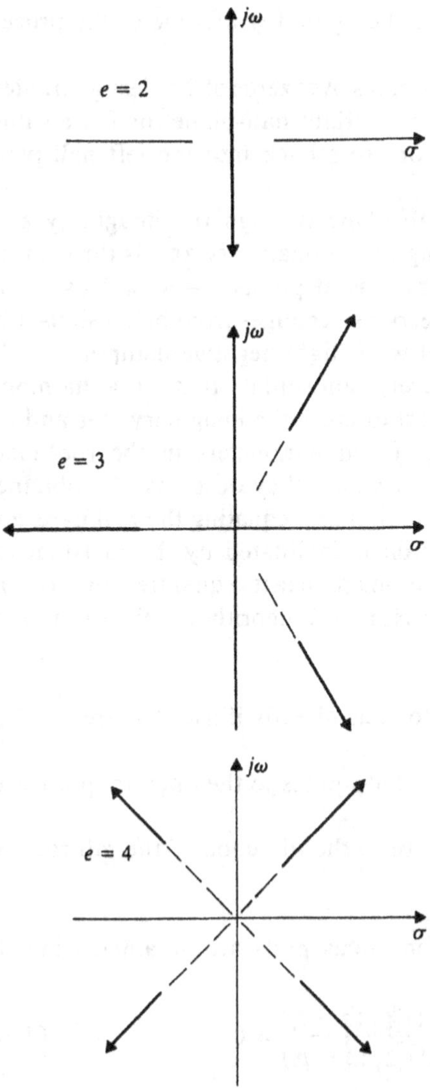

Figure 4.9 Asymptotes of root loci for several values of excess poles.

root loci always go from the poles to the zeros, but that those zeros (e in number) which are not in the finite part of the s plane lie at infinity.

Figure 4.9 shows that whenever the excess of poles over zeros is greater than 2, the root loci must eventually cross the imaginary axis into the right half of the s plane. Consequently, no system having an excess of two or more can be stable for all values of gain. Since the excess is two or more in most practical systems, the implication is that in practice there is a finite upper limit to the loop gain. The ratio of the loop gain at which a system is designed to operate to the gain at which it becomes unstable (expressed logarithmically), is known as the *gain margin* of the system. Gain margin is an important consideration in systems in which the loop gain may change during the operating lifetime of the process (due possibly to aging of components).

It must not be inferred that an excess of poles over zeros of 2 or less guarantees stability, since the root loci may cross into the right half-plane for finite values of gain, and remain there (when $e = 2$) or cross back into the left half-plane when K becomes large enough.

The root loci cross into the right half-plane through the imaginary axis. Except in the trivial case when the crossing of the imaginary axis is through the origin ($s = 0$), the loci cross the imaginary axis at points $s = \pm j\omega$. This means that the nature of the unforced dynamic response changes from being sinusoidal with slight positive damping to sinusoidal with slight negative damping. At the dividing line, the response becomes purely sinusoidal: that of a harmonic oscillator. The gains that cause the root loci to cross the imaginary axis and the frequencies at which they occur are significant parameters in the root-locus method. These frequencies and the gains at which they occur can be obtained by setting $s = j\omega$ into the characteristic equation and equating the real parts and the imaginary parts to zero. The calculation is facilitated by the fact that the gains at which the crossings occur also make exact equalities out of the inequalities that result from the Routh (or Hurwitz) algorithm. This was already illustrated in Example 4C.

The basic rules for drawing the root loci, as already illustrated are

The loci move continuously from the open-loop poles to the open-loop zeros or to infinity.

The loci approach infinity at lines which are in the direction of the eth roots of $-1 = e^{-j\pi}$ from the origin.

Many other rules for constructing root-locus plots are obtained from the basic root locus equation

$$1 + KG(s) = 1 + K\frac{\prod_{i=1}^{n_z}(s - z_i)}{\prod_{i=1}^{n_p}(s - p_i)} = 0 \tag{4.32}$$

Each factor $s - z_i$ or $s - p_i$ is represented in the complex plane by a vector (a "phasor" in electrical engineering parlance) from the zero z_i or pole p_i to the point x. If s is a point on the root locus, then (4.32) must hold, i.e.,

$$\frac{\prod_{i=1}^{n_z}(s - z_i)}{\prod_{i=1}^{n_p}(s - p_i)} = -\frac{1}{K} \tag{4.33}$$

which means that the product of the lengths of all the vectors from the zeros to the point s divided by the product of the lengths of all the vectors from the poles must come out to be $1/K$ and that the sum of all the angles of the vectors from the zeros to the point s minus the angles of the vectors from the poles must add to $-180°$. Rules obtained from this general principle, such as the directions of departure of the loci from the open-loop poles or of arrival at the open-loop zeros, can be found in various textbooks [11, 12] that concentrate on frequency-domain analysis.

The points at which the loci leave the real axis are known as *breakaway points.* To find these points consider the root-locus equation

$$1 + K\frac{\mathsf{N}(s)}{\mathsf{D}(s)} = 0$$

Multiply by the open-loop denominator to obtain the characteristic equation

$$\mathsf{P}(s) = \mathsf{D}(s) + K\mathsf{N}(s) = 0$$

The breakaway points are those at which $\mathsf{P}(s)$ has a multiple root. Thus if $s = a$ is a breakaway point we can write

$$\mathsf{P}(s) = (s - a)^2\mathsf{P}_1(s)$$

where $\mathsf{P}_1(s)$ is a polynomial of degree $k - 2$ obtained by multiplying all the factors except the factor $(s - a)^2$ arising because of the multiple root. ($\mathsf{P}_1(s)$ could conceivably have more than a double root at $s = a$, in which case $\mathsf{P}_1(s)$ could contain other $(s - a)$ factors. This is of no concern.)

The derivative of $\mathsf{P}(s)$ with respect to s is

$$\mathsf{P}'(s) = 2(s - a)\mathsf{P}_1(s) + (s - a)^2\mathsf{P}_1'(s)$$

Thus, at $s = a$

$$\mathsf{P}'(a) = 2(a - a)\mathsf{P}(a) + (a - a)^2\mathsf{P}'(s) = 0$$

In other words at a breakaway point $s = a$ the derivative of $\mathsf{P}(s)$ is zero. If $s = a$ is any other point on the root locus, we can write

$$\mathsf{P}(s) = (s - a)\mathsf{P}_1(s)$$

where $\mathsf{P}_1(a) \neq 0$. It thus follows that

$$\mathsf{P}'(s) = \mathsf{P}_1(s) + (s - a)\mathsf{P}_1'(s)$$

and hence

$$\mathsf{P}'(a) = \mathsf{P}_1(a) \neq 0$$

Thus the breakaway points on the real axis are distinguished from all other points on the real axis by the property that the derivative of $\mathsf{P}(s)$ goes to zero at the breakaway points. Note that $\mathsf{P}'(s)$ is a polynomial of degree $k - 1$ in s, and hence finding its roots poses a numerical problem only slightly less complicated than finding the roots of $\mathsf{P}(s)$ themselves. If the latter are to be found with the aid of a computer, it is hardly worth the trouble of finding the breakaway points by solving for the roots of

$$\mathsf{P}'(s) = 0 \tag{4.34}$$

The reader may wish to verify that $\mathsf{P}'(a) = 0$ at the breakaway points in the previous examples.

Nyquist diagram The earliest graphical method investigating the stability of linear systems was developed by H. Nyquist in 1932[1] and is based on the polar plot of the loop transmission transfer function. To understand Nyquist's method, recall that the condition for *instability* is that

$$1 + K\mathrm{G}(s) = 0 \qquad \text{or} \qquad \mathrm{G}(s) = -\frac{1}{K} \tag{4.35}$$

for some value of s in the *right* half of s plane. Conversely, if there does not exist a value s in the right half-plane for which $\mathrm{G}(s) = 1/K$, then we are assured that the system is stable.

For every point s in the right half-plane, there is a point $z = \mathrm{G}(s)$ in the z plane. (If $\mathrm{G}(s)$ is a rational function, then for each value of s there is *only* one value of $z = \mathrm{G}(s)$.) Thus the function $\mathrm{G}(s)$ "maps" the right half of the s plane into some region of the z plane. (Since $\mathrm{G}(s)$ is a continuous function and the right half-plane is a contiguous region, the map of the right half-plane by the function $\mathrm{G}(s)$ is also contiguous.) If the region of the z plane that is the map of the right half of the s plane under the function $\mathrm{G}(s)$ covers the point $-1/K$, the system is unstable; if the map does not cover the point $-1/K$, the system is stable. The two cases are depicted in Fig. 4.10.

The basic principle of the method of Nyquist is thus to determine whether or not the map of the right half of the s plane created by the function $\mathrm{G}(s)$ covers the point $-1/K$. Is it necessary to find $\mathrm{G}(s)$ for every s in the right half-plane? The answer, fortunately, is no. There is a theorem in complex variables which asserts that the map of the boundary of a region in the s plane is the boundary of the map of that region in the z plane. Thus, to find the map of the right half of the s plane under $\mathrm{G}(s)$ we need only find the map of the

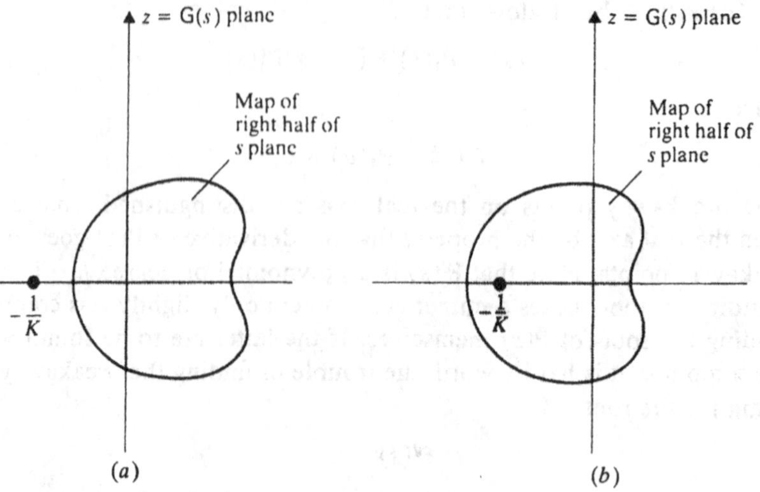

Figure 4.10 System with return difference $\mathrm{F}(s) = 1 + K\mathrm{G}(s)$ is unstable if map of right half of s plane covers the point $-1/K$. (a) Stable system; (b) Unstable system.

boundary of the right half of the s plane. The entire right half of the s plane is unbounded, of course. We get around that difficulty by finding the map of the large semicircular region bounded by the imaginary axis between $-j\Omega$ and $j\Omega$ in the semicircle of radius Ω in the right half-plane. Then we pass to the limit as $\Omega \to \infty$. If we are dealing with a proper rational function (i.e., the numerator degree is lower than the denominator degree) then as $\Omega \to \infty$, $G(s) \to 0$, so the whole semicircle maps into just one point: $G(s) = 0$.

To construct the map we start an excursion at the origin O and "walk" up the imaginary axis to the point A as shown in Fig. 4.11(a) at which $s = j\Omega$. The map of this portion of the imaginary axis may have the appearance of the curve $O' - A'$ in the $G(s)$ plane as shown in Fig. 4.11(b). Then we walk around the semicircle to the point B. The map of the semicircle $A - B$ is the arc $A' - B'$. Finally we return to the origin O upward along the imaginary axis along the path $B - O$ and obtain the corresponding arc $B' - O'$ in the $G(s)$ plane. The map of the entire right half-plane is obtained by letting $\Omega \to \infty$ which has the effect of shrinking the arc $A' - B'$ to a single point. Since we know that the semicircular arc maps into just one point, there is no need to bother with that arc. It is enough just to walk up the imaginary axis.

The map of the imaginary axis separates the map of the right half of the s plane from the map of the left half of the s plane. It is necessary of course to know which points on the $G(s)$ plane correspond to the points on the right half of the s plane, and which correspond to points on the left half of the s plane. We are aided in this process by the fact that the transformation $z = G(s)$ is "conformal": angles are locally preserved.[14] Thus if we take our excursion

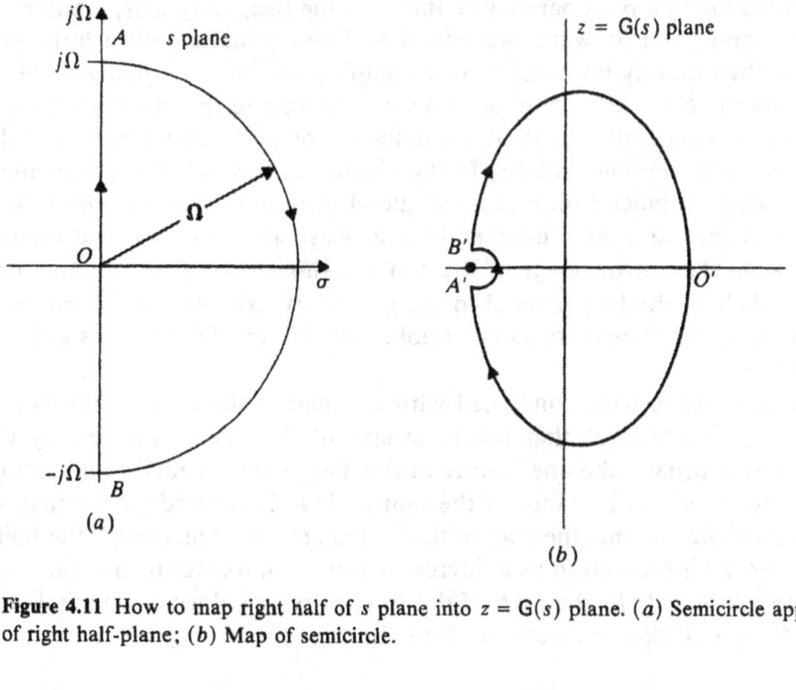

Figure 4.11 How to map right half of s plane into $z = G(s)$ plane. (a) Semicircle approximates all of right half-plane; (b) Map of semicircle.

along the imaginary axis with our right hand extended so that it lies over the right half-plane, the corresponding excursion over the map of the imaginary axis with the right hand outstretched is over the map of the right half-plane. This principle is sufficient to identify the map of the right half-plane in all cases, and is equivalent to Nyquist's "encirclement rule" which we shall give later on.

In Figs. 4.10 and 4.11 we drew the maps of the right half-plane (an infinite region) as a finite region, because the entire right half-plane outside a semicircular arc shrinks down to a single point. But what happens if $G(s)$ has poles on the imaginary axis? In that case, of course, the map of the region near a pole will result in very large values of z. Since our excursion along the imaginary axis is not permitted, we might consider an excursion along a line in the left half-plane parallel to and slightly left of the imaginary axis. This places the imaginary axis itself into the right half-plane—the region of instability which is where our previous classification of the region of instability would rightfully place it. On the other hand one might, with some justification, argue that a physical open-loop system is bound to have some damping present and hence that the open-loop poles are near but not exactly upon the imaginary axis. This means that an excursion up the imaginary axis is permitted, and that the poles encountered on the excursion are to our left. Each approach will result in a different Nyquist diagram. But there is no practical difference, because only the part of the Nyquist diagram that is remote from the open-loop poles is needed to assess the stability of a system. We adopt the approach of keeping the imaginary axis poles to our right. Thus, suppose for example, the loop transmission $G(s)$ has a pole at the origin and a pair of complex poles on the imaginary axis at $\omega = \omega_c$ (as well as poles and zeros elsewhere in the s plane). As we proceed along a path parallel to and near the imaginary axis, starting on the real axis and going upward, we find that the map starts with a large real number and then rapidly becomes a large complex number with phase angle of nearly 90°. As the excursion continues upward the map of the line continues to evolve in accordance with the total constellation of poles and zeros until the line brings us near the pole and $j\omega_c$. In the vicinity of this pole the phase angle goes from a large number B at a phase angle of +90° to a large number C at a phase angle of zero to a large number D at a phase angle of −90° and then to zero as $\omega \to \infty$. The mirror image of the contour shown in Fig. 4.12 is the map of the lower half of the line parallel to the imaginary axis. As the line in the s plane approaches the imaginary axis the points A', B', and D' move toward $\pm\infty$ as indicated.

Although we have been concerned with the map of the entire right half of the s plane, it is apparent that the boundary of the map produced by the imaginary axis is usually the one feature of the map that is needed to determine whether or not a system is stable. If the point $-1/K$ is covered by the map of the right half of the s plane, the map of the imaginary axis "encircles" the point $-1/K$ in a clockwise direction as ω increases from 0 to ∞. But if the map does not cover the point $-1/K$, the map of the imaginary axis does not encircle the point $-1/K$ in a clockwise direction. Thus, in most cases, only the map of the

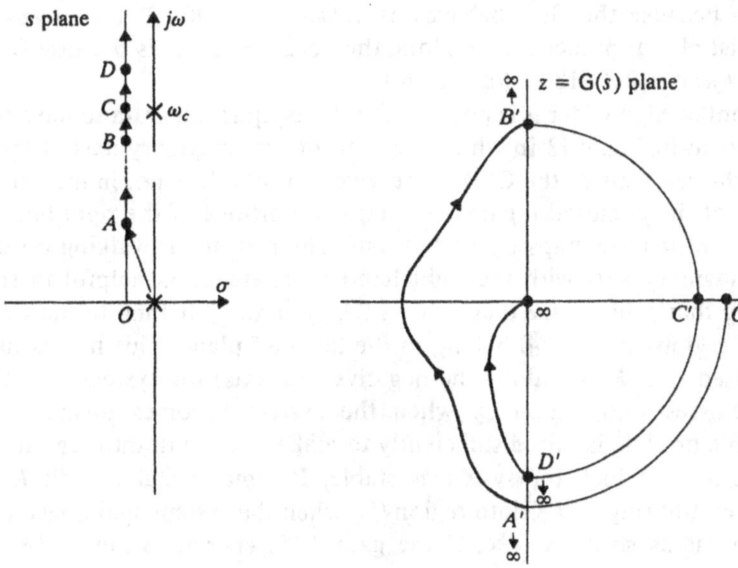

Figure 4.12 Nyquist diagram for transmission with poles on imaginary axis.

imaginary axis is drawn and this curve is called the Nyquist diagram. The customary stability criterion is thus familiarly stated as follows:

> **Nyquist stability criterion** A system having a return difference $1 + K\mathrm{G}(s)$ is stable if and only if the Nyquist diagram, i.e., the map of the imaginary axis, does not encircle the point $-1/K$ in the clockwise direction.

It must be noted that the encirclement test must be performed very carefully when the loop transmission $\mathrm{G}(s)$ itself has poles or zeros in the right half-plane, as discussed in various texts on complex variables and systems.[14] If $\mathrm{G}(s)$ has poles and/or zeros in the right half-plane it is safer to map the entire right half-plane by $\mathrm{G}(s)$ and check whether or not it covers the point $-1/K$.

The behavior of the Nyquist plot as $\omega \to 0$ depends on the order of the pole at the origin. If there is no pole at the origin $\mathrm{G}(0)$ is finite and is a real number. It is positive unless, for some perverse reason, the dc gain $\mathrm{G}(0)$ is defined to be negative. If there is a simple pole at the origin then as $\omega \to 0$, $\mathrm{G}(j\omega) \to C/j\omega$ which tends to infinity in magnitude and $-90°$ in phase. Similarly if there is a double pole at the origin then, as $\omega \to 0$, $\mathrm{G}(j\omega) \to C/(j\omega)^2 = -C/\omega^2$ which tends to infinity in magnitude and $-180°$ in phase. And so forth. The order of the pole at the origin is known as the system "type" and, as will be discussed in Sec. 4.7, governs the ability of the system to track an input in the form of a polynomial time function without steady state error.

As $\omega \to \infty$, the behavior of the Nyquist plot depends on the excess of poles over zeros. If the excess is one, the plot approaches the origin along the negative imaginary axis because the G(s) behaves as $C/j\omega \to 0\angle{-90°}$. If the excess is two, the Nyquist plot approaches zero along the negative real axis because G(s) behaves as $C/(j\omega)^2 \to 0\angle{-180°}$. And so forth.

For a system of high order it is possible for the Nyquist diagram to have the appearance shown in Fig. 4.13 in which the map of the imaginary axis of the s plane crosses the real axis of the G(s) plane several times. It is not immediately obvious which of the enclosed regions are maps of portions of the right half of the s plane and which are maps of the left half. The rule about walking up the map of the imaginary axis with the right hand outstretched is helpful in this case. Following that rule we see that regions ②, ③, and ⑤ belong to the right half-plane but regions ① and ④ belong to the left half-plane. This means that as K is increased ($-1/K \to 0$ along the negative real axis) the system is stable until $-1/K$ crosses into region ③ when the system becomes unstable. It remains unstable until K is raised sufficiently to make $-1/K$ fall into region ④, which is a region in which the system is stable. It remains stable until K is further increased to bring $-1/K$ into region ⑤, when the system again becomes unstable and remains so as $K \to \infty$. If the gain K is chosen to put $-1/K$ in

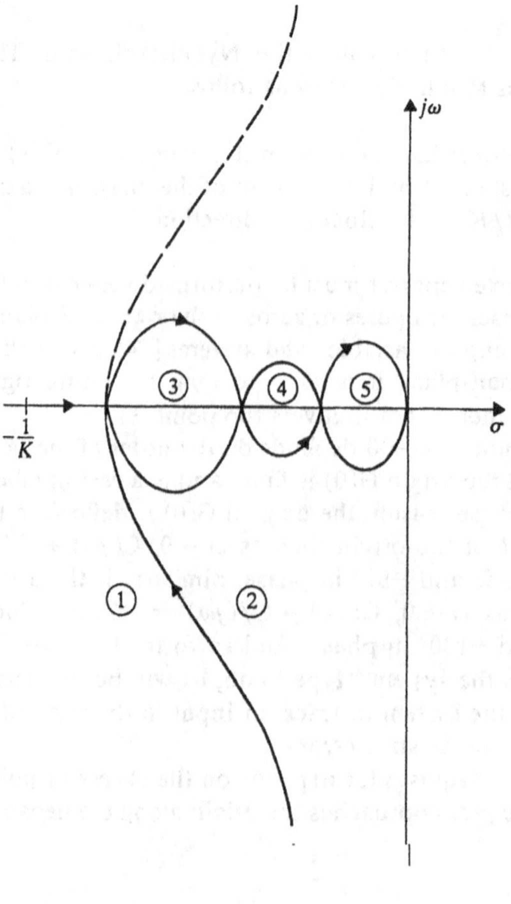

Figure 4.13 Nyquist diagram of a conditionally stable system. (System is stable if $-1/K$ is in regions 1 or 4.)

region ④ the system is said to be *conditionally stable.* A conditionally stable system is generally undesirable because of the danger that a reduction of gain as well as an increase can make the system unstable. Sometimes there is no way to avoid conditionally stable systems, but it is often possible to design a compensator to shape the Nyquist diagram to avoid having conditional stability. Methods that can be used to accomplish the required shaping are discussed in textbooks on frequency-domain methods of control system design.[11, 12]

The Nyquist diagram is used not only to assess the stability of a system (by determining whether the map of $G(s)$ covers the point $-1/K$, but also to investigate system "robustness" which is a measure of how much the system can change without becoming unstable. The further away the point $-1/K$ is from the map of the right half-plane, the more the system transmission (i.e., the map of the right half-plane under $G(s)$) can change without endangering stability. Hence it is desirable that this distance be substantial. A quantitative measure of this distance is the *gain margin* as discussed in Sec. 4.9 which deals with robustness in general.

Bode plots The Nyquist diagram can be regarded as a polar plot of the magnitude and phase of $G(s)$ when $s = j\omega$, that is, a polar plot of the magnitude and phase of $G(j\omega)$ with the frequency ω serving as a parameter. The same information can be presented in a pair of plots: one of the magnitude and the other of the phase of $G(j\omega)$, each as a function of frequency ω. These are known as the Bode plots of $G(s)$. In particular let

$$G(j\omega) = |G(j\omega)|\, e^{j\theta_G(\omega)}$$

where $|G(j\omega)|$ and $\theta_G(\omega)$ are known as the *magnitude* and *phase* functions of the loop transmission $G(s)$. Instead of plotting $|G(j\omega)|$ it is customary to plot

$$D(\omega) = 20 \log_{10} |G(j\omega)|$$

Regardless of the units of $G(s)$, the units of $D(\omega)$ are invariably *decibels* (abbreviated dB). Since there is no physical significance to the logarithm of a quantity that is not dimensionless, i.e., the ratio of two physical variables of the same type (e.g., voltage out/voltage in, etc.) it is not strictly proper to use the decibel notation unless $G(s)$ is a dimensionless ratio, i.e., unless the input and the output are the same physical type. But this improper usage is universally condoned and accepted.

The plot of $D(\omega)$ vs. ω is known as the Bode amplitude plot and the plot of $\theta_G(\omega)$ vs. ω is known as the Bode phase plot.

The Bode plot for a transfer function that has only real poles and zeros is particularly easy to construct graphically. In particular, consider a system having a loop transfer function

$$G(s) = G_0 \frac{\left(1 + \dfrac{s}{z_1}\right) \cdots \left(1 + \dfrac{s}{z_l}\right)}{\left(1 + \dfrac{s}{p_1}\right) \cdots \left(1 + \dfrac{s}{p_k}\right)} \tag{4.36}$$

This form of G(s) is especially convenient for Bode plots because each factor in the numerator and the denominator is unity at $s = 0$ and hence the dc gain, G(0) $= G_0$ is explicitly exhibited. When $s = j\omega$

$$\mathrm{G}(j\omega) = G_0 \frac{\left(1+\frac{j\omega}{z_1}\right)\cdots\left(1+\frac{j\omega}{z_l}\right)}{\left(1+\frac{j\omega}{p_1}\right)\cdots\left(1+\frac{j\omega}{p_k}\right)}$$

and hence

$$|\mathrm{G}(j\omega)| = |G_0| \frac{\left[1+\left(\frac{\omega}{z_1}\right)^2\right]^{1/2}\cdots\left[1+\left(\frac{\omega}{z_l}\right)^2\right]^{1/2}}{\left[1+\left(\frac{\omega}{p_1}\right)^2\right]^{1/2}\cdots\left[1+\left(\frac{\omega}{p_k}\right)^2\right]^{1/2}} \tag{4.37}$$

Thus

$$D(\omega) = 20\log|G_0| + 10\log\left[1+\left(\frac{\omega}{z_1}\right)^2\right] + \cdots + 10\log\left[1+\left(\frac{\omega}{z_l}\right)^2\right]$$
$$-10\log\left[1+\left(\frac{\omega}{p_1}\right)^2\right] - \cdots - 10\log\left[1+\left(\frac{\omega}{p_k}\right)^2\right] \tag{4.38}$$

and (for $G_0 > 0$)

$$\theta_G(\omega) = \tan^{-1}\left(\frac{\omega}{z_1}\right) + \cdots + \tan^{-1}\left(\frac{\omega}{z_l}\right) - \tan^{-1}\left(\frac{\omega}{p_1}\right) - \cdots - \tan^{-1}\left(\frac{\omega}{p_k}\right) \tag{4.39}$$

(If the dc gain G_0 is negative, a 180° phase shift must be added to (4.39).) These results may be interpreted as follows: (Fig. 4.14.)

The log-magnitude plot $D(\omega)$ is the sum of the log-magnitude plot of each contributing factor and the phase plot is the sum of the phase plots of each contributing factor.

With increasing frequency, the contribution of a zero is an increase in both the log-magnitude and the phase; the contribution of a pole is a decrease† in both log-magnitude and phase.

The contribution of a typical zero or pole is shown in Fig. 4.14. It is seen that at the frequency $\omega = z_i$ or $\omega = p_i$ the magnitude is exactly twice its value at dc ($\omega = 0$) and the phase shift is exactly 45°. As the frequency is further increased

† The phase relation is valid only when the contributing pole or zero is in the left half plane, that is, p_i or z_i is positive. If the zero or pole is in the *right* half plane, then z_i or p_i is negative, and the phase contribution is opposite. Bode[2] has called such poles or zeros *nonminimum phase.* Nonminimum phase *poles* are indicative of an unstable open loop system, of course. The effect of open loop zeros is more subtle, however, and is discussed in greater detail in Note 4.7.

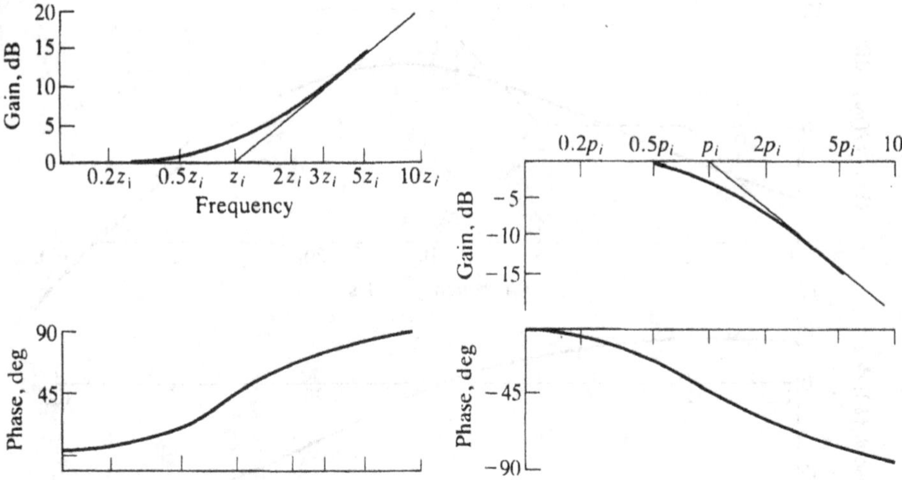

Figure 4.14 Bode plots for a zero and for a pole.

$$(a)\ \mathsf{G}(s) = 1 + \frac{s}{z_i} \qquad (b)\ \mathsf{G}(s) = \frac{1}{1 + (s/p_i)}$$

the contributions to log-magnitude plots are

$$D_i(\omega) \rightarrow 10 \log \left(\frac{\omega}{z_i}\right)^2 = 20 \log \left(\frac{\omega}{z_i}\right) \qquad \text{for a zero}$$

$$D_i(\omega) \rightarrow -10 \log \left(\frac{\omega}{p_i}\right)^2 = -20 \log \left(\frac{\omega}{p_i}\right) \qquad \text{for a pole}$$

Thus, if a logarithmic frequency scale is used D_i is asymptotic to a line having a slope of 20 (dB) for each tenfold increase in frequency, that is, "20 dB per decade." The slope is positive for a zero, and negative for a pole. The asymptote intersects the logarithmic frequency axis at $\omega = z_i$ or $\omega = p_i$. At these frequencies, known as the "corner" frequencies, the exact gain is $\pm 10 \log 2 = 3.010$ dB, so these are also known as the "3 dB" frequencies.

The phase contribution from each factor tends to ±90°. (Positive for a zero; negative for a pole.)

The log-magnitude and phase curves for the overall system are obtained by simply adding the curves of each contributing factor. Thus, for example, the log-magnitude and phase curve for

$$\mathsf{G}(s) = \frac{(1+s)\left(1+\frac{s}{5}\right)}{\left(1+\frac{s}{2}\right)\left(1+\frac{s}{10}\right)\left(1+\frac{s}{20}\right)}$$

has the appearance shown in Fig. 4.15. The maximum deviation from the straight line approximation is 3 dB.

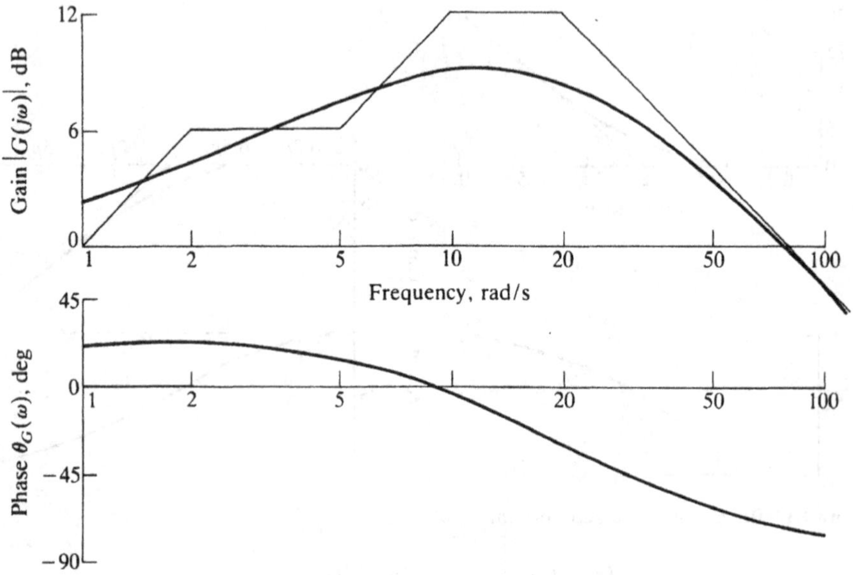

Figure 4.15 Bode plot for

$$G(s) = \frac{(s+1)(s/5+1)}{(s/2+1)(s/10+1)(s/20+1)}$$

A pole or a zero at the origin is treated slightly differently, because the log magnitude is not finite as $\omega \to 0$. A zero at the origin means that $D(\omega) \to -\infty$; a pole at the origin means that $D(\omega) \to +\infty$. These are the plots for a corner frequency of 0; in other words a zero at the origin contributes an increasing log-magnitude line with a constant slope of +20 dB/decade; a pole at the origin contributes a decreasing log-magnitude line with a slope of −20 dB/decade. Each passes through 0 dB at $\omega = 1$. The phase angle due to a pole is a constant of −90 degrees and the phase angle contribution of a zero is a constant +90 degrees.

The Bode plots for a transfer function G(s) that has complex poles or zeros is more complicated, because the straight-line approximation as illustrated in Fig. 4.13 is not applicable since a transfer function with a complex-conjugate pair of poles will include a factor of the form

$$G_i(s) = \frac{1}{1 + 2\zeta(s/\omega_0) + (s/\omega_0)^2} \tag{4.40}$$

The log-magnitude and phase functions corresponding to (4.40) are

$$\begin{aligned} D_i(\omega) &= -20 \log \left[1 + (4\zeta^2 - 2)\left(\frac{\omega}{\omega_0}\right)^2 + \left(\frac{\omega}{\omega_0}\right)^4 \right]^{1/2} \\ \theta_i(\omega) &= \tan^{-1}\left(\frac{-(\omega/\omega_0)\zeta}{1 - (\omega/\omega_0)^2}\right) \end{aligned} \tag{4.41}$$

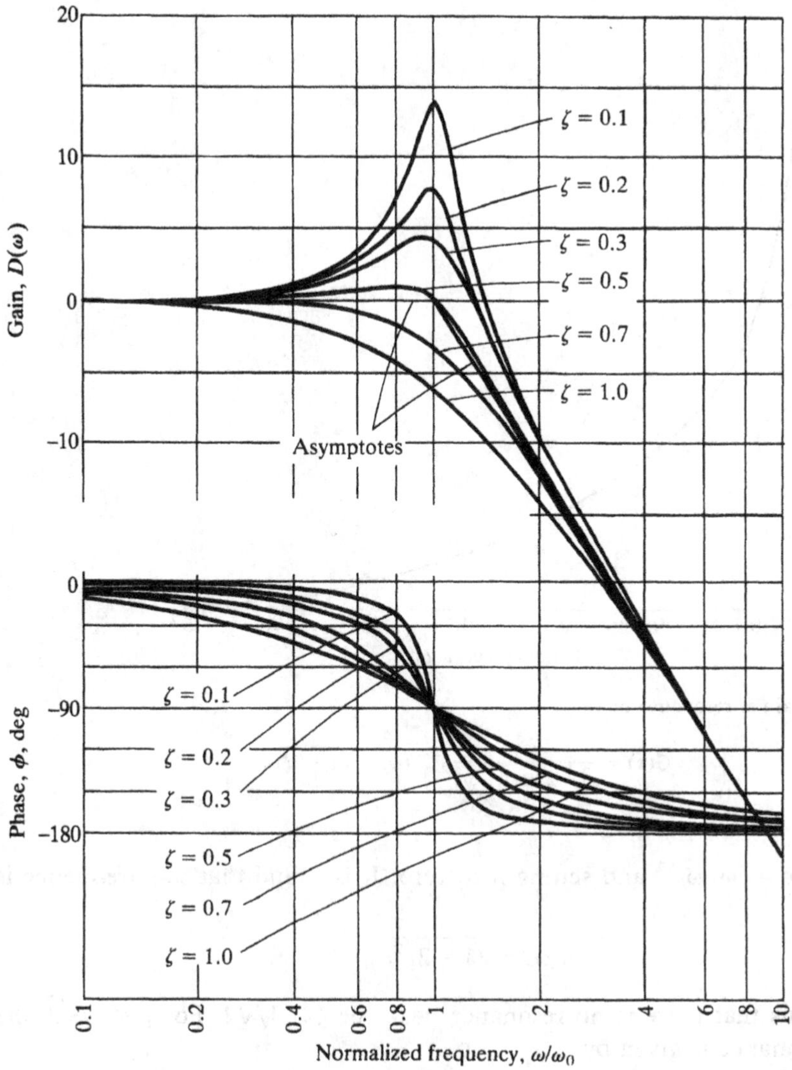

Figure 4.16 Bode plots for second-order system

$$G(s) = \frac{\omega_0^2}{s^2 + 2\zeta\omega_0 s + \omega_0^2}$$

The log-magnitude and phase curves vs. normalized frequency ω/ω_0 are shown in Fig. 4.16 for various damping factors ranging from $\zeta = 0.1$ (lightly damped) to $\zeta = 1.0$. It is seen that as the damping becomes very small, the log-magnitude becomes very large in the vicinity of the natural frequency $\omega \approx \omega_0$, and the phase shift rapidly changes from angles close to zero to angles close to 180°, crossing through exactly 90° at $\omega/\omega_0 = 1$.

The frequency ω_r at which the peak (often called a *resonance peak*) in the log-magnitude plot occurs can be found by taking the derivative of (4.41) with

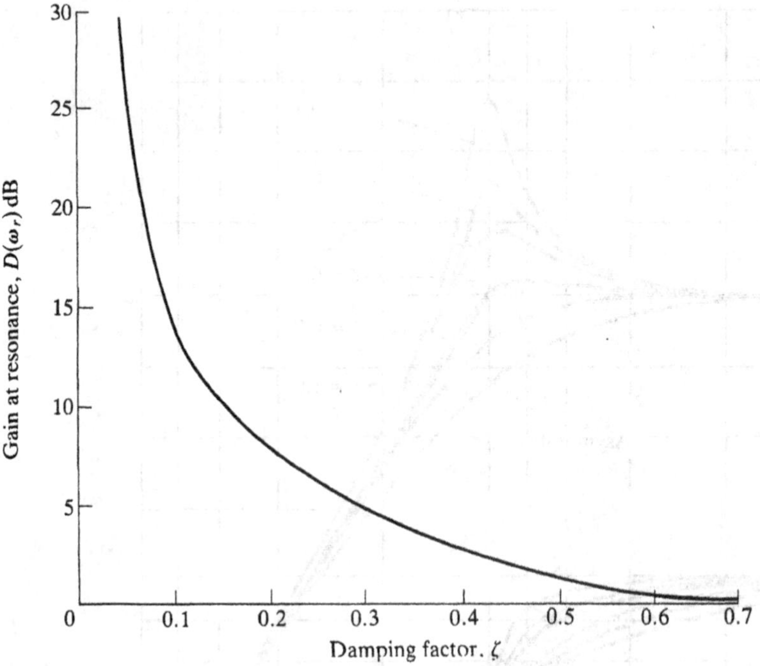

Figure 4.17 Gain at resonance of

$$G(s) = \frac{\omega_0^2}{s^2 + 2\zeta\omega_0 s + \omega_0^2} \qquad (\omega_r = \sqrt{1-2\zeta}\,\omega_0)$$

respect to $x = (\omega/\omega_0)^2$ and setting it to zero. It is found that this frequency is given

$$\omega_r = \sqrt{1 - 2\zeta^2}\,\omega_0$$

which means that there is no resonance peak for $\zeta > 1/\sqrt{2}$. For $\zeta < 1/\sqrt{2}$, the gain at resonance is given by

$$D_i(\omega_r) = -20 \log [4\zeta^2(1 - \zeta^2)] \qquad \text{(dB)} \tag{4.42}$$

a graph of which is shown in Fig. 4.17.

Example 4E Hydraulically actuated gun turret (continued) In Example 4B we found the transfer function between the input u and the output angle θ of the hydraulically actuated gun turret. Using the numerical data given in Example 2D for azimuth control:

$$K_v = 94.3 \qquad L_v = 1.0 \qquad J = 7900 \qquad K_m = 8.46 \times 10^6$$

$$\omega_m = 45.9 \quad \text{and} \quad K_{\Delta p} = 6.33 \times 10^{-6}$$

we find numerically that

$$H(s) = \frac{100\,980}{s(s^3 + 140.2s^2 + 10\,449s + 100\,980)} \tag{4E.1}$$

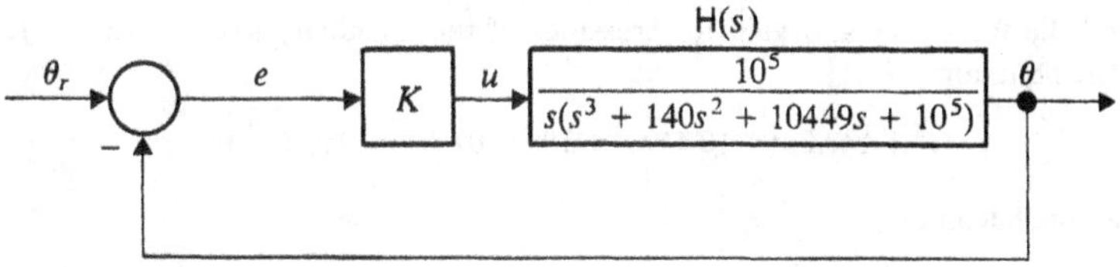

Figure 4.18 Closed-loop control of gun turret.

The root-locus equation for the closed-loop process (see Fig. 4.18) is

$$1 + \frac{100\,980K}{s(s^3 + 140.2s^2 + 10\,449s + 100\,980)} \tag{4E.2}$$

The open-loop poles of the process are at $s = 0$ and at the roots of the cubic factor $s^3 + 140.2s^2 + 10\,449s + 100\,980$. The latter are found numerically to be located at

$$s = -11.2$$

$$s = -64.5 \pm j69.6$$

Since there are four poles and no (finite) zeros of the transfer function, the root loci all go to ∞ parallel to lines at $\pm 45°$ and $\pm 135°$ angles from the real axis.

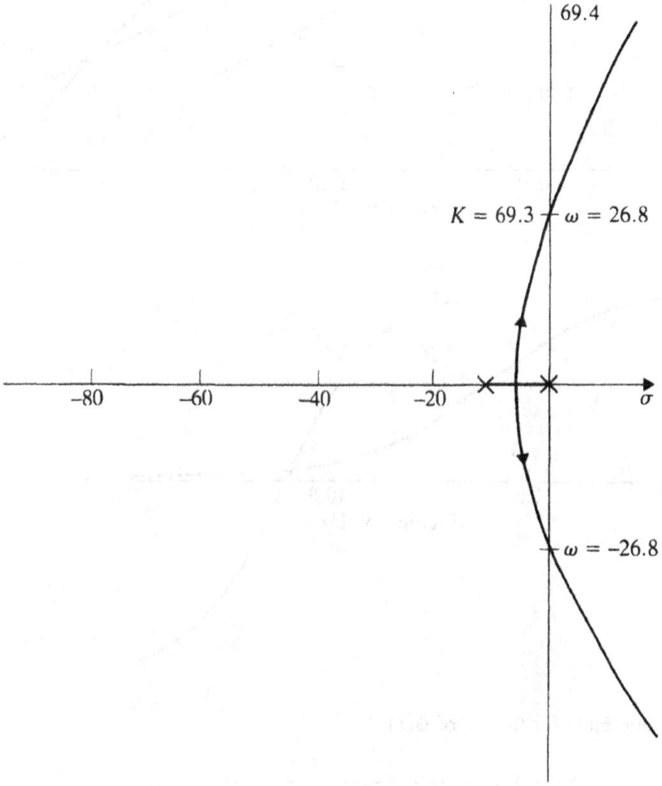

Figure 4.19 Root-locus plot for feedback control of hydraulically actuated gun turret.

To find the frequency and gain for crossings of the imaginary axis we set $s = j\omega$ in the characteristic equation

$$s^4 + 140.2s^3 + 10\,449s^2 + 100\,980s + 100\,980K = 0 \tag{4E.3}$$

with $s = j\omega$ this becomes

$$\omega^4 - j140.2\omega^3 - 10\,449\omega^2 + j100\,980\omega + 100\,980K = 0$$

or, on equating the real and the imaginary parts to zero,

$$\omega^4 - 10\,449\omega^2 + 100\,980K = 0 \tag{4E.4}$$

$$-140.2\omega^3 + 100\,980\omega = 0 \tag{4E.5}$$

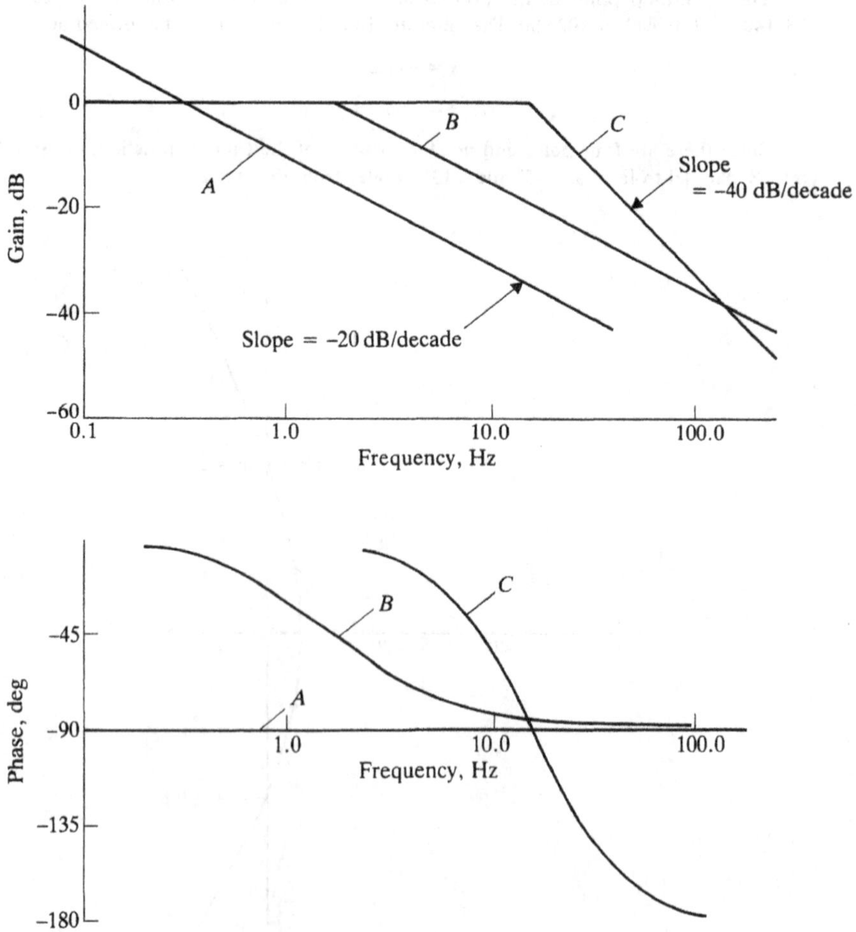

Figure 4.20(a) Bode plots for factors of $G(s)$.

$$G(s) = \frac{1}{s\left(1 + \dfrac{s}{11.2}\right)\left[1 + 2(0.68)\dfrac{s}{95} + \left(\dfrac{s}{95}\right)^2\right]}$$

The second equation gives $\omega = 0$, the starting point of the locus, and

$$\omega = \sqrt{100\,980/140.2} = 26.8$$

and this value of ω when substituted into (4E.5) gives $K = 69.4$. The same value of K could have been obtained by use of the Routh or the Hurwitz algorithm. (See Prob. 4.9.)

The root locus plot for this system is shown in Fig. 4.19.

The transfer function, in factored form, is

$$G(s) = \frac{100\,980}{s(s + 11.2)(s^2 + 129s + 9016)}$$

$$= \frac{1}{s(1 + s/11.2)[1 + 2(0.68)(s/95) + (s/95)^2]}$$

The Bode plots for each factor in G(s) are shown in Fig. 4.20(a); the composite is shown in Fig. 4.20(b).

The Nyquist plot corresponding to G(s) is shown in Fig. 4.21.

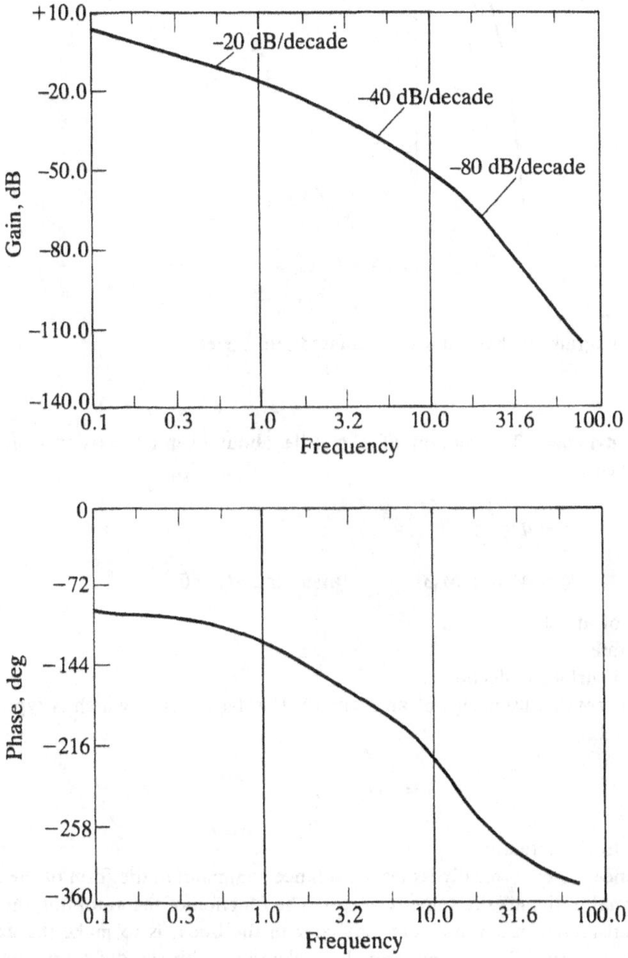

Figure 4.20(b) Bode plots for gun turret.

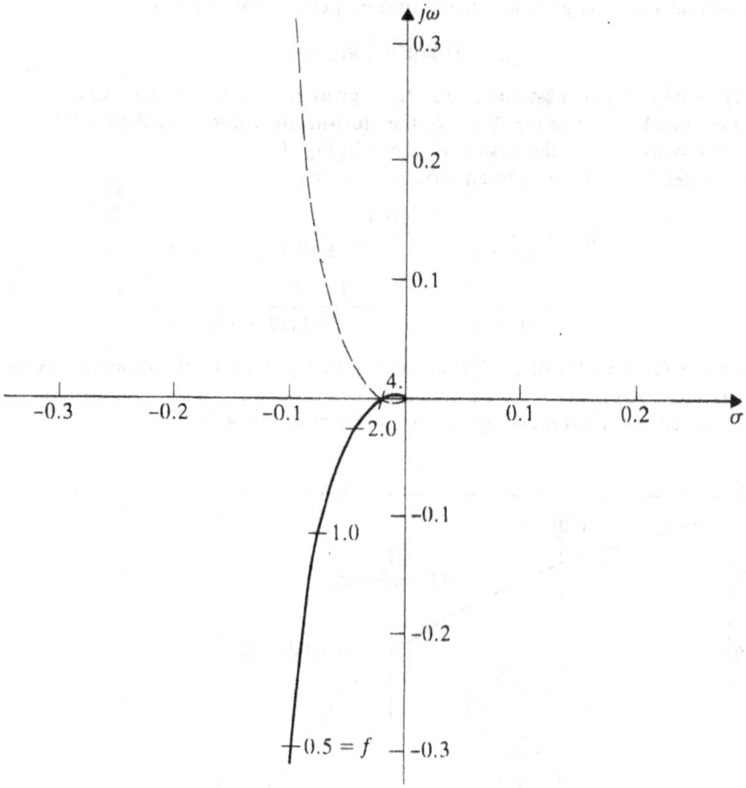

Figure 4.21 Nyquist diagram for hydraulically actuated gun turret.

Example 4F Missile dynamics The motion of a missile about its pitch axis was shown in Example 3F to be given by

$$\dot{\alpha} = q + \frac{Z_\alpha}{V}\alpha + \frac{Z_\delta}{V}\delta$$

$$\dot{q} = M_\alpha \alpha + M_\delta \delta \qquad \text{(assuming } M_q \approx 0\text{)} \tag{4F.1}$$

where α is the angle of attack
q is the pitch rate
δ is the control surface deflection

The control surface is rotated by means of an actuator, the dynamics of which is typical of a first-order lag: (Fig. 4.22)

$$\dot{\delta} = \frac{1}{\tau}(u - \delta) \tag{4F.2}$$

where u is the input to the actuator.

A missile guidance system typically issues a guidance command in the form of the desired acceleration a_{Nc} normal to the missile velocity vector. The function of the autopilot, the design of which shall be considered in several examples later in the book, is to make the achieved normal acceleration a_N "track" the commanded acceleration with good fidelity. It is thus appropriate to deal with the error e between the commanded and the achieved normal

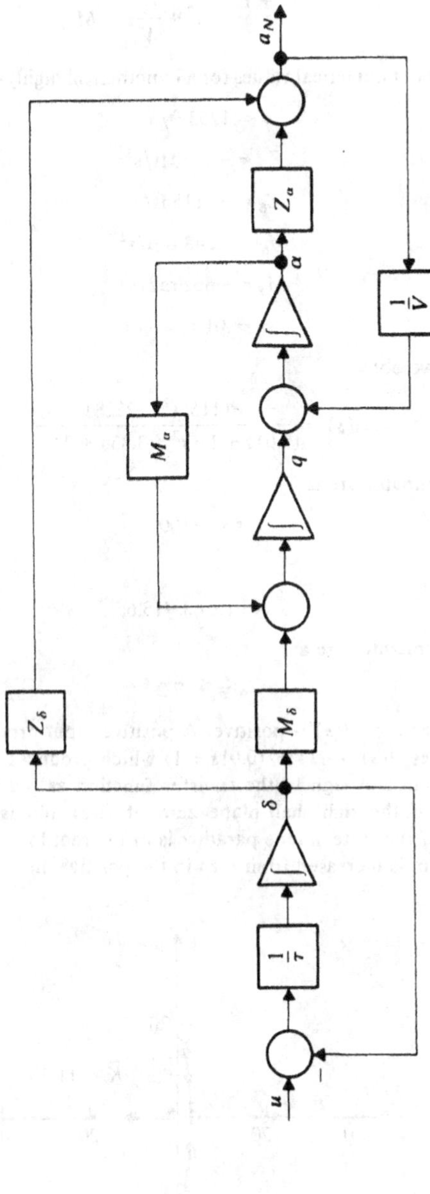

Figure 4.22 Missile attitude dynamics with normal acceleration as output.

acceleration. The latter is given by

$$a_N = Z_\alpha \alpha + Z_\delta \delta \tag{4F.3}$$

The transfer function from u to a_N is determined to be:

$$\mathrm{H}(s) = \frac{1}{\tau s + 1} \frac{Z_\delta s^2 + Z_\alpha M_\delta - Z_\delta M_\alpha}{s^2 + \dfrac{Z_\alpha}{V} s - M_\alpha} \tag{4F.4}$$

A representative set of numerical values for a hypothetical highly-maneuverable missile are:

$$V = 1253 \text{ ft/s}$$
$$Z_\alpha = -4170 \text{ ft/s}^2$$
$$Z_\delta = -1115 \text{ ft/s}^2$$
$$M_\alpha = -248 \text{ rad/s}^2$$
$$M_\delta = -662 \text{ rad/s}$$
$$\tau = .01 \text{ s}$$

For these values we obtain

$$\mathrm{H}(s) = \frac{-1115(s^2 - 2228)}{(0.01s + 1)(s^2 + 3.33s + 248)} \tag{4F.5}$$

The zeros of the denominator are at

$$s = -100$$

and at

$$s = -1.67 \pm j15.65$$

and the zeros of the numerator are at

$$s = \pm 47.2$$

Note that the dc gain of $\mathrm{H}(s)$ is positive: A positive input produces a positive response. But for high frequencies $\mathrm{H}(s) \to -1115/(0.01s + 1)$ which produces a negative response for a positive input. The change of sign in the transfer function as the frequency is increased is another consequence of the right half-plane zero of $\mathrm{H}(s)$ and is the source of apparent paradoxical behavior of the system. One paradox is in the root locus, shown in Fig. 4.23. It is observed that as the gain is increased from zero in the positive direction the one branch of the

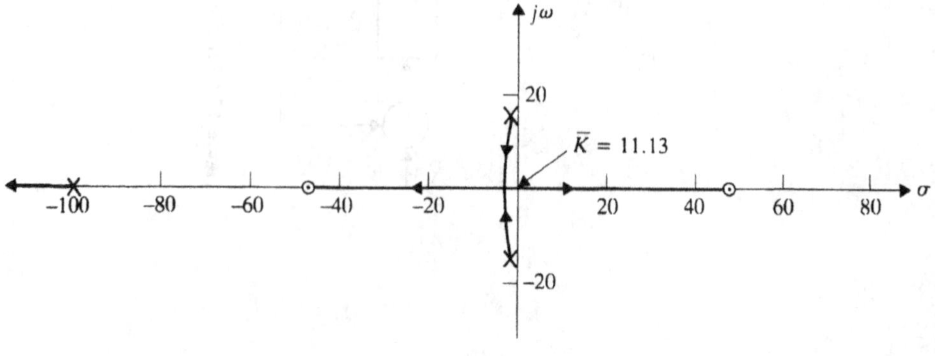

Figure 4.23 Root loci for missile dynamics. $\bar{K}$ increasing.

root locus crosses the imaginary axis at $s = 0$ and then continues to the real root as $s = 47.1$. This behavior is clear from the characteristic equation:

$$s^3 + (100.33 + \bar{K})s^2 + 581s + 24\,800 - 2228\bar{K} = 0$$

where

$$\bar{K} = 111\,500K$$

The coefficient of s^0 vanishes at

$$\bar{K} = \frac{24\,800}{2228} = 11.13$$

and hence there is a pole at the origin for this value of $\bar{K}$.

The graphical methods of Nyquist and Bode have an advantage over the algebraic Routh-Hurwitz methods: They are not restricted to rational transfer functions and thus not limited to systems characterized by ordinary differential equations. Thus they are applicable to systems characterized by partial-differential equations and pure delays. The following example provides a frequency-domain explanation for the instability exhibited (as found in Chap. 1) by a system with a delayed output.

Example 4G Pure delay In Chap. 1 we considered a system whose output $y(t)$ is a faithful, but delayed, version of the input $u(t)$

$$y(t) = u(t - T) \tag{4G.1}$$

The Laplace transform of the delayed input is

$$\mathrm{y}(s) = \int_0^\infty e^{-st}u(t - T)\,dt = \int_{-T}^\infty e^{-s(\tau+T)}u(\tau)\,d\tau \tag{4G.2}$$

On the assumption that $u(t)$ is zero for $t < 0$, (4G.2) becomes

$$\mathrm{y}(s) = e^{-sT}\int_0^\infty e^{-s\tau}u(\tau)\,d\tau = e^{-sT}\mathrm{u}(s)$$

Thus, the transfer function of a pure delay is

$$\mathrm{G}(s) = e^{-sT}$$

with $s = j\omega$

$$\mathrm{G}(j\omega) = e^{-j\omega T}$$

Thus

$$|\mathrm{G}(j\omega)| = 1$$

and

$$\theta_G(\omega) = -\omega T$$

The Nyquist diagram is thus a circle centered at the origin as shown in Fig. 4.24(a), and the closed-loop system, having the return difference

$$\mathrm{T}(s) = 1 + K\mathrm{G}(s) = 1 + Ke^{-sT} \tag{4G.3}$$

is unstable for $K > 1$, as was found in Chap. 1. The Bode diagram has a constant amplitude of 1 (0 dB) and a linearly decreasing phase (which does not look linear on a logarithmic frequency axis as shown in Fig. 4.24 for $T = 0.01$ s).

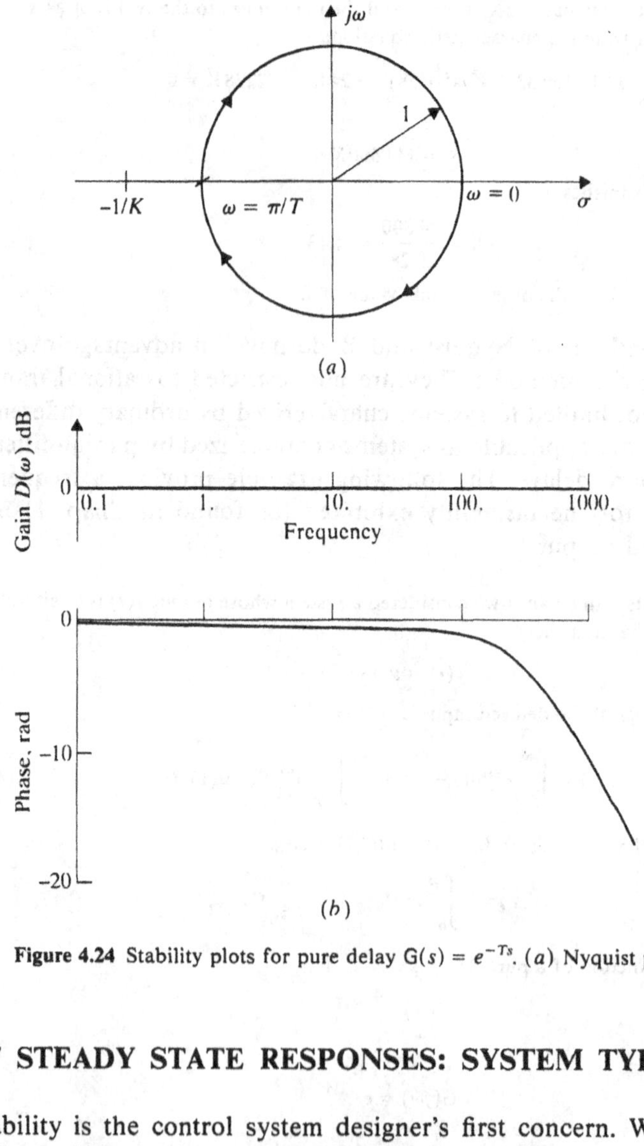

Figure 4.24 Stability plots for pure delay $G(s) = e^{-Ts}$. (*a*) Nyquist plot; (*b*) Bode plot.

4.7 STEADY STATE RESPONSES: SYSTEM TYPE

Stability is the control system designer's first concern. With stability assured, interest shifts to the nature of the response of the system to various types of reference inputs. (For the present we consider only single-input, single-output processes. The general multiple-input, multiple-output case is treated in Sec. 4.10.)

A system designed to follow a reference input, rather than merely to return to equilibrium, is generally known as a "tracking" system, and has the configuration shown in Fig. 4.25. The output of the system is y and the input to the open-loop plant is u. The difference between the desired reference input is called the *system error*

$$e = y_r - y$$

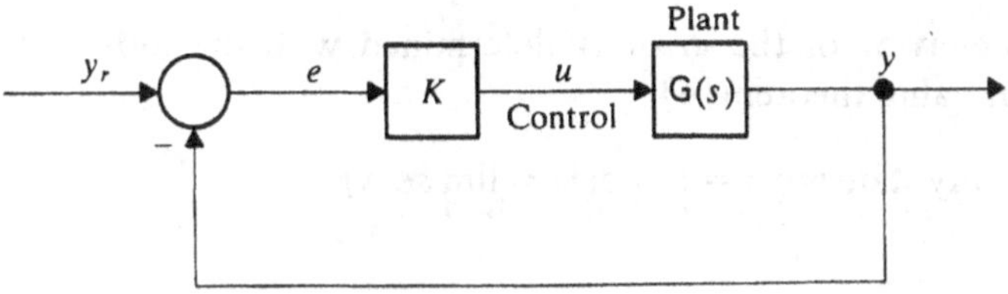

Figure 4.25 Error driven feedback control system.

In the simplest type of control system, the system error is multiplied by a gain K to produce the control input u. Since the input to the plant is proportional to the error, if the error becomes large it will produce a large input which, because of negative feedback, will tend to drive the error to zero.

To analyze the behavior of the system quantitatively, consider the transfer functions from the input to the error e and to the output y. Using the block-diagram algebra of Sec. 4.3, or any other convenient method, one can easily determine these transfer functions

$$\mathrm{H}_E(s) = \frac{1}{1 + K\mathrm{G}(s)} = \frac{\mathrm{e}(s)}{\mathrm{y}_r(s)} \tag{4.43}$$

$$\mathrm{H}_C(s) = \frac{K\mathrm{G}(s)}{1 + K\mathrm{G}(s)} = \frac{\mathrm{y}(s)}{\mathrm{y}_r(s)} \tag{4.44}$$

Note that the return difference $1 + K\mathrm{G}(s)$ is in the denominator of the transfer function (4.43) between the reference input and the error. To make the error small, we would like the return difference to be large. With the plant transfer function $\mathrm{G}(s)$ given, the return difference $1 + K\mathrm{G}(s)$ can be increased by increasing K. For reasons of stability, however, we usually can't make K arbitrarily large. Therefore it is not possible to design a system that can track *every* reference input with arbitrarily small errors. As a matter of fact, we might not be very happy with a system that does so. The reference input typically has rapid changes and noise which often one might not want to track with perfect fidelity.

One measure of the performance of a system is its steady state behavior when the reference input to the system is a polynomial time function:

$$y_r(t) = p_m(t) = C_1 + C_2 t + \cdots + \frac{C_{m+1}}{m!} t^m \tag{4.45}$$

It is useful to formalize this measure of performance with the following:

Definition A system is of "type m" if it can track a polynomial input of degree m with finite but nonzero steady state error.

We shall shortly discover that a system of type m can track a polynomial of degree $m - 1$ (or less) with zero steady state error, but that the error in tracking a polynomial reference input of degree $m + 1$ (or greater) becomes infinite.

The steady state behavior of the error is determined with the aid of the Laplace transform final-value theorem [4]:

$$\text{Steady state error} = \lim_{t\to\infty} e(t) = \lim_{s\to 0} s\mathsf{e}(s)$$

where, by (4.43)

$$\mathsf{e}(s) = \frac{1}{1 + K\mathsf{G}(s)}\mathsf{y}_r(s) \tag{4.46}$$

When the input $y_r(t)$ is the polynomial time function $p_m(t)$, then its Laplace transform

$$\begin{aligned}\mathsf{y}_r(s) = \mathsf{p}_m(s) &= \frac{C_1}{s} + \frac{C_2}{s^2} + \cdots + \frac{C_{m+1}}{s^{m+1}} \qquad (C_{m+1} \neq 0)\\ &= \frac{C_1 s^m + C_2 s^{m-1} + \cdots + C_{m+1}}{s^{m+1}}\end{aligned} \tag{4.47}$$

Thus, from (4.46),

$$s\mathsf{e}(s) = \frac{1}{1 + K\mathsf{G}(s)} \frac{C_1 s^m + \cdots + C_{m+1}}{s^m} \tag{4.48}$$

The limit as $s \to 0$ of $s\mathsf{e}(s)$ is infinite if $\mathsf{G}(0)$ is finite, because of the presence of the factor s^m in the denominator of $s\mathsf{e}(s)$ given by (4.48). The only way that, as $s \to 0$, $\lim s\mathsf{e}(s)$ can be finite is if $\mathsf{G}(s)$ has a pole of the proper order at $s = 0$, that is, if $\mathsf{G}(s)$ is of the form

$$\mathsf{G}(s) = \frac{\mathsf{N}(s)}{s^p \mathsf{D}(s)} \tag{4.49}$$

where neither $\mathsf{N}(s)$ nor $\mathsf{D}(s)$ have zeros at $s = 0$. When $\mathsf{G}(s)$ is of the form (4.49) then

$$\begin{aligned}s\mathsf{e}(s) &= \frac{1}{1 + K\dfrac{\mathsf{N}(s)}{s^p \mathsf{D}(s)}} \frac{C_1 s^m + \cdots + C_{m+1}}{s^m}\\ &= \frac{s^{p-m}\mathsf{D}(s)}{s^p \mathsf{D}(s) + K\mathsf{N}(s)}(C_1 s^m + \cdots + C_{m+1})\end{aligned} \tag{4.50}$$

From (4.50) we infer the following:

$$\begin{aligned}&\text{If } p > m, \quad \lim_{t\to\infty} e(t) = \lim_{s\to 0} s\mathsf{e}(s) = 0\\ &\text{If } p = m, \quad \lim_{t\to\infty} e(t) = \lim_{s\to 0} s\mathsf{e}(s) \text{ is finite but nonzero}\\ &\text{If } p < m, \quad \lim_{t\to\infty} e(t) = \lim_{s\to 0} s\mathsf{e}(s) \text{ is infinite}\end{aligned} \tag{4.51}$$

We thus conclude that:

The system type is equal to the order of the pole of $G(s)$ at $s = 0$.

Since a pole at the origin represents a perfect integrator, the system type is often defined as the number of cascaded integrators in the system.

It should be noted that the presence of a single integrator in the open-loop plant implies that it is stable, but not asymptotically stable, and that more than one integrator means that the open-loop plant is unstable. Thus we see that a closed-loop system, having the ability to track a polynomial input, cannot result when the open-loop plant is asymptotically stable! This should come as no surprise. The closed-loop system, after all, is error-driven, and we are insisting that the steady-state error go to zero. This means that the control input u also becomes zero in the steady state. But, at the same time we are demanding that the output be nonzero! To sustain a nonzero output with a zero input is not one of the properties of an asymptotically stable system.

It is emphasized that the system type is determined by the order of the pole at the origin in the open-loop process, and not the closed-loop process. A properly designed closed-loop process must invariably be asymptotically stable, and thus the closed-loop transfer function must have all its poles in the left half-plane and none on the imaginary axis, which includes the origin.

The steady state error that results when the polynomial input is the same degree as the system type is determined with the aid of (4.50)

$$\lim_{t\to\infty} e(t) = \lim_{s\to 0} se(s) = \begin{cases} \dfrac{C_1 D(0)}{D(0) + KN(0)} & \text{for } p = m = 0 \\[2ex] \dfrac{C_{m+1} D(0)}{KN(0)} & \text{for } p = m \geqq 1 \end{cases} \tag{4.52}$$

If we define the "fractional error," or "error ratio," as

$$r_m = \frac{1}{C_{m+1}} \lim_{t\to\infty} e(t)$$

then, from (4.52) we determine the fractional error for a type m system (when the input is a polynomial of degree m)

$$r_m = \begin{cases} \dfrac{1}{1 + KN(0)/D(0)} = \dfrac{1}{1 + KG(0)} & m = 0 \\[2ex] \dfrac{1}{KN(0)/D(0)} = \dfrac{1}{KG(0)} & m \geqq 1 \end{cases} \tag{4.53}$$

Thus the steady state error ratio decreases as the open-loop dc gain $KG(0)$ increases. Hence if the requisite number of integrators is not present to make the plant of the desired type, the steady state error ratio can be reduced (but not brought to zero) by making the loop gain $KG(0)$ high. But of course it can't be made arbitrarily high without compromising stability.

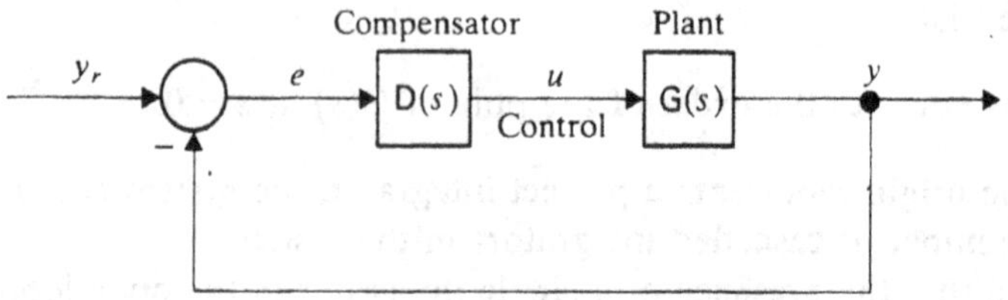

Figure 4.26 Error-driven feedback control system with compensator.

If closed-loop system tracking performance of a particular type is required, but the existing open-loop plant type is not sufficiently high, the required number of integrators is supplied by means of a "compensator": a dynamic system having the transfer function $\mathsf{D}(s)$ placed between the measured error and the input to the original plant, as shown in Fig. 4.26.

The most common type of compensator is the so-called PI (proportional + integral) compensator

$$\mathsf{D}(s) = \frac{K_1}{s} + K_2 = \frac{K_1 + K_2 s}{s}$$

The pole in $\mathsf{D}(s)$ raises the type of the open-loop system.

Before the era of digital computers, it was difficult to construct a perfect integrator: a device which maintains an absolutely constant output indefinitely in the absence of any input. The output of a physical device called an integrator would tend to drift, either to zero or, worse still, to infinity. The quality of an integrator was measured by the time it could be expected to hold a constant output. High-quality integrators needed to control processes whose natural time constants were of the order of several hours (typical in process control) were very expensive and marginally reliable, hence PI controls were bothersome. The digital computer now provides a means for realizing a perfect integrator, and the hardware problems of PI control have disappeared.

In principle, the compensator can provide a double pole at the origin thus raising the open-loop system type by 2. This is generally infeasible in practice, however, because the output of such a compensator is an increasing function of time even when the error is zero. This output is the input u to the original plant. Hence the physical input to the plant is constantly increasing in magnitude. Sooner or later a limit will be reached at which point the physical input to the plant will have to stop increasing: the control input "saturates." The input demanded by the compensator will not be physically attainable. This fact of saturation needs to be taken into account in the system design. The system should not be required to exhibit behavior of a type of which it is physically incapable.

In classical system designs, the integrators needed to raise the system type are frequently included in the plant model. The designer is concerned with shaping the dynamic behavior of that part of the compensator having no poles at the origin. With the state variable approach to be developed in this book, the integrators required in the compensator emerge in the normal design process.

4.8 DYNAMIC RESPONSE: BANDWIDTH

Another consideration in control system design is *dynamic response.* Not only must the closed-loop system be stable and reach the required steady state value eventually, but it cannot take forever to get to where it's going, and it should not be too oscillatory.

The dynamic characteristics of the system are typically defined in terms of the response to a unit-step input. The "step response" of a typical system is shown in Fig. 4.27. The parameters of major interest are as follows:

Overshoot Difference between peak value and steady state value;

Rise time Time for output to reach a specified fraction (usually $1 - e^{-1} = 0.632$) of the steady state value.

In addition to these parameters, often other parameters are of interest such as delay time (the time that it takes for the output to "get started"—to reach say 10 percent of its steady state value), "peak-time" (time to reach first peak in a system with positive overshoot).

There is no universal agreement on the definitions. For example, in some fields (e.g., process control) rise time is defined as the time it takes for the process to go from 10 to 90 percent of its final value. Or it may be defined as the time it takes to get to its steady state value the first time. (This is only meaningful in a system with overshoot.)

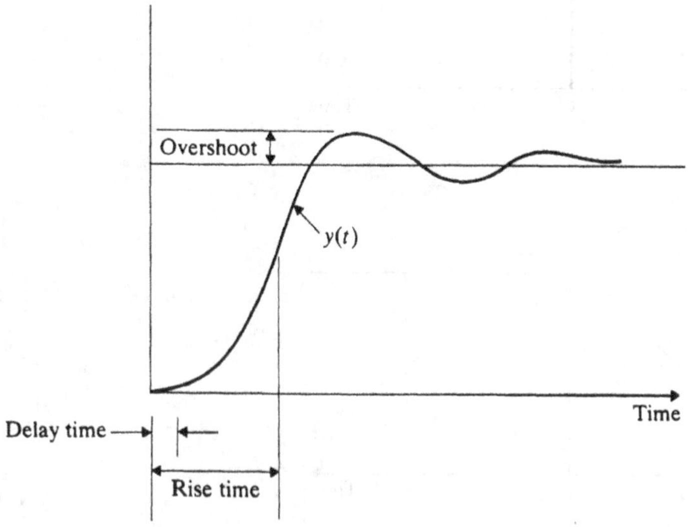

Figure 4.27 Characteristics of dynamic response.

Dynamic performance requirements of a system are typically specified in terms of the maximum permissible rise time and overshoot. These parameters are readily measured on a time-history plot of the output of an actual system or, for design purposes, on the output of a simulation of the actual system. But they are not readily calculated from the transfer function of the system. One way of avoiding the calculation problem is to define the response time as the *centroid* $\bar{t}$ of the impulse response of the system

$$\bar{t} = \frac{\int_0^\infty t h(t)\, dt}{\int_0^\infty h(t)\, dt} \tag{4.54}$$

The physical significance of the centroidal response time $\bar{t}$ is that it is the location of a single impulse which has the same effect as the actual system. An impulse located at $t = \bar{t}$ would give rise to a step occurring at that time, as shown in Fig. 4.28. This interpretation of the response time has a certain intuitive appeal.

The response time $\bar{t}$ cannot be determined by simply inspecting the step response of the system. Precise calculation of $\bar{t}$ from recorded data would require numerical integration of the step response: not a difficult task for a digital computer but not as easy as picking one or two points off a curve. The definition of response time by $\bar{t}$ has decided advantages with regard to analysis, however: It can be calculated directly from the transfer function $\mathsf{H}(s)$ without

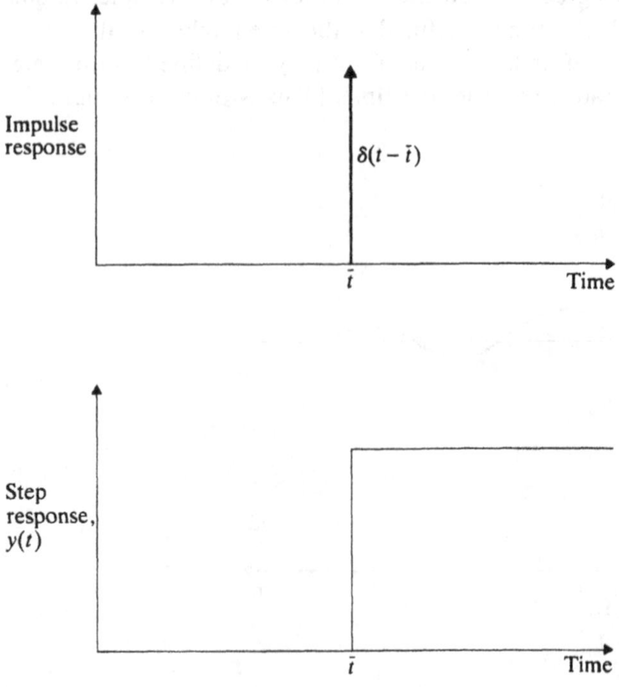

Figure 4.28 Definition of centroidal response time.

the need for first determining the expression for the step response or the impulse response. To see this, note that the transfer function is the Laplace transform of the impulse response

$$\mathrm{H}(s) = \int_0^\infty e^{-st} h(t)\, dt \tag{4.55}$$

Thus

$$\mathrm{H}(0) = \int_0^\infty h(t)\, dt \tag{4.56}$$

Now take the derivative of both sides of (4.55) with respect to s

$$\mathrm{H}'(s) = d\mathrm{H}(s)/ds = -\int_0^\infty t\, e^{-st} h(t)\, dt$$

Thus

$$-\mathrm{H}'(0) = \int_0^\infty t h(t)\, dt$$

Thus

$$\bar{t} = -\frac{\mathrm{H}'(0)}{\mathrm{H}(0)} \tag{4.57}$$

In words, the centroidal response time $\bar{t}$ is the ratio of the Laplace transforms of the derivative of the transfer function at the origin to the transfer function itself; $\bar{t}$ is thus a measure of how fast the transfer function decreases at the origin.

It is very easy to calculate $\bar{t}$ using (4.57). For this reason it is a useful definition notwithstanding the possible difficulty of determining it from measured data. In typical systems, moreover, $\bar{t}$ is very close to the rise time calculated using other definitions. In a first-order system, for example, with

$$\mathrm{H}(s) = \frac{\omega_0}{s + \omega_0} \tag{4.58}$$

it is found that (4.57) gives

$$\bar{t} = \frac{1}{\omega_0} \tag{4.59}$$

The step response corresponding to this transfer function is

$$a(t) = \mathscr{L}^{-1}\left[\frac{\omega_0}{s(s + \omega_0)}\right] = 1 - e^{-\omega_0 t}$$

The step response reaches $1 - e^{-1}$ of its final value at $t_{0.63}$ given by

$$\omega_0 t_{0.63} = 1 \tag{4.60}$$

Thus, for a first-order system, the centroidal response time and the 63 percent response time are exactly equal. (The response of a first-order system is the basis of the definition of the 63 percent response time.)

The centroidal response time is easy to calculate for a second-order system with

$$\mathsf{H}(s) = \frac{\omega_0^2}{s^2 + 2\zeta\omega_0 s + \omega_0^2} \tag{4.61}$$

Using (4.57) it is found that

$$\bar{t} = \frac{2\zeta}{\omega_0} \tag{4.62}$$

This means that the response time increases with the damping factor ζ. For two systems having equal natural frequencies, the system with the larger damping factor has the larger response time, which agrees with our intuition.

Calculation of the step response corresponding to $\mathsf{H}(s)$ as given by (4.61) and then solving for $t_{0.63}$ is a messy business. The step-response curves themselves have the appearance shown in Fig. 4.29. The 63 percent and the centroidal response times are both shown graphically vs. ζ in Fig. 4.30. They

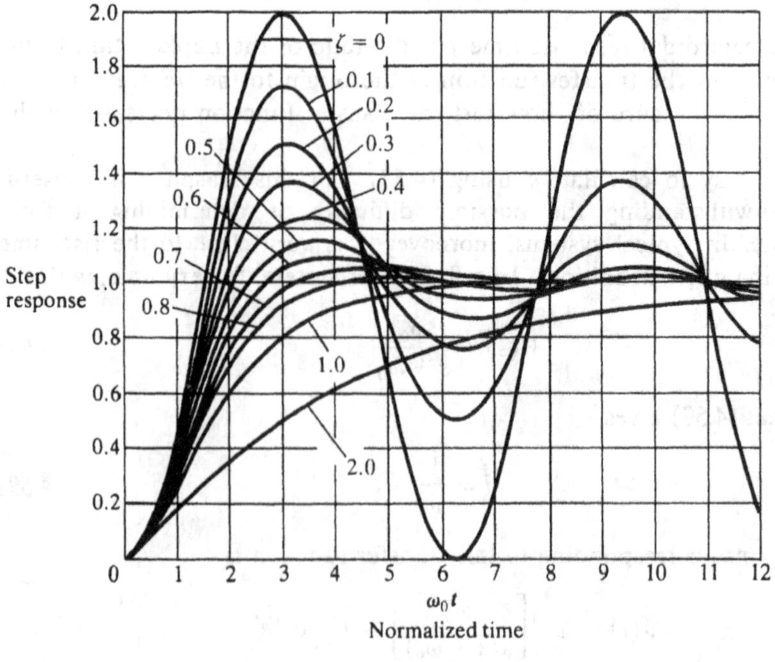

Figure 4.29 Step response of second-order system

$$\mathsf{G}(s) = \frac{\omega_0^2}{s^2 + 2\zeta\omega_0 s + \omega_0^2}$$

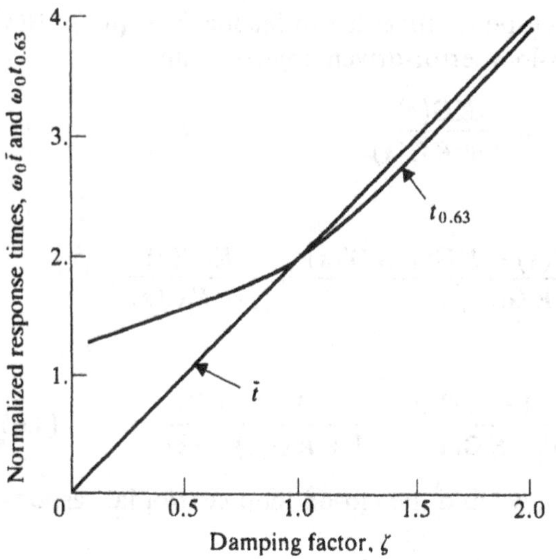

Figure 4.30 Response time of second-order

$$H(s) = \frac{\omega_0^2}{s^2 + 2\zeta\omega_0 s + \omega_0^2}$$

agree exactly at a damping factor $\zeta = 1$ and are within about 20 percent of each other for the range of damping factors of practical interest: $\zeta > 0.7$.

For very low damping factors, the centroidal rise time is much smaller than the 63 percent rise time, but neither gives a very good indication of system behavior.

Calculation of the centroidal rise time of systems in tandem (cascade) is instructive. Suppose a system $H(s)$ comprises two subsystems $H_1(s)$ and $H_2(s)$ in tandem

$$H(s) = H_1(s)H_2(s)$$

Then

$$H'(s) = H_1'(s)H_2(s) + H_2'(s)H_1(s)$$

Thus

$$\frac{H'(s)}{H(s)} = \frac{H_1'(s)H_2(s) + H_2'(s)H_1(s)}{H_1(s)H_2(s)} = \frac{H_1'(s)}{H_1(s)} + \frac{H_2'(s)}{H_2(s)} \tag{4.63}$$

Thus, on evaluating (4.63) at $s = 0$, we obtain

$$\bar{t} = \bar{t}_1 + \bar{t}_2 \tag{4.64}$$

i.e., the centroidal response time of a tandem combination of systems is the sum of the centroidal response times of each. This also agrees with our intuition: There is a lag in going through the first system, and the second system adds an additional lag, so we would expect that a formula like (4.64) will hold, at least approximately, for any reasonable definition of response time.

Finally, consider the effect on response time due to feedback. Suppose $\mathsf{H}(s)$ is the transfer function of a closed-loop error-driven control system

$$\mathsf{H}(s) = \frac{K\mathsf{G}(s)}{1 + K\mathsf{G}(s)}$$

Then

$$\mathsf{H}'(s) = \frac{[1 + K\mathsf{G}(s)]K\mathsf{G}'(s) - K\mathsf{G}(s)K\mathsf{G}'(s)}{[1 + K\mathsf{G}(s)]^2} = \frac{K\mathsf{G}'(s)}{[1 + K\mathsf{G}(s)]^2}$$

and

$$\frac{\mathsf{H}'(s)}{\mathsf{H}(s)} = \frac{K\mathsf{G}'(s)}{[1 + K\mathsf{G}(s)]^2} \frac{1 + K\mathsf{G}(s)}{K\mathsf{G}(s)} = \frac{1}{1 + K\mathsf{G}(s)} \frac{\mathsf{G}'(s)}{\mathsf{G}(s)} \tag{4.65}$$

Thus, on evaluating (4.65) at $s = 0$, we find the closed-loop centroidal response time given by

$$\bar{t}_{CL} = \frac{1}{1 + K\mathsf{G}(0)} \bar{t}_G \tag{4.66}$$

where $\bar{t}_G$ is the response time of the open-loop plant. We see that another benefit of feedback is the reduction of the system response time: The closed-loop response time is equal to the open-loop response time divided by the dc return difference.

The centroidal response time can readily be calculated from the state-space representation of the transfer function

$$\mathsf{H}(s) = C(sI - A)^{-1}B \tag{4.67}$$

where C is a $1 \times k$ matrix and B is a $k \times 1$ matrix, so the product $C(sI - A)^{-1}B$ is a 1×1 matrix. The dc transmission is

$$\mathsf{H}(0) = -CA^{-1}B$$

(We assume that the system (4.67) is asymptotically stable, which assures that A^{-1} exists.) To find $\mathsf{H}'(0)$ we write

$$\mathsf{H}(s) = C\Phi(s)B$$

where

$$\Phi(s) = (sI - A)^{-1}$$

is the resolvent of A. Then

$$\mathsf{H}'(0) = C\Phi'(0)B$$

Now

$$(sI - A)\Phi(s) = I$$

Differentiate with respect to s

$$\Phi(s)(sI - A)\Phi'(s) = 0$$

or

$$\Phi'(0) = A^{-1}\Phi(0) = -A^{-2}$$

and

$$\mathsf{H}'(0) = -CA^{-2}B$$

Finally

$$\bar{t} = \frac{\mathsf{H}'(0)}{\mathsf{H}(0)} = \frac{CA^{-2}B}{CA^{-1}B} \tag{4.68}$$

(It is legal to divide by $CA^{-1}B$ because $CA^{-1}B$ is a 1×1 matrix.)

One is tempted to cancel C and B in the numerator and denominator of (4.68) and thereby obtain an expression for $\bar{t}$ in terms of only the dynamics matrix A. This is not legal, of course. Even if it were legal, the final result wouldn't make any sense because the result would be $\bar{t} = A^{-1}$ which is equating a matrix to a scalar. Is there any way that something like this equation could make sense? The answer is yes, provided that an appropriate generalization is used. The quantity

$$\bar{t} = (\|A^{-1}\|/\|A\|)^{1/2} \tag{4.69}$$

has been studied by Bass.[15] For a further discussion of this, see Note 4.3.

In the frequency domain, the dynamic behavior of a system is characterized by its *bandwidth*. The reader has very likely formed an intuitive notion of bandwidth by reasoning as follows: Every input can be regarded as comprising components at various frequencies; a rapidly changing input means that it has a large high frequency content, while a smooth, slowly varying input has a relatively smaller high frequency content. Thus if the control system is to faithfully reproduce an input that is changing rapidly, it must be capable of faithfully reproducing inputs at high frequencies, i.e., to have a large bandwidth. On the other hand, if the input is slowly varying, the control system does not need a high bandwidth. A step change (i.e., a discontinuity) in the input has a large amount of high-frequency content: to reproduce it faithfully (with a short rise time) requires a high bandwidth. If a short rise time is not required then neither is a high bandwidth. Reasoning thus one reaches the conclusion that there is an inverse relationship between bandwidth and response time.

In communication systems where data is transmitted by modulation of a carrier, the bandwidth of a system is generally defined as the width of the resonance curve between the "half power" frequencies—i.e., the frequencies at which the gain is "3 dB down" from the value at resonance. In a control system, however, the oscillatory response characteristic of a resonant system is highly undesirable. If well designed, a closed-loop system will have a frequency response characteristic of the Bode plots of Fig. 4.16 with a damping factor ζ greater than about 0.4 or 0.5. The resonance peak, if any, is very small and the frequency response typically is very flat until a critical frequency is reached, at which point it begins falling off quite rapidly. A system of this type is called *low pass*—it passes low frequencies without substantial attenuation—and its bandwidth is usually defined as the frequency at which the gain is below the dc gain by a factor of $\sqrt{2}$. In other words

$$\left|\frac{\mathrm{H}(jW)}{\mathrm{H}(0)}\right| = \frac{1}{\sqrt{2}}$$

or

$$\frac{|\mathrm{H}(jW)|^2}{\mathrm{H}^2(0)} = \frac{1}{2} \tag{4.70}$$

where W is the bandwidth of the system.

In a first-order system

$$\mathrm{H}(s) = \frac{\omega_0}{s + \omega_0}$$

(4.70) becomes

$$\frac{\omega_0^2}{W^2 + \omega_0^2} = \frac{1}{\left(\dfrac{W}{\omega_0}\right)^2 + 1} = \frac{1}{2}$$

or

$$W = \omega_0 \tag{4.71}$$

We already knew this from the nature of the Bode plot for a first-order system. For a second-order system

$$\mathrm{H}(s) = \frac{\omega_0^2}{s^2 + 2\zeta\omega_0 s + \omega_0^2}$$

The bandwidth W is the solution of

$$\frac{1}{\left[1 - \left(\dfrac{W}{\omega_0}\right)^2\right]^2 + 4\zeta\left(\dfrac{W}{\omega_0}\right)^2} = \frac{1}{2}$$

or

$$\left(\frac{W}{\omega_0}\right)^2 = 1 - 2\zeta^2 + \sqrt{(1 - 2\zeta^2)^2 + 1} \tag{4.72}$$

A plot of the bandwidth of a second-order system vs. damping factor is shown in Fig. 4.31. For purposes of comparison, the reciprocal of the centroidal response time $\bar{t}$ is also shown. It is observed that for the useful range of damping factors ($\zeta > 0.4$) the reciprocal of $\bar{t}$ is a lower bound on the bandwidth:

$$W\bar{t} > 1$$

In other words the product of the bandwidth and the centroidal response is

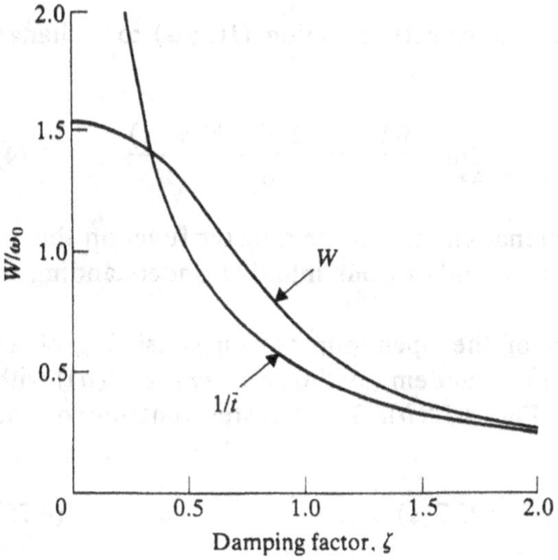

Figure 4.31 Bandwidth of second-order system

$$H(s) = \frac{\omega_0^2}{s^2 + 2\zeta\omega_0 + \omega_0^2}$$

always greater than 1—a sort of time-frequency uncertainty principle. Note, however that $W\bar{t}$ is never *much* greater than 1. Thus we can safely say that the product of bandwidth and response time, which is exactly 1 for a first-order system, is approximately 1 for properly damped second-order systems. The relationship

$$W\bar{t} \approx 1 \tag{4.73}$$

turns out to hold for a higher order as well, and reinforces our intuitive conception of the reciprocal relationship between bandwidth and response time.

4.9 ROBUSTNESS AND STABILITY (GAIN AND PHASE) MARGINS

In designing the control law for some process we deal not with the physical process per se but rather with a mathematical model of the process. The control law will be acceptable only if the mathematical model predicts the behavior of the physical process reasonably well. But no mathematical model can predict the behavior of a physical process *exactly*; there will always be some discrepancy between the actual behavior of the physical process and that predicted by the mathematical model. And the discrepancy may increase with time owing to normal aging and deterioration. Some uncertainty of the physical process is a reality that the control system engineer must contend with.

One of the well-known advantages of feedback is that it confers a degree of "robustness" or immunity to uncertainty or changes in the process. "Sensitivity" S of a process to a change in one of the parameters of the process is a way of quantifying the advantage of feedback. Suppose that the transfer function of the process is $H(s; \alpha)$ where α is some parameter that can change in the process. We define the *sensitivity* of the transfer function $H(s;\alpha)$ to a change in α by

$$S(\alpha) = \frac{1}{\mathsf{H}(s)}\frac{\partial \mathsf{H}(s)}{\partial \alpha} = \frac{1}{\mathsf{H}(s;\alpha)}\lim_{\Delta\alpha\to 0}\frac{\mathsf{H}(s;\alpha+\Delta\alpha)-\mathsf{H}(s;\alpha)}{\Delta\alpha} \tag{4.74}$$

The sensitivity is thus the fractional change in the transfer function due to a change in the parameter and corresponds to our intuitive understanding of sensitivity.

Let us compare the sensitivity of the open-loop system consisting of an amplifier of gain K and a plant $\mathsf{G}(s)$ in tandem (as shown in Fig. 4.32(*a*)) with the closed-loop system shown in Fig. 4.32(*b*). The transfer function of the open-loop system is

$$\mathsf{H}_0(s) = K\mathsf{G}(s) \tag{4.75}$$

and hence the sensitivity to a change in K is

$$S_0(K) = \frac{1}{\mathsf{H}_0}\frac{\partial \mathsf{H}_0}{\partial K} = \frac{1}{K}$$

The closed-loop transfer function, on the other hand, is

$$\mathsf{H}_c(s) = \frac{K\mathsf{G}(s)}{1+K\mathsf{G}(s)} \tag{4.76}$$

and the corresponding sensitivity is

$$S_c = \frac{1+\mathsf{G}(s)K}{\mathsf{G}(s)K}\,\frac{[1+\mathsf{G}(s)K]\mathsf{G}(s)-\mathsf{G}(s)K\mathsf{G}(s)}{[1+\mathsf{G}(s)K]^2} = \frac{1}{K(1+\mathsf{G}(s)K)} \tag{4.77}$$

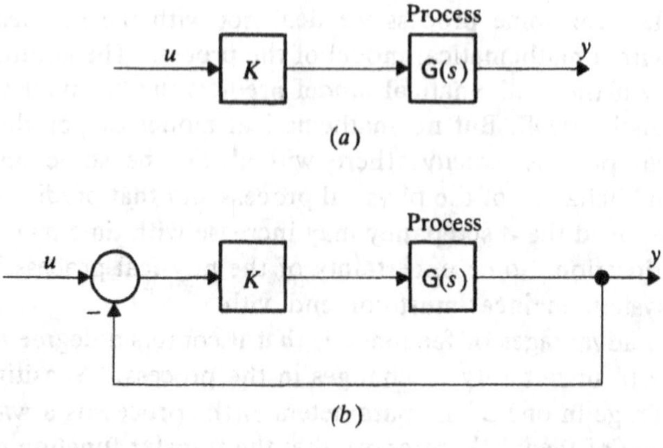

Figure 4.32 Open-loop and closed-loop processes for sensitivity comparisons. (*a*) Open-loop process $\mathsf{H}_0(s) = K\mathsf{G}(s)$; (*b*) Closed-loop process $\mathsf{H}_c(s) = K\mathsf{G}(s)/[1+K\mathsf{G}(s)]$.

The ratio of the closed-loop sensitivity to the open-loop sensitivity is

$$\frac{S_c}{S_0} = \frac{1}{1 + \mathsf{G}(s)K} \tag{4.78}$$

Feedback thus has the effect of reducing the sensitivity to gain variations by the reciprocal of the return difference $1 + \mathsf{G}(s)K$. The higher the return difference, the lower the sensitivity to parameter changes. Thus a high return difference not only speeds up the dynamic response, as shown in (4.66), but also tends to immunize the system to changes in parameters of the open-loop system. (In the above example, the sensitivity to changes in the gain K of the amplifier was computed. But the principle is equally valid for changes in other parameters of the plant $\mathsf{G}(s)$.)

From the standpoint of speed of response and immunity to parameter variations, we should like the return difference of the system to be large at all frequencies. There are various reasons, however, why this is not a practical goal. The most important reason is that the transfer function of every practical plant is "low pass," tending to zero (in magnitude) as frequency becomes infinite. If the amplifier has a fixed gain K, the loop transmission will tend to zero at high frequencies, and hence the return difference will tend to unity. Instead of an amplifier with a constant gain of K one might conceive of using a dynamic "compensator" $\mathsf{D}(s)$ with a gain that increases with frequency to counteract the decrease in the plant transfer function. Such compensators are feasible but not very desirable in practice because they amplify the inherent high-frequency noise in the system. Moreover, a *physical* compensator cannot sustain a transfer function that increases indefinitely with frequency. As with every physical device, the transfer function of any compensator must eventually "roll off" with frequency.

Thus, even with dynamic compensation, the loop transmission of a system must ultimately become zero, and the return difference must ultimately approach unity. The practical design problem is not how to keep the return difference large at all frequencies, but rather how to make the return difference tend to unity in a graceful, well-behaved manner.

The problem is phase shift. The decrease in amplitude of the loop gain is accompanied by a phase shift. It is possible for the loop gain to reach unity in *magnitude*, and to have a phase shift of 180° at some frequency. In this case the return difference is zero and the transfer function of the system becomes infinite: the system is unstable. In order for the system to be stable, it is not permissible for the return difference to go to zero at any frequency. Moreover, because of possible differences between the transfer function used for purposes of design and the true transfer function, it is imprudent to permit the return difference to come close to zero. In a practical design it is necessary to provide reasonable *stability margins.*

Gain margin and *phase margin* are the stability margins in common use. The gain margin is the amount that the loop gain can be changed, at the frequency at which the phase shift is 180°, without reducing the return difference to zero.

The phase margin is the amount of phase lag that can be added to the open-loop transfer function, at the frequency at which its magnitude is unity, without making the return difference zero. These margins can be illustrated on a Nyquist plot or a Bode plot for a typical transfer function, as shown in Fig. 4.33.

The Nyquist diagram corresponding to a typical loop transmission is shown in Fig. 4.33(a). The loop gain K is taken to be unity, as is customary in gain and phase margin analyses. The system as shown is stable. At the frequency ω_2 at which the phase shift is 180°, the magnitude of the loop transfer function $|G(j\omega_2)|$ is less than unity by the gain margin γ. This means that the loop transmission can be raised by an amount γ without causing the system to become unstable. In most instances the gain margin is expressed as a logarithmic ratio in $\text{dB} = 20 \log \gamma$. The Nyquist diagram of Fig. 4.33 also shows the phase margin ϕ which is the angle that the phasor $G(j\omega_1)$ makes with the negative real axis at the frequency ω_1 at which $|G(j\omega)|$ first reaches unity. If a phase lag less than the phase margin ϕ were added to each point on the plot of $G(j\omega)$, the Nyquist diagram would not encircle the $-1 + j0$ point and the closed-loop system would remain stable.

The gain and phase margins are shown in the Bode plot of the loop transmission in Fig. 4.33(b). Note again that the phase margin is the difference between 180° and the actual phase shift at the "gain crossover" frequency ω_1

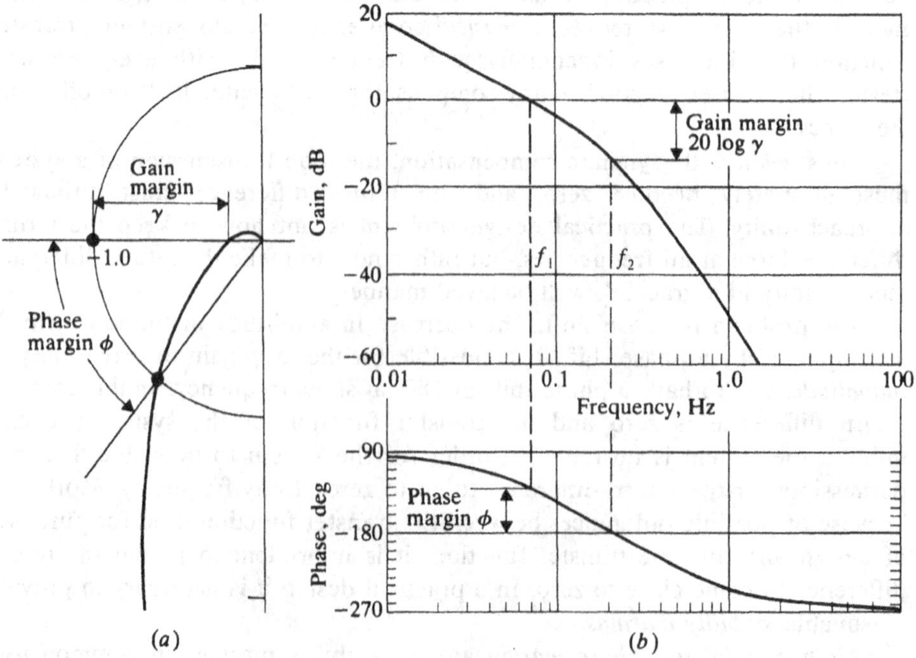

Figure 4.33 Gain and phase margins defined. (a) Nyquist diagram; (b) Bode plots.

and that the gain margin is the amount that the log magnitude plot falls below 0 dB at the "phase crossover" frequency ω_2.

The gain and phase margins are conveniently expressed in terms of the magnitude of the return difference

$$\mathsf{T}(j\omega) = 1 + \mathsf{G}(j\omega) \tag{4.79}$$

which is the phasor from the point $-1 + j0$ to the phasor $\mathsf{G}(j\omega)$ on the Nyquist plot, as shown in Fig. 4.34(*a*). It is seen directly from Fig. 4.34(*b*) that

$$\gamma = |\mathsf{T}(j\omega_2)| = |1 + \mathsf{G}(j\omega_2)| \qquad \theta_G(\omega_2) = 180° \tag{4.80}$$

And, by a simple geometric construction (see Fig. 4.34(*c*))

$$\phi = 2\sin^{-1}\left|\frac{\mathsf{T}(j\omega_1)}{2}\right| \qquad |\mathsf{G}(j\omega_1)| = 1 \tag{4.81}$$

Since the gain and phase margins are directly related to the variation with frequency ω of the return difference $\mathsf{T}(j\omega)$, it would not be very surprising to find that the return difference has an important role to play in the assessment of the robustness of a control system. The return difference retains its importance even in multiple-input, multiple-output systems, in which the concepts of gain and phase margin become problematic.

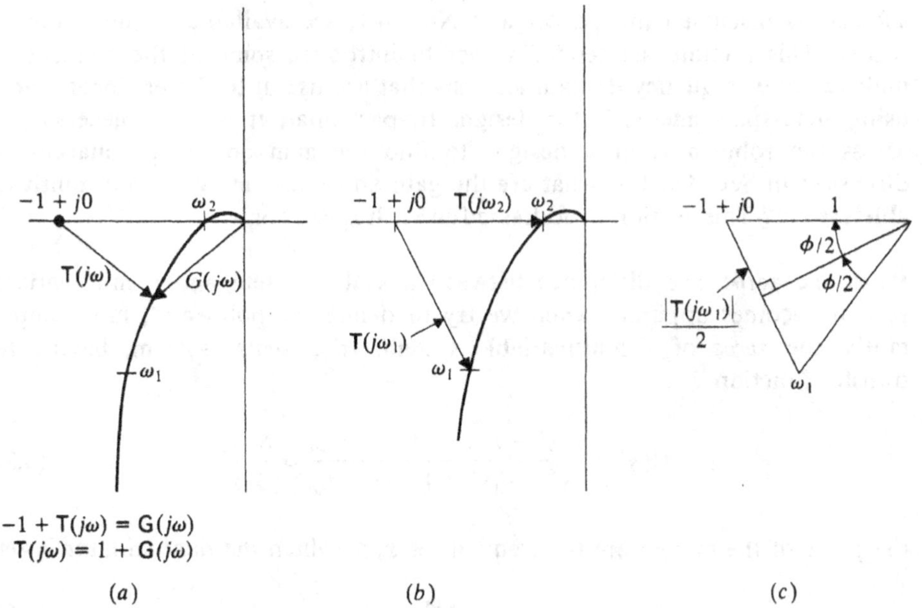

Figure 4.34 Use of return difference with phase and gain margins. (*a*) Return difference is phasor from $-1 + j0$ to phase $\mathsf{G}(j\omega)$; (*b*) Return differences at frequencies of gain and phase margin; (*c*) Construction for phase-margin formula.

4.10 MULTIVARIABLE SYSTEMS: NYQUIST DIAGRAM AND SINGULAR VALUES

Most of this chapter has been concerned with single-input, single-output "scalar" systems: The plant under examination has one input and one output and hence, in the frequency domain, it is characterized by a single transfer function. Most of the systems encountered in practice fall into this class. But there are many systems in which there are more than one control input and/or more than one output of concern. An example of a multiple-input, multiple-output (or, more simply, *multivariable*) system is the lateral channel of an aircraft (in which the inputs are the rudder and aileron deflections and in which possible outputs would be the roll and yaw angles). Another example is the distillation column discussed in Example 4A and earlier.

With state-space methods, the focus is on the *state* of the process more than on the inputs or the outputs. Hence there is less need for making a distinction between scalar systems and multivariable systems than there is with frequency-domain methods. (This is one of the advantages of state-space methods.) Nevertheless, a familiarity with some of the concepts of multivariable frequency-domain analysis cannot but be helpful to the user of state-space design methods.

The intention of this section is certainly not to provide an introduction to multivariable frequency-domain *design* methods. Several textbooks on this subject, as discussed in Sec. 4.1 and Note 4.1, are available to the interested reader. This section is intended rather to introduce some of the concepts of multivariable frequency-domain analysis that are useful to the engineer who is using state-space methods for design. In particular, it is often necessary to assess the robustness of a design—to find the gain and phase margins as discussed in Sec. 4.9. But what are the gain and phase margins in a multivariable system? This section is addressed to such questions.

Poles and zeros The difference between a scalar system and a multivariable system becomes apparent when we try to define the poles and, more importantly, the *zeros* of a multivariable system. In a scalar system, having the transfer function

$$\mathsf{H}(s) = \frac{b_0 s^k + b_1 s^{k-1} + \cdots + b_k}{s^k + a_1 s^{k-1} + \cdots + a_k} = \frac{\mathsf{N}(s)}{\mathsf{D}(s)} \tag{4.82}$$

the poles of the system are the frequencies s_p at which the denominator is zero

$$\mathsf{D}(s) = s^k + a_1 s^{k-1} + \cdots + a_k = 0 \tag{4.83}$$

and the zeros of the system are the frequencies at which the numerator is zero

$$\mathsf{N}(s) = b_0 s^k + b_1 s^{k-1} + \cdots + b_k = 0$$

A multivariable system, however, is represented by a *matrix* of transfer functions

$$\mathsf{H}(s) = \begin{bmatrix} \mathsf{H}_{11}(s) & \cdots & \mathsf{H}_{1m}(s) \\ \cdots & \cdots & \cdots \\ \mathsf{H}_{l1}(s) & \cdots & \mathsf{H}_{lm}(s) \end{bmatrix} \tag{4.84}$$

where m is the number of inputs to the system and l is the number of outputs. Every entry in the matrix represents the transfer function between one of the inputs and one of the outputs, and is the ratio of polynomials in s in the form of (4.82). Since each of these transfer functions has its own set of poles and zeros, how are the poles and zeros of the entire *system* to be defined? The issue is readily settled for poles: The poles of the system can be defined as the totality of the poles of all the transfer functions in the matrix (4.84). This definition of the system poles is equivalent to expressing the transfer functions as (possibly different) numerator polynomials in s all over a single common denominator polynomial. If the matrix $\mathsf{H}(s)$ of the system is obtained starting from a state-space model, as described in Sec. 3.5, this common denominator is the characteristic polynomial $|sI - A|$.

It is reasonable to define the poles of the system as the collection of all the poles of the transfer functions in $\mathsf{H}(s)$ because at any pole at least one transfer function becomes infinite and hence the matrix $\mathsf{H}(s)$ cannot be said to exist. On the other hand, the entire transfer matrix does not become zero at a zero of one of the elements in $\mathsf{H}(s)$. Thus it would not be appropriate to define the zeros of an entire system as the collection of the zeros of all the individual transfer function $\mathsf{H}_{ij}(s)$. On the other hand, it would not make much sense to define the zeros of the system as those frequencies at which the matrix $\mathsf{H}(s)$ becomes zero, for then unless all the transfer functions have a common factor (a very unusual condition), the system H would have no zeros at all!

What is needed here is a definition of the zeros of a multivariable system which is a natural generalization of the zeros of a scalar system. One way of interpreting the zeros of a scalar system is as the poles of the *inverse* system, the transfer function of which is

$$\mathsf{H}^{-1}(s) = \frac{\mathsf{D}(s)}{\mathsf{N}(s)}$$

If the multivariable system is "square," i.e., if there are exactly the same number of outputs as inputs, then $\mathsf{H}(s)$ will be a square matrix and, in general, will have an inverse except at isolated (complex) frequencies. It is appropriate to call these frequencies the zeros of the system. Since the condition for $\mathsf{H}(s)$ to have an inverse is that the determinant of the transfer matrix be nonzero

$$|\mathsf{H}(s)| = \begin{vmatrix} \mathsf{H}_{11}(s) & \cdots & \mathsf{H}_{1m}(s) \\ \cdots & \cdots & \cdots \\ \mathsf{H}_{l1}(s) & \cdots & \mathsf{H}_{lm}(s) \end{vmatrix} \neq 0$$

we can say that the zeros of a square system are the zeros of the determinant $|\mathsf{H}(s)|$.

A "nonsquare" system, i.e., one in which the number of inputs is not equal to the number of outputs, is more of a problem. One way of addressing the problem, when there are more inputs than outputs, is to define several more independent outputs (assuming that the order of the system is high enough to permit this). This is done by adding rows to the observation matrix C, independent of those already present, so that the number of rows in C equals the number of columns in B. Although these added outputs may not be of particular interest, this artifice permits use of any of the theory developed for square systems. This technique will not avail, however, when there are fewer inputs than outputs. Adding inputs that are not physically present is not permissible since the resulting mathematical model would no longer represent the physical process. A more general definition of the zeros of a nonsquare system avoids the need for the artifice of adding inputs or outputs. We define the zeros as those (complex) frequencies at which the *rank* of the transfer function matrix $\mathsf{H}(s)$ is reduced. Normally the transfer-function matrix will be of "full rank," i.e., $\text{rank}[\mathsf{H}(s)] = \min[l, m]$ except at the zeros of the system at which $\mathsf{H}(s)$ will drop to less than full rank. (See Note 4.4.)

The return difference and the multivariable Nyquist diagram In Sec. 4.9 we found that the phase and gain margins of a scalar system, traditional measures of system robustness, can be determined by an examination of the behavior of the return difference as a function of frequency. It turns out that the manner in which the return difference varies with frequency also provides a means of measuring robustness in a multivariable system. To develop the concept of robustness for multivariable systems we need to clarify the notion of return difference and find a useful and convenient way of characterizing its behavior with frequency.

In Sec. 4.3 (block-diagram algebra) we saw that it is generally possible to express the transfer function from the input of a feedback system to its output in the form of the product of a "forward transmission matrix" and the inverse of another matrix of the form $\mathsf{T}(s) = I + \mathsf{G}(s)$, where $\mathsf{G}(s)$ is the "loop transmission." The matrix $\mathsf{T}(s)$ was called the "return difference." This is the matrix that we investigate to assess the robustness of a multivariable system.

An important feature of the return difference $\mathsf{T}(s)$ is that it is the sum of an identity matrix (which of course is square) and the loop transmission $\mathsf{G}(s)$. In order for the sum of I and $\mathsf{G}(s)$ to be defined, $\mathsf{G}(s)$ must also be a square matrix of the same dimension as I. Hence the return difference $\mathsf{T}(s)$ is a square matrix and there is no problem defining its zeros: they are the zeros of the determinant

$$|\mathsf{T}(s)| = |I + \mathsf{G}(s)|$$

Since the inverse of the return-difference matrix appears as a factor in the transfer-function matrix of the closed-loop system, the poles of the latter will include the zeros of the return difference. Naturally we expect the return difference to be zero at some (complex) values of s. If the system is stable, these zeros will occur only in the left half-plane; if the system is unstable, however, one or more zeros of the return difference will occur in the right half-plane. The question that an analysis of robustness seeks to answer is how much a parameter of a stable system can be permitted to vary before the system becomes unstable. In a scalar system, we can estimate the robustness by using the classical stability margins which can be determined by examining the behavior of $\mathsf{T}(s)$. But by Nyquist's method, we only need to investigate $\mathsf{T}(s)$ for $s = j\omega$, that is, for s on the imaginary axis. In particular, the minimum value of the magnitude of the return difference indicates how close the Nyquist diagram comes to the "critical" $-1 + j0$ point (see Fig. 4.34) and is a pretty good measure of robustness: The larger the minimum value of the magnitude of the return difference, the more robust (i.e., tolerant of loop gain variations) the system.

The determinant of the return-difference matrix in a multivariable system plays the role of the return difference itself of a scalar system. It would be quite natural to assume that the robustness of a multivariable system can be determined by studying the Nyquist diagram for the determinant of the return difference of the system. A polar plot of the determinant of the return difference $|\mathsf{T}(j\omega)|$ as ω is varied from $-\infty$ to ∞ may be termed the *multivariable Nyquist diagram* for the closed-loop system. To make the plot resemble a scalar Nyquist diagram we can place the origin at the critical point $-1 + j0$. This is the same as writing

$$|\mathsf{T}(j\omega)| = 1 + \mathsf{G}(j\omega) \tag{4.85}$$

and obtaining the Nyquist plot for $\mathsf{G}(j\omega)$. In the multivariable case, however, $\mathsf{G}(j\omega)$ cannot be interpreted as the "open-loop" transmission.

Singular value analysis Although the multivariable Nyquist plot is fairly easy to obtain, especially with the aid of a computer, it often does not tell enough about the robustness of the system under investigation, because the determinant of a matrix is not always a good indicator of how near that matrix comes to being singular. And that is what we want to determine for the return difference matrix $\mathsf{T}(s)$.

To see why the determinant of a matrix may be a poor indicator of how near the matrix comes to being singular, consider the matrix

$$M = \begin{bmatrix} 1 & 0 \\ 1/\varepsilon & 1 \end{bmatrix} \tag{4.86}$$

This matrix has a determinant of unity independent of ε, yet is only an ε away from being singular: Replace the zero in the upper right-hand corner of M by ε and M is singular. The eigenvalues of a matrix are scarcely a better measure of the incipient singularity. In the case of M of (4.86), for example, the eigenvalues of M are both unity, and do not provide an inkling into the near-singularity of M.

A better measure of the near-singularity of a matrix is the set of *singular values* of the matrix, defined as the square-roots of the eigenvalues of $M^H M$ where M^H is the transpose of the complex conjugate of M. (These eigenvalues are always positive and real. See Note 4.5.) When M is a real matrix $M^H = M'$. But since the matrices we will be considering are in general complex functions of frequency (when $s = j\omega$), it is necessary to make use of the more general form. Thus, in a multivariable system, instead of investigating the behavior of the determinant of $\mathsf{T}(j\omega)$ or even of the eigenvalues of $\mathsf{T}(j\omega)$, it is more appropriate to investigate the singular values of $\mathsf{T}(j\omega)$, that is, the square roots of the eigenvalues of

$$\mathsf{S}(\omega) = \mathsf{T}^H(j\omega)\mathsf{T}(j\omega) = \mathsf{T}'(-j\omega)\mathsf{T}(j\omega) \tag{4.87}$$

(Note that $\mathsf{T}^H(j\omega) = \mathsf{T}'(-j\omega)$ because $\mathsf{T}(s)$ is a real function of s, that is, every element of $\mathsf{T}(s)$ is real when s is real.)

As an illustration of the advantage of singular values over eigenvalues as a measure of incipient singularity, we find the singular values of the matrix M in (4.86). Since M is real, we need the eigenvalues of

$$S = M'M = \begin{bmatrix} 1 & 1/\varepsilon \\ 0 & 1 \end{bmatrix} \begin{bmatrix} 1 & 0 \\ 1/\varepsilon & 1 \end{bmatrix} = \begin{bmatrix} 1 + 1/\varepsilon^2 & 1/\varepsilon \\ 1/\varepsilon & 1 \end{bmatrix}$$

The characteristic equation of S is

$$|\lambda I - S| = \begin{bmatrix} \lambda - 1 - 1/\varepsilon^2 & -1/\varepsilon \\ -1/\varepsilon & \lambda - 1 \end{bmatrix} = \lambda^2 - \lambda\left(2 + \frac{1}{\varepsilon^2}\right) + 1$$

and the eigenvalues are

$$\lambda_1 = 1 + \frac{1}{2\varepsilon^2} + \frac{1}{2\varepsilon^2}\sqrt{1 + 4\varepsilon^2}$$

$$\lambda_2 = 1 + \frac{1}{2\varepsilon^2} - \frac{1}{2\varepsilon^2}\sqrt{1 + 4\varepsilon^2}$$

The larger eigenvalue λ_1 tends to infinity as $\varepsilon \to 0$. But the smaller eigenvalue λ_2 tends to zero, and as a consequence, one of the singular values $\sigma_2 = \lambda^{1/2} \to 0$ as $\varepsilon \to 0$, thus providing an indication that M becomes singular as $\varepsilon \to 0$. Note also that the two singular values $\sigma_1 = \lambda_1^{1/2}$ and σ_2 separate more and more as $\varepsilon \to 0$. This is an indication of the "ill-conditioning" of M: Not only does it become singular as $\varepsilon \to 0$, but it also grows (i.e., one of its elements grows) to infinity.

The singular values of the return difference matrix $\mathsf{T}(j\omega) = I + \mathsf{G}(j\omega)$ can be used to estimate the gain and phase margins of a multivariable system. A number of theorems have been developed since the late 1970s for estimating these margins (see Note 4.6). A typical theorem is the following [16]:

Let $\underline{\sigma}[\mathsf{T}(j\omega)]$ denote the smallest of the singular values of the return difference at the frequency ω. Suppose that there is a constant $\alpha \leqq 1$ such that $\underline{\sigma}[\mathsf{T}(j\omega)] \geqq \alpha$ for all frequencies. Then there is a *guaranteed gain margin*

$$GM = \frac{1}{1 \pm \alpha} \tag{4.88}$$

and a *guaranteed phase margin*

$$PM = \pm 2 \sin^{-1}\left(\frac{\alpha}{2}\right) \tag{4.89}$$

A number of theorems related to the one cited above have been developed during the early 1980s. This subject is likely to continue to receive a great deal of attention by researchers of the decade.

One of the major problems of singular value analysis is that the stability margins that it guarantees are *extremely conservative*, because they allow for the simultaneous magnitude and phase variations of *any* of the gains at which the loop is closed. In other words, if the off-nominal return difference is

$$\tilde{\mathsf{T}}(s) = I + D\mathsf{G}(s)$$

the singular value analysis tends to seek out the least favorable variation of the matrix D that represents the departure of the plant from its nominal, or design, value. The matrix D in practice, however, is anything but arbitrary. The singular value analysis may predict very small gain or phase margins on the basis of unlikely, or even impossible, plant variations.

A more reasonable assessment of robustness would take into account only those variations in system parameters that can actually occur in reality. If the total range of variation of the system can be represented by a single parameter, say μ, which ranges between 0 and 1, that is, the plant transfer matrix is $\mathsf{G}(s, \mu)$ with $\mathsf{G}(s, 0) = \bar{\mathsf{G}}(s)$ being the "nominal" system and $\mathsf{G}(s, 1)$ being the transfer matrix for the largest possible variation of the physical parameters, then there is a general theorem [16] that asserts that the system is stable for all values of $0 \leqq \mu \leqq 1$ if, and only if, the multivariable Nyquist plot for $1 + \mathsf{G}(s, 0)$ deforms "smoothly" into the Nyquist plot for $1 + \mathsf{G}(s, 1)$ and at no time covers the critical point $-1 + j0$.

Example 4H Two-loop control of distillation column The distillation column introduced in Example 2G and discussed later in Examples 4A and 4C has two control inputs u_1 (the steam flow rate) and Δs (the vapor side-stream flow rate) and may be regarded as having two outputs Δz_1 and Δz_2, the positions of the "interphase fronts." This process thus seems a natural candidate for a two-loop control system, as shown in Fig. 4.35. The side-stream flow rate Δs is controlled by the displacement Δz_1 of the front between the water and the propanol, and the steam flow rate Δu_1 is controlled by the displacement Δz_2 of the front between the water and the glycol. A more general control in which each control input is a linear combination of the two control outputs would probably be employed in practice, but the resulting complexity of the overall system in this case would obscure the analysis of this example, the objective of which is to illustrate the multivariable Nyquist diagram and singular-value estimation of gain and phase margins.

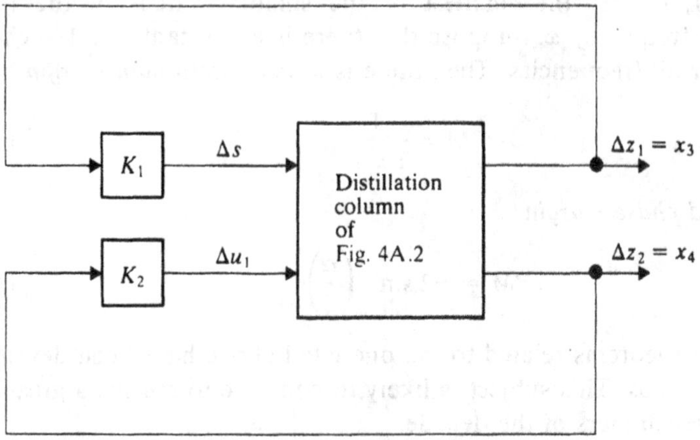

Figure 4.35 Two-loop control of distillation column.

Using the numerical data of Example 2G in the transfer matrix determined in Example 4A, we find that the transfer matrix for the plant is

$$\mathsf{H}(s) = \begin{bmatrix} \dfrac{3.04}{s} & \dfrac{-278.2}{s(s+6.02)(s+30.3)} \\ \dfrac{0.052}{s} & \dfrac{-206.6}{s(s+6.02)(s+30.3)} \end{bmatrix} \tag{4H.1}$$

with the diagonal matrix

$$K = \begin{bmatrix} K_1 & 0 \\ 0 & K_2 \end{bmatrix}$$

implied by the structure of Fig. 4.35, and we find the return-difference matrix

$$\mathsf{T}(s) = I + K\mathsf{H}(s) = \begin{bmatrix} 1 + \dfrac{3.04K_1}{s} & -\dfrac{278.2K_1}{s(s+6.02)(s+30.3)} \\ \dfrac{0.052K_2}{s} & 1 - \dfrac{206.6K_2}{s(s+6.02)(s+30.3)} \end{bmatrix} \tag{4H.2}$$

The closed-loop poles are determined by setting the determinant of the return difference to zero

$$0 = |\mathsf{T}(s)| = \left(1 + \frac{3.04K_1}{s}\right)\left(1 - \frac{206.6K_2}{s(s+6.02)(s+30.3)}\right) + \frac{(278.2)(0.052)K_1K_2}{s^2(s+6.02)(s+30.3)} \tag{4H.3}$$

From (4H.3) we obtain the characteristic equation of the system

$$\begin{aligned} \Delta(s) &= s^2(s+6.02)(s+30.3) + 3.04K_1 s(s+6.02)(s+30.3) - 206.6K_2 s - 613.9K_1K_2 \\ &= s^4 + (36.62 + 3.04K_1)s^3 + (182.4 + 110.4K_1)s^2 + (554.5K_1 - 206.2K_2)s - 613.9K_1K_2 \\ &= 0 \end{aligned} \tag{4H.4}$$

The ranges of K_1 and K_2 for which the closed-loop system is stable can be found using the Routh array or the Hurwitz matrix described in Sec. 4.5. In particular, the Hurwitz matrix for this example is

$$H = \begin{bmatrix} 36.32 + 3.04K_1 & 554.5K_1 - 206.6K_2 & 0. & 0. \\ 1. & 182.4 + 110.4K_1 & -613.6K_1K_2 & 0. \\ 0. & 36.32 + 3.04K_1 & 554.5K_1 - 206.6K_2 & 0. \\ 0. & 1. & 182.4 + 110.4K_1 & -613.6K_1K_2 \end{bmatrix}$$

By the Hurwitz matrix criterion, stability of the closed loop system is assured if

$$\begin{aligned} D_1 &= 36.32 + 3.04K_1 > 0 \\ D_2 &= \begin{vmatrix} 36.32 + 3.04K_1 & 554.5K_1 - 206.6K_2 \\ 1. & 182.4 + 110.4K_1 \end{vmatrix} > 0 \\ D_3 &= (554.5K_1 - 206.6K_2)D_2 + 613.6K_1K_2(36.32 + 3.04K_1)^2 > 0 \\ D_4 &= -613.6K_1K_2D_3 > 0 \end{aligned} \tag{4H.5}$$

Numerical analysis (Prob. 4.6) reveals that the four inequalities of (4H.5) are simultaneously satisfied in a region that is nearly rectangular and given by

$$0 \leqq K_1 \leqq 11.94$$
$$0 \leqq -K_2 \leqq \sim 32.1$$

(See Fig. 4.36.)

A "comfortable" operating point would be at

$$K_1 = 5 \qquad K_2 = -10$$

which provides a gain margin of over 2 for K_1 and over 3 for K_2. The return-difference matrix at this operating point is given by

$$\mathsf{T}(s) = \begin{bmatrix} 1 + \dfrac{15.2}{s} & -\dfrac{1392.5}{s(s + 6.02)(s + 30.3)} \\ -\dfrac{0.52}{s} & 1 + \dfrac{2066.}{s(s + 6.02)(s + 30.3)} \end{bmatrix}$$

$$= \begin{bmatrix} \dfrac{s + 15.2}{s} & \dfrac{-1392.5}{s(s + 6.02)(s + 30.3)} \\ \dfrac{-0.52}{s} & \dfrac{s^3 + 36.32s^2 + 182.4s + 2066.}{s(s + 6.02)(s + 30.3)} \end{bmatrix}$$

This is the matrix we will analyze using the methods of this section.

First we obtain the "multivariable Nyquist plot," i.e., the magnitude and phase of the determinant of the return difference:

$$\begin{aligned} |\mathsf{T}(s)| &= \frac{(s + 15.2)(s^3 + 36.32s^2 + 182.4s + 2066.) - (0.52)(1392.5)}{s^2(s + 6.02)(s + 30.3)} \\ &= \frac{s^4 + 5.52s^3 + 734.5s^2 + 4839s + 30\,686.}{s^4 + 36.32s^3 + 182.4s^2} \\ &= 1 + \mathsf{G}(s) \end{aligned}$$

where

$$\mathsf{G}(s) = \frac{15.2s^3 + 552.06s^2 + 4839.s + 30\,686.}{s^4 + 36.32s^3 + 182.4s^2} \tag{4H.6}$$

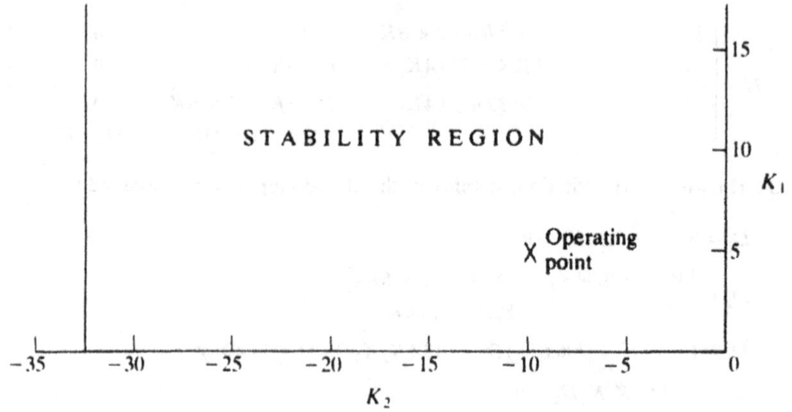

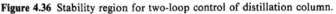

Figure 4.36 Stability region for two-loop control of distillation column.

The (multivariable) Nyquist plot for G(s) as given by (4H.6) is shown in Fig. 4.37. The plot starts with a phase shift of slightly more than 180° and crosses the 180° line (at a frequency of about 6 cycles per hour) at $\sigma \approx -4.7$. A scalar system with this Nyquist plot would be only conditionally stable (with a gain reduction margin of 4.7). This is somewhat misleading, because we know that the plant is open-loop stable and that both gains can be

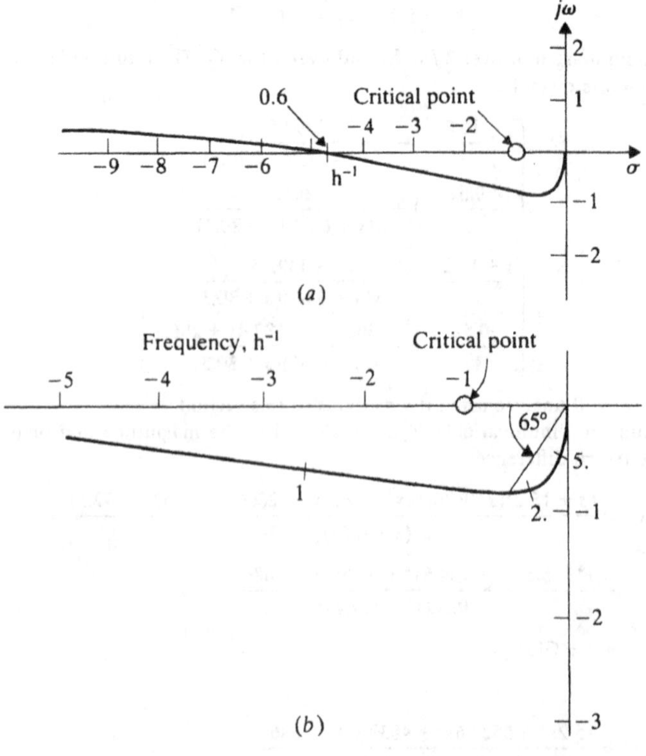

Figure 4.37 Multivariable Nyquist plot for two-loop control of distillation column. (a) Syste appears to be conditionally stable; (b) Apparent phase margin of 65°.

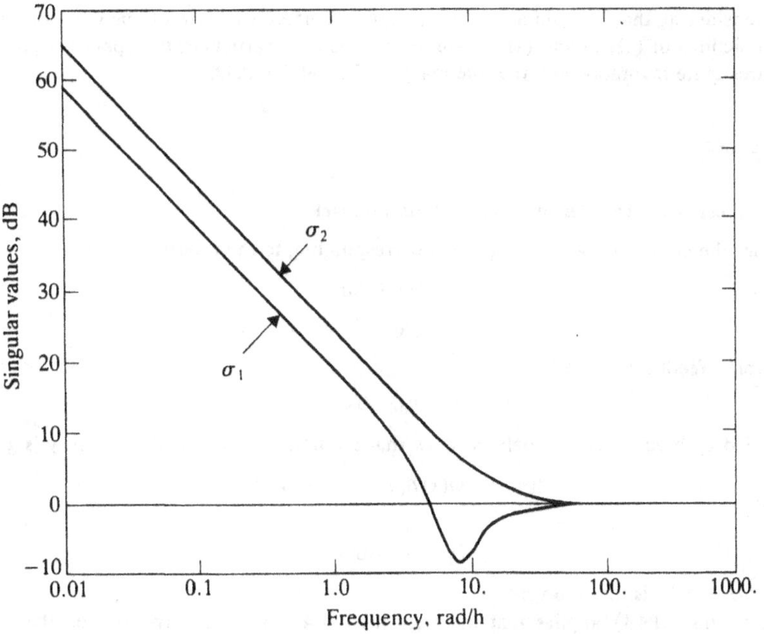

Figure 4.38 Singular values for two-loop control of distillation column.

reduced to zero (individually and simultaneously) without compromising stability. The Nyquist plot also seems to imply an infinite margin for gain increase and this again is misleading because we have already determined that there is a finite upper limit on one of the gains. Finally, we note that the Nyquist diagram suggests a phase margin of about 65°. But how is this phase margin to be interpreted?

If we examine how G(s) is computed from T(s) we can readily explain this apparent paradox. The characteristic equation (4H.3) contains K_1, K_2, and the product K_1K_2. We are trying to assess the effect of changes in K_1 and K_2 as if they resulted from the variation of a single parameter. It ought not to be too surprising to find that an apparent gain reduction margin in the Nyquist diagram is the effect of positive gain margins on K_1 and K_2.

The numerically computed plots (using the customary dB vs. log frequency scale) of the two singular values of the return difference are shown in Fig. 4.38. The smaller of the two singular values reaches a minimum of −8.16 dB at the frequency of 7.9 rad/h (1.26 cycles per hour). The corresponding value of $\alpha = 0.391$, and hence by (4.88) and (4.89) the guaranteed gain and phase margins, are

$$GM = \begin{cases} \dfrac{1}{1+0.391} = 0.718 \\[2ex] \dfrac{1}{1-0.391} = 1.64 \end{cases} \tag{4H.7}$$

and

$$PM = \begin{cases} 2\sin^{-1}\dfrac{0.391}{2} = 22.5° \\[2ex] -2\sin^{-1}\dfrac{0.391}{2} = -22.5° \end{cases} \tag{4H.8}$$

Considering that the gain and phase margin estimates are known to be very conservative, the predictions of (4H.7) and (4H.8) are not too bad. In particular, the upper margin of 1.64 compares quite favorably with the true margin of 11.94/5 = 2.39.

PROBLEMS

Problem 4.1 Closed-loop transfer function for state feedback

Consider the control process of Fig. P4.1, corresponding to the process

$$\dot{x} = Ax + Bu$$

$$y = Cx$$

"State-variable" feedback is used

$$u = u_0 - Gx$$

(*a*) Aided by block-diagram analysis, show that the transfer function from u_0 to y is given by

$$\mathsf{H}(s) = C\Phi(s)B[I + G\Phi(s)B]^{-1} \tag{P4.1}$$

where

$$\mathsf{y}(s) = \mathsf{H}(s)\mathsf{u}(s)$$

and $\Phi(s) = (sI - A)^{-1}$ is the resolvent.

(*b*) Show that (P4.1) implies that the transmission zeros of the process, i.e., the zeros of $|C\Phi(s)B|$ are not altered by state-variable feedback.

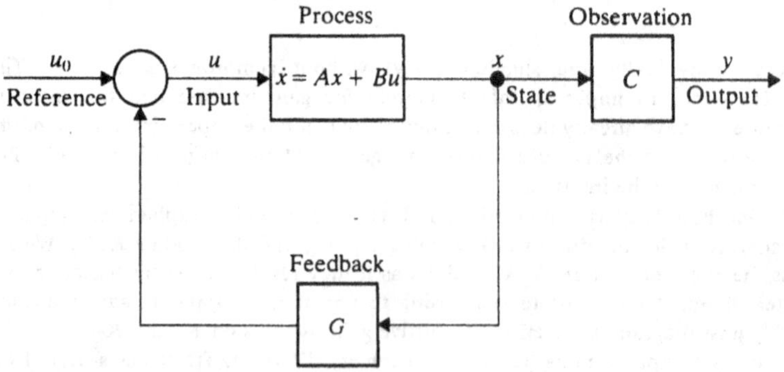

Figure P4.1 Control system with state-variable feedback.

Problem 4.2 Three-capacitance thermal system: feedback control

It is desired to control the temperature $y = v_3$ on the insulated rod of Prob. 2.3 et seq. For this purpose a temperature sensor (e.g., thermocouple) is attached to the rod, and the input power is varied in proportion to the difference $d = T_e - T_1$. In the electrical analog the feedback law is

$$e_0 = u = gd \qquad d = e_d - v_3$$

where e_d is the desired temperature and $v_0 = e_R$ is the ambient temperature.

(*a*) Find the closed loop transfer functions to the output from the reference input and from the disturbance

$$\mathsf{H}_1(s) = \mathsf{v}_3(s)/\mathsf{e}_d(s) \qquad \mathsf{H}_2(s) = \mathsf{v}_3(s)/\mathsf{v}_0(s)$$

respectively.

(*b*) Find the range of the feedback gain g for which the closed-loop system is stable. (Why is it not stable for all g?)

(*c*) Draw the root locus of the system.

(*d*) Draw the Nyquist and Bode diagrams.

(*e*) As a function of g, determine the steady state error $d_{ss} = e_d(\infty)$ when $v_0 \neq e_R$.

Use the following numerical data

$$R = 1 \qquad C = 2$$

Problem 4.3 Three-capacitance thermal system: PI control

The presence of a steady state error in the system of Prob. 4.2 makes the control law unsuited for precise temperature control. To improve the steady state performance, proportional + integral (PI) control is to be used. The transfer function of the "compensator" is to be

$$\frac{\mathrm{u}(s)}{\mathrm{d}(s)} = g_1 + \frac{g_2}{s}$$

instead of g as used in Prob 4.2.

(*a*) Find the range of gains g_1 and g_2 for which the closed-loop system is stable.

(*b*) For $g_2/g_1 = 1$, draw the root locus of the system with g_1 as the variable gain.

(*c*) Draw the Nyquist and Bode diagrams corresponding to part (*b*).

Problem 4.4 Aircraft lateral dynamics: modes and transfer functions

The aerodynamics data for a fighter aircraft are as follows

$$\frac{Y_\beta}{V} = -0.746 \quad \frac{Y_p}{V} = 0.006 \quad \frac{Y_r}{V} = 0.001 \quad \frac{g}{V} = 0.0369 \quad \frac{Y_A}{V} = 0.0012 \quad \frac{Y_R}{V} = 0.0092$$

$$L_\beta = -12.9 \quad L_p = -0.746 \quad L_r = 0.387 \quad L_A = 6.05 \quad L_R = 0.952$$

$$N_\beta = 4.31 \quad N_p = 0.024 \quad N_r = -0.174 \quad N_A = -0.416 \quad N_R = -1.76$$

(*a*) Using the state vector $x = [\beta, p, r, \phi]'$ as given in (2.41), write the A and B matrices.

(*b*) The eigenvalues for the lateral motion of an aircraft consist, typically, of two complex poles with relatively low damping, and a pair of real poles. The complex pair defines a mode called *dutch roll.* One real pole, relatively far from the origin, defines a mode called *roll subsidence,* and a real pole near the origin defines the *spiral* mode. (The latter is sometimes unstable—spiral divergence.) Using the data given above find the four modes for this aircraft.

(*c*) A stability augmentation system (SAS) is to be designed for this aircraft using two rate gyros, each of which measures one of the body rates p and r. Find the transfer functions

$$\frac{\mathrm{p}(s)}{\delta_A(s)} \quad \frac{\mathrm{r}(s)}{\delta_A(s)} \quad \frac{\mathrm{p}(s)}{\delta_R(s)} \quad \frac{\mathrm{r}(s)}{\delta_R(s)}$$

Is it apparent from these transfer functions why the ailerons are used for roll (p) control and the rudder is used for yaw (r) control?

(*d*) Find the transmission zeros of the process.

Problem 4.5 Aircraft longitudinal dynamics, simplified

The aerodynamic coefficients for an aircraft are approximated by

$$\frac{Z_\alpha}{V} = -1 \qquad \frac{Z_E}{V} = -0.1$$

$$M_q = -0.5 \quad M_\alpha = -5. \quad M_E = -9.$$

$$X_\alpha = -14. \quad X_E = -1.$$

All other coefficients are negligible.

(*a*) Find the open-loop poles (short-period and phugoid modes) of the aircraft.

(*b*) Find the transfer functions of the aircraft from δ_E to θ and Δu.

Problem 4.6 Distillation column

By use of the Hurwitz matrix, verify that the range of gains for stability of the closed-loop system of Example 4H is as shown in Fig. 4.36.

Problem 4.7 Double-effect evaporator

Consider the double-effect evaporator introduced in Example 2H, with the dynamics matrices as defined by (2H.5)-(2H.7).

(*a*) Find the open-loop poles (eigenvalues) of the system.

(*b*) The observed quantities ("outputs") are $y_1 = x_1$ (first-effect holdup) and $y_2 = x_4$ (second-effect holdup). What is the observation matrix? Find the transfer functions from the controls to the outputs.

Problem 4.8 Double-effect evaporator: feedback control

A two-loop control system is proposed for the double-effect evaporator of Prob. 4.7 in which the first-effect holdup x_1 is controlled by the steam-flow rate u_1 and the second-effect holdup x_4 is controlled by the first-effect bottoms flow rate u_2. The resulting system has the structure shown in Fig. P4.8.

(*a*) Find the range of gains g_1 and g_2 for which the closed-loop system is stable.

(*b*) Let the return difference for the process be given by

$$\mathsf{T}(s) = I + G\mathsf{H}(s)$$

where $G = [g_1, g_2]$ and $\mathsf{H}(s)$ is the 2×2 transfer-function matrix. Plot the singular values of the system as a function of frequency, with the loops opened at the input. Use a nominal value $G = \bar{G} = [-40, 40]$.

(*c*) Repeat part (*b*) with a nominal gain matrix $\bar{G} = [-20, 10]$.

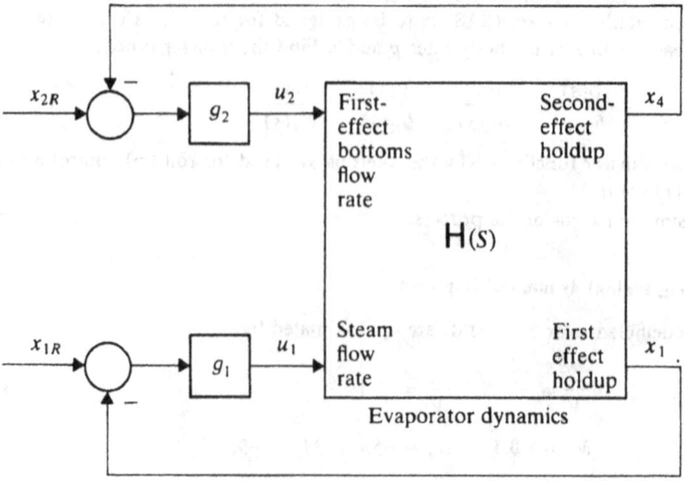

Figure P4.8 Control system for double-effect evaporator.

Problem 4.9 Gun turret: range of gain for stability

Using the Routh table or the Hurwitz matrix criterion, find the range of gains for which the turret control system of Example 4E is asymptotically stable.

Problem 4.10 Missile autopilot: Bode and Nyquist diagrams

Draw the Bode and Nyquist diagrams for the missile autopilot of Example 4F.

Problem 4.11 Missile autopilot: Acceleration and angular rate feedback

In addition to the feedback of the normal acceleration error $e = a_{NC} - a_N$, a a missile autopilot will frequently also make use of the pitch rate. This will thus result in a control law

$$u = K_1(a_{NC} - a_N) - K_2 q \tag{P4.11a}$$

(*a*) Using this control law, find the range of K_1 and K_2 for which the closed-loop system is stable.

(*b*) In order to implement the control law of (P4.11*a*) an additional sensor (a rate gyro to measure q) is needed. What are the benefits that such a control law might confer on the system that would justify the additional cost of the sensor?

NOTES

Note 4.1 Frequency-domain analysis

The frequency-domain approach to control system analysis and design which was developed during the 1940s and 1950s is often called the "classical" approach to distinguish it from the "modern" state-space approach which had its beginnings in the late 1950s and early 1960s.

Not all investigators agreed on the advantages of the state-space approach over the frequency-domain approach, and a significant minority remain unconvinced to the present day. The complaint of the frequency-domain advocates is that the reason for the use of feedback is the uncertainties in the dynamic process, and that when these uncertainties are present, the qualitative methods of frequency-domain analysis are more appropriate. Qualitative system properties such as bandwidth, stability margins, etc., were regarded as difficult to study by state-space methods. To answer the need for computational design tools for multivariable systems that would rival the state-space tools in power, the classicists developed such techniques as multivariable root loci, multivariable Nyquist plots, and various subsidiary techniques. Much of the theoretical results and most of the design software is the product of the efforts of investigators of the United Kingdom, led by H. H. Rosenbrock[17] of the University of Manchester and A. J. G. MacFarlane[18] of the University of Cambridge. I. Horowitz of the Weizmann Institute (Israel) is another leading exponent of the classical approach. Since state-space concepts have been included in many of the newer frequency-domain methods, it might be appropriate to call this work "neoclassical." Neoclassical frequency-domain activity of the western hemisphere is represented by the theoretical work of C. A. Desoer of the University of California (Berkeley) and G. Zames of McGill University.

Note 4.2 Aircraft dynamic modes

The terminology of aircraft dynamics stems from the 1930s and 1940s. The longitudinal modes are called "short-period" and "phugoid"; the lateral modes are "dutch-roll," "spiral," and "roll subsidence." The background of this terminology is given by Etkin.[19]

Note 4.3 Bandwidth of multivariable systems

R. W. Bass[15] has defined the bandwidth of a system in state-space form as

$$W = (\|A\| / \|A^{-1}\|)^{1/2}$$

and has proved that $\|\Phi(j\omega)\Phi^{-1}(0)\| \leqq 1/\sqrt{2}$ for $\omega \gg W$, where $\Phi(j\omega) = (j\omega I - A)^{-1}$, i.e., the resolvent of the system at $s = j\omega$. These inequalities are generalizations of the concept of bandwidth for scalar systems.

A shortcoming of this definition of bandwidth is that neither the control distribution matrix B nor the observation matrix C enter into the definition. Thus this definition depends only on the poles, and not on the zeros, of the system.

Note 4.4 Transmission zeros

It is said with considerable justification that state-space methods are concerned primarily with the poles of a system rather than with its zeros. This is surely one of the reasons that transmission zeros play a much larger role within the classical and neoclassical (frequency-domain) methodology than they do in the state-space methodology. In the latter the compensator is designed by the separation principle: first a "full-state" feedback law is designed assuming all the states are measured; then an observer is designed to estimate the missing states. In the first step only the A and B matrices are used; in the second step only the A and C matrices are used. The only place A, B, and C are brought together is when the full-state feedback law is combined with the observer to yield the required compensator. Since the transmission zeros depend on A, B, and C together, through $\mathsf{H}(s) = C(sI - A)^{-1}B$, the state-space approach obscures the transmission zeros. Since the behavior of a system depends not only on its poles, but also on its zeros, the absence of a clear connection in the state-space methodology between the compensator design and the transmission zeros is a shortcoming of this methodology and suggests a possible direction for future research.

As one might expect, transmission zeros receive the greatest attention in books emphasizing the frequency-domain methodology. In particular, see [17] and [18].

Note 4.5 Singular-value analysis

The singular values of a matrix are of particular importance in determining whether a matrix is relatively easy to invert ("well conditioned") or difficult to invert ("ill conditioned"). They are consequently of special interest in the branch of numerical analysis that is concerned with algorithms for the manipulation of large matrices. Singular-value analysis is prominent in books on numerical methods, such as Householder.[20]

Note 4.6 Robustness of multivariable control systems

The study of robustness of multivariable control systems by means of singular-value analysis is represented by the work of a number of investigators centered at the Massachusetts Institute of Technology beginning in the late 1970s with the doctoral research of M. G. Safonov.[21] A number of papers that make use of singular-value analysis as an analytical tool for multivariable control systems are included in the Special Issue on Linear Multivariable Control Systems of the *IEEE Transactions on Automatic Control*.[22]

Note 4.7 Nonminimum phase poles and zeros

It is readily seen that a pole or a zero at $s = -\sigma_0 + j\omega_0$ or at $s = +\sigma_0 + j\omega_0$ will have the same effect on the Bode amplitude characteristic of a system, but will have different effects on the phase characteristic. The phase associated with the pole or zero in the left half-plane is $\phi_1 = \tan^{-1}(\sigma_0/\omega + \omega_0)$ while the phase associated with the pole or zero in the right half-plane is $\phi_2 = \tan^{-1}(-\sigma_0/\omega + \omega_0)$; ϕ_1 is always less than 90 degrees while ϕ_2 is greater than 90 degrees. Thus of the two, the left half-plane pole or zero is the one of *minimum-phase*, a term first used by Bode.[2] A nonminimum phase pole is always indicative of an unstable system. Nonminimum phase zeros, on the other hand, can occur in a stable system, but if they do occur they are often a source of difficulty to the control system designer.

REFERENCES

1. Nyquist, H., "Regeneration Theory," *Bell System Technical Journal*, vol. 11, 1932, pp. 126–147.
2. Bode, H. W., *Network Analysis and Feedback Amplifier Design*, D. Van Nostrand, New York, 1945.
3. Black, H. S., "Inventing the Negative Feedback Amplifier," *IEEE Spectrum*, vol. 14, no. 1, January 1977, pp. 54–60.
4. Schwarz, R. J., and Friedland, B., *Linear Systems*, McGraw-Hill Book Co., New York, 1965.
5. Mason, S. J., "Feedback Theory: Some Properties of Signal Flow Graphs," *Proceedings of the IRE*, vol. 41, no. 9, September 1953.
6. Rynaski, E. J., "Flight Control Synthesis Using Robust Output Observers," *Proc. AIAA Guidance and Control Conference*, San Diego, CA, September 1982, pp. 825–831.
7. Routh, E. J., *A Treatise on the Stability of a Given State of Motion*, Macmillan & Co., London, 1877.
8. Hurwitz, A., "Über die Bedingungen, unter welchen einer Gleichung nur Wurzeln mit negativen reelen Teilen besitzt," *Math. Ann.*, vol. 146, 1895, pp. 273–284.
9. Parks, P. C., "A New Proof of the Routh-Hurwitz Stability Criterion Using the Second Method of Lyapunov," *Proc. Cambridge Philosophical Society*, vol. 58, pt. 4, 1962, pp. 694–702.
10. Lyapunov, M. A., "Le problème général de la stabilité du mouvement," *Ann. Fac. Sci. Toulouse*, vol. 9, 1907, pp. 203–474.
11. D'Azzo, J. J., and Houpis, C. H., *Linear Control System Analysis and Design: Conventional and Modern*, McGraw-Hill Book Co., New York, 1981.
12. Ogata, K., *Modern Control Engineering*, Prentice-Hall Inc., Englewood Cliffs, NJ, 1970.
13. Evans, W. R., "Graphical Analysis of Control Systems," *Trans. AIEE, Pt. II*, vol. 67, 1948, pp. 547–551.
14. Churchill, R. V., *Introduction to Complex Variables and Applications*, McGraw-Hill Book Co., New York, 1948.
15. Bass, R. W., "Robustified LQG Synthesis to Specifications," *Proc. 5th Meeting of Coord. Group On Modern Control Theory, Part II*, Dover, NJ, October 1983, pp. 11–93.
16. Lehtomaki, N. A., Sandell, N. R., Jr., and Athans, M., "Robustness Results on LQG Based Multivariable Control System Designs," *IEEE Trans. on Automatic Control*, vol. AC-26, no. 1, February 1981, pp. 75–93.
17. Rosenbrock, H. H., *Computer Aided Control System Design*, Academic Press, New York, 1974.
18. MacFarlane, A. J. G. (ed.), *Frequency Response Methods in Control Systems*, IEEE Press, New York, 1979.
19. Etkin, B., *Dynamics of Flight*, John Wiley & Sons, New York, 1959.
20. Householder, A. S., *The Theory of Matrices in Numerical Analysis*, Blaisdell Publishing Co., Waltham, MA, 1964.
21. Safonov, M. G., *Stability and Robustness of Multivariable Feedback Systems*, MIT Press, Cambridge, MA, 1980.
22. Special Issue on Linear Multivariable Control Systems, *IEEE Trans. on Automatic Control*, vol. AC-26, no. 1, February 1981.

CHAPTER

FIVE

CONTROLLABILITY AND OBSERVABILITY

5.1 INTRODUCTION

Some state-space concepts can be regarded as reinterpretations of older, frequency-domain concepts; others are distinctive to state-space methods. Controllability and observability are in this latter category.

The ideas of controllability and observability were introduced by R. E. Kalman in the mid 1950s as a way of explaining why a method of designing compensators for unstable systems by cancelling unstable poles (i.e., poles in the right half-plane) by zeros in the right half-plane is doomed to fail even if the cancellation is *perfect*. (It was already known that this method of compensation was not feasible because perfect cancellation is not possible in practice. See Note 5.1.) Kalman showed that a perfect pole-zero cancellation would result in an unstable system with a stable transfer function. The transfer function, however, is of lower order than the system, and the unstable modes are either not capable of being affected by the input (uncontrollable) or not visible in the output (unobservable).

In frequency-domain analysis it is tacitly assumed that the dynamic properties of a system are completely determined by the transfer function of the system. That this is not always the case is illustrated by the following example.

Example 5A Hypothetical system Figure 5.1 shows the block-diagram of a hypothetical system contrived specifically to illustrate the concepts of controllability and observability. There is no reason, however, why it would not represent some physical process.

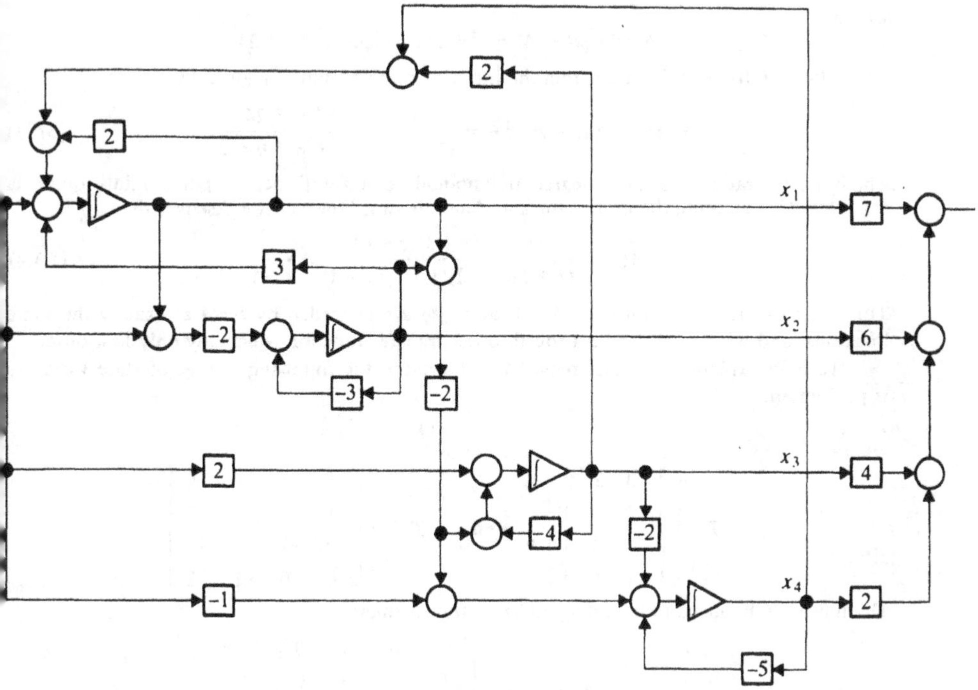

Figure 5.1 Hypothetical fourth-order system to illustrate controllability and observability.

The differential equations of the system, obtained by inspection of the block-diagram are

$$\begin{aligned}\dot{x}_1 &= 2x_1 + 3x_2 + 2x_3 + x_4 + u \\ \dot{x}_2 &= -2x_1 - 3x_2 - 2u \\ \dot{x}_3 &= -2x_1 - 2x_2 - 4x_3 + 2u \\ \dot{x}_4 &= -2x_1 - 2x_2 - 2x_3 - 5x_4 - u\end{aligned} \tag{5A.1}$$

and the observation equation is

$$y = 7x_1 + 6x_2 + 4x_3 + 2x_4 \tag{5A.2}$$

Thus, the matrices of the state-space representation are:

$$A = \begin{bmatrix} 2 & 3 & 2 & 1 \\ -2 & -3 & 0 & 0 \\ -2 & -2 & -4 & 0 \\ -2 & -2 & -2 & -5 \end{bmatrix} \qquad B = \begin{bmatrix} 1 \\ -2 \\ 2 \\ -1 \end{bmatrix} \qquad C = [7 \quad 6 \quad 4 \quad 2]$$

The resolvent corresponding to A is given by

$(sI - A)^{-1} =$

$$\frac{1}{\Delta(s)}\begin{bmatrix} s^3 + 12s^2 + 47s + 6 & 3s^2 + 21s + 36 & 2s^2 + 14s + 24 & s^2 + 7s + 12 \\ -2s^2 - 18s - 40 & s^3 + 7s^2 + 8s - 16 & -4s - 16 & -2s - 8 \\ -2s^2 - 12s - 10 & -2s^2 - 12s - 10 & s^3 + 6s^2 + 7s + 2 & -2s - 2 \\ -2s^2 - 6s - 4 & -2s^2 - 6s - 4 & -2s^2 - 6s - 4 & s^3 + 5s^2 + 8s + 4 \end{bmatrix}$$

where

$$\Delta(s) = |sI - A| = s^4 + 21s^3 + 35s^2 + 50s + 24$$

Thus the transfer function from the input u to the output y is given by

$$\mathsf{H}(s) = C(sI - A)^{-1}B = \frac{s^3 + 9s^2 + 26s + 24}{s^4 + 21s^3 + 35s^2 + 50s + 24} \tag{5A.3}$$

which is the ratio of a third-degree polynomial to a fourth-degree polynomial—quite as expected. On factoring the numerator and denominator, however, we discover that

$$\mathsf{H}(s) = \frac{(s+2)(s+3)(s+4)}{(s+1)(s+2)(s+3)(s+4)} = \frac{1}{s+1} \tag{5A.4}$$

Thus, three of the poles (at $s = -2, -3$, and -4) are cancelled by zeros at exactly the same locations, and what seems to be a fourth-order transfer function is actually only first-order.

To help explain this rather remarkable behavior, the following change of state variables is performed:

$$\bar{x} = Tx$$

where

$$T = \begin{bmatrix} 4 & 3 & 2 & 1 \\ 3 & 3 & 2 & 1 \\ 2 & 2 & 2 & 1 \\ 1 & 1 & 1 & 1 \end{bmatrix} \quad \text{and} \quad T^{-1} = \begin{bmatrix} 1 & -1 & 0 & 0 \\ -1 & 2 & -1 & 0 \\ 0 & -1 & 2 & -1 \\ 0 & 0 & -1 & 2 \end{bmatrix}$$

The matrix T happens to be a diagonalizing transformation

$$TAT^{-1} = \Lambda = \begin{bmatrix} -1 & 0 & 0 & 0 \\ 0 & -2 & 0 & 0 \\ 0 & 0 & -3 & 0 \\ 0 & 0 & 0 & -4 \end{bmatrix}$$

and the corresponding control and observation matrices are

$$\bar{B} = TB = \begin{bmatrix} 1 \\ 0 \\ 1 \\ 0 \end{bmatrix}, \quad \bar{C} = CT^{-1} = [1 \quad 1 \quad 0 \quad 0]$$

Hence the corresponding state equations are

$$\begin{aligned} \dot{\bar{x}}_1 &= -\bar{x}_1 + u \\ \dot{\bar{x}}_2 &= -2\bar{x}_2 \\ \dot{\bar{x}}_3 &= -3\bar{x}_3 + u \\ \dot{\bar{x}}_4 &= -4\bar{x}_4 \end{aligned} \tag{5A.5}$$

and the observation equation is

$$y = \bar{x}_1 + \bar{x}_2 \tag{5A.6}$$

A block-diagram representation of (5A.5) and (5A.6) is shown in Fig. 5.2. Clearly, the input u affects only the state variables $\bar{x}_1$ and $\bar{x}_3$; $\bar{x}_2$ and $\bar{x}_4$ are unaffected by the input. The output y depends only on $\bar{x}_1$ and $\bar{x}_2$; $\bar{x}_3$ and $\bar{x}_4$ do not contribute to the output. Thus, in the transformed coordinates, the system has four different subsystems. (In this case each subsystem is only first-order.)

$\bar{x}_1$: affected by the input; visible in the output
$\bar{x}_2$: unaffected by the input; visible in the output
$\bar{x}_3$: affected by the input; invisible in the output
$\bar{x}_4$: unaffected by the input; invisible in the output

Only the first subsystem $\bar{x}_1$ contributes to the transfer function $\mathsf{H}(s)$, which clearly is $1/(s+1)$.

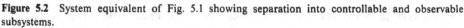

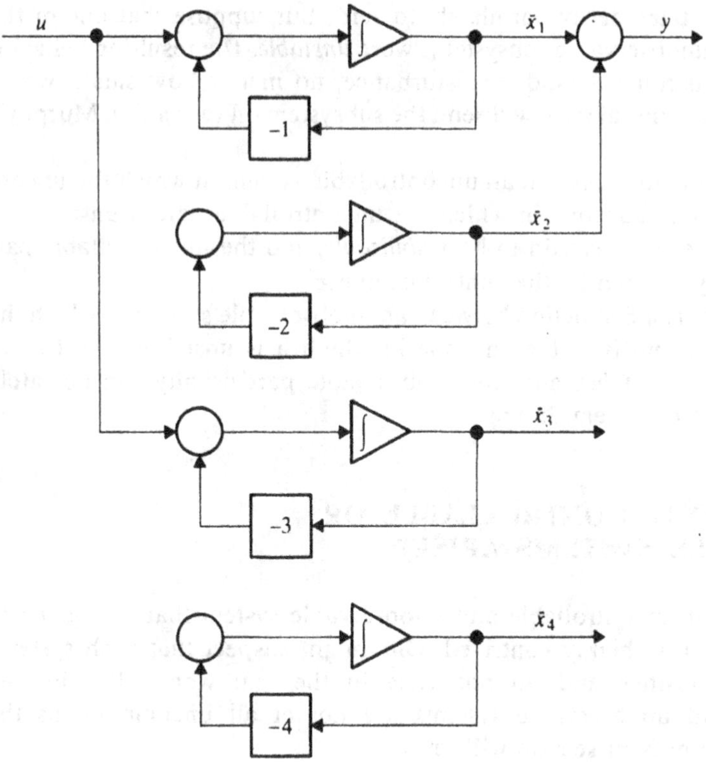

Figure 5.2 System equivalent of Fig. 5.1 showing separation into controllable and observable subsystems.

Example 5A is a microcosm of the general case. As Kalman has shown,[1] every system of the generic form

$$\dot{x} = Ax + Bu$$

$$y = Cx$$

can be transformed into the four subsystems of Fig. 5.2. The first subsystem is both controllable and observable: the second is uncontrollable but observable; the third is controllable but unobservable; and the fourth is neither observable nor controllable. The transfer function of the system is determined only by the controllable and observable subsystem. It thus follows that if the transfer function of a single-input, single-output system is of lower degree than the dimension of the state-space, then the system must contain an uncontrollable subsystem, or an unobservable subsystem, or possibly both. By convention, if a system contains an uncontrollable subsystem it is said to be *uncontrollable*; likewise, if it contains an unobservable subsystem it is said to be *unobservable.* (See Note 5.2.)

The system in the foregoing example is asymptotically stable: all its poles are in the left half-plane, so the consequences of the system being unobservable and uncontrollable are innocuous. Any initial conditions on the uncontrollable and unobservable states decay harmlessly to zero. But suppose that one of the uncontrollable or unobservable subsystems were *unstable.* The resulting behavior could well be disastrous: a random disturbance, no matter how small, which establishes a nonzero initial state will send the subsystem off to infinity. Murphy's law par excellence!

There is a distinction between an uncontrollable system in which the uncontrollable part is stable and one in which the uncontrollable part is unstable. A system of the former type is said to be *stabilizable*, and the uncontrollable part often can be safely ignored by the control engineer.

Similarly, there is a distinction between an unobservable system in which the unobservable subsystem is stable and one in which it is unstable. The former type is said to be *detectable*, and the unobservable part usually can be safely ignored in the control system design.

5.2 WHERE DO UNCONTROLLABLE OR UNOBSERVABLE SYSTEMS ARISE?

The example of an uncontrollable and unobservable system that was given in the previous section is highly contrived. One might suspect that such systems are academic curiosities and do not arise in the real world. But in fact uncontrollable and unobservable systems are not at all uncommon, as the illustrations of the present section will reveal.

Redundant state variables One common source of uncontrollable systems arises when redundant state variables are defined. Consider, for example, the dynamic system

$$\dot{x} = Ax + Bu$$

and suppose, for some reason, more state variables, proportional to those already present in the state vector x are defined:

$$z = Fx \tag{5.1}$$

where F is an $n \times k$ matrix. Then

$$\dot{z} = F\dot{x} = F(Ax + Bu)$$

is a true differential equation, so we can define a "metastate" vector

$$\mathbf{x} = \begin{bmatrix} x \\ z \end{bmatrix}$$

which satisfies the differential equation

$$\dot{\mathbf{x}} = \mathbf{A}\mathbf{x} + \mathbf{B}u \tag{5.2}$$

where

$$\mathbf{A} = \begin{bmatrix} A & 0 \\ FA & 0 \end{bmatrix} \quad \mathbf{B} = \begin{bmatrix} B \\ FB \end{bmatrix}$$

The system characterized by (5.2) has the block diagram shown in Fig. 5.3(a). There is a path from the input u to the state x and to the (redundant) state z; superficially the system seems to be controllable. But consider the change of variable

$$\bar{\mathbf{x}} = \mathbf{T}\mathbf{x} \tag{5.3}$$

where

$$\mathbf{T} = \begin{bmatrix} I_k & 0 \\ -F & I_n \end{bmatrix} \qquad \mathbf{T}^{-1} = \begin{bmatrix} I_k & 0 \\ F & I_n \end{bmatrix} \tag{5.4}$$

where I_l $(l = k, n)$ is an l-by-l identity matrix. (Multiply $\mathbf{T}$ by $\mathbf{T}^{-1}$ to verify (5.4).)

The dynamics matrix of the transformed system is given by

$$\bar{\mathbf{A}} = \mathbf{TAT}^{-1} = \begin{bmatrix} I_k & 0 \\ -F & I_n \end{bmatrix} \begin{bmatrix} A & 0 \\ FB & 0 \end{bmatrix} \begin{bmatrix} I_k & 0 \\ F & I_n \end{bmatrix} = \begin{bmatrix} A & 0 \\ 0 & 0 \end{bmatrix}$$

and the control matrix is given by

$$\bar{\mathbf{B}} = \mathbf{TB} = \begin{bmatrix} I_k & 0 \\ -F & I_n \end{bmatrix} = \begin{bmatrix} B \\ FB \end{bmatrix} = \begin{bmatrix} B \\ 0 \end{bmatrix}$$

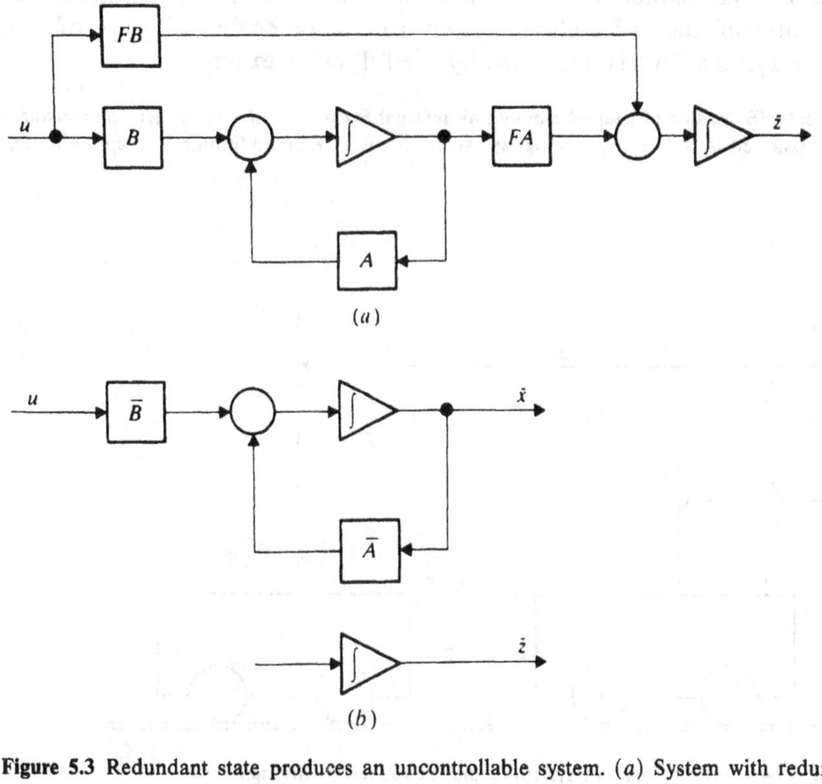

Figure 5.3 Redundant state produces an uncontrollable system. (a) System with redundant state $z = Fx$; (b) System of (a) after being transformed by $\bar{x} = Tx$.

Thus, in the transformed system,

$$\dot{\bar{x}} = \bar{A}\bar{x} + \bar{B}u$$

$$\dot{\bar{z}} = 0 \tag{5.5}$$

Differential equation (5.5) represents k integrators with no inputs connected to them (Fig. 5.3(b)) and hence the substate z is uncontrollable.

All the algebra used above is really quite superfluous. The transformation **T** of (5.3) merely asserts that

$$\bar{z} = z - Fx$$

and, by virtue of (5.1), $\bar{z} \equiv 0$, so surely (5.5) must hold.

Now of course no one would intentionally use more state variables than the minimum number needed to characterize the behavior of the dynamic process. In a complicated process with unfamiliar physics, however, the control system engineer may be tempted to write down differential equations for everything in sight and in so doing, may write down more equations than are necessary. This will invariably result in the model for an uncontrollable system.

Physically uncontrollable system Another instance of an uncontrollable system is one in which the only forces and torques are *internal* to the system. For example, as a consequence of Newton's law of action and reaction, the location of the center of mass of a closed system cannot be changed by use of forces within the system. This is illustrated by the following example.

Example 5B Motion of coupled masses with internal force Consider the system comprising two carts coupled by a (passive) spring, as shown in Fig. 5.4. In addition to the spring force, an

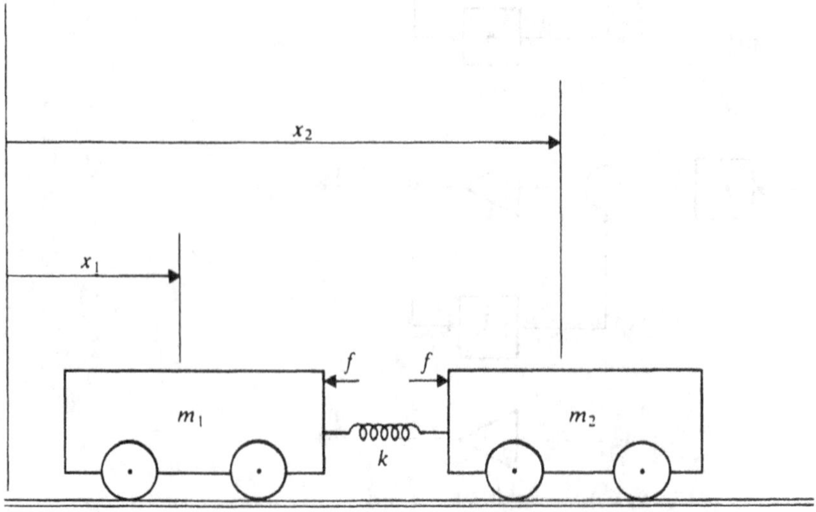

Figure 5.4 Center of mass of system cannot be moved by internal force.

active control force f is to be provided by some means within the system, so that whatever the force on cart 1, an equal and opposite reaction force from that source must push on cart 2. Thus the differential equations of the pair of carts are

$$\frac{dx_1}{dt} = \dot{x}_1 \tag{5B.1}$$

$$\frac{dx_2}{dt} = \dot{x}_2 \tag{5B.2}$$

$$\frac{d\dot{x}_1}{dt} = -\frac{k}{m_1}(x_1 - x_2) - \frac{f}{m_1} \tag{5B.3}$$

$$\frac{d\dot{x}_2}{dt} = -\frac{k}{m_2}(x_2 - x_1) + \frac{f}{m_2} \tag{5B.4}$$

From (5B.3) and (5B.4)

$$m_1 \frac{d\dot{x}_1}{dt} + m_2 \frac{d\dot{x}_2}{dt} = \frac{d}{dt}(m_1\dot{x}_1 + m_2\dot{x}_2) = 0$$

Thus

$$m_1\dot{x}_1 + m_2\dot{x}_2 = m\dot{x}_c = \text{const}$$

where

$$m = m_1 + m_2 \qquad x_c = \frac{m_1x_1 + m_2x_2}{m_1 + m_2} = \text{center of mass}$$

Thus the center of mass of the system cannot be moved by the internal force f.

This physical fact is formally illustrated by matrix analysis. For the original system (5B.1)-(5B.4) the state vector is

$$x = [x_1, x_2, \dot{x}_1, \dot{x}_2]'$$

and the corresponding matrices are

$$A = \begin{bmatrix} 0 & 0 & 1 & 0 \\ 0 & 0 & 0 & 1 \\ -k/m_1 & k/m_1 & 0 & 0 \\ k/m_2 & -k/m_2 & 0 & 0 \end{bmatrix} \qquad B = \begin{bmatrix} 0 \\ 0 \\ -1/m_1 \\ 1/m_2 \end{bmatrix} \tag{5B.5}$$

We make the change of state variables

$$x_c = \frac{m_1}{m}x_1 + \frac{m_2}{m}x_2$$

$$\delta = x_1 - x_2$$

$$\dot{x}_c = \frac{m_1}{m}\dot{x}_1 + \frac{m_2}{m}\dot{x}_2$$

$$\dot{\delta} = \dot{x}_1 - \dot{x}_2$$

Then

$$\bar{x} = \begin{bmatrix} x_c \\ \delta \\ \dot{x}_c \\ \dot{\delta} \end{bmatrix} = \begin{bmatrix} m_1/m & m_2/m & 0 & 0 \\ 1 & -1 & 0 & 0 \\ 0 & 0 & m_1/m & m_2/m \\ 0 & 0 & 1 & -1 \end{bmatrix} \begin{bmatrix} x_1 \\ x_2 \\ \dot{x}_1 \\ \dot{x}_2 \end{bmatrix}$$

Thus

$$T = \begin{bmatrix} m_1/m & m_2/m & 0 & 0 \\ 1 & -1 & 0 & 0 \\ 0 & 0 & m_1/m & m_2/m \\ 0 & 0 & 1 & -1 \end{bmatrix} \quad \text{and} \quad T^{-1} = \begin{bmatrix} 1 & m_2/m & 0 & 0 \\ 1 & -m_1/m & 0 & 0 \\ 0 & 0 & 1 & m_2/m \\ 0 & 0 & 1 & -m_1/m \end{bmatrix}$$

Thus we find that

$$\bar{A} = TAT^{-1} = \begin{bmatrix} 0 & 0 & 1 & 0 \\ 0 & 0 & 0 & 1 \\ 0 & 0 & 0 & 0 \\ 0 & -k(1/m_1 + 1/m_2) & 0 & 0 \end{bmatrix} \qquad \bar{B} = TB = \begin{bmatrix} 0 \\ 0 \\ 0 \\ -(1/m_1 + 1/m_2) \end{bmatrix}$$

Hence, as expected, the differential equations of the transformed system are

$$\frac{dx_c}{dt} = \dot{x}_c$$

$$\frac{d\delta}{dt} = \dot{\delta}$$

$$\frac{d\dot{x}_c}{dt} = 0$$

$$\frac{d\dot{\delta}}{dt} = -k\left(\frac{1}{m_1} + \frac{1}{m_2}\right)\delta - \left(\frac{1}{m_1} + \frac{1}{m_2}\right)f$$

The internal force f can change the distance δ between x_1 and x_2 but not the coordinates x_1 and x_2 independently. To do that, an external force is needed.

This example illustrates that the mathematical model must be consistent with the physics of the system. The A and B matrices must be exactly as given by (5B.5). If an error in calculation were made, for example, such that the fourth element in the B matrix were not $1/m_2$ but some other number, the system would seem to be controllable and one might try to move the center of mass by using the force f. But of course no matter how large we make f, the center of mass won't move.

Too much symmetry Another situation that results in an uncontrollable system arises when the system in question has too much symmetry. This typically arises in electrical networks that contain balanced bridges, and in mechanical systems which have similar symmetry. This is illustrated by the following example.

Example 5C Balanced bridges are uncontrollable The differential equations of the electrical network, or its thermal analog, shown in Fig. 5.5, were found in Chap. 2 to be

$$\dot{v}_1 = -\frac{1}{C_1}\left(\frac{1}{R_1} + \frac{1}{R_3}\right)v_1 + \frac{1}{C_1 R_3}v_2 + \frac{1}{C_1 R_1}e_0 \tag{5C.1}$$

$$\dot{v}_2 = \frac{1}{C_2 R_3}v_1 - \frac{1}{C_2}\left(\frac{1}{R_2} + \frac{1}{R_3}\right)v_2 + \frac{1}{C_2 R_2}e_0 \tag{5C.2}$$

Consider the difference voltage

$$\bar{v}_1 = v_1 - v_2$$

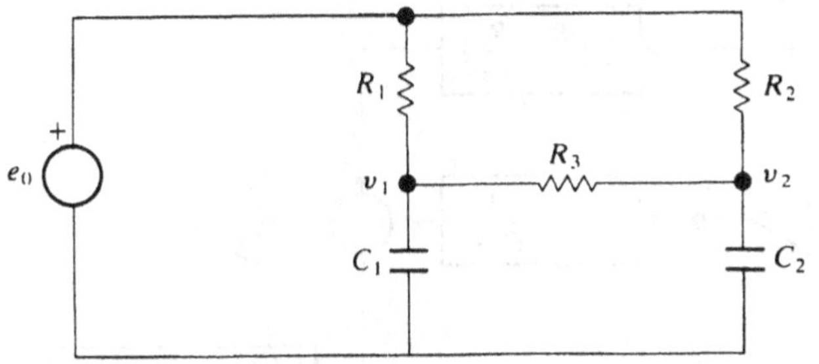

Figure 5.5 Electrical bridge network.

The time-derivative of $\bar{v}_1$ using (5C.1) and (5C.2) is

$$\frac{d\bar{v}_1}{dt} = -\left[\frac{1}{C_1}\left(\frac{1}{R_1}+\frac{1}{R_3}\right)+\frac{1}{C_2R_3}\right]v_1+\left[\frac{1}{C_1R_3}+\frac{1}{C_2}\left(\frac{1}{R_2}+\frac{1}{R_3}\right)\right]v_2$$
$$+\frac{R_2C_2-R_1C_1}{C_1C_2R_1R_2}e_0 \tag{5C.3}$$

If the bridge is *balanced*, i.e.,

$$R_1C_1 = R_2C_2 \tag{5C.4}$$

then the coefficient of the input voltage e_0 vanishes. And moreover, the bracketed coefficients of v_1 and v_2 become equal. Thus (5C.3) reduces to

$$\frac{d\bar{v}_1}{dt} = -\frac{R_1+R_2+R_3}{C_1R_1R_3}\bar{v}_1$$

This means that the voltage $\bar{v}_1 = v_1 - v_2$ between the terminals of R_3 cannot be influenced by the input e_0; the voltage $\bar{v}_1$ decays from whatever initial voltage it starts with to zero with the time constant $\tau = C_1R_1R_3/(R_1+R_2+R_3)$ irrespective of the input.

If the only observation is the voltage $\bar{v}_1 = v_1 - v_2$, then the system is also *unobservable.* To see this we define the transformed state

$$\bar{v}_1 = v_1 - v_2$$
$$\bar{v}_2 = v_2$$

To this definition of transformed voltages there corresponds the transformation matrix

$$T = \begin{bmatrix} 1 & -1 \\ 0 & 1 \end{bmatrix} \quad \text{and} \quad T^{-1} = \begin{bmatrix} 1 & 1 \\ 0 & 1 \end{bmatrix}$$

The transformed differential equations are

$$\dot{\bar{v}}_1 = -\left[\frac{1}{C_1}\left(\frac{1}{R_1}+\frac{1}{R_2}\right)+\frac{1}{C_2R_3}\right]\bar{v}_1+\left(\frac{1}{C_2R_2}-\frac{1}{C_1R_1}\right)\bar{v}_2+\left(\frac{R_2}{C_1}-\frac{R_1}{C_2}\right)\frac{e_0}{R_1R_2}$$
$$\dot{\bar{v}}_2 = \frac{1}{C_2R_3}\bar{v}_1-\frac{1}{C_2R_2}\bar{v}_2+\frac{R_1}{C_1}\frac{e_0}{R_1R_2}$$

and the observation is given by

$$y = \bar{v}_1$$

When the bridge is balanced there is no path from $\bar{v}_2$ to the output, hence $\bar{v}_2$ cannot be observed. (See Fig. 5.6.)

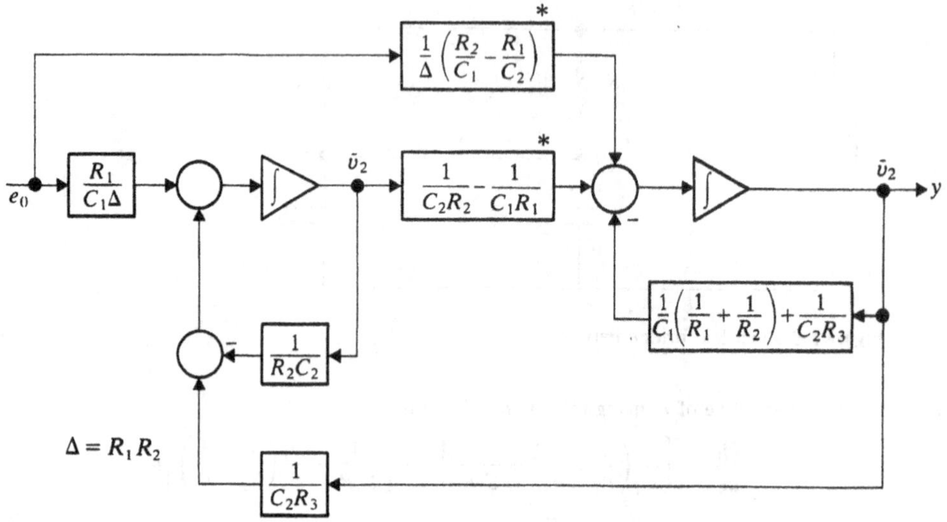

Figure 5.6 Block diagram showing that balanced bridge is neither controllable nor observable. (Elements with * open when bridge is balanced.)

When numerical values are inserted for the physical parameters in the systems of Examples 5B and 5C there is no way of distinguishing between the qualitative nature of the uncontrollability of the two systems: they are both simply uncontrollable. But physically there is a very important distinction between the two systems. The two-mass mechanical system is uncontrollable for *every* value of the parameters (masses, spring rates); the only way to control the position of the center of mass is to add an external force. This necessitates a *structural* change to the system. The balanced bridge, however, is uncontrollable only for one specific relationship between the parameters, namely the balance condition (5C.4). In other words, the system is *almost always controllable.* (As a practical matter, it will be difficult to control v_1 and v_2 independently when (5C.4) is nearly true. This raises the issue of degree of controllability, a topic discussed in Note 5.3.)

It is important for the control system engineer to recognize this distinction, particularly when dealing with an unfamiliar process for which the state-space representation is given only by numerical data. A numerical error in calculating the elements of the A and B matrices, or an experimental error in measuring them, may make an uncontrollable system seem controllable. A control system designed with this data may seem to behave satisfactorily in simulation studies based on the erroneous design data, but will fail in practice. On the other hand, a process that appears to be uncontrollable (or nearly uncontrollable), but which is not structurally uncontrollable, may be rendered more tractable by changing some parameter of the process—by "unbalancing the bridge."

Example 5D How not to control an unstable system (inverted pendulum) There are many ways of designing perfectly fine control systems for unstable processes such as the inverted pendulum of Examples 2E and 3D. These will be discussed at various places later on in this

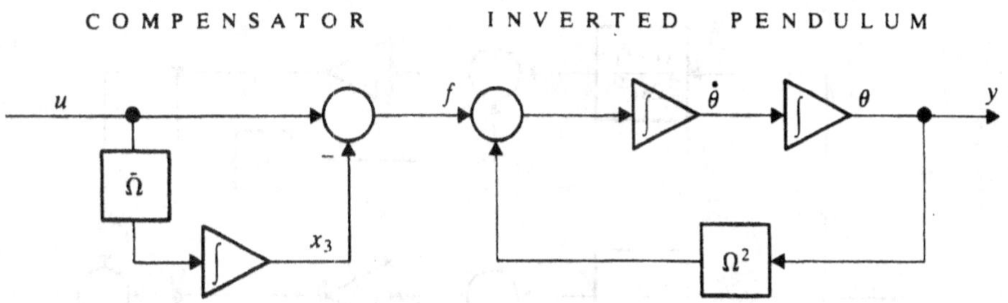

Figure 5.7 Unstabilizable compensation of inverted pendulum.

book. But one way guaranteed to be disastrous is to try to cancel the unstable pole with a zero in the compensator. The reason for the disaster is the subject of this example.

Consider the inverted pendulum of Example 3D with the output being the measured position. The transfer function from the input (force) to the output (position) is

$$\mathsf{H}(s) = \frac{\mathsf{y}(s)}{\mathsf{f}(s)} = \frac{1}{s^2 - \Omega^2} = \frac{1}{(s+\Omega)(s-\Omega)} \tag{5D.1}$$

This is obviously unstable. A much better transfer function would be

$$\mathsf{H}(s) = \frac{1}{s(s+\Omega)} \tag{5D.2}$$

which is stable and, because of the pole at the origin, would be a "type-one" system, with zero steady state error. Thus, one might be tempted to "compensate" the unstable transfer function by means of a compensator having the transfer function (Fig. 5.7)

$$\mathsf{G}(s) = \frac{s - \bar{\Omega}}{s} = 1 - \frac{\bar{\Omega}}{s} \tag{5D.3}$$

with

$$\bar{\Omega} = \Omega$$

Of course it will not be possible to make $\bar{\Omega}$ precisely equal to Ω so the compensation will not be perfect. But that is not the trouble, as we shall see.

The compensator transfer function (5D.3) represents "proportional plus integral" compensation which is quite customary in practical process control systems. The transfer function of the compensated system is now

$$\mathsf{H}_c(s) = \mathsf{G}(s)\mathsf{H}(s) = \frac{s - \bar{\Omega}}{s(s^2 - \Omega^2)} \to \mathsf{H}(s) \quad \text{as} \quad \bar{\Omega} \to \Omega \tag{5D.4}$$

A block diagram representation of this system is shown in Fig. 5.7, and the state-space equations corresponding to this representation are

$$\begin{aligned} \dot{x}_1 &= x_2 \\ \dot{x}_2 &= \Omega^2 x_1 - x_3 + u \\ \dot{x}_3 &= \bar{\Omega} u \end{aligned} \tag{5D.5}$$

where x_3 is the state of the integrator in the compensator. The matrices of the process (5D.5) are

$$A = \begin{bmatrix} 0 & 1 & 0 \\ \Omega^2 & 0 & -1 \\ 0 & 0 & 0 \end{bmatrix} \qquad B = \begin{bmatrix} 0 \\ 1 \\ \bar{\Omega} \end{bmatrix}$$

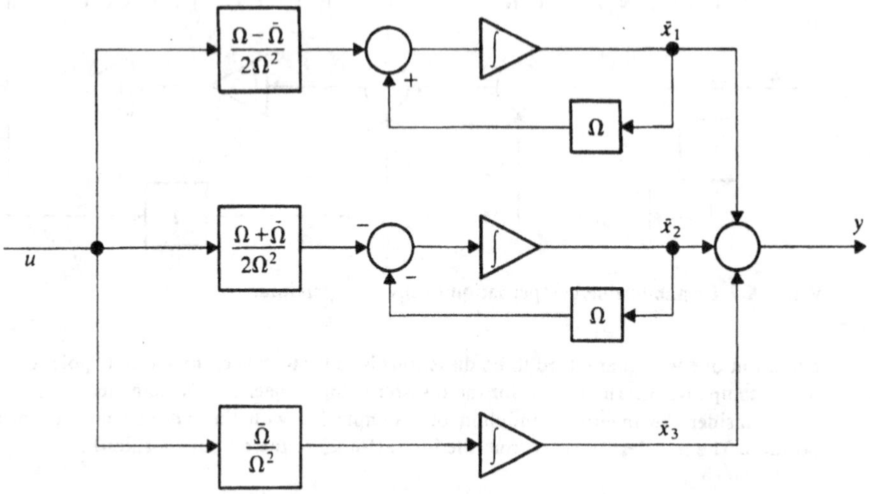

Figure 5.8 Partial fraction representation of Fig. 5.7.

The A matrix can be transformed to diagonal form by the transformation matrix

$$T = \frac{1}{2\Omega^2}\begin{bmatrix} \Omega^2 & \Omega & -1 \\ \Omega^2 & -\Omega & -1 \\ 0 & 0 & 2 \end{bmatrix} \qquad T^{-1} = \begin{bmatrix} 1 & 1 & 1 \\ \Omega & -\Omega & 0 \\ 0 & 0 & \Omega^2 \end{bmatrix}$$

We find that

$$\bar{A} = TAT^{-1} = \begin{bmatrix} \Omega & 0 & 0 \\ 0 & -\Omega & 0 \\ 0 & 0 & 0 \end{bmatrix}$$

$$\bar{B} = TB = \frac{1}{2\Omega^2}\begin{bmatrix} \Omega - \bar{\Omega} \\ -(\Omega + \bar{\Omega}) \\ 2\bar{\Omega} \end{bmatrix}$$

The state-space representation of the transformed system is as shown in Fig. 5.8. This block-diagram corresponds directly to the partial-fraction expansion of (5D.4):

$$H_c(s) = \frac{\bar{\Omega}/\Omega^2}{s} + \frac{(\Omega - \bar{\Omega})/2\Omega^2}{s - \Omega} + \frac{-(\Omega + \bar{\Omega})/2\Omega^2}{s + \Omega} \tag{5D.6}$$

Note carefully what happens when $\bar{\Omega} \to \Omega$. In the block-diagram the connection between the control input u and the *unstable* state x_1 is broken, rendering the system uncontrollable and *unstabilizable*. In (5D.6) the residue at the unstable pole vanishes. But now we understand that the vanishing of a residue at a pole of a transfer function does not imply that the subsystem giving rise to the pole disappears, but rather that it becomes "invisible."

If the original inverted pendulum could have arbitrary initial conditions, the transformed system (5D.5) could also have arbitrary initial conditions and hence the inverted pendulum would most assuredly not remain upright, regardless of how the loop were closed between the measurement y and the control input u.

More reasons for unobservability The foregoing examples were instances of uncontrollable systems. Instances of unobservable systems are even more abundant. An unobservable system results any time a state variable is not measured

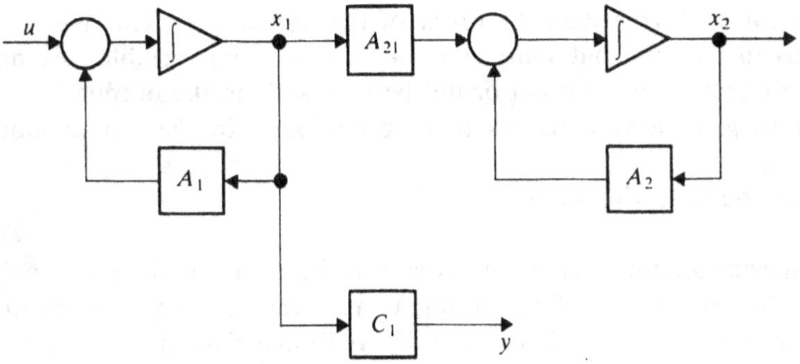

Figure 5.9 Systems in tandem that are unobservable.

directly and is not fed back to those state variables that are measured. Thus, any system comprising two subsystems in tandem (as shown in Fig. 5.9, in which none of the states of the right-hand subsystem can be measured) is unobservable. The transfer function from the inputs to the outputs obviously depends only on the left-hand subsystem.

Physical processes which have the structure shown in Fig. 5.9 are not uncommon. A mass m acted upon by a control force f is unobservable if only its velocity, and not its position, can be measured. This means that no method of velocity feedback can serve as a means of controlling position. In this regard it is noted that the integral of the measured velocity is not the same as the actual position. A control system shown in Fig. 5.10 will not be effective in controlling the position x of the mass, no matter how well it controls the velocity $\dot{x}$; any initial position error will remain in the system indefinitely.

In addition to the obvious reasons for unobservability there are also some of the more subtle reasons such as symmetry, as was illustrated by Example 5C.

5.3 DEFINITIONS AND CONDITIONS FOR CONTROLLABILITY AND OBSERVABILITY

In Secs. 5.1 and 5.2 we found that uncontrollable and/or unobservable systems were characterized by the property that the transfer function from the input to

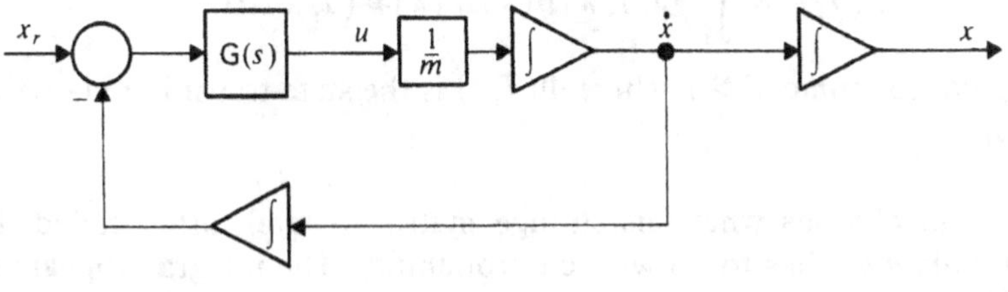

Figure 5.10 Position of mass cannot be observed and cannot be controlled using only velocity feedback.

the output is of lower degree than the order of the dynamic system. We were able to trace this to the fact that some combinations of state variables are not capable of being affected by the input or not being visible in the output.

It is useful to give these concepts more precision with the aid of more precise definitions.

We start with the following basic:

Definition of controllability A system is said to be controllable if and only if it is possible, by means of the input, to transfer the system from *any* initial state $x(t) = x_t$ to *any* other state $x_T = x(T)$ in a *finite* time $T - t \geqq 0$.

The emphasized words "any" and "finite" are essential to the definition. If it is only possible to make the system go from *some* states to *some* other states, then the system is not controllable. Moreover, if it takes an infinite amount of time to go from the arbitrary initial state to the arbitrary final state the system is likewise not controllable.

(In some texts, a system is called *completely controllable* when it is possible to transfer it from *any* state to *any* other state in finite time.[2] A system is not *completely* controllable when it is possible only to transfer it from some states.)

In the definition of controllability the initial time t is not specified and the final time is not fixed. This is done to accommodate time-varying systems, in which it may happen that the possibility of reaching x_T from x_t depends on the initial time t. (See Note 5.2.) In a time-invariant system, however, no generality is lost in taking the initial time t to be zero.

The terminal time T must be finite in order for the system to be controllable. In time-varying systems it may be necessary to restrict T to be greater than some fixed time, say $\bar{T}$. But in time-invariant systems, as we shall see, the only restriction on T is that it be greater than zero. (In fact, if the use of impulsive inputs is permitted, it is possible in a controllable system to go from any state to any other state in zero time, i.e., instantaneously. (See Note 5.2.) As a practical matter, it is possible in a controllable system to go from any state to any other state in an arbitrarily short time if we are willing to use a sufficiently large input.)

Controllability theorem A system is controllable if and only if the matrix

$$P(T, t) = \int_t^T \Phi(T, \lambda)B(\lambda)B'(\lambda)\Phi'(T, \lambda)\, d\lambda \tag{5.6}$$

is nonsingular for some $T > t$, where $\Phi(T, t)$ is the state-transition matrix of the system.

It is not at all obvious what this strange matrix integral, often called the *controllability grammian*, has to do with controllability. The integral appears to have been fetched from out of the sky. Later in the book, we will encounter integrals of this type quite often.

Since there is no obvious connection between the controllability grammian and getting from the state x_t to the state x_T, we should not be surprised that the proof of the controllability theorem is not entirely obvious. And it isn't. The nonobvious part of the theorem is the necessary condition, namely that if the integral $P(T, t)$ is singular for all finite $T > t$, then there are some states that can't be reached by any input.

To prove that the existence of an inverse of the controllability grammian guarantees the ability of going from any state x_t to any other state x_T, we recall from Chap. 3 (Eq. (3.21)) that

$$x_T = \Phi(T, t)x_t + \int_t^T \Phi(T, \lambda)B(\lambda)u(\lambda)\,d\lambda \tag{5.7}$$

Now suppose that $P(T, t)$ is nonsingular (i.e., has an inverse) on the interval $[t, T]$ for some finite T. Then an input that forces the process from x_t to x_T is given by

$$u(\lambda) = B'(\lambda)\Phi'(T, \lambda)P^{-1}(T, t)[x_T - \Phi(T, t)x_t] \qquad \text{for } t \leqq \lambda \leqq T \tag{5.8}$$

To verify this just substitute (5.8) into (5.7):

$$x_T = \Phi(T, t)x_t + \int_t^T \Phi(T, \lambda)B(\lambda)B'(\lambda)\Phi'(T, \lambda)\,d\lambda\, P^{-1}(T, t)\{x_T - \Phi(T, t)x_t\} \tag{5.9}$$

By (5.6) the integral in (5.9) is $P(T, t)$, so (5.9) becomes

$$x_T = \Phi(T, t)x_t + x_T - \Phi(T, t)x_t$$

which is an identity. This verifies that the input (5.8) does indeed transfer the system from x_t at time t to x_T at time T.

Note that the input given by (5.8) requires the inverse $P^{-1}(T, t)$ which exists only if $P(T, t)$ is nonsingular for some t and $T > t$. If $P(T, t)$ is singular for all $T > t$, then the input (5.8) cannot be used to transfer x_t to x_T. But perhaps we can use some other input. The answer is no, and demonstrating this constitutes the second part of the proof. We want to show that there are some states that can't be reached if $P(T, t)$ is singular for all finite $T > t$. Consider some time T for which $P(T, t)$ is singular. Then there must be some *nonzero* vector v such that

$$v'P(T, t)v = 0$$

Thus, by the definition (5.6)

$$\int_t^T v'\Phi(T, \lambda)B(\lambda)B'(\lambda)\Phi'(T, \lambda)v\,d\lambda = 0 \tag{5.10}$$

The integrand can be written

$$\int_t^T z'(\lambda)z(\lambda)\,d\lambda = 0 \tag{5.11}$$

where

$$z(\lambda) = B'(\lambda)\Phi'(T, \lambda)v \tag{5.12}$$

Since the integrand is a sum of squares: $z_1^2(\lambda) + \cdots + z_n^2(\lambda)$ it must be nonnegative. The only way that an integral, over a positive interval $T > t$, of a nonnegative quantity can be zero, is when the integrand itself is *identically zero* over the entire interval. Thus, for a singular grammian we have found a vector v for which

$$z(\lambda) = B'(\lambda)\Phi'(T, \lambda)v \equiv 0 \qquad \text{for all } \lambda \text{ in the interval } [t, T] \tag{5.13}$$

Then, as we shall see, it is impossible in the time interval $[t, T]$ to get from the origin to any state in the direction of v. Suppose the contrary: that it is possible to go to a state cv (where c is a scalar) in the direction of v. Then, with $x_t = 0$, by (5.7) we must have

$$cv = \int_t^T \Phi(T, \lambda)B(\lambda)u(\lambda)\, d\lambda \tag{5.14}$$

On premultiplying both sides of (5.14) by v' we find

$$cv'v = \int_t^T v'\Phi(T, \lambda)B(\lambda)u(\lambda)\, d\lambda \tag{5.15}$$

The left-hand side of (5.15) is clearly nonzero. But the integrand of the right-hand side is $z'(\lambda)u(\lambda)$ and in (5.13) we have found that $z(\lambda) \equiv 0$ in the entire interval $[t, T]$. Thus, the right-hand side of (5.15) is zero, independent of the input. This is a contradiction. We are forced to conclude that no input can transfer the system from the origin to a state cv in the interval $[t, T]$. If the controllability grammian is singular for all t and T there will always be some states that we will not be able to reach in any finite time. Hence the controllability grammian must be nonsingular for some t and $T > t$, in order that the system be controllable. This completes the proof.

The controllability theorem and its proof are the contributions of R. E. Kalman. See Note 5.1.

For a time-invariant system, the controllability grammian is given by

$$P(T - t) = \int_t^T e^{A(T-\lambda)}BB'\, e^{A'(T-\lambda)}\, d\lambda = \int_0^{T-t} e^{At}BB'\, e^{A't}\, dt$$

or simply

$$P(T) = \int_0^T e^{At}BB'\, e^{A't}\, dt \tag{5.16}$$

The matrix used for the controllability test is sometimes written

$$\tilde{P}(T) = \int_0^T e^{-At}BB'\, e^{-A't}\, dt \tag{5.17}$$

which is not the same matrix as $P(T)$ but which can easily be shown to have the same rank as $P(T)$.

Matrices having the form of the controllability grammian (5.6) in the general case, or in the form of (5.16) in the time-invariant case, sometimes need to be evaluated for optimum control and estimation problems, as will be discussed from Chap. 9 onward. But to evaluate the integrals merely for the purpose of testing controllability of a system seems a great deal of effort to achieve a simple objective. A simpler test would be most welcome. For time-varying systems there does not seem to be a simpler alternative. But for time-invariant systems, several simpler alternatives are available. We have already used one of the alternatives in the examples: transform the matrix to diagonal form (or block-diagonal form) and find whether or not any subsystem cannot be reached by the input. It may not always be easy to do this, however. A still simpler criterion, based on the rank of the matrix $[B, AB, \ldots, A^{k-1}B]$, will be given in the next section. We postpone a discussion of this condition, however, until addressing the topic of observability. We shall see that a close similarity exists between the concept of controllability and the concept of observability, which make it desirable to treat the two concepts together.

Just as the output y is not considered in the definition of controllability, the input u is generally not considered in defining observability. Thus, we deal with the unforced system

$$\dot{x} = A(t)x$$

with the observation given by

$$y(t) = C(t)x(t)$$

We use the following:

Definition of observability An unforced system is said to be observable if and only if it is possible to determine *any* (arbitrary initial) state $x(t) = x_t$ by using only a finite record, $y(\tau)$ for $t \leqq \tau \leqq T$, of the output.

This definition seems to square with our intuitive concept of what ought to constitute an observable system. Note that the definition requires ability to determine the initial state no matter where that state might be in the state-space. If only *some*, but not all, initial states can be determined, then the system is not observable.

The general condition for observability is given by the following:

Observability theorem A system is observable if and only if the matrix

$$M(T, t) = \int_t^T \Phi'(\lambda, t)C'(\lambda)C(\lambda)\Phi(\lambda, t)\, d\lambda \tag{5.18}$$

is nonsingular for some $T > t$, where $\Phi(T, t)$ is the state-transition matrix of the system.

The matrix $M(T, t)$ for testing observability is often called the *observability grammian*, and bears a strong resemblance to the controllability grammian (5.6): in place of the transition matrix $\Phi(T, \lambda)$ in (5.6), its transpose appears in (5.18); in place of the control matrix B in (5.6), the transpose of the observation matrix C appears in (5.18). Because of the close resemblance between controllability and observability, these are frequently referred to as *dual* concepts.

To prove the observability theorem we use the fact that the output y is given by

$$y(\lambda) = C(\lambda)\Phi(\lambda, t)x_t \qquad (\lambda \geqq t) \tag{5.19}$$

when the system starts in the state x_t. Multiply both sides of (5.19) by $\Phi'(\lambda, t)C'(\lambda)$ and integrate over the interval $[t, T]$ to obtain

$$\int_t^T \Phi'(\lambda, t)C'(\lambda)y(\lambda)\, d\lambda = \left(\int_t^T \Phi'(\lambda, t)C'(\lambda)C(\lambda)\Phi(\lambda, t)\, d\lambda\right)x_t \tag{5.20}$$

The integral on the right-hand side of (5.20) is recognized as the observability grammian $M(T, t)$ of (5.18). Thus, if the observability grammian is nonsingular, we can solve (5.20) for x_t:

$$x_t = M^{-1}(T, t) \int_t^T \Phi'(\lambda, t)C'(\lambda)y(\lambda)\, d\lambda \tag{5.21}$$

This formula furnishes an actual procedure for finding the initial state x_t, given $y(t)$ over the interval of the integral. Of course, it may not be the only way to determine x_t. Perhaps another way can be found to determine x_t that does not entail the inverse of the observability grammian. The answer, as we already suspect, is no. The reason why the answer is no is a consequence of an argument like that used for establishing the dual result for controllability: if the observability grammian $M(T, t)$ is singular then there exists a vector w for which the function

$$q(\lambda) = C(\lambda)\Phi(\lambda, t)w \equiv 0 \qquad \text{for all } \lambda \text{ in the interval } [t, T]$$

This function $q(\lambda)$, which is identically zero over the interval $[t, T]$ is precisely the output of the system when the initial state is w. It thus follows that if the initial state is w or anywhere on the line cw it will yield an output of zero and there will be no way of determining that initial state. If the observability grammian is singular for every t and T, there will always be some initial state which will produce zero outputs for intervals of any length, and hence the system is not observable.

For time-invariant systems the observability grammian of (5.18) may be written

$$M(T - t) = \int_t^T e^{A'(\lambda - t)}C'C\, e^{A(\lambda - t)}\, d\lambda = \int_0^{T-t} e^{A'\tau}C'C\, e^{A\tau}\, d\tau$$

or simply

$$M(T) = \int_0^T e^{A'\tau} C' C e^{A\tau} \, d\tau \tag{5.22}$$

Other forms of the observability grammian are also used, such as

$$\tilde{M}(T) = \int_0^T e^{-A't} C' C e^{-At} \, dt$$

Again, as with controllability, these matrices are not equal to $M(t)$ but have the same rank.

Also, as is the case with controllability, it is not necessary to evaluate the observability grammian to test for observability. There is a simpler algebraic test which is the subject of the next section.

5.4 ALGEBRAIC CONDITIONS FOR CONTROLLABILITY AND OBSERVABILITY

In the previous section we have seen that the necessary and sufficient condition for controllability of a time-invariant system is that the controllability grammian $P(T)$, given by (5.16), be nonsingular for some finite time T.

The algebraic criterion equivalent to this is expressed by the following:

Algebraic controllability theorem The time-invariant system $\dot{x} = Ax + Bu$ is controllable if and only if the rank $r(Q)$ of the *controllability test matrix*

$$Q = [B \quad AB \quad \cdots \quad A^{k-1}B] \tag{5.23}$$

is equal to k, the order of the system.

Note that Q is a matrix having k rows and kl columns, where l is the number of inputs. The rank of Q thus cannot be greater than k. But the rank of Q can be smaller than k. If so, the system is not controllable.

To prove the algebraic controllability theorem we note that if $P(T)$ is singular, then, by (5.13), there is a nonzero vector v such that the function

$$z(t) = B' e^{A't} v \equiv 0 \qquad \text{for} \qquad 0 \leqq t \leqq T \tag{5.24}$$

Since the function $z(t)$ is *identically* zero (flat), all its derivatives must also be identically zero. Thus we must have

$$\begin{aligned} \dot{z}(t) &= B'A' e^{A't} v \equiv 0 \\ \ddot{z}(t) &= B'(A')^2 e^{A't} v \equiv 0 \\ &\cdots\cdots\cdots\cdots\cdots\cdots \\ z^{(k-1)}(t) &= B'(A')^{k-1} e^{A't} v \equiv 0 \end{aligned} \tag{5.25}$$

We can keep going with this process but there is no need to do so.

We can arrange (5.24) and (5.25) in the following form:

$$\underbrace{\begin{bmatrix} B' \\ B'A' \\ \vdots \\ B'(A')^{k-1} \end{bmatrix}}_{Q'} e^{At} v \equiv 0 \qquad (5.26)$$

(k columns; kl rows)

The long matrix in (5.26), is Q'. Let its columns be denoted by $q_1, q_2, \ldots, q_k$:

$$Q' = [q_1 \quad q_2 \quad \cdots \quad q_k]$$

Also let

$$e^{At}v = \begin{bmatrix} \alpha_1(t) \\ \vdots \\ \alpha_k(t) \end{bmatrix} \qquad \text{a } k\text{-dimensional vector}$$

Thus (5.26) becomes

$$\alpha_1(t)q_1 + \alpha_2(t)q_2 + \cdots + \alpha_k(t)q_k \equiv 0 \qquad (5.27)$$

In other words, the columns of Q' are linearly *dependent*, which implies that the rank of the matrix Q' must be less than the order k of the system. We have thus established that if the controllability grammian is singular the rank of the matrix Q' is less than k. Since the rank of any matrix is equal to the rank of its transpose, we also can say that the singularity of the controllability grammian implies that the rank of Q is less than k.

To prove the converse, we expand e^{At} in a power series in t:

$$e^{At} = I + At + A^2t^2/2! + \cdots + A^{k-1}t^{k-1}/(k-1)! + A^kt^k/k! + \cdots \qquad (5.28)$$

By the Cayley–Hamilton theorem, however,

$$A^k = -a_1A^{k-1} - a_2A^{k-2} - \cdots - a_kI \qquad (5.29)$$

where $a_1, a_2, \ldots, a_k$ are the coefficients of the characteristic polynomial of A. Thus, by repeated use of (5.29), any power of A greater than $k-1$ can be expressed as a polynomial of degree $k-1$ in A. Thus (5.28) can be written as

$$e^{At} = If_1(t) + Af_2(t) + \cdots + A^{k-1}f_k(t)$$

where $f_1(t), f_2(t), \ldots, f_k(t)$ are the time functions obtained by substituting the expressions for powers of A higher than $k-1$ into (5.28) and collecting terms.

(The proof doesn't depend on what these functions are.) Thus

$$e^{At}B = [Bf_1(t) + ABf_2(t) + \cdots + A^{k-1}Bf_k(t)]$$
$$= Q\begin{bmatrix} f_1(t) \\ \vdots \\ f_k(t) \end{bmatrix}$$

Thus the controllability grammian can be expressed as

$$P(T) = Q\int_0^T \begin{bmatrix} f_1^2(t) & \cdots & f_1(t)f_k(t) \\ \cdots & \cdots & \cdots \\ f_1(t)f_k(t) & \cdots & f_k^2(t) \end{bmatrix} dt\, Q' \qquad (5.30)$$
$$= QGQ'$$

If we knew the functions $f_1, \ldots, f_k$ we would have an expression for $P(T)$ in terms of the grammian matrix G appearing in (5.30) between Q and Q'. But no matter what this matrix is, the rank of $P(T)$ cannot be greater than the rank of Q, since the rank of the product of matrices cannot exceed the rank of any of its factors. Thus, if the rank of Q is less than k, then the rank of $P(T)$ must surely be less than k, which means, of course, that $P(T)$ must be singular.

This completes the proof of the algebraic controllability theorem.

Since Q is a constant matrix it has constant rank. Thus, if Q is singular, then $P(T)$ is singular for *every* T. Similarly, if $P(T)$ is nonsingular for any $T > 0$, it must be nonsingular for every $T > 0$. This means that if a system is controllable, there is an input that will transfer the system from any starting state to any other state in an arbitrarily short time. The shorter the time, the larger the needed input, of course.

From the manner in which the algebraic controllability theorem was established using the controllability grammian, we can immediately assert the dual:

Algebraic observability theorem The (unforced) time invariant system

$$\dot{x} = Ax$$

with the observation vector

$$y = Cx$$

is observable if and only if the rank $r(N)$ of the *observability test matrix*

$$N = [C' \quad A'C' \quad \cdots \quad (A')^{k-1}C'] \qquad (5.31)$$

is equal to k, the order of the system.

Intuitively, one might conjecture that the ranks of the controllability grammian P and the corresponding matrix Q are related to each other and to the dimension of the subspace of states that can be reached. It turns out that this conjecture is quite correct. In fact, the ranks of $P(T)$, and Q, and the dimension of the "controllable subspace" are all *equal.* Likewise, the ranks of the observability grammian $M(T)$ and N and the dimension of the "observable subspace" are all equal. Complete proofs of all these facts is beyond the scope of this book. (See Note 5.2.) But we can gain insight into why this happens by considering transformations of state variables. In particular, suppose

$$\bar{x} = Tx$$

so that the matrices for the transformed system are

$$\bar{A} = TAT^{-1} \qquad \bar{B} = TB \qquad \bar{C} = CT^{-1} \tag{5.32}$$

Then the controllability test matrix for the transformed system is

$$\bar{Q} = [\bar{B} \quad \bar{A}\bar{B} \quad \cdots \quad \bar{A}^{k-1}\bar{B}]$$

But from (5.32)

$$\bar{A}^i\bar{B} = TA^iT^{-1}TB = TA^iB$$

Thus

$$\bar{Q} = [TB \quad TAB \quad \cdots \quad TA^{k-1}B] = TQ$$

Since the rank of a product cannot exceed the rank of either factor

$$r(\bar{Q}) \leqq r(Q)$$

But

$$Q = T^{-1}\bar{Q}$$

So

$$r(Q) \leqq r(\bar{Q})$$

Thus we conclude that

$$r(Q) = r(\bar{Q})$$

In other words, the rank of the controllability matrix is invariant to a change of state variable. Suppose that the transformed system is block-diagonal, i.e.,

$$\bar{A} = \begin{bmatrix} \bar{A}_1 & 0 \\ 0 & \bar{A}_2 \end{bmatrix} \begin{matrix} \updownarrow k_1 \\ \updownarrow k_2 \end{matrix} \qquad (k_1 + k_2 = k)$$

and, moreover, that the subsystem (of order k_1) corresponding to $\bar{A}_1$ is controllable, but that no input at all goes to the subsystem (of order k_2) corresponding to $\bar{A}_2$. Thus, we have

$$\bar{B} = \begin{bmatrix} \bar{B}_1 \\ 0 \end{bmatrix}$$

The controllability test matrix is

$$Q = \begin{bmatrix} \bar{B}_1 & \bar{A}_1\bar{B}_1 & \cdots & \bar{A}_1^{k_1-1}\bar{B} & \bar{A}_1^{k_1}\bar{B} & \cdots & \bar{A}_1^{k_1+k_2}\bar{B} \\ 0 & 0 & & 0 & 0 & & 0 \end{bmatrix} \begin{matrix} \updownarrow k_1 \\ \updownarrow k_2 \end{matrix}$$

The upper left submatrix $[\bar{B}_1 \quad \bar{A}_1\bar{B}_1 \quad \cdots \quad \bar{A}_1^{k_1-1}\bar{B}_1]$ is the controllability test matrix of the controllable subsystem, so it is of rank k_1. Thus the matrix $\bar{Q}$ is at least of rank k_1. But it cannot be of rank greater than k_1 because it contains k_2 rows of zero elements. Thus the *rank deficiency* $k_2 = k - k_1$ of $\bar{Q}$ is precisely equal to the dimension of the subspace which receives no input. Because the controllability matrix of the original system has the same rank as $\bar{Q}$, the dimension of the subspace that is uncontrollable remains equal to k_2.

The very same transformation concept applies to the relationship between the dimension of the "unobservable" subspace and the rank of the observability test matrix.

The block-diagonal matrix $\bar{A}$ can be the matrix of the Jordan canonical form. In that case, as discussed in Chap. 4, the state variables are the "normal modes" of the system. All the normal modes which can be controlled can be identified with subsystem 1 and all those normal modes which cannot be controlled can be identified with subsystem 2. It thus follows immediately that the rank of the controllability test matrix Q is equal to the number of controllable normal modes. Similarly, the number of observable normal modes is equal to the rank of the observability test matrix N.

These concepts are illustrated by the following examples.

Example 5E Hypothetical system (continued from Example 5A) The test matrix Q for controllability of the hypothetical system of Example 5A is

$$Q = [B \quad AB \quad A^2B \quad A^3B] = \begin{bmatrix} 1 & -1 & 1 & -1 \\ -2 & 4 & -10 & 28 \\ 2 & -6 & 18 & -54 \\ -1 & 3 & -9 & 27 \end{bmatrix}$$

The sum of the elements in each column of Q is zero so Q is clearly singular. Moreover, the sum of the elements of the first two rows minus the fourth row are also zero. Thus, only two rows of Q are linearly independent and the rank of Q is thus 2.

The test matrix N for observability of the system is

$$N = [C' \quad A'C' \quad (A')^2C' \quad (A')^3C'] = \begin{bmatrix} 7 & -10 & 16 & 28 \\ 6 & -9 & 15 & -27 \\ 4 & -6 & 10 & -18 \\ 2 & -3 & 5 & -9 \end{bmatrix}$$

It is similarly verified that the rank of N is 2.

Thus, there are two observable modes and two controllable modes. This is clear from Fig. 5.2. But since one of the controllable modes is also an observable mode, one mode remains that is neither observable nor controllable.

Example 5F Coupled masses with an internal force The controllability test matrix for the system of Example 5B is

$$Q = [B \quad AB \quad A^2B \quad A^3B] = \begin{bmatrix} 0 & -1/m_1 & 0 & km/m_1^2m_2 \\ 0 & 1/m_2 & 0 & -km/m_1m_2^2 \\ -1/m_1 & 0 & km/m_1^2m_2 & 0 \\ 1/m_2 & 0 & -km/m_1m_2^2 & 0 \end{bmatrix} \qquad (m = m_1 + m_2)$$

The third and fourth columns of Q are proportional to the first and second, respectively, so the rank of Q is only 2, as expected from Example 5B. The two uncontrollable state variables are the position of the center of mass x_c and its derivative $\dot{x}_c$.

Example 5G Distillation column A schematic diagram corresponding to the simplified model of the distillation column as developed by Gilles and Retzbach[3] (Example 2G) is shown in Fig. 4.2 on page 120. It is observed that there is a path from the input Δu_1 (steam flow rate) to each of the state variables. Nevertheless, the process cannot be controlled by u_1 alone, because x_2 and x_3 are both integrators and thus give the appearance of redundant state variables. It is also evident that the process is not controllable from the input Δu_2 (vapor side stream flow rate): there is not even a path from u_2 to x_1 and to x_2. But, by using both inputs, the process is controllable.

These facts can be verified by use of the algebraic controllability criterion. First, consider the single input u_1. The corresponding control matrix is

$$B_1 = \begin{bmatrix} b_{11} \\ 0 \\ 0 \\ 0 \end{bmatrix}$$

Hence, the controllability matrix Q_1 corresponding to u_1 is given by

$$Q_1 = [B_1 \quad AB_1 \quad A^2B_1 \quad A^3B_1]$$

$$= \begin{bmatrix} 1 & a_{11} & a_{11}^2 & a_{11}^3 \\ 0 & a_{21} & a_{21}(a_{11}+a_{22}) & a_{21}[a_{11}^2 + a_{22}(a_{11}+a_{22})] \\ 0 & 0 & a_{21}a_{32} & a_{32}a_{21}(a_{11}+a_{22}) \\ 0 & 0 & a_{21}a_{42} & a_{42}a_{21}(a_{11}+a_{22}) \end{bmatrix} b_{11}$$

The upper left-hand 3-by-3 submatrix is triangular and thus has a nonzero determinant (unless $a_{21} = 0$). Thus the rank of Q_1 is at least 3. But

$$|Q_1| = b_{11}^4 a_{21} \begin{vmatrix} a_{21}a_{32} & a_{32}a_{21}(a_{11}+a_{22}) \\ a_{21}a_{42} & a_{42}a_{21}(a_{11}+a_{22}) \end{vmatrix} = 0$$

Thus, the rank of $Q_1 < 4$. Thus, we conclude that the rank of $Q_1 = 3$, which means that the process is not controllable using only Δu_1. The control matrix for the input Δu_2 is

$$B_2 = \begin{bmatrix} 0 \\ 0 \\ b_{32} \\ b_{42} \end{bmatrix}$$

and the corresponding controllability matrix is

$$Q_2 = \begin{bmatrix} 0 & 0 & 0 & 0 \\ 0 & 0 & 0 & 0 \\ b_{32} & 0 & 0 & 0 \\ b_{42} & 0 & 0 & 0 \end{bmatrix}$$

which has a rank of 1.

The controllability matrix using both inputs is a 4×8 matrix whose columns are the columns of Q_1 and Q_2, interlaced. If the system is controllable, the resulting matrix must have four linearly independent columns, for example, the first three columns of Q_1 and the first column of Q_2. The determinant of the matrix formed from these columns is

$$\Delta = \begin{vmatrix} b_{11} & b_{11}a_{11} & b_{11}a_{11}^2 & 0 \\ 0 & b_{11}a_{21} & b_{11}(a_{11}+a_{22}) & 0 \\ 0 & 0 & b_{11}a_{21}a_{32} & b_{32} \\ 0 & 0 & b_{11}a_{21}a_{42} & b_{42} \end{vmatrix}$$

$$= b_{11}^3 a_{21}^2 \begin{vmatrix} a_{32} & b_{32} \\ a_{42} & b_{42} \end{vmatrix}$$

Except for specific values of a_{32}, a_{42}, b_{32}, and b_{42} the determinant $\Delta \neq 0$ and hence, the controllability matrix has a full rank of 4 and the process is, in general, controllable using both inputs. From the numerical data given with Example 2G, it is seen that a_{32} and a_{42} are of the same magnitude, while b_{32} and b_{42} are very much different in magnitude. Thus, Δ is not even approximately zero and the process is easily controllable using both inputs.

It is very important to recognize that the algebraic controllability and observability tests are only valid for time-invariant systems. That they are not generally valid for time-varying systems is vividly illustrated by a simple, but practical example, in which the state vector x is a constant:

$$\dot{x} = 0 \tag{5.33}$$

hence the dynamics matrix A is zero. If the observation matrix C is constant then the observability test matrix

$$N = [C' \quad 0 \quad 0 \quad \cdots \quad 0] \tag{5.34}$$

N has rank k if and only if C has rank k, i.e., that there are as many independent components of the observation vector as there are components in x. If C is time-varying, the observability test matrix is still given by (5.34) which would imply that x is unobservable, unless C is of rank k. But in fact x may be observable even if the observation vector y is a scalar, if C is time varying. Consider the scalar observation

$$y(t) = C(t)x = c'(t)x = c_1(t)x_1 + \cdots + c_k(t)x_k \tag{5.35}$$

At k different time instants $t_1, t_2, \ldots, t_k$ we have

$$y(t_1) = c'(t_1)x$$
$$y(t_2) = c'(t_2)x$$
$$\cdots\cdots\cdots\cdots$$
$$y(t_k) = c'(t_k)x$$

or

$$\begin{bmatrix} y(t_1) \\ \vdots \\ y(t_k) \end{bmatrix} = \begin{bmatrix} c'(t_1) \\ \vdots \\ c'(t_k) \end{bmatrix} x$$

If the time instants t_i are chosen such that the matrix multiplying x is nonsingular, then

$$x = \begin{bmatrix} c'(t_1) \\ \vdots \\ c'(t_k) \end{bmatrix}^{-1} \begin{bmatrix} y(t_1) \\ \vdots \\ y(t_k) \end{bmatrix}$$

Determination of an unknown constant vector by looking at it at different times, or "from different angles" is a standard procedure in the calibration of instruments, and the selection of suitable time instants, depending on the nature of $c(t)$, or the design of a suitable $c(t)$, is an important issue in practical calibration procedures. See Note 5.4.

5.5 DISTURBANCES AND TRACKING SYSTEMS: EXOGENOUS VARIABLES

In order to use state-space methods on design problems in which there are reference inputs and/or disturbances, it is frequently desirable to represent these inputs and disturbances by additional state variables.

The particular dynamic process we might wish to control would be of the form

$$\dot{x} = Ax + Bu + Fx_d \tag{5.36}$$

where x_d is a disturbance vector (which may or may not be subject to direct measurement).

In addition, we might wish to require that the state x track a reference state x_r.

To formulate the problem purely in terms of state variables, it is often expedient to assume that x_d and x_r satisfy known differential equations:

$$\dot{x}_d = A_d x_d \tag{5.37}$$

$$\dot{x}_r = A_r x_r \tag{5.38}$$

These supplementary states are surely not subject to control by the designer, so that these are *unforced* differential equations. The system comprising x, x_d, and x_r is necessarily uncontrollable. (Fig. 5.11.)

In general, we are concerned with the error defined by

$$e = x - x_r \tag{5.39}$$

The differential equation for the error using (5.36) and (5.38) becomes

$$\begin{aligned} \dot{e} = \dot{x} - \dot{x}_r &= A(e + x_r) + Fx_d + Bu - A_r x_r \\ &= Ae + (A - A_r)x_r + Fx_d + Bu = Ae + Ex_0 + Bu \end{aligned} \tag{5.40}$$

where

$$E = [A - A_r \mid F] \tag{5.41}$$

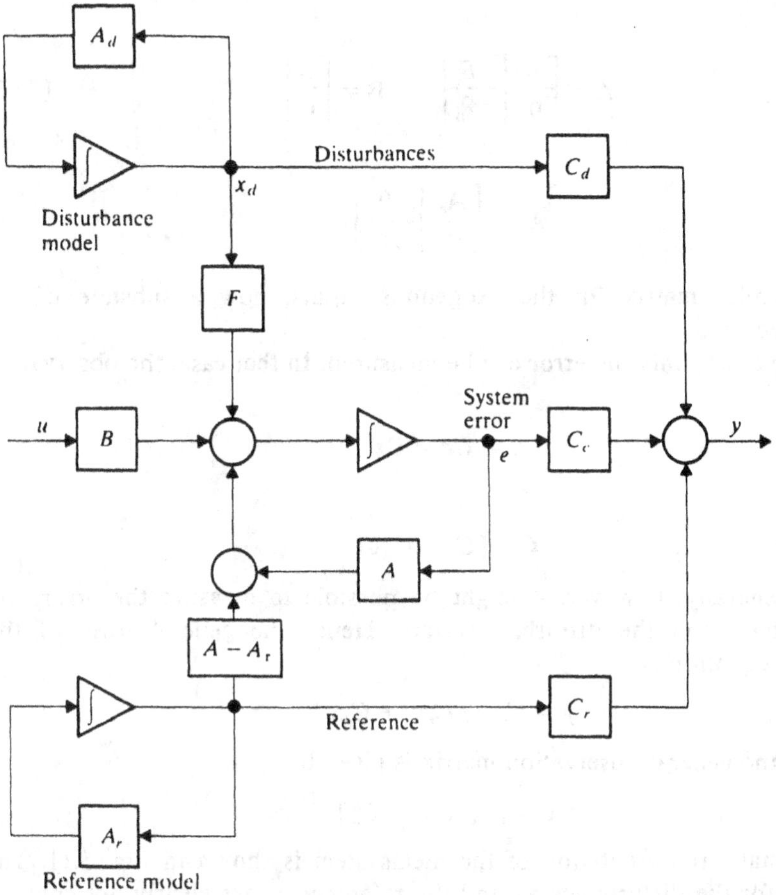

Figure 5.11 State-space representation of system with disturbances and reference input. (Models for disturbances and reference state are uncontrollable.)

and

$$x_0 = \begin{bmatrix} x_r \\ \hdashline x_d \end{bmatrix} \tag{5.42}$$

The vector x_0 represents the "exogenous" inputs to the system. To the differential equation of the error is adjoined the equations for the reference and disturbance states to produce a system of order $2k + 1$ having the "metastate" vector

$$\mathbf{x} = \begin{bmatrix} e \\ \hdashline x_0 \end{bmatrix} \begin{matrix} \updownarrow & k \\ \updownarrow & k+1 \end{matrix} \tag{5.43}$$

and satisfying the "metastate equation"

$$\dot{\mathbf{x}} = \mathbf{A}\mathbf{x} + \mathbf{B}u \tag{5.44}$$

where

$$\mathbf{A} = \left[\begin{array}{c|c} A & E \\ \hline 0 & A_0 \end{array}\right] \qquad \mathbf{B} = \left[\begin{array}{c} B \\ \hline 0 \end{array}\right] \tag{5.45}$$

where

$$\mathbf{A}_0 = \left[\begin{array}{c|c} A_r & 0 \\ \hline 0 & A_d \end{array}\right]$$

is the dynamics matrix for the exogenous inputs, now a substate of the metastate vector $\mathbf{x}$.

In some cases, only the error can be measured. In that case, the observation equation is

$$y = Ce = \mathbf{C}\mathbf{x}$$

where

$$\mathbf{C} = [C \mid 0 \quad 0]$$

More generally, however, it might be possible to measure the error, the reference state, and the disturbance state. Hence the general form of the observation equation is

$$y = C_e e + C_r x_r + C_d x_d$$

and hence, the general observation matrix is given by

$$\mathbf{C} = [C_e \mid C_r \quad C_d]$$

A schematic representation of the metasystem is shown in Fig. 5.11. The subsystems for the disturbance x_d and the reference x_r are clearly not controllable. With C_d and C_r present, the system is likely to be observable. But even if only C_e is present, the system may be observable because there is a path from x_r to the output through the subsystem that generates the error.

The very natural way in which an uncontrollable system arises when exogenous disturbances and reference inputs are modeled does not alter the fact that such systems *are* uncontrollable and hence that design techniques based on the premise of a controllable system cannot be applied willy-nilly to the metasystem. This doesn't imply that these methods are useless for this type of metasystem (or other types of uncontrollable systems) but rather that it is necessary to be cautious in their use.

PROBLEMS

Problem 5.1 Exogenous variables: controllability and observability

Consider the metasystem (5.44) with $\mathbf{A}$ and $\mathbf{B}$ as given by (5.45)

(*a*) Using the algebraic controllability test (5.23), show that the metasystem is *not* controllable. (This result is intuitively obvious.)

(*b*) Assume that only the state x (and not x_0) is measured, i.e., $y = Cx$, and that the original system $\dot{x} = Ax + Bu$ is observable (i.e., $[C', A'C', \ldots, (A')^{k-1}C']$ is of rank k). Discuss the conditions under which the metasystem is observable.

Problem 5.2 Two-car train

Consider the two-car train of Probs. 2.5 and 3.9.
(*a*) Is it controllable using only one motor?
(*b*) Is it controllable using both motors?
(*c*) Is it observable if only the position z_1 of the first car is measurable?
(*d*) Is it observable if the velocity v_1 of the first car is measurable?
(*e*) Is it observable if the velocities of both cars are measurable?

Problem 5.3 Aircraft lateral dynamics: controllability

Consider the lateral aircraft dynamics of Prob. 4.4.
(*a*) Is the dynamic process controllable using only the ailerons?
(*b*) Is the dynamic process controllable using only the rudder?

Problem 5.4 Inverted pendulum on cart: observability

Consider the inverted pendulum on a motor-driven cart described in Probs. 2.1 and 3.6. Determine whether or not it is observable with the following sets of observations:
(*a*) Cart displacement: $y = x_1$;
(*b*) Pendulum angle: $y = x_3$;
(*c*) Cart velocity: $y = x_2$;
(*d*) Cart velocity and pendulum angle: $y_1 = x_2$, $y_2 = x_3$.

Problem 5.5 Double-effect evaporator: controllability

Determine whether or not the evaporator of Example 2H is controllable from each of the following combinations of inputs:
(*a*) u_1 only;
(*b*) u_1 and u_2;
(*c*) u_1 and u_3;
(*d*) u_2 and u_3.
If the system is not controllable for any of the above cases, explain why not and, if possible, identify the states that are not controllable. *Hint:* Refer to Fig. 2.21.

Problem 5.6 Double-effect evaporator: observability

Determine whether or not the evaporator of Example 2H is observable from each of the following combinations of outputs:
(*a*) x_1 and x_4;
(*b*) x_3 and x_5;
(*c*) x_3, x_4, and x_5.
If, in any case, the system is not observable, explain why not and, if possible, identify the unobservable states.

NOTES

Note 5.1 Background of controllability and observability

In 1954 Bergen and Ragazzini[4] presented a method of compensating a sampled-data system by solving for the transfer function of the compensator given the desired closed-loop transfer function. They recognized that this method of compensation entailed cancellation of undesirable poles and zeros of the plant and substitution of more desirable ones. A mathematically exact cancellation would not be possible with real hardware. Thus they developed rules governing the incorporation of "nonminimum phase" (Note 4.7) poles and zeros into the specification of the desired closed-loop transfer function.

Kalman observed that the problem of nonminimum phase pole-zero cancellation would be present even if the cancellation were mathematically perfect, because the resulting system would turn out to be uncontrollable. With J. E. Bertram he presented a state-space design procedure[5] making use of state variable feedback in which the concept of controllability is hinted at. By 1960, Kalman had fully elucidated the concept of controllability and the dual concept of observability.[6]

Note 5.2 Varieties of controllability and observability

In this book we say a system is *controllable* if it is possible to find an input which brings it to the origin (or any other state) from any state in a finite time. (Kalman called such a system *completely controllable*.[1]) If a system is not controllable, it can be divided into two subsystems, one of which (if it exists) is controllable and the other is uncontrollable. If the uncontrollable subsystem is stable, the entire system is said to be *stabilizable*. The set of stabilizable systems thus includes the controllable systems as a subset: every controllable system is stabilizable, but not every stabilizable system is controllable. Similar distinctions apply with regard to observability. A system that is not observable (*completely* observable, in Kalman's terminology) can be divided into two subsystems, one of which (if it exists) is observable and the other is not. If the unobservable subsystem is stable, the entire system is said to be *detectable*. Thus the observable systems are a subset of the detectable systems.

These definitions and concepts are adequate for time-invariant systems, a category that includes most systems considered in this book. When time-varying systems are considered, however, the situation becomes more complicated. In a time-invariant system controllability is independent of the initial time. If this is true in a time-varying system, the system is said to be *uniformly* controllable. The dual of uniform controllability is uniform observability. With regard to the latter, it is noted that our definition of observability requires the ability to determine the present state based on *future* outputs. In a time-invariant system this is equivalent to the ability to determine the present state on the basis of *past* outputs. These are not necessarily equivalent, however, in time-varying systems. Thus we have another concept, namely *reconstructability*, which is the ability to determine the present state from past inputs.

A reasonably comprehensive treatment of observability, controllability, and various derivative concepts can be found in [2].

Note 5.3 Degree of controllability

By the definition of this chapter, a system is either controllable or it is not. In the real world, however, it may not be possible to make such sharp distinctions. An electrical bridge network, for example, is uncontrollable (or unobservable) for one discrete combination of its parameters. Since exact mathematical balancing is not possible, every practical bridge network is controllable and observable. If the balance condition is close to being satisfied, however, it will be very difficult to control or to observe all the state variables of the bridge. The problem with the standard definitions of controllability and observability is that they can lead to discontinuous functions of the system parameters: an arbitrarily small change in a system parameter can cause an abrupt change in the rank of the matrix by which controllability or observability is determined. It would be desirable to have definitions which can vary continuously with the parameters of the system and thus can reflect the degree of controllability of the system. Kalman et al.[6] recognized the need and suggested using the determinant of the corresponding test matrix or grammian as a measure of the degree of controllability or observability. Friedland,[7] noting that basing the degree of controllability or observability on the determinant of the test matrix suffers from sensitivity to the scaling of the state variables, suggested using the ratio of the smallest of the singular values to the largest as a preferable measure. Moore[8] subsequently elaborated upon this suggestion.

Note 5.4 Application to calibration

The development of an analytical technique for the determination of a constant vector b based on time-varying measurement signal $y(t) = C(t)b$ and generalizations of this technique is considered by Friedland.[9]

REFERENCES

1. Kalman, R. E., "Mathematical Description of Linear Dynamic Systems," *SIAM J. on Control*, ser. A., vol. 1, no. 2, 1963, pp. 152–192.
2. Wonham, W. M., *Linear Multivariable Control: a Geometric Approach*, 2d ed., Springer-Verlag, New York, 1979.
3. Gilles, E. D., and Retzbach, B., "Reduced Models and Control of Distillation Columns with Sharp Temperature Profiles," *IEEE Trans. on Automatic Control*, vol. AC-28, no. 5, May 1983, pp. 628–630.
4. Bergen, A. R., and Ragazzini, J. R., "Sampled-Data Processing Techniques for Feedback Control Systems," *Trans. AIEE, Pt. II*, vol. 73, November 1954, pp. 236–244.
5. Kalman, R. E., "On the General Theory of Control Systems," *Proc. First International Congress, IFAC, Moscow, USSR*, 1960, pp. 481–492.
6. Kalman, R. E., Ho, Y. C., and Narendra, K. S., "Controllability of Linear Dynamic Systems," *Contributions to Differential Equations*, vol. 1, no. 2, 1963, pp. 189–213.
7. Friedland, B., "Controllability Index Based On Conditioning Number," *Trans. ASME (J. Dynamic Systems, Measurement & Control)*, vol. 97G, no. 4, December 1975, pp. 444–445.
8. Moore, B. C., "Principal Component Analysis in Linear Systems: Controllability, Observability, and Model Reduction," *IEEE Trans. on Automatic Control*, vol. AC-26, no. 1, February 1981, pp. 17–32.
9. Friedland, B., "On The Calibration Problem," *IEEE Trans. on Automatic Control*, vol. AC-22, no. 6, December 1977, pp. 899–905.

CHAPTER

SIX

SHAPING THE DYNAMIC RESPONSE

6.1 INTRODUCTION

At last we have arrived at the point of using state-space methods for control system design. In this chapter we will develop a simple method of designing a control system for a process in which all the state variables are accessible for measurement—the method known as *pole-placement.* We will find that in a controllable system, with all the state variables accessible for measurement, it is possible to place the closed-loop poles anywhere we wish in the complex s plane. This means that we can, in principle, completely specify the closed-loop dynamic performance of the system. In principle, we can start with a sluggish open-loop system and force it to behave with alacrity; in principle, we can start with a system that has very little open-loop damping and provide any amount of damping desired. Unfortunately, however, what can be attained in principle may not be attainable in practice. Speeding the response of a sluggish system requires the use of large control signals which the actuator (or power supply) may not be capable of delivering. The consequence is generally that the actuator saturates at the largest signal that it can supply. In some instances the system behavior may be acceptable in spite of the saturation. But in other cases the effect of saturation is to make the closed-loop system unstable. It is usually not possible to alter open-loop dynamic behavior very drastically without creating practical difficulties.

Adding a great deal of damping to a system having poles near the imaginary axis is also problematic, not only because of the magnitude of the control signals needed, but also because the control system gains are very sensitive to the location of the open-loop poles. Slight changes in the open-loop pole location may cause the closed-loop system behavior to be very different from that for which it is designed.

We will first address the design of a regulator. Here the problem is to determine the gain matrix G in a linear feedback law

$$u = -Gx \tag{6.1}$$

which shapes the dynamic response of the process in the absence of disturbances and reference inputs. Afterward we shall consider the more general problem of determining the matrices G and G_0 in the linear control law

$$u = -Gx - G_0 x_0 \tag{6.2}$$

where x_0 is the vector of exogenous variables. The reason it is necessary to separate the exogenous variables from the process state x, rather than deal directly with the metastate

$$\mathbf{x} = \begin{bmatrix} x \\ \hline x_0 \end{bmatrix} \tag{6.3}$$

introduced in Chap. 5, is that in developing the theory for the design of the gain matrix, we must assume that the underlying process is *controllable.* Since the exogenous variables are not true state variables, but additional inputs that cannot be affected by the control action, they cannot be included in the state vector when using a design method that requires controllability.

The assumption that all the state variables are accessible to measurement in the regulator means that the gain matrix G in (6.1) is permitted to be any function of the state x that the design method requires. In most practical instances, however, the state variables are not all accessible for measurement. The feedback control system design for such a process must be designed to use only the measurable output of the process

$$y = Cx$$

where y is a vector of lower dimension than x. In some cases it may be possible to determine the gain matrix G_y for a control law of the form

$$u = -G_y y \tag{6.4}$$

which produces acceptable performance. But more often it is not possible to do so. It is then necessary to use a more general feedback law, of the form

$$u = -G\hat{x} \tag{6.5}$$

where $\hat{x}$ is the state of an appropriate dynamic system known as an "observer." The design of observers is the subject of Chap. 7. And in Chap. 8, we shall show that when a feedback law of the form of (6.5) is used with a properly designed observer, the dynamic properties of the overall system can be specified at will, subject to practical limitations on control magnitude and accuracy of implementation.

6.2 DESIGN OF REGULATORS FOR SINGLE-INPUT, SINGLE-OUTPUT SYSTEMS

The present section is concerned with the design of a gain matrix

$$G = g' = [g_1, g_2, \ldots, g_k] \tag{6.6}$$

for the single-input, single-output system

$$\dot{x} = Ax + Bu \tag{6.7}$$

where

$$B = b = \begin{bmatrix} b_1 \\ b_2 \\ \vdots \\ b_k \end{bmatrix} \tag{6.8}$$

With the control law $u = -Gx = -g'x$ (6.7) becomes

$$\dot{x} = (A - bg')x$$

Our objective is to find the matrix $G = g'$ which places the poles of the closed-loop dynamics matrix

$$A_c = A - bg' \tag{6.9}$$

at the locations desired. We note that there are k gains $g_1, g_2, \ldots, g_k$ and k poles for a kth order system, so there are precisely as many gains as needed to specify each of the closed-loop poles.

One way of determining the gains would be to set up the characteristic polynomial for A_c:

$$|sI - A_c| = |sI - A + bg'| = s^k + \bar{a}_1 s^{k-1} + \cdots + \bar{a}_k \tag{6.10}$$

The coefficients $\bar{a}_1, \bar{a}_2, \ldots, \bar{a}_k$ of the powers of s in the characteristic polynomial will be functions of the k unknown gains. Equating these functions to the numerical values desired for $\bar{a}_1, \ldots, \bar{a}_k$ will result in k simultaneous equations the solution of which will yield the desired gains $g_1, \ldots, g_k$.

This is a perfectly valid method of determining the gain matrix g', but it entails a substantial amount of calculation when the order k of the system is higher than 3 or 4. For this reason, we would like to develop a direct formula for g in terms of the coefficients of the open-loop and closed-loop characteristic equations.

If the original system is in the companion form given in (3.90), the task is particularly easy, because

$$A = \begin{bmatrix} -a_1 & -a_2 & \cdots & -a_{k-1} & -a_k \\ 1 & 0 & \cdots & 0 & 0 \\ 0 & 1 & \cdots & 0 & 0 \\ \cdots & \cdots & \cdots & \cdots & \cdots \\ 0 & 0 & \cdots & 1 & 0 \end{bmatrix} \tag{6.11}$$

$$bg' = \begin{bmatrix} 1 \\ 0 \\ 0 \\ \vdots \\ 0 \end{bmatrix} [g_1, g_2, \ldots, g_k] = \begin{bmatrix} g_1 & g_2 & \cdots & g_k \\ 0 & 0 & \cdots & 0 \\ \cdots & \cdots & \cdots & \cdots \\ 0 & 0 & \cdots & 0 \end{bmatrix}$$

Hence

$$A_c = A - bg' = \begin{bmatrix} -a_1 - g_1 & -a_2 - g_2 & \cdots & -a_k - g_k \\ 1 & 0 & \cdots & 0 \\ 0 & 1 & \cdots & 0 \\ \cdots & \cdots & \cdots & \cdots \\ 0 & 0 & \cdots & 0 \end{bmatrix}$$

The gains $g_1, \ldots, g_k$ are simply added to the coefficients of the open-loop A matrix to give the closed-loop matrix A_c. This is also evident from the block-diagram representation of the closed-loop system as shown in Fig. 6.1. Thus for a system in the companion form of Fig. 6.1, the gain matrix elements are given by

$$a_i + g_i = \hat{a}_i \qquad i = 1, 2, \ldots, k$$

or

$$g = \hat{a} - a \tag{6.12}$$

where

$$a = \begin{bmatrix} a_1 \\ \vdots \\ a_k \end{bmatrix} \qquad \hat{a} = \begin{bmatrix} \hat{a}_1 \\ \vdots \\ \hat{a}_k \end{bmatrix} \tag{6.13}$$

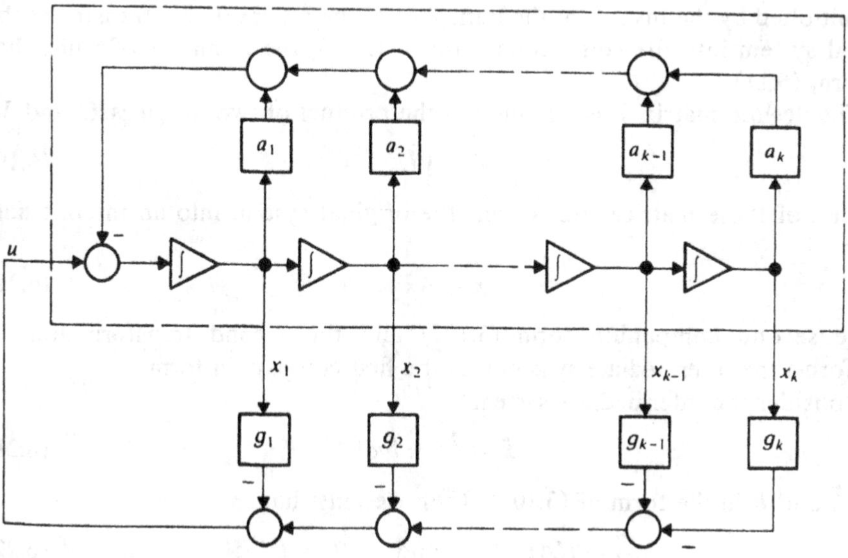

Figure 6.1 State variable feedback for system in first companion form.

are vectors formed from the coefficients of the open-loop and closed-loop characteristic equations, respectively.

The dynamics of a typical system are usually not in companion form. It is necessary to transform such a system into companion form before (6.12) can be used. Suppose that the state of the transformed system is $\bar{x}$, achieved through the transformation

$$\bar{x} = Tx \tag{6.14}$$

Then, as shown in Chap. 3,

$$\dot{\bar{x}} = \bar{A}\bar{x} + \bar{b}u \tag{6.15}$$

where

$$\bar{A} = TAT^{-1} \qquad \text{and} \qquad \bar{b} = Tb$$

For the transformed system the gain matrix is

$$\bar{g} = \hat{a} - \bar{a} = \hat{a} - a \tag{6.16}$$

since $\bar{a} = a$ (the characteristic equation being invariant under a change of state variables). The desired control law in the original system is

$$u = -g'x = -g'T^{-1}\bar{x} = -\bar{g}'\bar{x} \tag{6.17}$$

From (6.17) we see that

$$\bar{g}' = g'T^{-1}$$

Thus the gain in the original system is

$$g = T'\bar{g} = T'(\hat{a} - a) \tag{6.18}$$

In words, the desired gain matrix for a general system is the difference between the coefficient vectors of the desired and actual characteristic equation, premultiplied by the inverse of the transpose of the matrix T that transforms the general system into the companion form of (3.90), the A matrix of which has the form (6.11).

The desired matrix T is obtained as the product of two matrices U and V:

$$T = VU \tag{6.19}$$

The first of these matrices transforms the original system into an intermediate system

$$\dot{\tilde{x}} = \tilde{A}\tilde{x} \tag{6.20}$$

in the second companion form (3.107) and the second transformation U transforms the intermediate system into the first companion form.

Consider the intermediate system

$$\dot{\tilde{x}} = \tilde{A}\tilde{x} + \tilde{b}u \tag{6.21}$$

with $\tilde{A}$ and $\tilde{b}$ in the form of (3.107). Then we must have

$$\tilde{A} = UAU^{-1} \qquad \text{and} \qquad \tilde{b} = Ub \tag{6.22}$$

The desired matrix U is precisely the inverse of the controllability test matrix Q of Sec. 5.4. To prove this fact, we must show that

$$U^{-1}\tilde{A} = AU^{-1} \tag{6.23}$$

or

$$Q\tilde{A} = AQ \tag{6.24}$$

Now, for a single-input system

$$Q = [b, Ab, \ldots, A^{k-1}b]$$

Thus, with $\tilde{A}$ given by (3.107), the left-hand side of (6.23) is

$$Q\tilde{A} = [b, Ab, \ldots, A^{k-1}b]\begin{bmatrix} 0 & 0 & \cdots & -a_k \\ 1 & 0 & \cdots & -a_{k-1} \\ 0 & 1 & \cdots & -a_{k-2} \\ \cdots & \cdots & \cdots & \cdots \\ 0 & 0 & \cdots & -a_1 \end{bmatrix}$$

$$= [Ab, A^2b, \ldots, A^{k-1}b, -a_kb - a_{k-1}Ab - \cdots - a_kA^{k-1}b] \tag{6.25}$$

The last term in (6.25) is

$$(-a_kI - a_{k-1}A - \cdots - a_kA^{k-1})b \tag{6.26}$$

Now, by the Cayley-Hamilton theorem, (see Appendix):

$$A^k = -a_1A^{k-1} - a_2A^{k-2} - \cdots - a_kI$$

so (6.26) is A^kb. Thus the left-hand side of (6.24) as given by (6.25) is

$$Q\tilde{A} = [Ab, A^2b, \ldots, A^kb] = A[b, Ab, \ldots, A^{k-1}b] = AQ$$

which is the desired result.

If the system is not controllable, then Q^{-1} does not exist and there is no general method of transforming the original system into the intermediate system (6.21); in fact it is not possible to place the closed-loop poles anywhere one desires. Thus, controllability is an essential requirement of system design by pole placement. If the system is *stabilizable* (i.e., the uncontrollable part is asymptotically stable, as discussed in Chap. 5) a stable closed-loop system can be achieved by placing the poles of the controllable subsystem where one wishes and accepting the pole locations of the uncontrollable subsystem. In order to apply the formula of this section, it is necessary to first separate the uncontrollable subsystem from the controllable subsystem.

The control matrix $\tilde{b}$ of the intermediate system is given by

$$\tilde{b} = Ub \tag{6.27}$$

We now show that

$$\tilde{b} = \begin{bmatrix} 1 \\ 0 \\ \vdots \\ 0 \end{bmatrix} \tag{6.28}$$

Multiply (6.28) by Q to obtain

$$Q\tilde{b} = [b, Ab, \ldots, A^{k-1}b]\begin{bmatrix} 1 \\ 0 \\ \vdots \\ 0 \end{bmatrix} = b$$

which is the same as (6.27), since $Q^{-1} = U$.

The final step is to find the matrix V that transforms the intermediate system (6.21) into the final system (6.15). We must have

$$\bar{x} = V\tilde{x} \tag{6.29}$$

For the transformation (6.28) to hold, we must have

$$\bar{A} = V\tilde{A}V^{-1}$$

or

$$V^{-1}\bar{A} = \tilde{A}V^{-1} \tag{6.30}$$

The matrix V^{-1} that satisfies (6.30) is the transpose of the upper left-hand k-by-k submatrix of the (triangular Toeplitz) matrix appearing in (3.103)

$$V^{-1} = \begin{bmatrix} 1 & a_1 & \cdots & a_{k-1} \\ 0 & 1 & \cdots & a_{k-2} \\ \cdots & \cdots & \cdots & \cdots \\ 0 & 0 & \cdots & 1 \end{bmatrix} = W \tag{6.31}$$

To prove this, we note that the left-hand side of (6.30) is

$$V^{-1}\bar{A} = \begin{bmatrix} 1 & a_1 & \cdots & a_{k-1} \\ 0 & 1 & \cdots & a_{k-2} \\ \cdots & \cdots & \cdots & \cdots \\ 0 & 0 & \cdots & 1 \end{bmatrix}\begin{bmatrix} -a_1 & -a_2 & \cdots & -a_k \\ 1 & 0 & \cdots & 0 \\ \cdots & \cdots & \cdots & \cdots \\ 0 & 0 & \cdots & 0 \end{bmatrix}$$

$$= \begin{bmatrix} 0 & 0 & \cdots & 0 & -a_k \\ 1 & a_1 & \cdots & a_{k-2} & 0 \\ 0 & 1 & \cdots & a_{k-3} & 0 \\ \cdots & \cdots & \cdots & \cdots & \cdots \\ 0 & 0 & \cdots & 1 & 0 \end{bmatrix} \tag{6.32}$$

(Note that the zeros in the first row of $V^{-1}\bar{A}$ are the result of the difference of two terms $a_1 - a_1$, $a_2 - a_2$, etc.) and the right-hand side of (6.30) is

$$\tilde{A}V^{-1} = \begin{bmatrix} 0 & 0 & \cdots & -a_k \\ 1 & 0 & \cdots & -a_{k-1} \\ 0 & 1 & \cdots & -a_{k-2} \\ \cdots & \cdots & \cdots & \cdots \\ 0 & 0 & \cdots & -a_k \end{bmatrix} \begin{bmatrix} 1 & a_1 & \cdots & a_{k-1} \\ 0 & 1 & \cdots & a_{k-2} \\ 0 & 0 & \cdots & a_{k-3} \\ \cdots & \cdots & \cdots & \cdots \\ 0 & 0 & \cdots & 1 \end{bmatrix}$$

$$= \begin{bmatrix} 0 & 0 & \cdots & 0 & -a_k \\ 1 & a_1 & \cdots & a_{k-2} & 0 \\ 0 & 1 & \cdots & a_{k-3} & 0 \\ \cdots & \cdots & \cdots & \cdots & \cdots \\ 0 & 0 & \cdots & 1 & 0 \end{bmatrix}$$

which is the same as (6.32). Thus (6.30) is proved.

We also need

$$\bar{b} = V\tilde{b}$$

We will show that

$$\bar{b} = \tilde{b}$$

Consider

$$\tilde{b} = V^{-1}\bar{b}$$

with

$$b = V^{-1}\bar{b} = \begin{bmatrix} 1 & a_1 & \cdots & a_{k-1} \\ 0 & 1 & \cdots & a_{k-2} \\ \cdots & \cdots & \cdots & \cdots \\ 0 & 0 & \cdots & 1 \end{bmatrix} \begin{bmatrix} 1 \\ 0 \\ \vdots \\ 0 \end{bmatrix} = \begin{bmatrix} 1 \\ 0 \\ \vdots \\ 0 \end{bmatrix}$$

Thus $\tilde{b}$ and $\bar{b}$ are the same.

The result of this calculation is that the transformation matrix T whose transpose is needed in (6.18) is the inverse of the product of the controllability test matrix and the triangular matrix (6.31).

The above results may be summarized as follows. The desired gain matrix g, by (6.18) and (6.19), is given by

$$g = (VU)'(\hat{a} - a) \tag{6.33}$$

where

$$V = W^{-1} \qquad \text{and} \qquad U = Q^{-1}$$

Thus

$$VU = W^{-1}Q^{-1} = (QW)^{-1}$$

Hence (6.33) becomes

$$g = [(QW)']^{-1}(\hat{a} - a) \tag{6.34}$$

where Q is the controllability test matrix, W is the triangular matrix defined by (6.31), $\hat{a}$ is the vector of coefficients for the desired (closed-loop) characteristic polynomial, and a is the vector of coefficients of the open-loop system.

The basic pole-placement formula (6.34) was first stated by Bass and Gura.[1] It can be derived by other methods as discussed in Note 6.1.

Now that we have a specific formula for the gains of a controllable, single-input system that will place the poles at any desired location, several questions arise: If the closed-loop poles can be placed anywhere, where *should* they be placed? How can the technique be extended to multiple input systems? We shall address these questions and others after considering several examples.

Example 6A Instrument servo A dc motor driving an inertial load constitutes a simple instrument servo for keeping the load at a fixed position.

As shown in Chap. 2 (Example 2B), the state-space equations for the motor-driven inertia are

$$\dot{\theta} = \omega \tag{6A.1}$$

$$\dot{\omega} = -\alpha\omega + \beta u \tag{6A.2}$$

where θ is the angular position of the load, ω is the angular velocity, u is the applied voltage, and α and β are constants that depend on the physical parameters of the motor and load:

$$\alpha = -K^2/JR \qquad \beta = K/JR$$

If the desired position θ_r is a constant then we can define the servo error

$$e = \theta - \theta_r$$

Then

$$\dot{e} = \dot{\theta} - \dot{\theta}_r = \omega \qquad (\theta_r = \text{const}) \tag{6A.3}$$

and (6A.3) replaces (6A.1) to give

$$\begin{bmatrix} \dot{e} \\ \dot{\omega} \end{bmatrix} = \begin{bmatrix} 0 & 1 \\ 0 & -\alpha \end{bmatrix}\begin{bmatrix} e \\ \omega \end{bmatrix} + \begin{bmatrix} 0 \\ \beta \end{bmatrix} u \tag{6A.4}$$

The angular position measurement can be instrumented by a potentiometer on the motor shaft and the angular velocity by a tachometer. Thus, the closed-loop system would have the configuration illustrated in Fig. 6.2. Note that the position gain is shown multiplying the negative of the system error which in turn is added to the control signal. This is consistent with the convention normally used for servos, wherein the position gain multiplies the difference $\theta_r - \theta$ between the reference and the actual positions. The quantity e defined above (6A.3) is the negative of the system error as normally defined in elementary texts.

The characteristic polynomial of the system is

$$|sI - A| = \begin{vmatrix} s & -1 \\ 0 & s+\alpha \end{vmatrix} = s^2 + \alpha s$$

Thus

$$a = \begin{bmatrix} \alpha \\ 0 \end{bmatrix}$$

The controllability test matrix Q and the matrix W are given respectively by

$$Q = [b, Ab] = \begin{bmatrix} 0 & \beta \\ \beta & -\alpha\beta \end{bmatrix} \qquad W = \begin{bmatrix} 1 & \alpha \\ 0 & 1 \end{bmatrix}$$

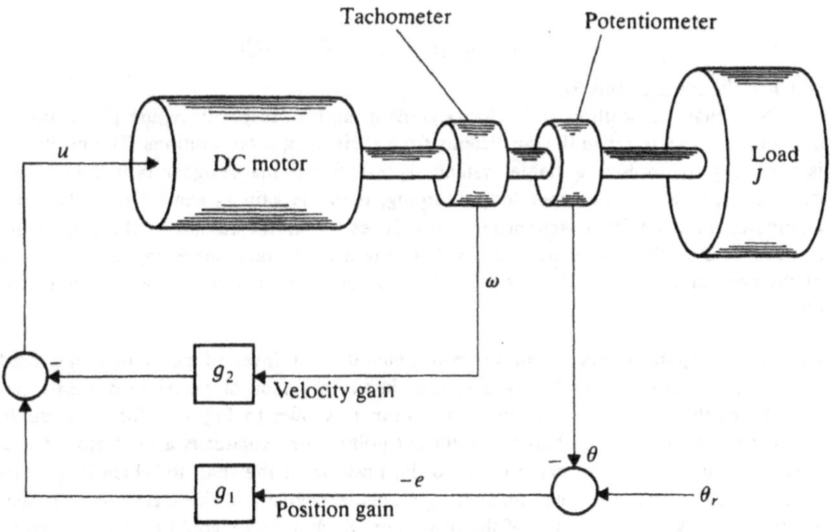

Figure 6.2 Implementation of an instrument servo.

Thus

$$QW = \begin{bmatrix} 0 & \beta \\ \beta & 0 \end{bmatrix} = (QW)'$$

and

$$[(QW)']^{-1} = \begin{bmatrix} 0 & 1/\beta \\ 1/\beta & 0 \end{bmatrix}$$

Thus the desired gain matrix, by the Bass-Gura formula (6.34), is

$$g = \begin{bmatrix} 0 & 1/\beta \\ 1/\beta & 0 \end{bmatrix} \begin{bmatrix} \bar{a}_1 - \alpha \\ \bar{a}_2 \end{bmatrix} = \begin{bmatrix} \bar{a}_2/\beta \\ (\bar{a}_1 - \alpha)/\beta \end{bmatrix} \tag{6A.5}$$

where $\bar{a}_1$ and $\bar{a}_2$ are the coefficients of the desired characteristic polynomial.

While the above calculation illustrates the general procedure, the gains could have been more easily computed directly. For a control law of the form

$$u = -g_1 e - g_2 \omega$$

(6A.4) becomes

$$\dot{e} = \omega$$

$$\dot{\omega} = -g_1 \beta e - (\alpha + \beta g_2)\omega$$

which has the closed-loop matrix

$$A_c = \begin{bmatrix} 0 & 1 \\ -g_1\beta & -(\alpha + g_2\beta) \end{bmatrix}$$

with the characteristic equation

$$|sI - A_c| = s^2 + (\alpha + g_2\beta)s + g_1\beta$$

Thus

$$\bar{a}_1 = \alpha + g_2\beta \qquad \bar{a}_2 = g_1\beta$$

or

$$g_1 = \bar{a}_2/\beta \qquad g_2 = (\bar{a}_1 - \alpha)/\beta$$

which is the same as (6A.5).

Note that the position and velocity gains g_1 and g_2, respectively, are proportional to the amounts we wish to move the coefficients from their open-loop positions. The position gain g_1 is necessary to produce a stable system: $\bar{a}_2 > 0$. But if the designer is willing to settle for $a_1 = \alpha$, i.e., to accept the open-loop damping, then the gain g_2 can be zero. This of course eliminates the need for a tachometer and reduces the hardware cost of the system. It is also possible to alter the system damping without the use of a tachometer, by using an estimate $\hat{\omega}$ of the angular velocity ω. This estimate is obtained by means of an observer as discussed in Chap. 7.

Example 6B Stabilization of an inverted pendulum An inverted pendulum can readily be stabilized by a closed-loop feedback system, just as a person of moderate dexterity can do it.

A possible control system implementation is shown in Fig. 6.3, for a pendulum constrained to rotate about a shaft at its bottom point. The actuator is a dc motor. The angular position of the pendulum, being equal to the position of the shaft to which it is attached, is measured by means of a potentiometer. The angular velocity in this case can be measured by a "velocity pick-off" at the top of the pendulum. Such a device could consist of a coil of wire

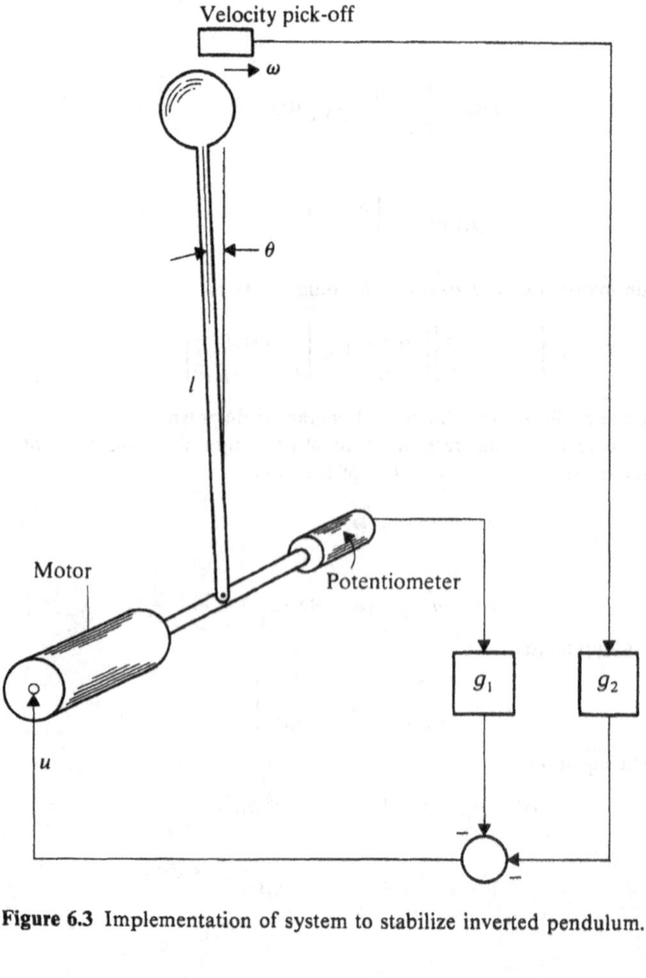

Figure 6.3 Implementation of system to stabilize inverted pendulum.

in a magnetic field created by a small permanent magnet in the pendulum bob. The induced voltage in the coil is proportional to the linear velocity of the bob as it passes the coil. And since the bob is at a fixed distance from the pivot point the linear velocity is proportional to the angular velocity. The angular velocity could of course also be measured by means of a tachometer on the dc motor shaft.

As determined in Prob. 2.2, the dynamic equations governing the inverted pendulum in which the point of attachment does not translate is given by

$$\begin{aligned} \dot{\theta} &= \omega \\ \dot{\omega} &= \Omega^2\theta - \alpha\omega + \beta u \end{aligned} \tag{6B.1}$$

where α and β are given in Example 6A, with the inertia J being the total reflected inertia:

$$J = J_m + ml^2$$

where m is the pendulum bob mass and l is the distance of the bob from the pivot. The natural frequency Ω is given by

$$\Omega^2 = \frac{mgl}{J + ml^2} = \frac{g}{l + J/ml}$$

(Note that the motor inertia J_m affects the natural frequency.)

Since the linearization is valid only when the pendulum is nearly vertical, we shall assume that the control objective is to maintain $\theta = 0$. Thus we have a simple regulator problem.

The matrices A and b for this problem are

$$A = \begin{bmatrix} 0 & 1 \\ \Omega^2 & -\alpha \end{bmatrix} \qquad b = \begin{bmatrix} 0 \\ \beta \end{bmatrix}$$

The open-loop characteristic polynomial is

$$|sI - A| = \begin{vmatrix} s & -1 \\ -\Omega^2 & s + \alpha \end{vmatrix} = s^2 + \alpha s - \Omega^2$$

Thus

$$\begin{aligned} a_1 &= \alpha \\ a_2 &= -\Omega^2 \end{aligned}$$

The open-loop system is unstable, of course.

The controllability test matrix and the W matrix are given respectively by

$$Q = \begin{bmatrix} 0 & \beta \\ \beta & -\alpha\beta \end{bmatrix} \qquad W = \begin{bmatrix} 1 & \alpha \\ 0 & 1 \end{bmatrix}$$

(which are the same as they were for the instrument servo). And

$$[(QW)']^{-1} = \begin{bmatrix} 0 & 1/\beta \\ 1/\beta & 0 \end{bmatrix}$$

Thus the gain matrix required for pole placement using (6.34), is

$$g = \begin{bmatrix} 0 & 1/\beta \\ 1/\beta & 0 \end{bmatrix} \begin{bmatrix} (\bar{a}_1 - \alpha) \\ \bar{a}_2 + \Omega^2 \end{bmatrix} = \begin{bmatrix} (\bar{a}_2 + \Omega^2)/\beta \\ (\bar{a}_1 - \alpha)/\beta \end{bmatrix}$$

Example 6C Control of spring-coupled masses The dynamics of a pair of spring-coupled masses, shown in Fig. 3.7(a), were shown in Example 3I to have the matrices

$$A = \begin{bmatrix} 0 & 1 & 0 & 0 \\ 0 & 0 & 1 & 0 \\ 0 & 0 & 0 & 1 \\ 0 & 0 & -K/\bar{M} & 0 \end{bmatrix} \qquad B = \begin{bmatrix} 0 \\ 0 \\ 0 \\ 1 \end{bmatrix}$$

The system has the characteristic polynomial

$$D(s) = s^4 + (K/\bar{M})s^2$$

Hence $$a_1 = a_3 = a_4 = 0, \qquad a_2 = K/\bar{M}.$$

The controllability test and W matrices are given, respectively, by

$$Q = \begin{bmatrix} 0 & 0 & 0 & 1 \\ 0 & 0 & 1 & 0 \\ 0 & 1 & 0 & -K/\bar{M} \\ 1 & 0 & -K/\bar{M} & 0 \end{bmatrix} \qquad W = \begin{bmatrix} 1 & 0 & K/\bar{M} & 0 \\ 0 & 1 & 0 & K/\bar{M} \\ 0 & 0 & 1 & 0 \\ 0 & 0 & 0 & 1 \end{bmatrix} \tag{6C.1}$$

Multiplying we find that

$$QW = (QW)' = (QW)^{-1} = \begin{bmatrix} 0 & 0 & 0 & 1 \\ 0 & 0 & 1 & 0 \\ 0 & 1 & 0 & 0 \\ 1 & 0 & 0 & 0 \end{bmatrix} \tag{6C.2}$$

(This rather simple result is not really as surprising as it may at first seem. Note that A is in the first companion form but using the right-to-left numbering convention. If the left-to-right numbering convention were used the A matrix would already be in the companion form of (6.11) and would not require transformation. The transformation matrix T given by (6C.2) has the effect of changing the state variable numbering order from left-to-right to right-to-left, and vice versa.)

The gain matrix g is thus given by

$$g = \begin{bmatrix} 0 & 0 & 0 & 1 \\ 0 & 0 & 1 & 0 \\ 0 & 1 & 0 & 0 \\ 1 & 0 & 0 & 0 \end{bmatrix} \begin{bmatrix} \bar{a}_1 \\ \bar{a}_2 - K/\bar{M} \\ \bar{a}_3 \\ \bar{a}_4 \end{bmatrix} = \begin{bmatrix} \bar{a}_4 \\ \bar{a}_3 \\ \bar{a}_2 - K/\bar{M} \\ \bar{a}_1 \end{bmatrix}$$

A suitable pole "constellation" for the closed-loop process might be a Butterworth pattern as discussed in Sec. 6.5. To achieve this pattern the characteristic polynomial should be of the form

$$\bar{D}(s) = s^4 + (1 + \sqrt{3})\Omega s^3 + (2 + \sqrt{3})\Omega^2 s^2 + (1 + \sqrt{3})\Omega^3 s + \Omega^4$$

Thus

$$\bar{a}_1 = (1 + \sqrt{3})\Omega$$
$$\bar{a}_2 = (2 + \sqrt{3})\Omega^2$$
$$\bar{a}_3 = (1 + \sqrt{3})\Omega^3$$
$$\bar{a}_4 = \Omega^4$$

Thus the gain matrix g is given by

$$g = \begin{bmatrix} \Omega^4 \\ (1 + \sqrt{3})\Omega^3 \\ (2 + \sqrt{3})\Omega^2 - K/\bar{M} \\ (1 + \sqrt{3})\Omega \end{bmatrix}$$

6.3 MULTIPLE-INPUT SYSTEMS

If the dynamic system under consideration

$$\dot{x} = Ax + Bu$$

has more than one input, that is, B has more than one column, then the gain matrix G in the control law

$$u = -Gx$$

has more than one row. Since each row of G furnishes k gains that can be adjusted, it is clear that in a controllable system there will be more gains available than are needed to place all of the closed-loop poles. This is a benefit: the designer has more flexibility in the design than in the case of a single-input system; it is possible to specify all the closed-loop poles and still be able to satisfy other requirements. How should these other requirements be specified? The answer to this question may well depend on the circumstances of the particular application. One possibility might be to set some of the gains to zero. For example, it is sometimes possible to place the closed-loop poles at locations desired with a gain matrix which has a column of zeros. This means that the state variable corresponding to that column is not needed in the generation of any of the control signals in the vector u, and hence there is no need to measure (or estimate) that state variable. This simplifies the resulting control system structure. If all the state variables, except those corresponding to columns of zeros in the gain matrix, are accessible for measurement then there is no need for an observer to estimate the state variables that cannot be measured. A very simple and robust control system is the result.

Another possible method of selecting a particular structure for the gain matrix is to make each control variable depend on a different group of state variables which are physically more closely related to that control variable than to the other control variables.

Still another possibility arises in systems which have a certain degree of structural symmetry and in which it is desired to retain the symmetry in the closed-loop system by an appropriate feedback structure.

The following example illustrates one method of selecting the gain matrix.

Example 6D Distillation column For the distillation column of Example 4A, having the block-diagram of Fig. 4.2, we saw in Example 5G that both inputs are needed in order for the system to be controllable, because there are redundant poles at the origin (due to the integrators) from either Δu_1 or Δu_2. If there were only one integrator present, it is easy to see that the system would be controllable from Δu_1 alone. This suggests a gain structure in which Δu_1 depends on x_1, x_2, and x_3, and Δu_2 depends on x_4. This gives four adjustable gains for the closed-loop fourth-order system and we would expect to be able to locate the closed-loop poles at whatever locations are desired.

Thus we use a gain matrix of the form

$$G = \begin{bmatrix} g_1 & g_2 & g_3 & 0 \\ 0 & 0 & 0 & g_4 \end{bmatrix} \tag{6D.1}$$

With the A and B matrices as given by (2G.5) it is found that the closed-loop dynamics matrix is

$$A_c = A - BG = \begin{bmatrix} a_{11} - b_{11}g_1 & -b_{11}g_2 & -b_{11}g_3 & 0 \\ a_{21} & a_{22} & 0 & 0 \\ 0 & a_{32} & 0 & -b_{32}g_4 \\ 0 & 0 & 0 & -b_{42}g_4 \end{bmatrix}$$

Thus the closed-loop characteristic equation is

$$|sI - A_c| = \left|\begin{array}{ccc|c} s - a_{11} + b_{11}g_1 & b_{11}g_2 & b_{11}g_3 & 0 \\ -a_{21} & s - a_{22} & 0 & 0 \\ 0 & -a_{32} & s & 0 \\ \hline 0 & 0 & 0 & s + b_{42}g_4 \end{array}\right|$$

$$= (s + b_{42}g_4)\begin{vmatrix} s - a_{11} + b_{11}g_1 & b_{11}g_2 & b_{11}g_3 \\ -a_{21} & s - a_{22} & 0 \\ 0 & -a_{32} & s \end{vmatrix}$$

$$= (s + b_{42}g_4)(s^3 + \bar{a}_1 s^2 + \bar{a}_2 s + \bar{a}_3) \tag{6D.2}$$

Note that $|sI - A_c|$ factors into two terms, a first-order term giving a pole at $s = -b_{42}g_4$ and a third-order term. The third-order term is the same as would result for a third-order system having dynamics and control matrices given respectively by

$$A_3 = \begin{bmatrix} a_{11} & 0 & 0 \\ a_{21} & a_{22} & 0 \\ 0 & a_{32} & 0 \end{bmatrix} \qquad B_3 = \begin{bmatrix} b_{11} \\ 0 \\ 0 \end{bmatrix}$$

with a state variable feedback of the form

$$G_3 = [g_1, g_2, g_3]$$

Thus we can adjust g_1, g_2, and g_3 to achieve any desired location of the roots of the third-order factor in (6D.2) and use g_4 to adjust the location of the pole at $s = -b_{42}g_4$.

Note that if the gains are real numbers, as they must be in a physical system, then one pole must be the real pole at $s = -b_{42}g_4$, and hence one of the poles arising from the cubic factor in (6D.2) must also be real when the gains g_1, g_2, and g_3 are real. Thus, by using a gain matrix having the structure of (6D.1), we do not have freedom to place the closed-loop poles anywhere in the complex plane. This is not a contradiction of controllability, because (6D.1) is not the most general form that the gain matrix G can take: four of the possible gains have been set to zero. Since a very satisfactory transient response can be achieved, however, with two real poles, the gain matrix structure of (6D.1) is, in the practical sense, perfectly acceptable.

6.4 DISTURBANCES AND TRACKING SYSTEMS: EXOGENOUS VARIABLES

In the previous section we considered the design of regulators in which the performance objective is to achieve a specified closed-loop dynamic behavior (pole locations) of the system in response to arbitrary initial disturbances. A more general design objective is to control the system error not only for initial disturbances, but also for persistent disturbances, and also to track reference inputs.

In Chap. 5 the general problem was set up by defining the system error

$$e = x - x_r \tag{6.35}$$

where x_r is assumed to satisfy a differential equation

$$\dot{x}_r = A_r x_r \tag{6.36}$$

In addition to the reference input we also have a disturbance x_d, so that the error is given by

$$\dot{e} = Ae + (A - A_r)x_r + Fx_d + Bu = Ae + Bu + Ex_0 \tag{6.37}$$

In Chap. 5 we defined the metastate

$$\mathbf{x} = \begin{bmatrix} e \\ \hline x_0 \end{bmatrix}$$

which makes it possible to regard the design problem, including reference and disturbance inputs, as a regulator problem. As was shown in Chap. 5, however, the metasystem is not controllable, so it is not possible to apply the pole-placement design technique to the metasystem. (Since controllability is not a requirement for formulating the optimum control problem, as discussed in Chap. 9, we will be able to use the metasystem formulation in connection with optimum control system design.)

Instead of working with the metasystem, we work directly with the error differential equation (6.37). The exogenous vector x_0 is treated as an input just like u. The design problem is really to arrange matters so that the control input u counteracts the effects of the exogenous variables. The control that we seek should be effective not only for a specific exogenous input, but rather for an entire class of inputs. Only the characteristics of the *class* are known to the designer; the specific member of the class is determined by measurements on x_0 while the process is in operation.

Since we are limiting our attention to linear systems, we consider only a linear control law, which now takes the general form

$$u = -Ge - G_0 x_0 = -Ge - G_r x_r - G_d x_d \tag{6.38}$$

The closed-loop system using a control of the form (6.38) has the appearance shown in Fig. 6.4. Note the presence of two paths in addition to the feedback

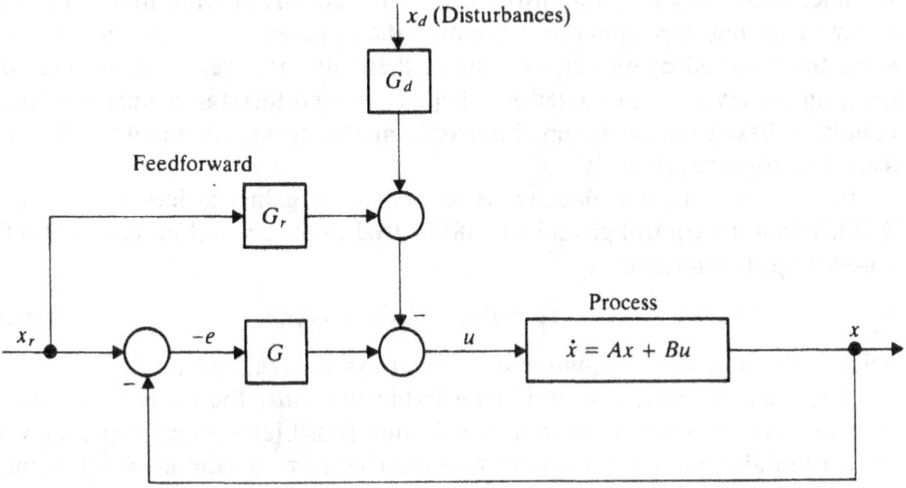

Figure 6.4 Schematic of feedback system for process with reference state and disturbance input.

loop in which the system error appears. There is a "feedforward" path with a gain G_r and a path through the gain G_d, the purpose of which is to counteract the effect of these disturbances.

The design is based, as already mentioned, on the assumption that the exogenous input vector $x_0 = [x_r', x_d']'$ as well as the system error e are accessible for measurement during the operation of the control system (i.e., in "real time"). Since x_r is a reference input, one might think that it is always accessible. The instrumentation, however, might be such that only the system error can be measured; it may be difficult (hence costly in terms of hardware) to measure x_d independent of the system error. It is noted that reference input x_r appears in (6.37) through the difference $A - A_r$ between the dynamics matrix A and the matrix A_r used to model the reference input. If $A_r = A$, that is, if the reference input can be generated as the solution of the unforced differential equation of the open-loop process, then no feedforward path is needed. If the open-loop process is stable, then the only reference inputs that can be generated are decaying exponentials which go to zero in the steady state. Thus if we need to track steps, ramps, etc., in the steady state, we cannot have $A_r = A$ for an asymptotically stable open-loop system. On the other hand if A has a pole at the origin of order ν, then by proper choice of initial conditions x_r can include a polynomial in time of order $\nu - 1$; we can still have $A_r = A$ and hence not require feedforward. (Recall from Chap. 5 that the presence of a pole of order ν in the open-loop system makes it a "type ν" system. We thus see again that a type ν system can follow reference inputs containing polynomials of degree up to and including $\nu - 1$ without use of feedforward.)

Sometimes the disturbance x_d can be measured easily, sometimes not. In a temperature control system, for example, in which x_d is the ambient temperature of the environment of the process, it is not too difficult to accomplish this measurement with an extra thermometer. In an aircraft autopilot design, on the other hand, in which the disturbances may consist of wind-induced forces, it may be all but impossible to instrument the required measurements. In cases where the required quantities, or some of them, are not accessible for measurement, an observer, as discussed in Chap. 7, is used to infer estimates of these quantities, based on the assumed dynamic model, using the quantities that are accessible for measurement.

For the present, our objective is to design the gain matrices G and G_0 in (6.38). When the control given by (6.38) is used in the general process (6.37) the closed-loop dynamics are

$$\dot{e} = Ae + Ex_0 - B(Ge + G_0 x_0) \tag{6.39}$$

which is the differential equation of a linear system excited by x_0.

If it were possible, it would be desirable to choose the gains G and G_0 to keep the system error zero. But this is not possible: system errors may be present initially that cannot instantly be reduced to zero. And even when initial errors are zero, there are usually not enough control variables (i.e., columns in the B matrix) to make the coefficients of x_0 vanish, as they must in order that the error be zero for *any* x_0 and e.

More reasonable performance objectives are the following:

(*a*) The closed-loop system should be asymptotically stable.
(*b*) A linear combination of the error state variables (rather than the entire state vector) is to be zero in the steady state.

In order for the closed-loop system to be asymptotically stable the closed-loop dynamics matrix $A_c = A - BG$ must have its characteristic roots in the left half-plane. If the system is controllable, this can be accomplished by a suitable choice of the gain matrix.

The steady-state condition is characterized by a constant error state vector, i.e., in the steady state

$$\dot{e} \equiv 0$$

which, from (6.39), means that

$$(A - BG)e = (BG_0 - E)x_0$$

If the closed-loop system is asymptotically stable, $A - BG = A_c$ has no characteristic roots at the origin, and hence its inverse exists. Thus the steady state error is given by

$$e = (A - BG)^{-1}(BG_0 - E)x_0 \tag{6.40}$$

As noted before it is not reasonable to expect that e be zero. Instead we require that

$$y = Ce \equiv 0 \tag{6.41}$$

where C is a singular matrix of suitable dimension. We'll see shortly what a "suitable" dimension is. If (6.41) holds, then from (6.40)

$$C(A - BG)^{-1}(BG_0 - E)x_0 = 0 \tag{6.42}$$

Remember that we want (6.42) to hold for any x_0. This can be achieved if and only if the coefficient matrix multiplying x_0 vanishes:

$$C(A - BG)^{-1}(BG_0 - E) = 0 \tag{6.43}$$

The matrix G_0 which satisfies (6.43) will meet the requirement of (6.41). We note that (6.43) can be written

$$C(A - BG)^{-1}BG_0 = C(A - BG)^{-1}E \tag{6.44}$$

We examine the possibility of solving (6.44) for the required gain matrix G_0. Here is where the dimension of C becomes significant. Suppose that the dimension of y is j. Then C is a $j \times k$ matrix, $(A - BG)^{-1}$ is a $k \times k$ matrix, and B is a $k \times m$ matrix, where m is the number of control variables. The product of the three matrices multiplying G_0 is thus a $j \times m$ matrix. If $j > m$, then (6.44) is "overdetermined": there are too many conditions to be satisfied by G_0 and, except for special values of E, no solution to (6.44) for G_0 exists. If $j < m$, then (6.44) is "underdetermined": G_0 is not uniquely specified by (6.44). This poses no problem; it only means that G_0 can be chosen to satisfy not only (6.41), but also to satisfy other conditions.

Analytically the "cleanest" case is when the number of inputs m is equal to the dimension of y. (If y is regarded as the system output, then we can say that the process is "square," having the same number of inputs as outputs.) In this case, when the matrix multiplying G_0 is nonsingular, the desired gain matrix is given by

$$G_0 = [C(A - BG)^{-1}B]^{-1}C(A - BG)^{-1}E \tag{6.45}$$

The big, messy matrix

$$B^{\#} = [C(A - BG)^{-1}B]^{-1}C(A - BG)^{-1} \tag{6.46}$$

that multiplies E in (6.45) has the property that

$$B^{\#}B = I \tag{6.47}$$

A matrix having this property is called a left inverse (or left "pseudoinverse") of B. Matrices of this type are encountered frequently in linear systems analysis. In terms of $B^{\#}$, (6.45) can be written

$$G_0 = B^{\#}E \tag{6.48}$$

Under what circumstances does the matrix $C(A - BG)^{-1}B$ possess an inverse? One might think that the existence of an inverse depends on the stabilizing gain matrix G. In fact, this is not the case. The existence of an inverse depends only on the open-loop dynamics: it can be shown that $C(A - BG)^{-1}B$ possesses an inverse if and only if

$$\lim_{s \to 0} \mathrm{H}_0(s) = |C(sI - A)^{-1}B| \neq 0 \tag{6.49}$$

If A is nonsingular (6.49) reduces to the requirement that $|CA^{-1}B| \neq 0$. The reason that invertability of $C(A - BG)^{-1}B$ is independent of G is related to the fact that state-variable feedback does not alter the transmission zeros of a process, as discussed in Prob. 4.1. (See also Note 6.2.)

The specific value of the inverse, however, does in general depend on G. Nevertheless, one can safely choose any feedback gain matrix G without being concerned about the possibility that this choice of gain will compromise the invertability of $C(A - BG)^{-1}B$.

In most cases, the reference state x_r does not need to have all of its components specified. In other words, the error that the control system must be designed to reduce to zero may be of lower dimension that the state vector. The other components of the state vector may be unspecified. In these cases, the component of the exogenous vector corresponding to the reference state may be of lower dimension that the state x and the corresponding submatrix of E will not be $A - A_d$ but a different matrix with fewer than k columns. Rather than try to express this in general terms, we illustrate it by the example that follows.

Example 6E Temperature control Consider the temperature control problem having the electrical network analog shown in Fig. 6.5. The voltage u may be regarded as the analog of the temperature of a heater and x_0 as the ambient temperature. Since there is only one heater (i.e., one input) then we can in general control only a single quantity, perhaps v_1 or v_2, or a linear combination of the two, such as their average $y = (v_1 + v_2)/2$.

The dynamic equations, in state-space form, are (see Example 2C)

$$\begin{bmatrix} \dot{v}_1 \\ \dot{v}_2 \end{bmatrix} = \begin{bmatrix} -a_{11} & a_{12} \\ a_{21} & -a_{22} \end{bmatrix} \begin{bmatrix} v_1 \\ v_2 \end{bmatrix} + \begin{bmatrix} b_1 \\ b_2 \end{bmatrix} u + \begin{bmatrix} e_1 \\ e_2 \end{bmatrix} x_0 \tag{6E.1}$$

with

$$a_{11} = -\frac{1}{C_1}[R_3^{-1} + D_1(R_2^{-1} + R_0^{-1} + R^{-1})] \qquad a_{12} = \frac{1}{C_1}[R_3^{-1} + R_2^{-1}D_1]$$

$$b_1 = R_0^{-1}D_1 \qquad e_1 = R^{-1}D_1$$

$$a_{21} = \frac{1}{C_2}[R_3^{-1} + R_1^{-1}D_2], \qquad a_{22} = -\frac{1}{C_2}[R_3^{-1} + D_2(R_1^{-1} + R_0^{-1} + R^{-1})]$$

$$b_2 = R_0^{-1}D_2, \qquad e_2 = R^{-1}D_2$$

where

$$D_1 = [1 + R_1(R_0^{-1} + R_2^{-1} + R^{-1})]^{-1} \qquad D_2 = [1 + R_2(R_0^{-1} + R_1^{-1} + R^{-1})]^{-1}$$

We assume that the desired state is

$$x_{1d} = \bar{v}_1 = \text{const}$$

$$x_{2d} = \bar{v}_2 = \text{const}$$

Thus

$$A_0 = 0 \tag{6E.2}$$

We take as the output matrix

$$C = [c_1, c_2] \tag{6E.3}$$

Assume the feedback gain matrix is

$$G = [g_1, g_2]$$

Then

$$A_c^{-1} = (A - BG)^{-1} = \begin{bmatrix} -a_{11} - b_1 g_1 & a_{12} - b_1 g_2 \\ a_{21} - b_2 g_1 & -a_{22} - b_2 g_1 \end{bmatrix}^{-1}$$

$$= \frac{\begin{bmatrix} -a_{22} - b_2 g_2 & -a_{12} + b_1 g_2 \\ -a_{21} + b_2 g_1 & -a_{11} - b_1 g_1 \end{bmatrix}}{(a_{11} + b_1 g_1)(a_{22} + b_2 g_2) - (a_{21} - b_2 g_1)(a_{12} - b_1 g_2)} \tag{6E.4}$$

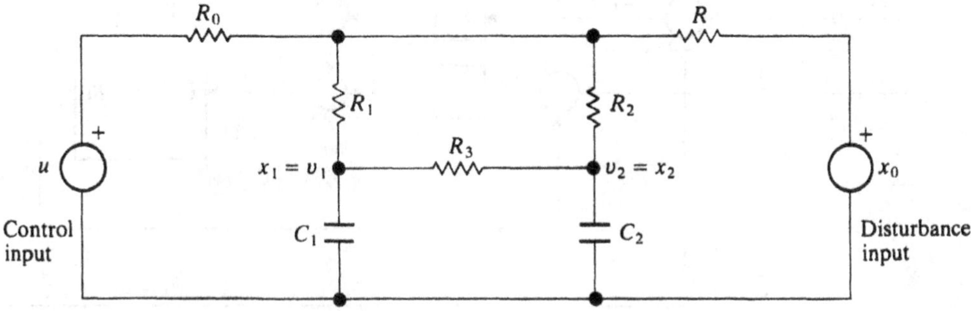

Figure 6.5 Electric network analog of temperature control problem.

and

$$C(A-BG)^{-1}B = \frac{[c_1 \quad c_2]\begin{bmatrix}-a_{22}b_1 - a_{12}b_2 \\ -a_{21}b_1 - a_{11}b_2\end{bmatrix}}{a_{11}a_{22} - a_{21}a_{12} + g_1(b_1a_{22} + b_2a_{12}) + g_2(b_1a_{21} + b_2a_{11})} \tag{6E.5}$$

Note that the numerator of (6E.5) is independent of the feedback gains, irrespective of c_1 and c_2.

In this example $C(A - BG)^{-1}B$ is a scalar (i.e., a 1-by-1 matrix) given by

$$p = C(A-BG)^{-1}B = \frac{-c_1(a_{22}b_1 + a_{12}b_2) - c_2(a_{21}b_1 + a_{11}b_2)}{a_{11}a_{22} - a_{21}a_{12} + g_1(b_1a_{22} + b_2a_{12}) + g_2(b_1a_{21} + b_2a_{11})}$$

Thus, from (6.47)

$$B^{\#} = [q_1, q_2]/p$$

where

$$q_1 = -c_1(a_{22} + b_2g_2) + c_2(-a_{21} + b_2g_1)$$

$$q_2 = c_1(-a_{12} + b_1g_2) - c_2(a_{11} + b_1g_1)$$

and

$$G_0 = B^{\#}E = [e_1q_1 + e_2q_2]/p \qquad \text{(a scalar)}$$

The implementation of the control law is illustrated in Fig. 6.6. It is noted that even though the performance criterion $y = Ce = c_1(v_1 - \bar{v}_1) + c_2(v_2 - \bar{v}_2)$ is a scalar combination of the two errors e_1 and e_2, the feedforward signal $g_{r1}\bar{v}_1 + g_{r2}\bar{v}_2$ is not expressible as a function of the difference between $\bar{v}_1$ and $\bar{v}_2$; both $\bar{v}_1$ and $\bar{v}_2$ are required in the control law implementation.

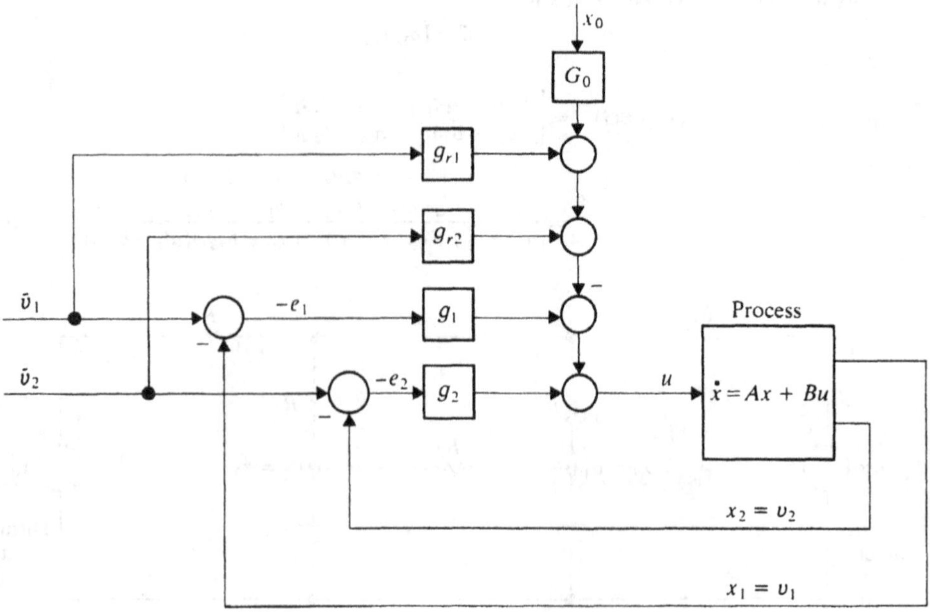

Figure 6.6 Control law implementation.

6.5 WHERE SHOULD THE CLOSED-LOOP POLES BE PLACED?

Having determined that the closed-loop poles of a controllable system can be placed anywhere, it is natural to ask where the poles *should* be placed. To assert that they should be placed to meet the performance requirements is begging the question, which is how to relate the performance requirements to the gain matrix G that is used in the implementation of the feedback law. A systematic method of selecting the gains by minimizing a quadratic performance integral is given in Chap. 9. That method has many advantages but it is by no means the only method available. Among the concerns that the designer might wish to address are those to be discussed in this section.

The control law for a regulator $u = -Gx$ implies that for a given state x the larger the gain, the larger the control input. There are limits on the control input in practical systems: The actuator which supplies the control u cannot be arbitrarily large without incurring penalties of cost and weight. Other reasons for limiting the control may be to avoid the potential damaging effects of stresses on the process that large inputs might cause. If the control signal generated by the linear feedback law $u = -Gx$ is larger than possible or permissible for reasons of safety, the actuator will "saturate" at a lower input level. The effect of occasional control saturation is usually not serious: in fact a system which never saturates is very likely overdesigned, having a larger and less efficient actuator than is needed to accomplish even the most demanding tasks. On the other hand, if the control signals produced by the linear control law are so large that the actuator is almost always saturated, it is not likely that the system behavior will be satisfactory, unless the actuator saturation is explicitly accounted for in an intentionally nonlinear (e.g., "bang-bang") control law design. If such a design is not intended, the gain matrix should be selected to avoid excessively large control signals for the range of states that the control system can encounter during operation.

The effect of control system gain on pole locations can be appreciated by considering the Bass-Gura formula (6.34) for a single-input system. (Qualitatively, similar considerations apply to multiple-input systems.) Note first that the gains are proportional to the amounts that the poles are to be moved, i.e., to the distance that the coefficients of the characteristic polynomial must move between the open-loop and the closed-loop system. The less the poles are moved, the smaller the gain matrix. Thus, large system gains are avoided by limiting the changes in the coefficients of the characteristic equation. It is also noted that the control system gains are inversely proportional to the controllability test matrix. The less controllable the system, the larger the gains that are needed to effect a change in the system poles. There is nothing surprising about this.

The inference that may be reasonably drawn from this is that the designer should not attempt to alter the dynamic behavior of the open-loop process more than is required. One reason for trying to alter the behavior of a process is to stabilize it, if it is unstable, or to increase its stability by moving its poles into the interior of the left half of the s plane. Although stability is the most important consideration it is not the only consideration. Speed of response (i.e., bandwidth) is also important. Fast response—high bandwidth—of the closed-loop system is often sought after, since the errors in following rapidly changing inputs will be smaller. There may be instances, however, in which the bandwidth of the closed-loop system is intentionally not made as high as it can be. If the reference input contains a good deal of noise, it might be desirable to reduce the bandwidth to prevent the system from becoming excessively agitated by following the noise.

Another reason for limiting the bandwidth of the closed-loop system is the uncertainty of the high-frequency dynamics of the process. A mechanical system, for example, has resonance effects (modes) due to the elasticity of the structural members. The dynamic model used for design ignores many if not all of these effects: their magnitudes are small; the exact frequencies are not easy to determine; the effort required to include them in the model is not justified. Other types of processes (thermal, chemical, etc.) also have uncertain behavior at high frequencies. If the uncertain high-frequency poles are included within the bandwidth of the closed-loop process, these resonances may be excited and result in unexpected high-frequency oscillation, or even instability. A prudent design requires that the loop transmission be well below unity at the frequencies where these resonances may occur.

The bandwidth of a system is governed primarily by its *dominant* poles, i.e., the poles with real parts closest to the origin. To see this, visualize the partial-fraction expansion of the transfer function of the system. Terms corresponding to poles whose (negative) real parts are farthest from the origin have relatively high decay rates (damping) and hence, after an initial transient period, they will contribute less to the total response than terms corresponding to poles with real parts close to the origin. (While this behavior is typical of physical processes, there is no theoretical reason why the residues at poles with high damping cannot be much greater than the residues at the poles with less damping. If the highly damped poles have large residues, their effects may persist simply because they start out much larger.)

In order for the transient to decay as rapidly as is required by the poles that are far from the origin, it is necessary to change the energy in the system rapidly; this would require the use of large control inputs. If there are some poles that are far from the origin and others that are close to the origin, the maximum control amplitudes will be governed by the former, but the system speed of response is slowed by the latter. This behavior suggests that the feedback gains are such that the available control is not efficiently used. Efficient use of the control signal would require that all the closed-loop poles be about the same distance from the origin.

Having reasoned that it is imprudent to try to move the open-loop poles farther than is necessary (obviously it is necessary to move them to the left half-plane if the open-loop process is unstable) and inefficient to make some poles much more highly damped (farther from the origin) than the other poles, one might seek to *optimize* the closed-loop pole locations. How to accomplish this in general is the subject of Chap. 9. One result of optimization theory that can be used here concerns "asymptotic pole location": As control effort becomes increasingly less "expensive," the closed-loop poles tend to radiate out from the origin along the spokes of a wheel in the left half-plane as given by the roots of

$$\left(\frac{s}{\omega_0}\right)^{2k} = (-1)^{k+1} \tag{6.50}$$

where k is the number of poles in the left half-plane. (Fig. 6.7.) Poles located in

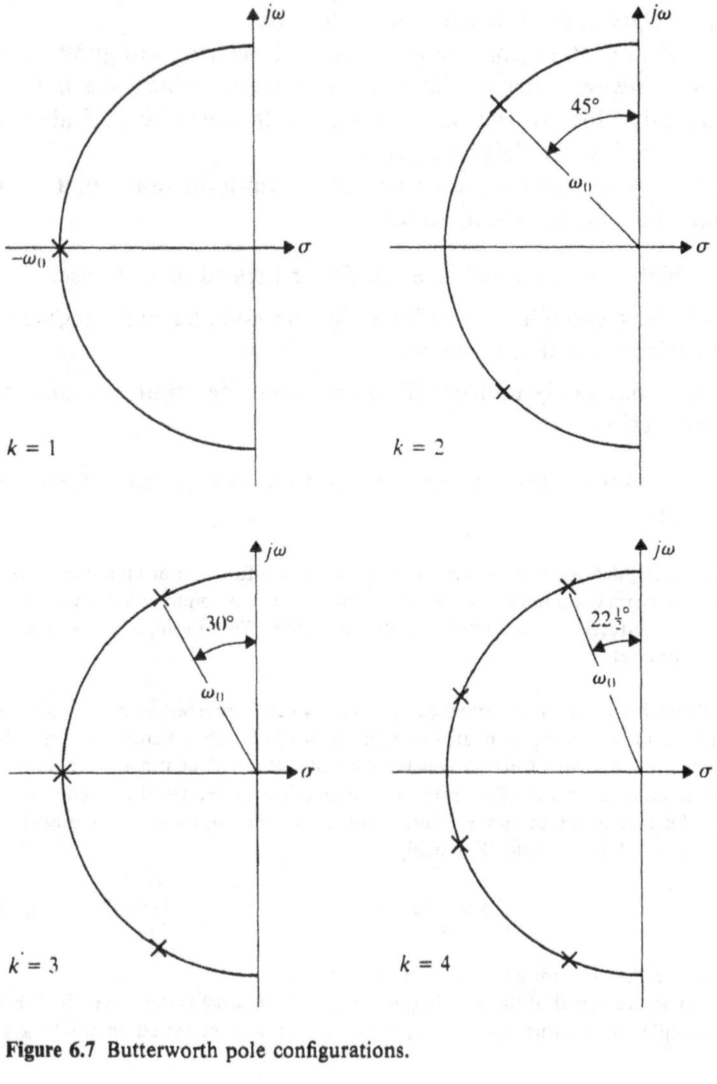

Figure 6.7 Butterworth pole configurations.

accordance with (6.50) are said to have a "Butterworth configuration," a term that originated in communication networks.

The polynomial having as its factors the zeros of (6.50) in the *left half-plane only* are known as *Butterworth polynomials* $B_k(z)$, $z = s/\omega_0$, the first few of which are:

$$B_1(z) = z + 1$$

$$B_2(z) = z^2 + \sqrt{2}z + 1$$

$$B_3(z) = z^3 + 2z^2 + 2z + 1$$

$$B_4(z) = z^4 + 2.613z^3 + (2 + \sqrt{2})z^2 + 2.613z + 1$$

Some of the properties of transfer functions having Butterworth polynomials for their denominators are given in Note 6.3 and Prob. 6.10.

In the absence of any other consideration, a Butterworth configuration is often suitable. Note, however that as the order k becomes high, one pair of poles come precariously close to the imaginary axis. It might be desirable to move these poles farther into the left half-plane.

The considerations that govern the choice of closed-loop poles that were discussed above may be summarized as follows:

Select a bandwidth high enough to achieve the desired speed of response.

Keep the bandwidth low enough to avoid exciting unmodeled high-frequency effects and undesired response to noise.

Place the poles at approximately uniform distances from the origin for efficient use of the control effort.

These broad guidelines allow plenty of latitude for special needs of individual applications.

Example 6F Missile autopilot As noted in Example 4F, the usual function of an autopilot in a missile is to make the normal component of acceleration a_N track a commanded acceleration signal a_{NC} which is produced by the missile guidance system. This example illustrates the design of such an autopilot.

Open-loop dynamics A high-performance missile, when provided with a suitable autopilot, is capable of achieving a relatively high bandwidth. This bandwidth may be comparable to that of the actuator which drives the control surface. Thus it may be necessary to include the dynamics of the actuator in order to have an adequate model of the process. We assume this to be the case in the present example, and use the first-order dynamic model for the actuator that was used in Example 4F, namely

$$\dot{\delta} = \frac{1}{\tau}(u - \delta) \tag{6F.1}$$

where u is the input to the actuator and τ is its time constant.

The complete dynamic model of the missile (airframe and actuator) is thus given by (4F.1) and (4F.2) in Example 4F. In this application, however, we are interested in tracking an acceleration command and hence prefer to use the acceleration error

$$e = a_{NC} - a_N \tag{6F.2}$$

as a state variable instead of the angle of attack. The derivative of the acceleration error is

$$\dot{e} = \dot{a}_{NC} - \dot{a}_N$$

Now, although the commanded acceleration is not constant, we can approach the design problem on the assumption that it is: $\dot{a}_{NC} \approx 0$. (A better design might be achieved by making use of the actual rate of change of normal acceleration command, a signal that might be available from the missile guidance system.)

In addition to approximating the commanded acceleration by a constant, we also assume that the aerodynamic coefficients Z_α and Z_δ and the missile speed V are approximately constant. Using all these approximations

$$\dot{e} = -\dot{a}_N = -Z_\alpha \dot{\alpha} - Z_\delta \dot{\delta} \tag{6F.3}$$

But, from (3F.1) and (3F.4),

$$\dot{\alpha} = q + \frac{a_N}{V} = q + \frac{1}{V}(a_{NC} - e)$$

Thus, by (6F.2) and (6F.1), we obtain from (6F.3)

$$\dot{e} = -Z_\alpha \left[q + \frac{1}{V}(a_{NC} - e) \right] + \frac{Z_\delta}{\tau}(\delta - u) \tag{6F.4}$$

The angle of attack α, by (4F.3), is

$$\alpha = \frac{1}{Z_\alpha}(a_N - Z_\delta \delta) = \frac{1}{Z_\alpha}(a_{NC} - e - Z_\delta \delta) \tag{6F.5}$$

Thus the differential equation for the pitch rate, using (4F.1), is

$$\dot{q} = \frac{M_\alpha}{Z_\alpha}(a_{NC} - e - Z_\delta \delta) + M_q q + M_\delta \delta \tag{6F.6}$$

A single third-order vector-matrix equation defining the system is obtained from (6F.1), (6F.4), and (6F.6). Defining the state vector by

$$x = [e, q, \delta]' \tag{6F.7}$$

we obtain the state-space equations

$$\dot{x} = Ax + Bu + Ea_{NC} \tag{6F.8}$$

where

$$A = \begin{bmatrix} Z_\alpha/V & -Z_\alpha & Z_\delta/\tau \\ -M_\alpha/Z_\alpha & M_q & \bar{M}_\delta \\ 0 & 0 & -1/\tau \end{bmatrix} \quad B = \begin{bmatrix} -Z_\delta/\tau \\ 0 \\ 1/\tau \end{bmatrix} \quad E = \begin{bmatrix} -Z_\alpha/V \\ M_\alpha/Z_\alpha \\ 0 \end{bmatrix} \tag{6F.9}$$

where

$$\bar{M}_\delta = M_\delta - \frac{M_\alpha}{Z_\alpha} Z_\delta \tag{6F.10}$$

The following numerical data were obtained for a representative highly maneuverable tactical missile:

$V = 1253$ ft/s (Mach 1.1)

$Z_\alpha = -4170$ ft/s^2 (per radian of angle of attack)

$$Z_\delta = -1115 \text{ ft/s}^2 \text{ (per radian of surface deflection)}$$
$$M_\alpha = -248 \text{ rad/s}^2 \text{ (per radian of angle of attack)}$$
$$M_q \approx 0$$
$$M_\delta = -662 \text{ rad/s}^2 \text{ (per radian of surface deflection)}$$
$$\tau = .01 \text{ s}$$

The characteristic equation of this system (with $M_q = 0$) is

$$\left(s + \frac{1}{\tau}\right)\left(s^2 + \frac{Z_\alpha}{V} s - M_\alpha\right) = 0 \tag{6F.11}$$

and, using the numerical data given above, (6F.11) becomes

$$(s + 100)(s^2 + 3.33s + 248) = 0$$

with roots at

$$s = -100 \qquad \text{(due to actuator)}$$

and at

$$s = -1.67 \pm j15.65 \qquad \text{(due to airframe)}$$

as shown in Fig 6.8. The open loop thus has very little damping and a natural frequency ω of approximately 15.65 rad/s = 2.49 Hz.

Design considerations If the damping factor were raised to a more suitable value (say $\zeta \approx 0.707$) the natural frequency of 2.49 Hz would result in a time constant of about 0.4 s. A shorter closed-loop time constant would be desirable for a high-performance missile: about 0.2 s would be more appropriate. Thus we should seek a natural frequency of $\omega \approx 30$ and $\zeta \approx 0.707$. This suggests a quadratic factor in the closed-loop characteristic polynomial of

$$s^2 + 30\sqrt{2}s + (30)^2 \tag{6F.12}$$

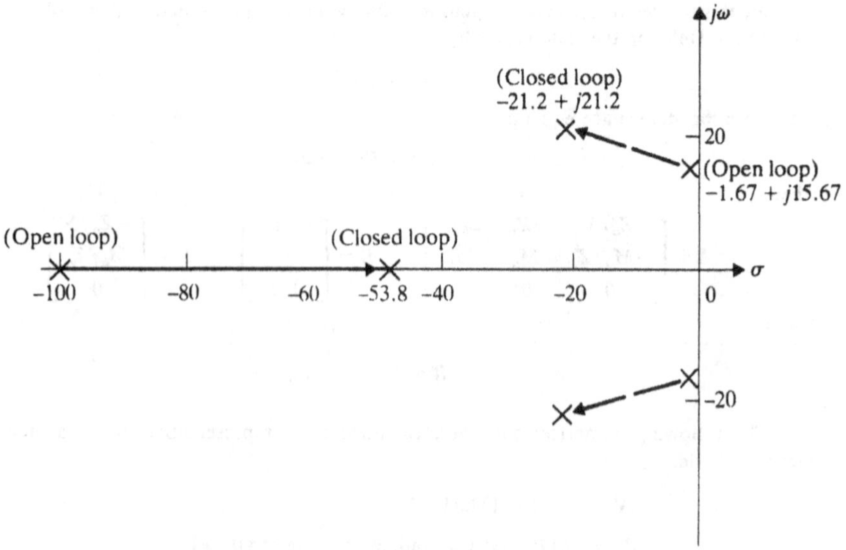

Figure 6.8 Open- and closed-loop poles for missile autopilot.

The location of the real pole at $s = -100$ due to the actuator is satisfactory: it is far enough away from the origin so as not to add substantially to the autopilot lag. We shall shortly discover, however, that to keep a closed-loop pole at $s = -100$ entails measuring (or estimating) and feeding back the actuator output δ. To simplify the implementation of the autopilot it might be desirable to permit the open-loop actuator pole to move to a different location provided that the overall system performance is not degraded. This is a design option we wish to explore.

The autopilot design will be done in two steps, as described earlier in the text. First we will design a regulator for a commanded normal acceleration of zero, then we will compute the feedforward gain to eliminate the steady state error for a nonzero commanded acceleration.

Regulator design To apply the Bass-Gura formula we need the open-loop characteristic equation: From (6F.11) this is

$$s^3 + 103.33s^2 + 581.s + 24\,800. = 0$$

Thus the open-loop coefficient vector is

$$a = \begin{bmatrix} 103.33 \\ 581. \\ 24\,800. \end{bmatrix}$$

thus

$$W = \begin{bmatrix} 1 & 103.33 & 581. \\ 0 & 1 & 103.33 \\ 0 & 0 & 1 \end{bmatrix}$$

We also find

$$Q = \begin{bmatrix} 111\,500. & -11.5 \times 10^6 & 8.77 \times 10^8 \\ 0. & -66.2 \times 10^3 & 6.64 \times 10^6 \\ 100. & -10^4 & 10^6 \end{bmatrix}$$

and

$$QW = \begin{bmatrix} 11\,500. & 0. & -0.248 \times 10^9 \\ 0. & -66\,204. & 0.198 \times 10^6 \\ 100. & 333.0 & 24\,800. \end{bmatrix}$$

from which:

$$(QW)^{-1} = \begin{bmatrix} 0.8657 \times 10^{-6} & 0.4544 \times 10^{-4} & 0.9035 \times 10^{-2} \\ 0.1090 \times 10^{-7} & -0.1517 \times 10^{-4} & -0.1215 \times 10^{-4} \\ -0.3637 \times 10^{-8} & 0.2040 \times 10^{-7} & 0.4055 \times 10^{-5} \end{bmatrix}$$

For any choice of closed-loop poles, the feedback gain matrix is given by:

$$g = G' = \begin{bmatrix} g_1 \\ g_2 \\ g_3 \end{bmatrix} = [(QW)']^{-1} \begin{bmatrix} \hat{a}_1 - a_1 \\ \hat{a}_2 - a_2 \\ \hat{a}_3 - a_3 \end{bmatrix} \tag{6F.13}$$

As discussed earlier, practical implementation is simplified by omitting the feedback from the control surface deflection. This is achieved by having $g_3 = 0$. From (6F.13), this requirement is satisfied by making

$$0.9035 \times 10^{-2}(\hat{a}_1 - a_1) - 0.1215 \times 10^{-4}(\hat{a}_2 - a_2) + 0.4055 \times 10^{-5}(\hat{a}_3 - a_3) = 0 \tag{6F.14}$$

We already decided that one factor of the characteristic polynomial be given by (6F.12). Thus the complete characteristic polynomial is chosen to be

$$(s+\omega_c)(s^2+30\sqrt{2}s+900) = s^3+\hat{a}_1 s^2+\hat{a}_2 s+\hat{a}_3$$

where

$$\begin{aligned}\hat{a}_1 &= \omega_c + 30\sqrt{2}\\ \hat{a}_2 &= 30\sqrt{2}\omega_c + 900\\ \hat{a}_3 &= 900\omega_c\end{aligned} \tag{6F.15}$$

with ω_c as yet undetermined. Equations (6F.14) and (6F.15) constitute four linear equations in the four unknowns $\hat{a}_1$, $\hat{a}_2$, $\hat{a}_3$, and ω_c. These are solved to yield

$$\begin{aligned}\hat{a}_1 &= 96.24\\ \hat{a}_2 &= 3182.\\ \hat{a}_3 &= 48\,419.\\ \omega_c &= 53.8\end{aligned}$$

The location of the real pole at $s = -\omega_c = -53.8$ is satisfactory, so no feedback gain from the surface deflection is necessary. Thus the gain matrix contains only two nonzero elements:

$$G = [-0.6366\times10^{-4}, -0.3929\times10^{-1}, 0] \tag{6F.16}$$

Feedforward gain Having decided that no feedback of the control surface deflection is necessary, and having adjusted the gains from the acceleration error $a_{NC} - a_N$ and the pitch rate q to provide the desired closed-loop poles, it now remains to set the feedforward gain G_0 to eliminate the steady-state error for a step input of acceleration.

The C matrix for the scalar error is

$$C = [1 \quad 0 \quad 0]$$

and the closed-loop A matrix is

$$A_c = A - BG = \begin{bmatrix} 3.767 & -8550.3 & -111\,500.\\ -0.0595 & 0. & -595.7\\ 0.006\,366 & 3.929 & -100.\end{bmatrix}$$

and

$$A_c^{-1} = -\begin{bmatrix} 0.048\,33 & 8.613 & -105.20\\ -0.000\,201 & 0.006\,88 & 0.1834\\ -0.000\,005 & 0.000\,82 & 0.0105\end{bmatrix}$$

Thus

$$CA_c^{-1} = [-0.048\,33 \quad -8.613 \quad 105.20]$$

and

$$CA_c^{-1}B = 5130.$$

Hence

$$B^{\#} = (CA_c^{-1}B)^{-1}CA_c^{-1} = [-9.42\times10^{-6} \quad -1.68\times10^{-3} \quad 2.05\times10^{-2}]$$

and, finally,

$$G_0 = B^{\#}E = -1.313\times10^{-4} \tag{6F.17}$$

The autopilot can be implemented as shown in Fig. 6.9. A body-mounted accelerometer measures the actual normal acceleration and a rate gyro measures the actual body pitch rate.

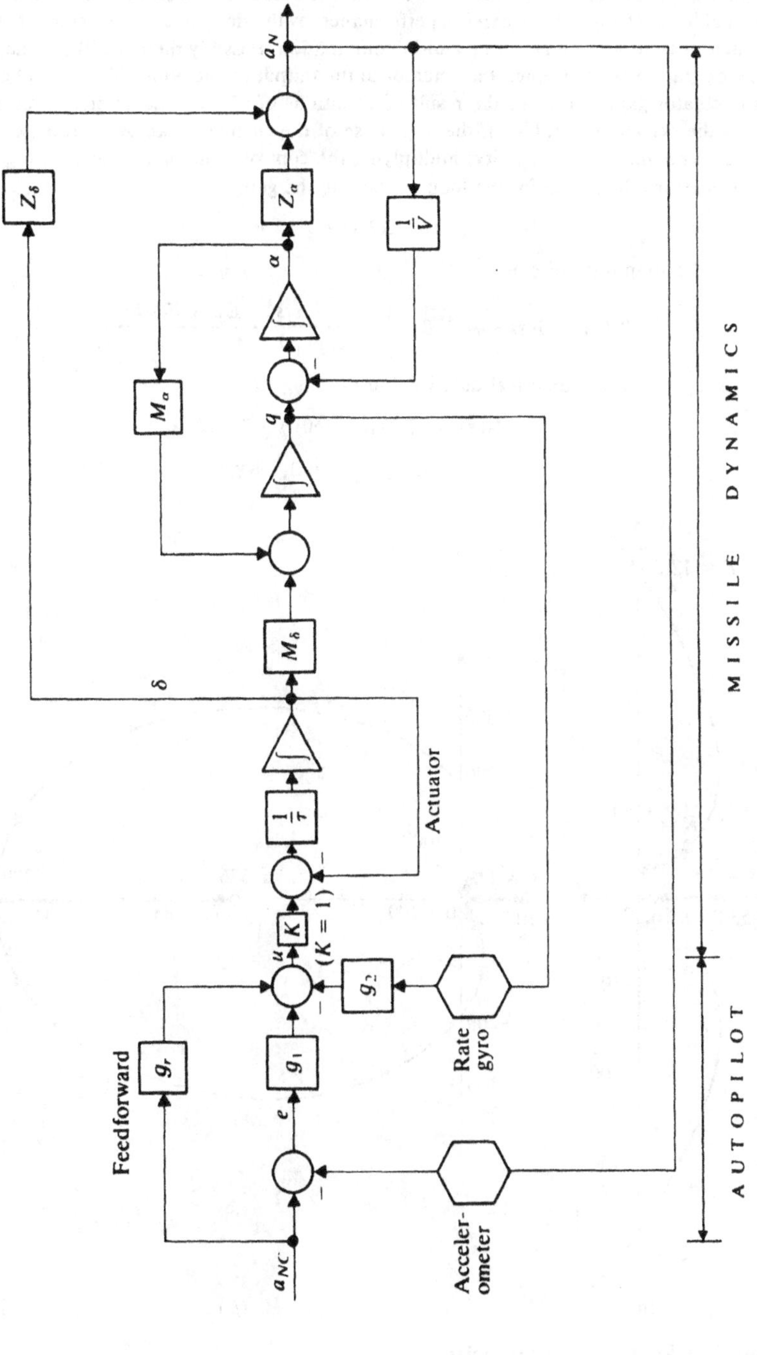

Figure 6.9 Dynamics of missile with autopilot.

Robustness of design The "robustness" of the design, i.e., its ability to withstand parameter variations, is of interest. It is not likely that the gain of the accelerometer or the gyro will vary by more than a fraction of a percent. The actuator and airframe dynamics are much more liable to change. In a careful performance evaluation, one would study the effect of parameter variations one at a time and in combination. Possibly the most likely change would be the dc transmission through the actuator to the output acceleration. This could be the result of an actuator gain change or the result of variations of airframe parameters from the values used in the design. Regardless of the true cause of the change, it can be represented by a gain K (with a nominal value of unity) multiplying the control signal u as shown in Fig. 6.9.

The return difference for the loop containing the gain K is

$$1 + KG(sI - A)^{-1}B$$

The forward loop transmission

$$\mathsf{G}_0(s) = G(sI - A)^{-1}B = \frac{G(s^3I + E_1s^2 + E_2s + E_3)B}{|sI - A|} = \frac{\mathsf{N}(s)}{\mathsf{D}(s)}$$

Using the above numerical data we find that

$$\mathsf{N}(s) = -7.09s^2 + 2601.s + 23\,608$$
$$= 7.09(-s + 376.)(s + 8.86)$$

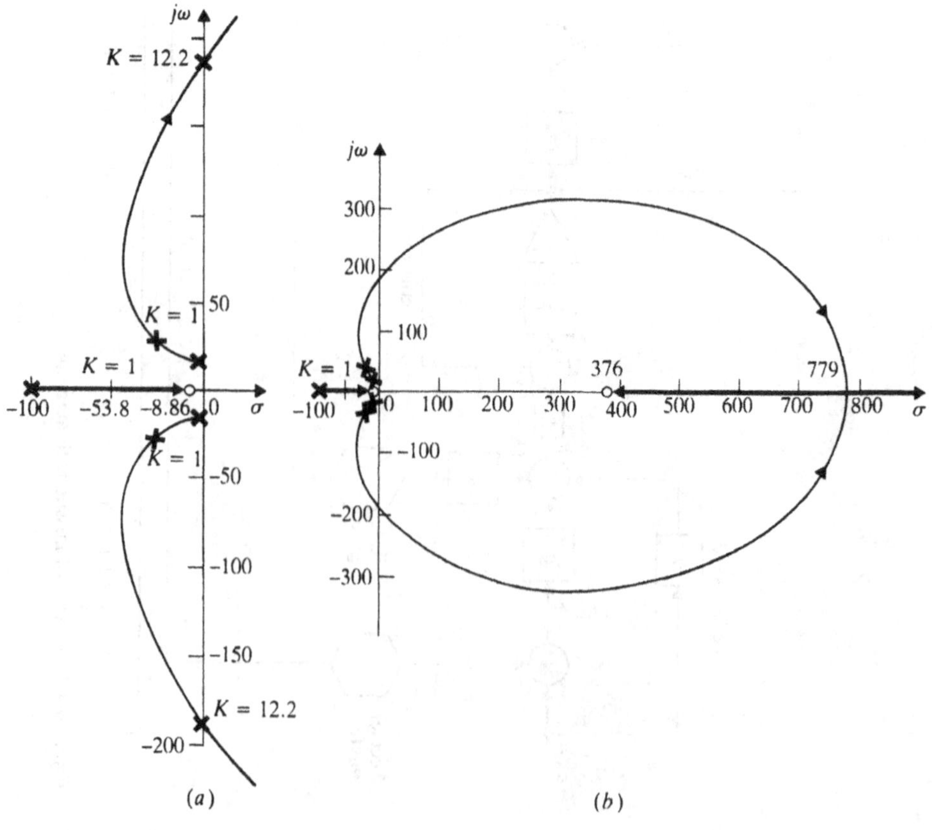

Figure 6.10 Root-locus plot for missile autopilot

$$\mathsf{G}(s) = \frac{7.09(-s + 376)(s + 8.86)}{(s + 53.8)(s^2 + 30\sqrt{2}s + 900)}$$

which results in apparent zeros at $s = 376.$ and $s = -8.86$. (These are not zeros of the open-loop process, but are created by the use of the sensors of acceleration and pitch rate.)

The root locus has the appearance shown in Fig. 6.10. The root locus starts at the open-loop poles at $s = -100$ and $s = -1.67 \pm j15.65$ and goes to the apparent zeros. At a nominal gain of $K = 1$, the loci pass through the poles for which the operation was designed ($s = -15\sqrt{2} \pm j15\sqrt{2}$, and $s = -53.8$) and then continue toward the imaginary axis and ultimately into the right half-plane. Because of the nonminimum phase zero at $s = 376$, the locus has a branch that goes out along the *positive* real axis as $K \to \infty$, as was discussed in Chap. 5.

The range of gain K for which the system is stable can be found using the Routh or Hurwitz algorithm of Chap. 5 and is

$$-1.14 < K < 12.21$$

The gain margin is thus 12.2 (or 22 dB) which is more than ample. The frequency at which the root locus crosses the imaginary axis is found to be $\omega = 187$. The right half-plane root-locus plot is shown in Fig. 6.10(b). It should be noted that the loci, after crossing the imaginary axis,

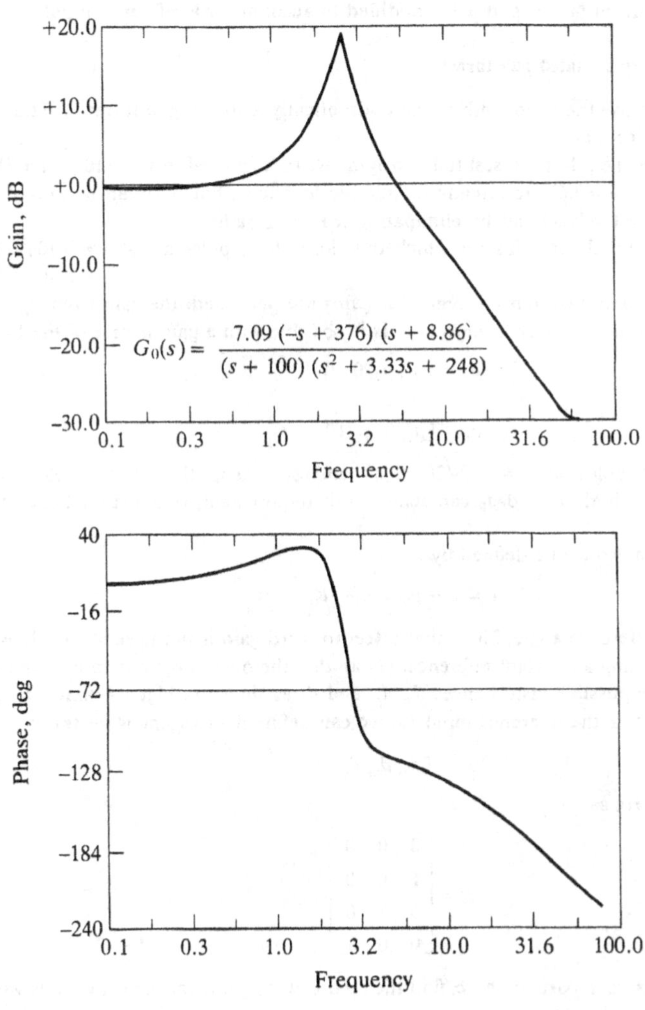

Figure 6.11 Bode plot for open-loop transmission for missile autopilot.

bend over toward the positive real axis and reach it at some positive real value of $s > 376$, the positive zero of $N(s)$. Then one branch goes to the zero and the other goes to $+\infty$.

The Bode plot for the open-loop transmission $G_0(s)$ is shown in Fig. 6.11.

PROBLEMS

Problem 6.1 Inverted pendulum on cart: state variable feedback

Consider the inverted pendulum on the motor-driven cart of Prob. 2.1 with numerical data as given in Prob. 3.6. It is desired to place the dominant poles (in a Butterworth configuration) at

$$s = -4 \quad \text{and} \quad s = -2 \pm j2\sqrt{3}$$

and to leave the pole at $s = -25$ unchanged.

(*a*) Find the gain matrix that produces this set of closed-loop poles.

(*b*) It is desired to move the cart from one position to another without causing the pendulum to fall. How must the control law of part *a* be modified to account for a reference input x_r?

Problem 6.2 Hydraulically actuated gun turret

It is desired to increase the bandwidth of the hydraulically actuated gun turret of Example 4E by use of state-variable feedback.

The dominant poles, i.e., those closest to the origin, are to be moved to $s = -10\sqrt{2}(1 \pm j1)$. The other poles (at $s = -64.5 \pm j69.6$) are already in suitable locations, but they can be moved in the interest of simplifying the feedback law by eliminating feedback paths.

(*a*) Determine the regulator gains for which the closed-loop poles are at $s = -10\sqrt{2}(1 \pm j1)$ and at $s = -64.5 \pm j69.6$.

(*b*) For simplicity, only two nonzero regulator gains are permitted: the gain from $x_1 = \theta$ and one other gain, either from $x_2 = \omega$ or from $x_3 = p$. Is it possible with a gain matrix of the form

$$g_A = [g_1, g_2, 0, 0]$$

or

$$g_B = [g_1, 0, g_3, 0]$$

to place the dominant poles at $s = -10\sqrt{2}(1 \pm j1)$ and still keep the "fast" poles at their approximate locations? If both g_A and g_B can achieve this requirement, which is the better choice? Explain.

(*c*) Let the tracking error e be defined by

$$e = \theta - \theta_0 = x_1 - \theta_0$$

where θ_0 is a constant reference angle. Show that a feedforward gain is not needed to achieve zero steady state error in tracking a constant reference. (Note that the open-loop system is "type 1.")

(*d*) There are three possible disturbances d_q, d_p, and d_τ as shown in Fig. 4.4. Since, by part *a*, it is unnecessary to estimate the reference input θ_0, we can define the exogenous vector as

$$x_0 = [d_\tau, d_p, d_q]'$$

and the distribution matrix as

$$E = \begin{bmatrix} 0 & 0 & 0 \\ 1 & 0 & 0 \\ 0 & 1 & 0 \\ 0 & 0 & 1 \end{bmatrix}$$

For each of the sets of gains in parts *a* and *b*, find the feedforward gains for the exogenous variables which will ensure zero steady state error.

Problem 6.3 Two-car train

It is desired to bring the two-car train of Probs. 2.5 and 3.9 to rest at the origin using only the motor on car 1. Find the gain matrix $G = g'$ in the control law $u = -g'x$ which places the poles at $s = -1 \pm j1$ and at $s = -100 \pm j100$.

Problem 6.4 Two-car train (continued)

Modify or redesign the control law obtained in Prob. 6.3, so that the train maintains a constant velocity $V = \text{const.}$

Problem 6.5 Aircraft longitudinal motion with simplified dynamics

The speed variations in aircraft longitudinal motion are often "trimmed" by a separate throttle control so that Δu can be assumed negligible. Thus we can use a simplified dynamic model in which the state variables are

$$x_1 = \alpha \qquad x_2 = q \qquad x_3 = \theta$$

Using these state variables and the aerodynamic coefficients of Prob. 4.5, find the gains that place the closed-loop poles in the Butterworth pattern: $s = -2$, $s = -1 \pm j\sqrt{3}$.

Problem 6.6 Constant-altitude autopilot

The altitude h of an aircraft is given by

$$\dot{h} = V\gamma = V(\theta - \alpha)$$

where γ is the flight-path angle. (See Fig. P6.6.) Hence the aircraft altitude can be maintained constant by keeping the flight-path angle $\gamma = \theta - \alpha$ zero. Add a state variable

$$x_4 = (h - h_0)/V$$

where h_0 is a reference altitude, to the state variables used in Prob. 6.5.

(a) Draw the block diagram of the closed-loop system.

(b) Find the gains for which the closed-loop poles lie in the Butterworth pattern:

$$s = 2.5(-1/2 \pm j\sqrt{3}/2) \qquad s = 2.5(-\sqrt{3}/2 \pm j1/2)$$

Problem 6.7 Aircraft lateral dynamics: turn coordination

When an aircraft executes a perfectly coordinated turn the sideslip angle β is zero. (When this occurs, the net force vector acting on the aircraft lies in the vertical plane of the aircraft so the occupant has the same kinesthetic sensation as when the aircraft is flying without banking.)

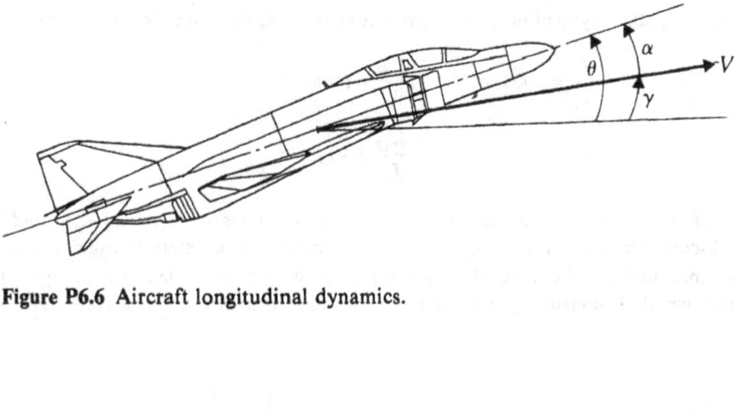

Figure P6.6 Aircraft longitudinal dynamics.

(*a*) The rudder is often used for turn coordination. We may thus assume a control law for the rudder, using (2.41)

$$\frac{Y_R}{V}\delta_R = -\frac{Y_A}{V}\delta_A - \frac{g}{V}\phi + \left(1 - \frac{Y_r}{V}\right)r - \frac{Y_p}{V}p - \left(\frac{Y_\beta}{V} + \frac{1}{T}\right)\beta \qquad \text{(P6.7}a\text{)}$$

when (P6.7*a*) is substituted into the first equation of (2.41) we obtain

$$\dot{\beta} = -\frac{1}{T}\beta \qquad \text{(P6.7}b\text{)}$$

Hence any sideslip that may be initially present will be reduced to zero with a time constant of T. When (P6.7*a*) is substituted into the next three equations of (2.41) a third-order system with a single control δ_A and a disturbance β is obtained. The poles of that system may be placed by use of the Bass-Gura formula. Using the data of Prob. 4.4, find the control law for the ailerons that makes the sideslip decay time constant $T = 0.2$ s and places the remaining poles at $s = -1$ and $s = -1 \pm j3$. Combine the result with (P6.7*a*) to obtain the entire control law.

(*b*) As the aircraft makes a constant-radius turn the bank angle ϕ becomes constant. Thus if a constant radius turn is desired, a constant bank angle ϕ_0 is commanded. Modify the control law of part *a* so that the aircraft error $e = \phi - \phi_0$ is reduced to zero in the steady state. (Let e be a state variable in place of ϕ.)

Problem 6.8 Three-capacitance thermal system

A state-variable feedback control law is to be designed for the thermal control system considered in Prob. 3.7, et seq.

(*a*) Find the control gains that place the regulator poles in a third-order Butterworth configuration of radius 2, i.e., the characteristic equation of the closed-loop system is to be

$$D(s) = \left(\frac{s}{2}\right)^3 + 2\left(\frac{s}{2}\right)^2 + 2\left(\frac{s}{2}\right) + 1 = 0$$

(*b*) It is desired to keep point 3 (i.e., v_3) at a constant temperature $\bar{v}$ in the presence of an external temperature v_0. Let the state be defined as $x = [v_1, v_2, e]'$ where $e = v_3 - \bar{v}$, and the exogenous vector as $x_0 = [\bar{v}, v_0]'$. Find the matrix E for the system, and, using the gain matrix from part *a*, find the feedforward gain matrix $G_0 = B^{\#}E$.

(*c*) Draw a block diagram of the control law showing the feedback and feedforward paths. Does anything seem unusual about this structure?

Problem 6.9 Two-axis gyro: gains by pole placement

A control law such as shown in Fig. 2.15 is to be designed for a two-axis gyro described in Example 2F (et seq.). The design will be accomplished in a number of steps which will encompass several problems.

The present problem is to design a deterministic control law under the assumption that all the state variables are measurable. The dynamic model to be used for the design is summarized in Example 3. The following data, typical of a small gyro, may be used for numerical calculations:

$$\frac{H}{J_d} = 3000\ \text{sec}^{-1} \qquad \frac{K_Q}{J_d} = 60\ \text{sec}^{-2}$$

$$\frac{B}{J_d} \approx 0 \qquad \frac{K_D}{J_d} = 30\ \text{sec}^{-2}$$

For this stage of the design it is assumed that the state variables δ_x, δ_y, ω_{xB}, ω_{yB}, *and* the external angular velocity components ω_{xE}, ω_{yE} are all measurable. (The external angular velocity components are *not* measurable, of course. If they were, there would be no need for the gyro!) In subsequent problems we shall consider the design of observers to measure those state variables, namely ω_{xB}, ω_{yB}, ω_{xE}, ω_{yE} that cannot be measured, using only observations given the measurements of δ_x and δ_y.

A linear control law of the form

$$u = \begin{bmatrix} u_x \\ u_y \end{bmatrix} = \begin{bmatrix} \tau_x/J_d \\ \tau_y/J_d \end{bmatrix} = -G_\delta \begin{bmatrix} \delta_x \\ \delta_y \end{bmatrix} - G_\omega \begin{bmatrix} \omega_{xB} \\ \omega_{yB} \end{bmatrix} - G_0 \begin{bmatrix} \omega_{xE} \\ \omega_{yE} \end{bmatrix}$$

The matrices G_δ and G_ω are the "regulator" gains, to keep the gyro wheel stable in the absence of external angular velocity components, and G_0 is the gain matrix for the exogenous inputs, in this case the external angular velocity components ω_{xE} and ω_{yE}.

(*a*) Considerations of symmetry suggest that the regulator gain matrices should be of the form

$$G_\delta = \begin{bmatrix} g_1 & g_2 \\ -g_2 & g_1 \end{bmatrix} \qquad G_\omega = \begin{bmatrix} g_3 & g_4 \\ -g_4 & g_3 \end{bmatrix}$$

This means that there are four parameters for a fourth-order system, and a unique design can be achieved by pole placement. Determine the regulator closed-loop characteristic equation in terms of g_1, g_2, g_3, g_4. Does this place any restriction on the closed-loop pole locations?

(*b*) Using the theory developed in Sec. 6.4, find the matrix G_0 that maintains δ_x, δ_y at zero in the steady state, given that the exogenous input angular velocity components ω_{xE} and ω_{yE} are constants.

(*c*) In the steady state with $\omega_{xE} \neq 0$ and $\omega_{yE} \neq 0$ the control vector $u = [\tau_x/J_d, \tau_y/J_d]'$ is not zero. How does it depend on the input angular velocity components? Does this suggest a method for determining the input angular velocity?

Problem 6.10 Properties of Butterworth filters

Let

$$\mathrm{H}(s) = \frac{1}{B_k(s/\omega_0)}$$

where $B_k(z)$ is a Butterworth polynomial of order k.

(*a*) Show that

$$|\mathrm{H}(j\omega)| = \frac{1}{[1 + (\omega/\omega_0)^{2k}]^{1/2}}$$

(*b*) Sketch the amplitude plot corresponding to $\mathrm{H}(j\omega)$.

(*c*) Explain why the Butterworth polynomial is said to have a "maximally flat" amplitude response as compared with other systems of the same order.

NOTES

Note 6.1 Bass-Gura formula

The Bass-Gura formula[1] was originally derived by a method that closely resembles that used in this book. A simpler but less intuitive derivation may be found in Chap. 3 of Kailath's book[2] which contains several other formulas for the feedback gains.

Note 6.2 Zeros of closed-loop system

The matrix of transfer functions for the m-input, m-output system

$$\dot{x} = Ax + Bu \qquad y = Cx \qquad (l = m)$$

is

$$\mathrm{H}(s) = C(sI - A)^{-1}B$$

In accordance with the definition of transmission zeros given in Sec. 4.10, the transmission zeros of $\mathsf{H}(s)$ are the zeros of $|C(sI - A)^{-1}B|$.

As revealed in the analysis of Prob 4.1, the transmission zeros of a system in which state-variable feedback is used are not altered by the use of such feedback, i.e., the transmission zeros of $\mathsf{H}_c(s) = C(sI - A + BG)^{-1}B$ are the zeros of $\mathsf{H}(s)$.

For $B^{\#}$ as given by (6.46) to exist, it is necessary that $|C(A - BG)B| \neq 0$, which is the same as requiring that $\mathsf{H}_c(s)$ have no transmission zeros at the origin ($s = 0$). Since the transmission zeros of $\mathsf{H}_c(s)$ coincide with those of $\mathsf{H}(s)$, however, we conclude that the necessary and sufficient condition for $B^{\#}$ to exist is that $\mathsf{H}(s)$ have no transmission zeros at the origin.

Note 6.3 Butterworth polynomials

Butterworth polynomials have found extensive application in communication networks for their "maximally flat" frequency response characteristics. (See Problem 6.10.) They have also occurred in control system design by classical methods. (See [3], for example.) That the optimum closed-loop pole locations tend to a Butterworth configuration as the control cost decreases was first pointed out by Kalman[4] and subsequently studied in considerable detail by Kwakernaak.[5] (See Note 9.4 for further discussion of asymptotic behavior.)

REFERENCES

1. Bass, R. W., and Gura, I., "High-Order System Design Via State-Space Considerations," *Proc. Joint Automatic Control Conf.*, Troy, NY, June 1965, pp. 311-318.
2. Kailath, T., *Linear Systems*, Prentice-Hall Inc., Englewood Cliffs, NJ, June 1980.
3. Chang, S. S. L., *Synthesis of Optimum Control Systems*, McGraw-Hill Book Co., New York, June 1961.
4. Kalman, R. E., "When Is a Linear Control System Optimal?," *Trans. ASME (J. Basic Engineering)*, vol. 86D, no. 1, March 1964, pp. 51-60.
5. Kwakernaak, H., and Sivan, R., *Linear Optimal Control Systems*, Wiley-Interscience, New York, 1972.

CHAPTER

SEVEN

LINEAR OBSERVERS

7.1 THE NEED FOR OBSERVERS

In Chap. 6 we studied methods for shaping the dynamic response of a closed-loop system by selecting the feedback gains to "place" the resulting poles at desired locations. In order to place the poles at arbitrary locations, it is generally necessary to have all the state variables available for feedback. There are many systems, of course, such as those illustrated in Examples 4D and 4E, in which acceptable performance can be achieved by feeding back only those state variables that are accessible to measurement. But often it is not possible to achieve acceptable performance using only those state variables that can be measured. Must we abandon the hope of controlling such systems? Fortunately not. If the system is observable, it is possible to estimate those state variables that are not directly accessible to measurement using the measurement data from those state variables that are accessible. And by use of these state-variable estimates rather than their measured values one can usually achieve acceptable performance. State-variable estimates may in some circumstances be even preferable to direct measurements, because the errors introduced by the instruments that provide these measurements may be larger than the errors in estimating these variables.

A dynamic system whose state variables are the estimates of the state variables of another system is called an *observer* of the latter system. This term was introduced into linear system theory by D. Luenberger in 1963[1, 2, 3] (see Note 7.1). Luenberger showed that, for any observable linear system, an observer can be designed having the property that the estimation error (i.e., the difference between the state of the actual system and the state of the observer) can be made to go to zero as fast as one may desire. The design technique is equivalent to pole placement in feedback system design.

Several years before Luenberger's introduction of observers, R. E. Kalman, with the collaboration of R. Bucy, published two famous papers[4, 5] on linear filtering and prediction. These papers defined a state estimator that is optimum with respect to the process noise and observation noise. This state estimator (now called a *Kalman filter*) has the structure of a linear observer, so in a sense a linear observer may be regarded as a suboptimum Kalman filter. Alternatively, a Kalman filter may be regarded as an optimum observer. The latter viewpoint is adopted in Chap. 11, which deals with Kalman filters.

Although observers are useful for estimating the state of a linear system having a known external input, their main use is in estimating the state variables that cannot be measured but are needed for implementation of feedback control. In this chapter, however, we focus on the observer, treating the input u as a known quantity. In the next chapter we show how the linear observer can be combined with a linear control law to produce a complete compensator, the goal of the linear system designer.

7.2. STRUCTURE AND PROPERTIES OF OBSERVERS

Suppose we have a dynamic system in our usual state-space representation

$$\dot{x} = Ax + Bu \tag{7.1}$$

for which we have already designed a control law $u = -Gx$ under the assumption that x is accessible for measurement. But instead of being able to measure the state x we can only measure

$$y = Cx \tag{7.2}$$

where the dimension m of the observation vector y is less than the dimension of x. (We really ought to say that the rank of C is less than the dimension of the state vector. This will cover the possibility that there are as many sensors as state variables, or even more, but that not all of them are independent.)

If we could invert the matrix C we would be able to solve for $x(t)$ directly, given $y(t)$. But we are assuming that C is a singular matrix, so we can't determine $x(t)$ using only $y(t)$; we must use not only present observation data but also past observation data: $y(\tau)$ for $\tau < t$. One procedure for finding $x(t)$ given $y(\tau)$ for $\tau < t$ is to find the state $x(\tau)$ at some earlier time $\tau < t$, using the integral formula (5.21), and then extrapolate to the present time using (7.1). If it were practical to implement this procedure, and if there were no errors in the measurement $y(t) = Cx(t)$, it would be possible to determine $x(t)$ exactly using only a finite amount of past data. But it is not practical to use this procedure because of the complexity entailed in the numerical evaluation of the integral. And it wouldn't be worthwhile, because of the errors inevitably present in the measurement of y. The presence of these errors means that we can only obtain an estimate $\hat{x}(t)$ of $x(t)$ and never $x(t)$ itself. (This is true even when all the state variables *can* be measured.)

A better procedure for obtaining an estimate $\hat{x}(t)$ of $x(t)$ is to make the estimate be the output of a dynamic system

$$\dot{\hat{x}} = \hat{A}\hat{x} + \hat{B}u + Ky \tag{7.3}$$

excited by the measurement y and the input u, selecting the matrices $\hat{A}$, $\hat{B}$, and K to make the error

$$e = x - \hat{x} \tag{7.4}$$

acceptably small. This is Luenberger's method.

A differential equation can be obtained for the error e by using (7.1)-(7.4). Specifically

$$\begin{aligned}\dot{e} = \dot{x} - \dot{\hat{x}} &= Ax + Bu - \hat{A}(x - e) - \hat{B}u - KCx \\ &= \hat{A}e + (A - KC - \hat{A})x + (B - \hat{B})u\end{aligned} \tag{7.5}$$

If we demand that the error go to zero asymptotically, independent of x and u, then the coefficients of x and u in (7.5) must be zero and $\hat{A}$ must be the dynamics matrix of a stable system. This means that

$$\hat{A} = A - KC \tag{7.6}$$

and

$$\hat{B} = B \tag{7.7}$$

Thus we cannot pick $\hat{A}$, $\hat{B}$, and K arbitrarily. In fact there is no choice at all in the selection of $\hat{B}$: it must be the control matrix B. And once the matrix K is selected, the matrix $\hat{A}$ is determined. The only matrix we are free to select is the matrix K. These restrictions are incorporated into the observer by writing the defining equation (7.3) as

$$\begin{aligned}\dot{\hat{x}} &= (A - KC)\hat{x} + Bu + Ky \\ &= A\hat{x} + Bu + K(y - C\hat{x})\end{aligned} \tag{7.8}$$

This expression for the observer shows that it has the same form as the original process (7.1), except that it has an additional input $K(y - C\hat{x})$. The quantity

$$r = y - C\hat{x} = C(x - \hat{x}) = Ce$$

is the difference between the actual measurement y and the estimated measurement. It is often called the *residual.* If the error $e = x - \hat{x}$ is forced to zero then the residual will likewise tend to zero.

A block-diagram representation of the observer, as given by (7.8), is shown in Fig. 7.1.

The observer is in the form of a feedback system in which the residual r has the role of the "error" typical of a feedback system. The closed-loop dynamics matrix is seen to be $\hat{A} = A - KC$ which is the same matrix that appears in the

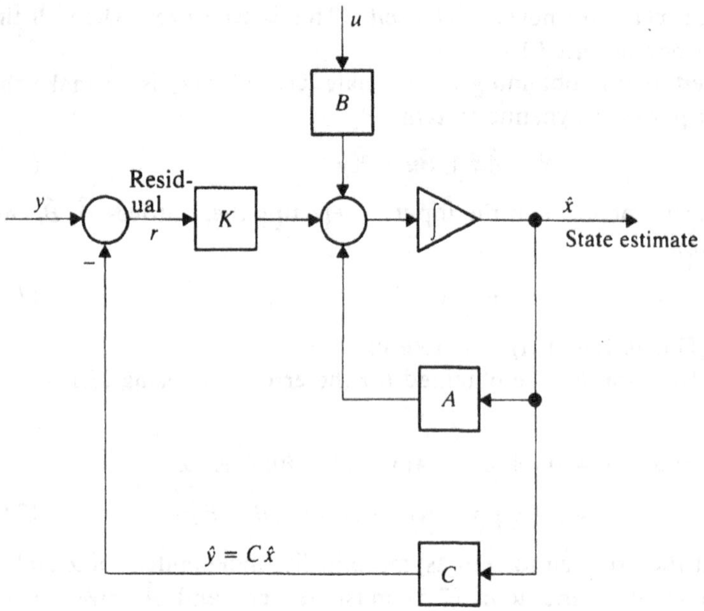

Figure 7.1 Block diagram of linear observer.

differential equation (7.5) for the propagation of the system error. When (7.6) and (7.7) are satisfied, (7.5) reduces to the homogeneous equation

$$\dot{e} = \hat{A}e \tag{7.9}$$

In order for the error to approach zero asymptotically it is necessary that $\hat{A}$ be a stability matrix, i.e., that the eigenvalues of the closed-loop matrix $\hat{A} = A - KC$ all be in the left half-plane. Determination of the feedback matrix K that accomplishes this is a pole-placement task similar to the task that the designer is faced with in shaping the response of a system with full-state variable feedback.

Just as the eigenvalues of the full-state feedback control system can be placed at arbitrary locations if the controllability test matrix Q is of rank k, so also the eigenvalues of $\hat{A} = A - KC$ can be placed at arbitrary locations if the observability test matrix

$$N = [C' \quad A'C' \quad \cdots \quad (A')^{k-1}C'] \tag{7.10}$$

is of rank k. If there is only a single output, then, as shown in the next section, the observer gain matrix K becomes a column vector and is uniquely determined by the desired eigenvalues of $\hat{A}$.

The presence of more than one output provides more flexibility: it is possible to place all the eigenvalues and do other things. Or, alternatively, some of the observer gains can be set to zero to simplify the resulting observer structure.

The transfer function from the observation vector y and control u to the state estimate $\hat{x}$ is found by calculating the Laplace transform of (7.8):

$$(sI - \hat{A})\hat{x}(s) = Bu(s) + Ky(s)$$

or

$$\hat{x}(s) = (sI - \hat{A})^{-1}Bu(s) + (sI - \hat{A})^{-1}Ky(s) \tag{7.11}$$

where $\hat{A} = A - KC$.

7.3 POLE-PLACEMENT FOR SINGLE-OUTPUT SYSTEMS

When there is only one output variable the output equation is

$$y = c_1x_1 + \cdots + c_kx_k = [c_1, \ldots, c_k]\begin{bmatrix} x_1 \\ \vdots \\ x_k \end{bmatrix}$$

Thus C is a row vector

$$C = c' = [c_1, c_2, \ldots, c_k]$$

and the gain matrix K is a column vector:

$$K = k = \begin{bmatrix} k_1 \\ \vdots \\ k_k \end{bmatrix} \tag{7.12}$$

In this case

$$\hat{A} = A - kc' \tag{7.13}$$

and the objective is to find the gain matrix K such that the eigenvalues of $\hat{A}$ have the desired locations. To obtain a formula for the gain matrix k we note that any matrix and its transpose, in particular, $\hat{A}$ and $\hat{A}'$, have the same eigenvalues. Thus, assigning the eigenvalues of $\hat{A}'$ is equivalent to assigning the eigenvalues of $\hat{A}$. But

$$\hat{A}' = A' - ck' \tag{7.14}$$

Compare (7.14) to the closed-loop control matrix

$$A_c = A - bg'$$

of (6.9) for the single-input full-state feedback problem. It is seen that they are of the same form: In particular, we have the following counterparts:

Full-state feedback	Observer
A	A'
b	c
g	k

We thus conclude that determination of the gain matrix k for the observer, with A' and c given, is the same problem as the determination of the gain matrix g with A and b given. We can thus translate the Bass-Gura formula (6.34) for the feedback gain g into the formula for the observer gain matrix:

$$k = [(NW)']^{-1}(\hat{a} - a) \tag{7.15}$$

where

$$N = [C', A'C', \ldots, (A')^{k-1}C'] \tag{7.16}$$

is the observability test matrix, and

$$\hat{a} = \begin{bmatrix} \hat{a}_1 \\ \vdots \\ \hat{a}_k \end{bmatrix} \qquad a = \begin{bmatrix} a_1 \\ \vdots \\ a_k \end{bmatrix} \tag{7.17}$$

where $\hat{a}_1, \ldots, \hat{a}_k$ are the coefficients of the desired characteristic equation:

$$|sI - \hat{A}| = s^k + \hat{a}_1 s^{k-1} + \cdots + \hat{a}_k = 0 \tag{7.18}$$

and $a_1, \ldots, a_k$ are the coefficients of the original characteristic equation:

$$|sI - A| = s^k + a_1 s^{k-1} + \cdots + a_k = 0$$

and

$$W = \begin{bmatrix} 1 & a_1 & \cdots & a_{k-1} \\ 0 & 1 & \cdots & a_{k-2} \\ \cdots & \cdots & \cdots & \cdots \\ 0 & 0 & \cdots & 1 \end{bmatrix} \tag{7.19}$$

Example 7A Instrument servo In Example 6A we considered the design of an instrument servo: a control system for an electric motor driving an inertial load. A control law was obtained for the voltage input to the motor as a function of the error $e = \theta - \theta_r$ (where θ is the measured angular position and θ_r is the desired reference position) and the angular velocity ω of the motor. There is no way of implementing the servo without measuring the error, but is it necessary to measure the angular velocity? One way of avoiding the measurement of angular velocity would be to make the feedback gain from the angular velocity zero and accept the transient response that can be achieved. If that solution is not acceptable, however, the alternative is to use a linear observer as described in this chapter.

The structure of the observer is defined by (7.8):

$$\dot{\hat{x}} = A\hat{x} + Bu + K(y - C\hat{x}) \tag{7A.1}$$

with the state vector given by

$$\hat{x} = [\hat{e}, \hat{\omega}]'$$

with

$$A = \begin{bmatrix} 0 & 1 \\ 0 & -\alpha \end{bmatrix} \qquad B = \begin{bmatrix} 0 \\ \beta \end{bmatrix}$$

The quantity that can be measured is the system error e. Thus

$$y = e = [1, 0]\begin{bmatrix} e \\ \omega \end{bmatrix}$$

Thus the observation matrix is

$$C = [1, 0]$$

In detail the differential equations corresponding to (7A.1) are

$$\begin{aligned} \dot{\hat{e}} &= \hat{\omega} + k_1(e - \hat{e}) \\ \dot{\hat{\omega}} &= -\alpha\hat{\omega} + \beta u + k_2(e - \hat{e}) \end{aligned} \tag{7A.2}$$

and have the structure shown in Fig. 7.2. Note that the input u must not be omitted.

Having determined the observer structure, it only remains to determine the gain matrix

$$K = k = \begin{bmatrix} k_1 \\ k_2 \end{bmatrix}.$$

By (7.14) this gain matrix is

$$k = [(NW)']^{-1}(\hat{a} - a) \tag{7A.3}$$

where N is the observability test matrix, and W is the triangular Toeplitz matrix defined by (7.19). In this example

$$N = [C' \quad A'C'] = \begin{bmatrix} 1 & 0 \\ 0 & 1 \end{bmatrix}$$

and, from Example 6A

$$W = \begin{bmatrix} 1 & \alpha \\ 0 & 1 \end{bmatrix}$$

Thus

$$(NW)' = \begin{bmatrix} 1 & 0 \\ \alpha & 1 \end{bmatrix} \qquad [(NW)']^{-1} = \begin{bmatrix} 1 & 0 \\ -\alpha & 1 \end{bmatrix}$$

The vector formed from the open-loop characteristic polynomial

$$D(s) = s^2 + \alpha s$$

is

$$a = \begin{bmatrix} \alpha \\ 0 \end{bmatrix}$$

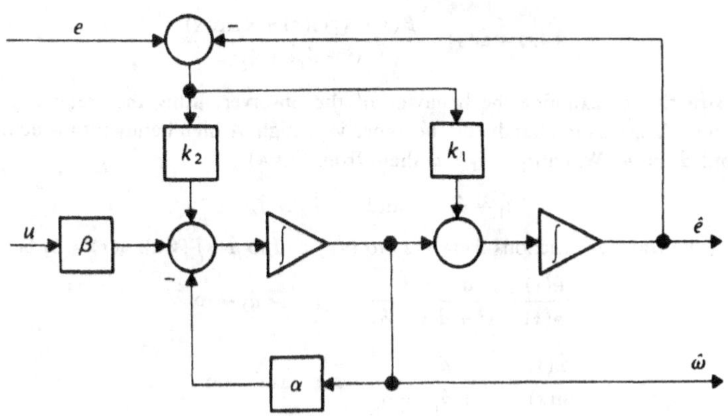

Figure 7.2 Observer for instrument servo.

Suppose that the observer characteristic polynomial is

$$\hat{D}(s) = s^2 + \hat{a}_1 s + \hat{a}_2$$

Then

$$\hat{a} = \begin{bmatrix} \hat{a}_1 \\ \hat{a}_2 \end{bmatrix}$$

Thus, finally, the observer gain matrix is

$$k = \begin{bmatrix} k_1 \\ k_2 \end{bmatrix} = \begin{bmatrix} 1 & 0 \\ -\alpha & 1 \end{bmatrix} \begin{bmatrix} \hat{a}_1 - \alpha \\ \hat{a}_2 \end{bmatrix} = \begin{bmatrix} \hat{a}_1 - \alpha \\ \hat{a}_2 - \alpha(\hat{a}_1 - \alpha) \end{bmatrix} \tag{7A.4}$$

Using (7.11) we find the transfer functions from the observation y and control u to the state estimates

$$\hat{A} = A - KC = \begin{bmatrix} 0 & 1 \\ 0 & -\alpha \end{bmatrix} - \begin{bmatrix} k_1 \\ k_2 \end{bmatrix} [1 \quad 0]$$

Thus

$$(sI - \hat{A})^{-1} = \begin{bmatrix} s + k_1 & -1 \\ k_2 & s + \alpha \end{bmatrix}^{-1} = \frac{1}{(s + k_1)(s + \alpha) + k_2} \begin{bmatrix} s + \alpha & 1 \\ -k_2 & s + k_1 \end{bmatrix}$$

Note that the denominator of $(sI - \hat{A})^{-1}$ is

$$s^2 + (k_1 + \alpha)s + \alpha k_1 + k_2 = s^2 + \hat{a}_1 s + \hat{a}_2$$

using k_1 and k_2 given by (7A.4). This should come as no surprise. Thus

$$(sI - \hat{A})^{-1} B = \frac{1}{s^2 + \hat{a}_1 s + \hat{a}_2} \begin{bmatrix} \beta \\ \beta(s + k_1) \end{bmatrix}$$

and

$$(sI - \hat{A})^{-1} K = \frac{1}{s^2 + \hat{a}_1 s + \hat{a}_2} \begin{bmatrix} (s + \alpha)k_1 + k_2 \\ k_2 s \end{bmatrix} = \begin{bmatrix} k_1 s + \hat{a}_2 \\ k_2 s \end{bmatrix}$$

Thus

$$\begin{aligned} \hat{x}_1(s) = \hat{e}(s) &= \frac{\beta u(s) + (k_1 s + \hat{a}_2) e(s)}{s^2 + \hat{a}_1 s + \hat{a}_2} \\ \hat{x}_2(s) = \hat{\omega}(s) &= \frac{\beta(s + k_1) u(s) + k_2 s e(s)}{s^2 + \hat{a}_1 s + \hat{a}_2} \end{aligned} \tag{7A.5}$$

It is instructive to examine the behavior of the observer, using the frequency-domain characterization (7A.5), as the bandwidth becomes very high. A high bandwidth is achieved by making $\hat{a}_1$ and $\hat{a}_2$ large. When $\hat{a}_2 \gg \hat{a}_1 \gg \alpha$ then, from (7A.4),

$$k_1 \approx \hat{a}_1 \quad \text{and} \quad k_2 \approx \hat{a}_2$$

It is seen that the transfer functions from $e(s)$ to $\hat{e}(s)$ and to $\hat{\omega}(s)$ (with $u(s) = 0$) become

$$\frac{\hat{e}(s)}{e(s)} = \frac{\hat{a}_1 s + \hat{a}_2}{s^2 + \hat{a}_1 s + \hat{a}_2} \to 1 \qquad \text{as } \hat{a}_2 \to \infty$$

$$\frac{\hat{\omega}(s)}{e(s)} = \frac{\hat{a}_2 s}{s^2 + \hat{a}_1 s + \hat{a}_2} \to s \qquad \text{as } \hat{a}_2 \to \infty$$

In words, the transfer function from the measured error e to the estimated error $\hat{e}$ approaches unity and the transfer function from e to $\hat{\omega}$ approaches that of a differentiator.

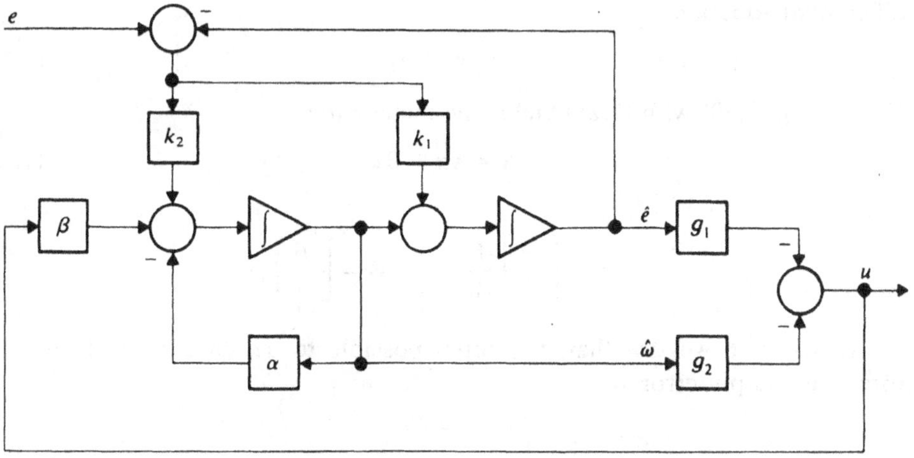

Figure 7.3 Compensator synthesized using observer.

When the bandwidth of the observer is not infinite, the estimate $\hat{e}$ is a filtered version of the measured e and the estimate $\hat{\omega}$ is a smoothed derivative of ω. The filtering/smoothing action of the observer may be desirable if the measured error is noisy.

Where does the input u that is needed in the synthesis of the observer come from? In the control problem u is a function of the state x. Since only one component e of the state x can be measured, does this mean that we can't determine u? The answer is that the control u that is used in the observer, as well as the input to the plant, is computed using the *estimated state*

$$u = -G\hat{x} = -g_1\hat{e} - g_2\hat{\omega}$$

The justification for this is given in the next chapter. But, to anticipate those results, we show the total structure of the compensator in Fig. 7.3. The control gains g_1 and g_2 that multiply the estimated states $\hat{e}$ and $\hat{\omega}$ are obtained as discussed in Example 6A.

7.4 DISTURBANCES AND TRACKING SYSTEMS: EXOGENOUS VARIABLES

In the previous two chapters we have considered systems in which the open-loop process, using the error as the state vector, has other inputs in addition to the control input u. These are the disturbance input x_d and the reference input x_r. The dynamics of the error are

$$\dot{e} = Ae + Bu + Ex_0 \tag{7.20}$$

where

$$x_0 = \begin{bmatrix} x_d \\ \hdashline x_r \end{bmatrix}$$

designates the complete set of exogenous inputs, whether due to the reference state or the disturbance state. As earlier, we assume that x_0 satisfies a known differential equation

$$\dot{x}_0 = A_0 x_0 \tag{7.21}$$

Combining (7.20) with (7.21) yields the metasystem

$$\dot{\mathbf{x}} = \mathbf{A}\mathbf{x} + \mathbf{B}u \tag{7.22}$$

with

$$\mathbf{A} = \left[\begin{array}{c|c} A & E \\ \hline 0 & A_0 \end{array}\right] \qquad \mathbf{B} = \left[\begin{array}{c} B \\ \hline 0 \end{array}\right]$$

In Chap. 6 we saw that it is often possible to design a control law that forces the output error

$$y = Ce$$

to zero in the steady state, even when the total exogenous vector x_0 does not go to zero. The control law

$$u = -Ge - G_0 x_0 \tag{7.23}$$

where

$$G_0 = [G_d, G_r]$$

requires knowledge of the exogenous vector x_0. This knowledge can be acquired by direct measurement of x_0, which sometimes may be feasible, or estimation of x_0. This section is concerned with the latter, and the approach is to design an observer for the metastate $\mathbf{x}$ which includes the error e and the exogenous vector x_0. In the event that all the components of the error vector e can be measured, and only x_0 needs to be estimated, then a reduced-order observer as discussed in Sec. 7.5 may be used.

We assume that the observation vector y may depend on both the error e and the exogenous vector x_0:

$$y = Ce + Dx_0 = \mathbf{C}\mathbf{x} \tag{7.24}$$

where

$$\mathbf{C} = [C \quad D]$$

The general observer equation (7.8) as applied to the metasystem (7.22) and (7.24) gives

$$\dot{\hat{\mathbf{x}}} = \mathbf{A}\mathbf{x} + \mathbf{B}u + \mathbf{K}(y - \mathbf{C}\hat{\mathbf{x}}) \tag{7.25}$$

Separating (7.25) into equations for the system error estimate $\hat{e}$ and the error in estimating the exogenous state x_0:

$$\begin{aligned} \dot{\hat{e}} &= A\hat{e} + Bu + E\hat{x}_0 + K_e(y - C\hat{e} - D\hat{x}_0) \\ \dot{\hat{x}}_0 &= A_0\hat{x}_0 + K_0(y - C\hat{e} - D\hat{x}_0) \end{aligned} \tag{7.26}$$

A block-diagram representation of (7.26) is shown in Fig. 7.4.

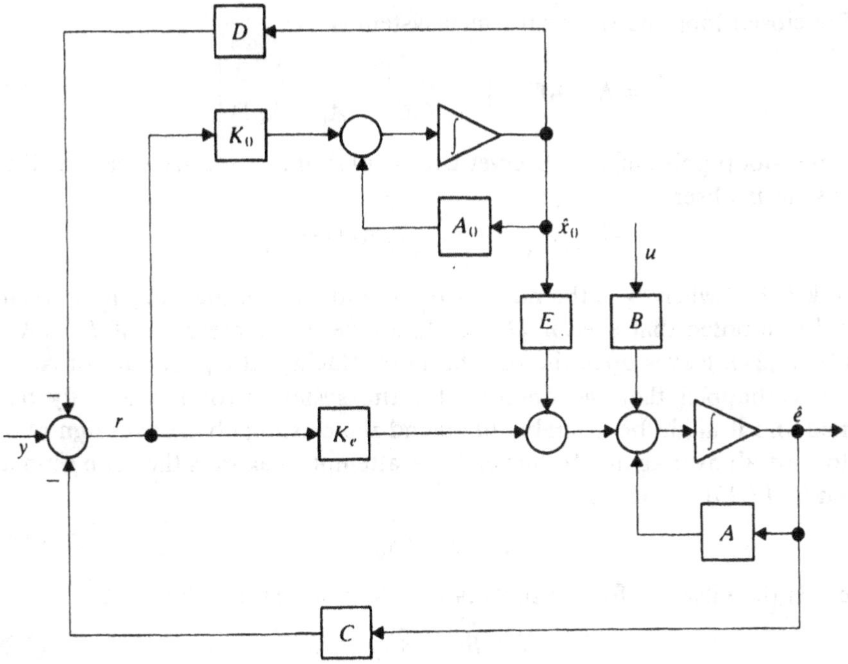

Figure 7.4 Block diagram of observer including estimation of exogenous vector.

A notable feature of the observer of Fig. 7.4 is that there is no feedback from the estimate of the system error $\hat{e}$ to the estimate of the exogenous state $\hat{x}_0$. The entire structure of the portion of the observer that estimates $\hat{x}_0$ is in parallel with the gain K_e. We shall see in the next chapter that when x_0 is not directly measurable, the estimate $\hat{x}_0$ is used in place of x_0. This means only that the input to $\hat{e}$ contains another term $G_0\hat{x}_0$ in addition to the term $E\hat{x}_0$ already shown in Fig. 7.4, but there is still no feedback from $\hat{e}$ to $\hat{x}_0$. Thus, if x_0 is a constant (i.e., $A_0 = 0$) and it does not appear in the measurement, then all that is left in the observer of x_0 is the path through K_0, as shown in Fig. 7.4, which is a bank of integrators, one for each component of $\hat{x}_0$, in parallel with the path through K_e. This means that in the determination of the error $\hat{e}$ there is a path proportional to the *integral* of the residual $r = y - C\hat{e}$ in addition to the path through K_e which is proportional to the residual itself. Because of the integral path, it is possible for the residual r to become zero without $\hat{x}_0$ going to zero, and a control signal based on $\hat{x}_0$ (instead of x_0, which we are assuming is not measurable) will also not go to zero. Thus it is possible to produce a constant control signal even when the system error can be driven to zero. The presence of a constant control signal in the absence of system error is a characteristic of a type 1 (or higher type) system which, in classical control system design is achieved by means of integral control, and is achieved automatically by the use of an observer to estimate the unmeasurable exogenous state vector x_0.

The closed-loop matrix for the metasystem is

$$\hat{\mathbf{A}} = \mathbf{A} - \mathbf{KC} = \begin{bmatrix} A - K_e C & E - K_e D \\ -K_0 C & A_0 - K_0 D \end{bmatrix} \tag{7.27}$$

The closed-loop poles of the observer can be placed at arbitrary locations if the metasystem is observable, i.e., if

$$\mathbf{N} = [\mathbf{C}', \mathbf{A}'\mathbf{C}', \ldots, (\mathbf{A}')^{k+\nu-1}\mathbf{C}']$$

has rank $k + \nu$, where ν is the number of components in the exogenous vector. It should be noted that even if D and A_0 are zero, the presence of E in $\hat{\mathbf{A}}$, as given by (7.27), leaves open the possibility of placing all eigenvalues of $\hat{\mathbf{A}}$.

It may happen that the observer for the system error has already been designed and it might be desirable to amend the existing observer design rather than to start all over again. To this end, we attempt to express the state estimate $\hat{e}$, given by (7.26) in the form

$$\hat{e} = \tilde{e} + V\hat{x}_0 \tag{7.28}$$

where $\tilde{e}$ is the observer for the process with $x_0 = 0$, that is,

$$\dot{\tilde{e}} = A\tilde{e} + Bu + \tilde{K}(y - C\tilde{e}) \tag{7.29}$$

where $\tilde{K}$ is the gain matrix for the observer in the absence of x_0. We assume that (7.26) and (7.29) hold and endeavor to find the matrix V which makes (7.28) hold as well.

If (7.28) holds then

$$\begin{aligned}\dot{\hat{e}} &= \dot{\tilde{e}} + V\dot{\hat{x}}_0 \\ &= A\tilde{e} + Bu + \tilde{K}(y - C\tilde{e}) + V(A_0\hat{x}_0 + K_0[y - C(\tilde{e} + V\hat{x}_0) - D\hat{x}_0]) \\ &= A\tilde{e} + Bu + (\tilde{K} + VK_0)(y - C\tilde{e}) + [VA_0 - VK_0(CV + D)]\hat{x}_0 \end{aligned} \tag{7.30}$$

But, from (7.26) and (7.28)

$$\begin{aligned}\dot{\hat{e}} &= A(\tilde{e} - V\hat{x}_0) + Bu + E\hat{x}_0 + K_e[y - C\tilde{e} - (CV + D)\hat{x}_0] \\ &= A\tilde{e} + Bu + K_e(y - C\tilde{e}) + [AV + E - K_e(CV + D)]\hat{x}_0 \end{aligned} \tag{7.31}$$

We thus have two expressions for $\hat{e}$. In order for them both to hold for all residuals $y - C\tilde{e}$ and all disturbance estimates, we must have

$$\tilde{K} + VK_0 = K_e \tag{7.32}$$

and

$$VA_0 - VK_0(CV + D) = AV + E - K_e(CV + D)$$

or, upon use of $VK_0 = K_e - \tilde{K}$ from (7.32),

$$VA_0 - (A - \tilde{K}C)V = E - \tilde{K}D \tag{7.33}$$

This is a linear matrix equation of the form $VA + BV = C$. The existence of a solution to (7.33) depends on the eigenvalues of A_0 and $A - \tilde{K}C$. (See Note

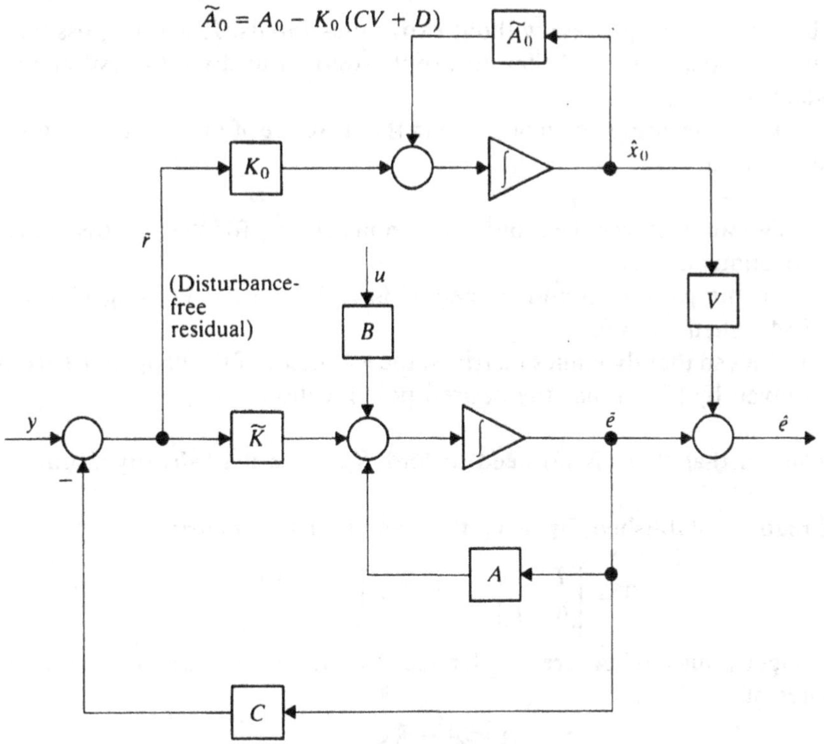

Figure 7.5 Alternate form of observer in which estimate of exogenous input is obtained using "disturbance-free" residual.

7.2.) In the special case in which x_0 is a constant, however, (that is, $A_0 = 0$), a solution to (7.33) exists if $A - \tilde{K}C$ is nonsingular. If the undisturbed process is observable,† then $\tilde{K}$ can be chosen to place the eigenvalues of $A - \tilde{K}C$ in the left half-plane, thereby guaranteeing that the inverse of $A - \tilde{K}C$ exists. In this special, but very important, case

$$V = -(A - \tilde{K}C)^{-1}(E - \tilde{K}D) \tag{7.34}$$

Finally, consider the equation of (7.26) for estimating the disturbance:

$$\begin{aligned}\dot{\hat{x}}_0 &= A_0\hat{x}_0 + K_0[y - C(\tilde{e} + V\hat{x}_0) - D\hat{x}_0] \\ &= [A_0 - K_0(CV + D)]\hat{x}_0 + K_0(y - C\tilde{e}) \end{aligned} \tag{7.35}$$

The observer structure implicit in (7.28), (7.29), and (7.35) is illustrated in the block diagram of Fig. 7.5.

Note that the input to the estimate of the exogenous vector is the residual

$$r = y - C\tilde{e}$$

† It is only necessary for the undisturbed process to be "detectable"; see Chap. 5.

of the observer for the process without exogenous inputs. It is thus possible to design that observer first, and then to use its residual to drive the estimator of the exogenous inputs.

In summary, the design of an observer in the structure of Fig. 7.5, the following procedure is used:

Step 1. Design an observer (i.e., find the gain matrix $\tilde{K}$) for the process without exogenous inputs.

Step 2. Using the gain $\tilde{K}$ found in step 1, find the matrix V using (7.33), or (7.34) when $A_0 = 0$.

Step 3. Find K_0 so that dynamics matrix of the estimator of the exogenous vector, as given by (7.35), has the desired pole locations.

It is noted that there is no need to find K_e, since it is already defined by (7.32).

It is readily established, by using the transformation matrix

$$\mathbf{T} = \begin{bmatrix} I & V \\ 0 & I \end{bmatrix} \qquad \mathbf{T}^{-1} = \begin{bmatrix} I & -V \\ 0 & I \end{bmatrix}$$

that the eigenvalues of $\hat{\mathbf{A}}$ are as defined by (7.27) and are located at the eigenvalues of

$$\tilde{A} = A - KC$$

and the eigenvalues of

$$\tilde{A}_0 = A_0 - K_0(CV + D)$$

Example 7B Inverted pendulum with disturbance In Example 6B we considered the design of a control system for a motor-driven inverted pendulum. If the damping provided by the motor is inadequate, feedback of both position and velocity is needed. Position feedback is required in any case, because the open-loop process is unstable.

We assume that in addition to the control input u a constant disturbance (perhaps due to the wind) is also present. Thus the complete dynamic model, accounting for the disturbance, is

$$\dot{\theta} = \omega$$

$$\dot{\omega} = \Omega^2\theta - \alpha\omega + u + d$$

$$\dot{d} = 0$$

Thus,

$$\mathbf{A} = \left[\begin{array}{cc|c} 0 & 1 & 0 \\ \Omega^2 & -\alpha & 1 \\ \hline 0 & 0 & 0 \end{array}\right] \qquad \mathbf{B} = \left[\begin{array}{c} 0 \\ \beta \\ \hline 0 \end{array}\right]$$

Thus, for this example

$$E = \begin{bmatrix} 0 \\ 1 \end{bmatrix} \qquad D = 0$$

Observers for this process will be designed in two ways: First, an observer will be designed directly for the metasystem, and then an observer in the form of Fig. 7.5 will be designed.

It is assumed that only the angular position θ is observed thus

$$y = [1 \quad 0 \mid 0]\begin{bmatrix}\theta \\ \omega \\ \hline d\end{bmatrix}$$

Thus

$$C = [1 \quad 0] \qquad \mathbf{C} = [C, 0]$$

The observability test matrix for the metasystem is

$$\mathbf{N} = [\mathbf{C}', \mathbf{A}'\mathbf{C}', \mathbf{A}'^2\mathbf{C}'] = \begin{bmatrix} 1 & 0 & \Omega^2 \\ 0 & 1 & -\alpha \\ 0 & 0 & 1 \end{bmatrix}$$

The open-loop characteristic equation is

$$|sI - \mathbf{A}| = \begin{vmatrix} s & -1 & 0 \\ -\Omega^2 & s+\alpha & -1 \\ 0 & 0 & s \end{vmatrix} = s(s^2 + \alpha s - \Omega^2)$$

Therefore

$$a_1 = \alpha \qquad a_2 = -\Omega^2 \qquad a_3 = 0$$

and hence

$$\mathbf{W} = \begin{bmatrix} 1 & \alpha & -\Omega^2 \\ 0 & 1 & \alpha \\ 0 & 0 & 1 \end{bmatrix} \qquad \mathbf{NW} = \begin{bmatrix} 1 & \alpha & 0 \\ 0 & 1 & 0 \\ 0 & 0 & 1 \end{bmatrix} \qquad (\mathbf{NW})^{-1} = \begin{bmatrix} 1 & -\alpha & 0 \\ 0 & 1 & 0 \\ 0 & 0 & 1 \end{bmatrix}$$

Thus the gain matrix is

$$\mathbf{K} = \begin{bmatrix} 1 & 0 & 0 \\ -\alpha & 1 & 0 \\ 0 & 0 & 1 \end{bmatrix}\begin{bmatrix} \hat{a}_1 - \alpha \\ \hat{a}_2 + \Omega^2 \\ \hat{a}_3 \end{bmatrix} = \begin{bmatrix} \hat{a}_1 - \alpha \\ \hat{a}_2 + \Omega^2 - \alpha(\hat{a}_1 - \alpha) \\ \hat{a}_3 \end{bmatrix} = \begin{bmatrix} k_1 \\ k_2 \\ k_3 \end{bmatrix}$$

The observer dynamics are given by

$$\begin{aligned} \dot{\hat{\theta}} &= \hat{\omega} + k_1(y - \hat{\theta}) \\ \dot{\hat{\omega}} &= \Omega^2\hat{\theta} - \alpha\hat{\omega} + \beta u + \hat{d} + k_2(y - \hat{\theta}) \\ \dot{\hat{d}} &= k_3(y - \hat{\theta}) \end{aligned} \tag{7B.1}$$

and has the block-diagram representation of Fig. 7.6.

To apply the alternate method of design, we first determine the gain matrix for the disturbance-free observer. For this calculation we need

$$N = [C', A'C'] = \begin{bmatrix} 1 & 0 \\ 0 & 1 \end{bmatrix} \qquad \text{and} \qquad W = \begin{bmatrix} 1 & \alpha \\ 0 & 1 \end{bmatrix}$$

Thus

$$(NW)^{-1} = \begin{bmatrix} 1 & -\alpha \\ 0 & 1 \end{bmatrix} \qquad [(NW)^{-1}]' = \begin{bmatrix} 1 & 0 \\ -\alpha & 1 \end{bmatrix}$$

Hence

$$\tilde{K} = [(NW)^{-1}]'\begin{bmatrix} \hat{a}_1 - a_1 \\ \hat{a}_2 - a_2 \end{bmatrix} = \begin{bmatrix} 1 & 0 \\ -\alpha & 1 \end{bmatrix}\begin{bmatrix} \hat{a}_1 - \alpha \\ \hat{a}_2 + \Omega^2 \end{bmatrix} = \begin{bmatrix} \hat{a}_1 - \alpha \\ \hat{a}_2 + \Omega^2 - \alpha(\hat{a}_1 - \alpha) \end{bmatrix} = \begin{bmatrix} \tilde{k}_1 \\ \tilde{k}_2 \end{bmatrix}$$

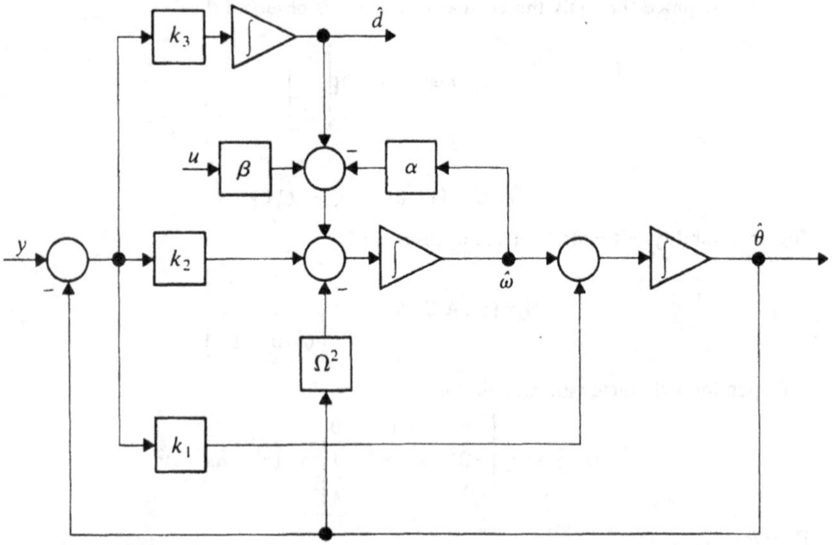

Figure 7.6 Observer for inverted pendulum with unknown disturbance.

Note that $\tilde{k}_1 = k_1$ and $\tilde{k}_2 = k_2$ in this case. The closed-loop matrix of the disturbance-free observer is

$$\tilde{A} = A - KC = \begin{bmatrix} 0 & 1 \\ \Omega^2 & -\alpha \end{bmatrix} - \begin{bmatrix} \tilde{k}_1 \\ \tilde{k}_2 \end{bmatrix} [1 \quad 0] = \begin{bmatrix} -\tilde{k}_1 & 1 \\ \Omega^2 - \tilde{k}_2 & -\alpha \end{bmatrix}$$

Thus

$$\tilde{A}^{-1} = (A - \tilde{K}C)^{-1} = \frac{1}{\hat{a}_2} \begin{bmatrix} -\alpha & -1 \\ -\Omega^2 + \tilde{k}_2 & -\tilde{k}_1 \end{bmatrix}$$

Hence the "correction matrix" is

$$V = -(A - \tilde{K}C)^{-1} E = \frac{1}{\hat{a}_2} \begin{bmatrix} \alpha & 1 \\ \Omega^2 - \tilde{k}_2 & \tilde{k}_1 \end{bmatrix} \begin{bmatrix} 0 \\ 1 \end{bmatrix} = \frac{1}{\hat{a}_2} \begin{bmatrix} 1 \\ \tilde{k}_1 \end{bmatrix}$$

and

$$CV + D = [1 \quad 0] \begin{bmatrix} 1/\hat{a}_2 \\ k_1/\hat{a}_2 \end{bmatrix} = 1/\hat{a}_2$$

Thus equation (7.35) for the disturbance estimator is

$$\dot{\hat{d}} = -(k_d/\hat{a}_2)\hat{d} + k_d(y - \tilde{\theta}) \tag{7B.2}$$

The disturbance-estimation gain k_d can be selected to be any positive number. The equation for the disturbance-free observer is

$$\begin{aligned} \dot{\tilde{\theta}} &= \tilde{\omega} + \tilde{k}_1(y - \tilde{\theta}) \\ \dot{\tilde{\omega}} &= \Omega^2 \tilde{\theta} - \alpha\tilde{\omega} + k_2(y - \tilde{\theta}) \end{aligned} \tag{7B.3}$$

and (7.28) becomes

$$\begin{aligned} \hat{\theta} &= \tilde{\theta} + \frac{1}{\hat{a}_2}\hat{d} \\ \hat{\omega} &= \tilde{\omega} - \frac{\tilde{k}_1}{\hat{a}_2}\hat{d} \end{aligned} \tag{7B.4}$$

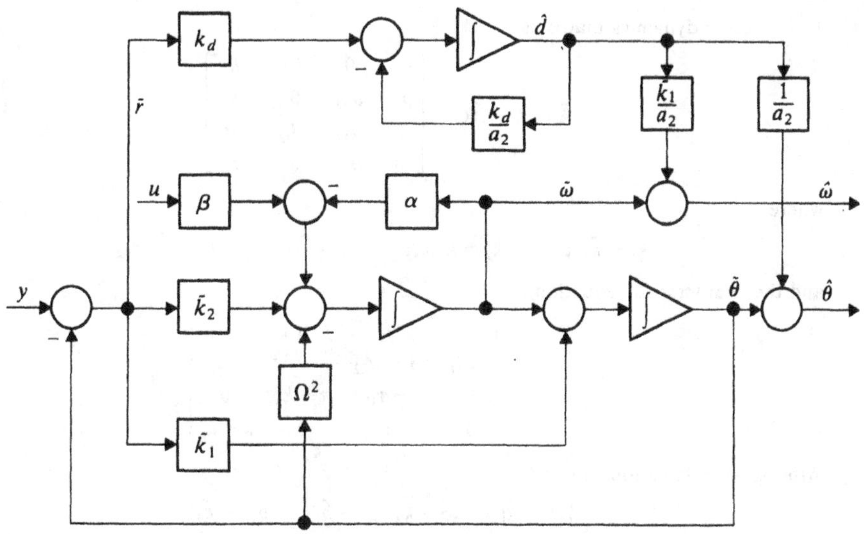

Figure 7.7 Alternate form of observer for inverted pendulum with unknown disturbance.

A block diagram showing the implementation of (7B.2) through (7B.4) is given in Fig. 7.7. In this case the alternate form of the observer is no simpler than the direct form of (7B.1) and calculation of the gains of (7B.1) is no more complicated than calculation of the gains for the alternate form. But this is not always the case, as the following example will show.

Example 7C Distillation column The distillation column considered earlier in Examples 2G, 4A, and 6D has two disturbances represented by the vector

$$d = \begin{bmatrix} \Delta F_A \\ \Delta X_{FA1} \end{bmatrix}$$

Appending these two disturbances to the fourth-order dynamic model would result in a sixth-order metasystem. Determination of the observer gain would be a formidable problem. Using the alternate form of the observer, however, reduces the sixth-order design to a fourth-order design for the disturbance-free observer, and a supplementary second-order design.

First, we consider the observer design for the disturbance-free process. Following Gilles and Retzbach[6] we choose an observer gain matrix of the form

$$\bar{K} = \begin{bmatrix} 0 & \bar{k}_{12} \\ 0 & \bar{k}_{22} \\ \bar{k}_{31} & 0 \\ 0 & \bar{k}_{42} \end{bmatrix} \tag{7C.1}$$

Using the dynamic model of Chap. 2, Example 2G, the disturbance-free observer is given by

$$\dot{\tilde{x}}_1 = a_{11}\tilde{x}_1 + \bar{k}_{12}(y_2 - c_{42}\tilde{x}_4) + b_{11}u_1$$

$$\dot{\tilde{x}}_2 = a_{21}\tilde{x}_1 + a_{22}\tilde{x}_2 + \bar{k}_{22}(y_2 - c_{42}\tilde{x}_4)$$

$$\dot{\tilde{x}}_3 = a_{32}\tilde{x}_2 + \bar{k}_{31}(y_1 - c_{31}\tilde{x}_3) + b_{32}u_2$$

$$\dot{\tilde{x}}_4 = a_{42}\tilde{x}_2 + \bar{k}_{42}(y_2 - c_{42}\tilde{x}_4) + b_{42}u_2$$

The observer dynamics matrix is

$$\tilde{A} = A - \tilde{K}C = \begin{bmatrix} a_{11} & 0 & 0 & k_1 \\ a_{21} & a_{22} & 0 & k_2 \\ 0 & a_{32} & -k_3 & 0 \\ 0 & a_{42} & 0 & -k_4 \end{bmatrix} \tag{7C.2}$$

where

$$k_1 = \tilde{k}_{12}c_{42} \qquad k_2 = \tilde{k}_{22}c_{43} \qquad k_3 = \tilde{k}_{31}c_{31} \qquad k_4 = \tilde{k}_{42}c_{42}$$

and the characteristic equation is

$$|sI - \tilde{A}| = \begin{vmatrix} s - a_{11} & 0 & 0 & -k_1 \\ -a_{21} & s - a_{22} & 0 & -k_2 \\ 0 & -a_{32} & s + k_3 & 0 \\ 0 & -a_{42} & 0 & s + k_4 \end{vmatrix} = 0$$

After some calculation, we find that

$$|sI - \tilde{A}| = -(s + k_3)(s^3 + \hat{b}_1 s^2 + \hat{b}_2 s + \hat{b}_3)$$

where

$$\hat{b}_1 = -a_{11} - a_{22} + k_4$$

$$\hat{b}_2 = a_{11}a_{22} - (a_{11} + a_{22})k_4 - a_{42}k_2$$

$$\hat{b}_3 = a_{11}a_{22}k_4 + a_{11}a_{42}k_2 - a_{21}a_{42}k_1$$

One of the desired observer poles is determined at $s = -k_3$. The three other observer poles determine $\hat{b}_1$, $\hat{b}_2$, $\hat{b}_3$. Given these values we can solve for k_1, k_2, and k_4:

$$\begin{aligned} k_4 &= \hat{b}_1 + a_{11} + a_{12} \\ k_2 &= -\frac{1}{a_{42}}[\hat{b}_2 - a_{11}a_{22} + (a_{11} + a_{22})k_4] \\ k_1 &= -\frac{1}{a_{21}a_{42}}[\hat{b}_3 - a_{11}a_{22}k_4 - a_{11}a_{42}k_2] \end{aligned} \tag{7C.3}$$

Having determined the gains for the disturbance-free observer, we now turn to the calculation of the observer for the disturbances. In this example

$$A_0 = 0 \qquad \text{and} \qquad D = 0$$

Hence (7.35) becomes

$$\dot{\hat{x}}_0 = -K_0 C V \hat{d} + K_0(y - C\tilde{e})$$

There are two observations and two disturbances, so K_0 is a 2×2 matrix; C is a 2×4 matrix; and V is a 4×2 matrix given by

$$V = -(A - \tilde{K}C)^{-1}E$$

where, as given in Chap. 2, E is a 4×2 matrix.

The completion of the observer design is the subject of Prob. 7.7.

7.5 REDUCED-ORDER OBSERVERS

The observer defined by (7.8) in Sec. 7.2 is the same order as the system that produces the observed output. It does not depend on the number of outputs.

But suppose that there is one output for every state variable:

$$y = Cx$$

where C is a nonsingular matrix. In this case there is no need for a dynamic observer; we can get x directly from y:

$$x = \hat{x} = C^{-1}y \tag{7.36}$$

Is it reasonable that if all the state variables can be measured, then a dynamic observer is not needed, but if even one state variable cannot be measured, then we should need an observer of the same order as the state of the system? On intuition one might suspect that it should be possible to get by using an observer of order $k - l$, where k is the order of the system and l is the number of independent outputs. This intuition is quite correct, as we shall see in this section.

In many applications it is possible to group the state variables into two sets: those that can be measured directly and those that depend indirectly on the former. The state vector is partitioned accordingly:

$$x = \begin{bmatrix} x_1 \\ \hline x_2 \end{bmatrix}$$

with

$$\begin{aligned} \dot{x}_1 &= A_{11}x_1 + A_{12}x_2 + B_1u \\ \dot{x}_2 &= A_{21}x_1 + A_{22}x_2 + B_2u \end{aligned} \tag{7.37}$$

The observation is given by

$$y = C_1x_1 \tag{7.38}$$

The standard observer (7.8) for (7.37) and (7.38) is

$$\dot{\hat{x}}_1 = A_{11}\hat{x}_1 + A_{12}\hat{x}_2 + B_1u + K_1(y - C_1\hat{x}_1) \tag{7.39}$$

$$\dot{\hat{x}}_2 = A_{21}\hat{x}_1 + A_{22}\hat{x}_2 + B_2u + K_2(y - C_1\hat{x}_1) \tag{7.40}$$

But why take the trouble to implement the observer equation (7.39) for $\hat{x}_1$ when we can solve for x_1 directly from (7.38)?

$$x_1 = \hat{x}_1 = C_1^{-1}y \tag{7.41}$$

In this case the observer (7.40) for those states that cannot be measured directly becomes

$$\dot{\hat{x}}_2 = A_{21}C_1^{-1}y + A_{22}\hat{x}_2 + B_2u \tag{7.42}$$

which is a dynamic system of the same order as the number of state variables that cannot be measured directly.

The dynamic behavior of the reduced-order observer is governed by the eigenvalues of A_{22} which is a submatrix of the open-loop dynamics matrix A, a matrix over which the designer has no control. If the eigenvalues of A_{22} are suitable, then (7.42) could be a satisfactory observer. Since there is no assurance that the eigenvalues of A_{22} are suitable, however, we need a more general system for the reconstruction of $\hat{x}_2$. (For the time being $\hat{x}_1$ is still obtained using (7.41); later we will generalize the estimation of x_1 also.)

A suitably general structure for the estimation of x_2 is given by

$$\hat{x}_2 = Ly + z \tag{7.43}$$

where z is the state of a $(k-l)$th-order system†

$$\dot{z} = Fz + \bar{G}y + Hu \tag{7.44}$$

As we did for the full-order observer, we define the estimation error

$$e = x - \hat{x} = \begin{bmatrix} x_1 - \hat{x}_1 \\ \hline x_2 - \hat{x}_2 \end{bmatrix} = \begin{bmatrix} e_1 \\ \hline e_2 \end{bmatrix} \tag{7.45}$$

By (7.41)

$$e_1 = x_1 - \hat{x}_1 = 0 \tag{7.46}$$

so we are concerned only with e_2, the differential equation for which is

$$\begin{aligned} \dot{e}_2 = \dot{x}_2 - \dot{\hat{x}}_2 &= A_{21}x_1 + A_{22}x_2 + B_2u - L\dot{y} - \dot{z} \\ &= A_{21}x_1 + A_{22}x_2 + B_2u - L[C_1(A_{11}x_1 + A_{12}x_2 + B_1u)] - Fz - \bar{G}y - Hu \end{aligned} \tag{7.47}$$

But, from (7.43)

$$z = \hat{x}_2 - Ly = x_2 - e_2 - Ly = x_2 - e_2 - LC_1x_1 \tag{7.48}$$

So (7.47) becomes

$$\begin{aligned} \dot{e}_2 = Fe_2 &+ (A_{21} - LC_1A_{11} - \bar{G}C_1 + FLC_1)x_1 \\ &+ (A_{22} - LC_1A_{12} - F)x_2 + (B_2 - LC_1B_1 - H)u \end{aligned} \tag{7.49}$$

In order for the error to be independent of x_1, x_2, and u, the matrices multiplying x_1, x_2, and u must vanish:

$$F = A_{22} - LC_1A_{12} \tag{7.50}$$

$$H = B_2 - LC_1B_1 \tag{7.51}$$

$$\bar{G}C_1 = A_{21} - LC_1A_{11} + FLC_1 \tag{7.52}$$

Then (7.49) becomes

$$\dot{e}_2 = Fe_2$$

and hence, for asymptotic stability, the eigenvalues of F must lie in the left half of the s plane.

Comparing (7.50) with (7.6) we see that selecting the matrix L in (7.50) to place the eigenvalues of F is the same type of problem as selecting the gain

† The overbar on the matrix $\bar{G}$ is used here to distinguish this matrix from the control gain matrix G.

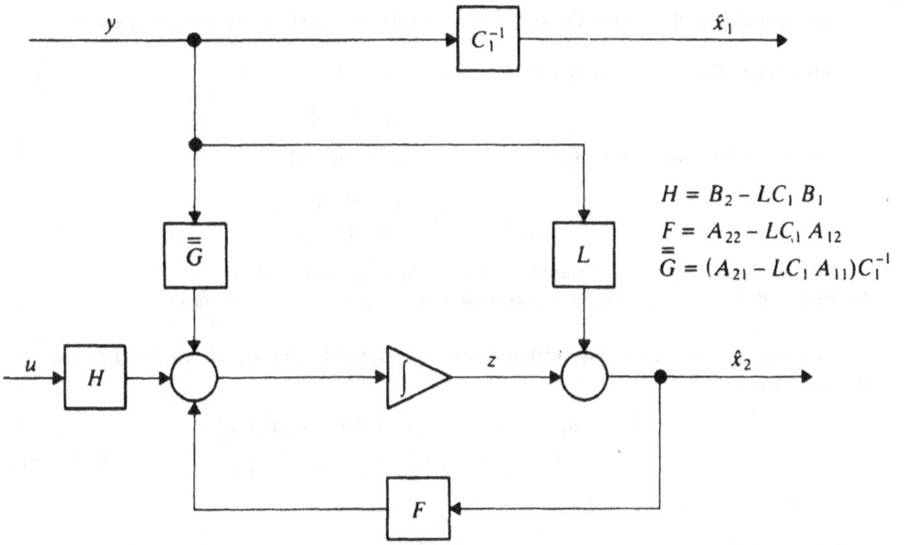

Figure 7.8 Reduced-order observer for observation $y = C_1 x_1$ with C_1 nonsingular.

matrix K to place the eigenvalues of $\hat{A}$. The submatrix A_{22} has the role of A in (7.6) and the product $C_1 A_{12}$ has the role of C in (7.6). In order to place the poles of F, it is necessary that the rank of the corresponding observability test matrix

$$N_1 = [A'_{12}C'_1, A'_{22}A'_{12}C'_1, \ldots, (A'_{22})^{k-l-1}A'_{12}C'_1]$$

be of rank $k - l$. Luenberger has shown [2] that this requirement is satisfied, if the full-state observability test matrix (7.10) is of rank k.

Having selected the matrix L to place the reduced-order observer poles, the matrix H is determined from (7.51) and the matrix $\bar{G}$ is determined from (7.52):

$$\bar{G} = (A_{21} - LC_1 A_{11})C_1^{-1} + FL \tag{7.53}$$

Using (7.53) in (7.44) gives

$$\dot{z} = F\hat{x}_2 + (A_{21} - LC_1 A_{11})C_1^{-1}y + Hu \tag{7.54}$$

This equation, together with (7.41) and (7.43) define the reduced-order observer. A block-diagram representation of these equations is given in Fig. 7.8.

It should be noted that when $L = 0$, the reduced-order observer reduces to (7.42). Thus the reduced-order observer of Fig. 7.8 can be regarded as a generalization of (7.42) to be used in the event that the dynamics matrix A_{22} does not have suitable eigenvalues.

Example 7D Temperature control In Example 6E we considered the design of a temperature control system. In metastate form the equations are

$$\begin{bmatrix} \dot{v}_1 \\ \dot{v}_2 \\ \dot{x}_0 \end{bmatrix} = \begin{bmatrix} -a_{11} & a_{12} & e_1 \\ a_{21} & -a_{22} & e_2 \\ 0 & 0 & 0 \end{bmatrix} \begin{bmatrix} v_1 \\ v_2 \\ x_0 \end{bmatrix} + \begin{bmatrix} b_1 \\ b_2 \\ 0 \end{bmatrix} u \tag{7D.1}$$

In this example we determine the structure and gains for each of the following cases:

A. Full-order observer assuming only x_1 is measured;

$$y = v_1 \qquad C = [1 \quad 0 \quad 0]$$

B. Full-order observer assuming both v_1 and v_2 are measured;

$$\begin{aligned} y &= v_1 \\ y_2 &= v_2 \end{aligned} \qquad C = \begin{bmatrix} 1 & 0 & 0 \\ 0 & 1 & 0 \end{bmatrix}$$

C. Reduced-(second-)order observer, assuming only v_1 is measured;
D. Reduced-(first-)order observer, assuming both v_1 and v_2 are measured.

Case A Full-order observer with one measured variable The equations for the observer in this case are

$$\begin{aligned} \dot{\hat{v}}_1 &= -a_{11}\hat{v}_1 + a_{12}\hat{v}_2 + e_1\hat{x}_d + b_1 u + k_1(y - \hat{x}_1) \\ \dot{\hat{v}}_2 &= a_{21}\hat{v}_1 - a_{22}\hat{v}_2 + e_2\hat{x}_d + b_2 u + k_2(y - \hat{x}_1) \\ \dot{\hat{x}}_0 &= k_3(y - \hat{x}_1) \end{aligned} \tag{7D.2}$$

The gain matrix

$$K = \begin{bmatrix} k_1 \\ k_2 \\ k_3 \end{bmatrix}$$

is chosen such that the eigenvalues of $A - KC$ are in suitable locations. The open-loop characteristic polynomial is

$$\begin{aligned} |sI - A| &= \begin{vmatrix} s + a_{11} & -a_{21} & -e_1 \\ -a_{21} & s + a_{22} & -e_2 \\ 0 & 0 & s \end{vmatrix} \\ &= s[s^2 + (a_{11} + a_{22})s + a_{11}a_{22} - a_{12}a_{21}] \end{aligned}$$

Thus the vector of coefficients is

$$a = \begin{bmatrix} a_1 \\ a_2 \\ a_3 \end{bmatrix} = \begin{bmatrix} a_{11} + a_{22} \\ a_{11}a_{22} - a_{12}a_{21} \\ 0 \end{bmatrix}$$

and

$$W = \begin{bmatrix} 1 & a_{11} + a_{22} & a_{11}a_{22} - a_{12}a_{21} \\ 0 & 1 & a_{11} + a_{22} \\ 0 & 0 & 1 \end{bmatrix}$$

The observability test matrix is

$$N = [C' \quad A'C' \quad (A')^2C'] = \begin{bmatrix} 1 & -a_{11} & a_{11}^2 + a_{12}a_{21} \\ 0 & a_{12} & -a_{11}a_{12} - a_{22}a_{12} \\ 0 & e_1 & -a_{11}e_1 + a_{12}e_2 \end{bmatrix}$$

Then

$$(NW) = \begin{bmatrix} 1 & a_{22} & 0 \\ 0 & a_{12} & 0 \\ 0 & e_1 & a_{22}e_1 + a_{12}e_2 \end{bmatrix}$$

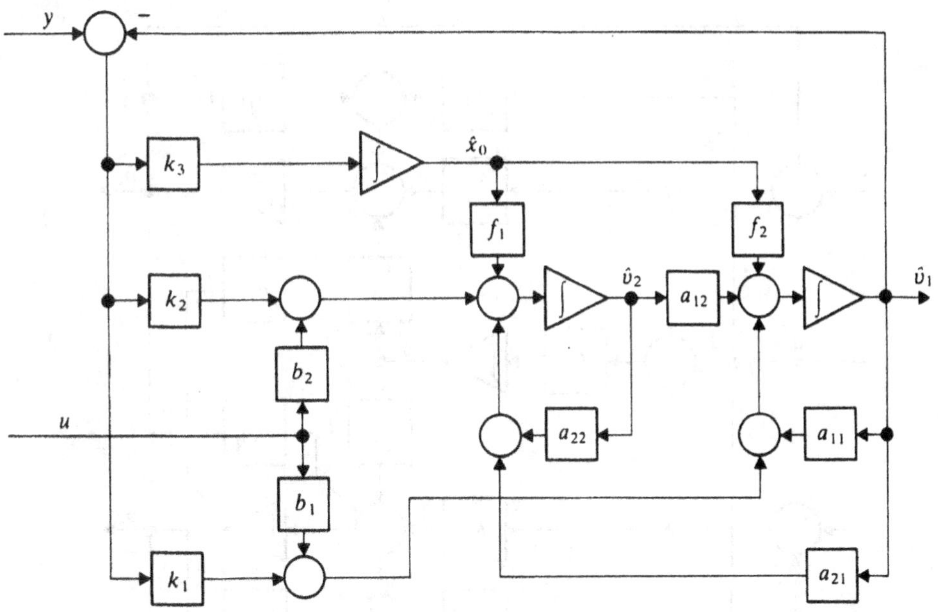

Figure 7.9 Full-order observer for temperature control with one measurement.

Hence

$$K = [(NW)^{-1}]' \begin{bmatrix} a_1 - (a_{11} + a_{22}) \\ a_2 - a_{11}a_{22} + a_{21}a_{12} \\ a_3 \end{bmatrix}$$

$$= \begin{bmatrix} 1 & 0 & 0 \\ -\dfrac{a_{22}}{a_{12}} & \dfrac{1}{a_{12}} & \dfrac{-e_1}{a_{12}(a_{22}e_1 + a_{12}e_2)} \\ 0 & 0 & \dfrac{1}{a_{22}e_1 + a_{12}e_2} \end{bmatrix} \begin{bmatrix} \hat{a}_1 - (a_{11} + a_{22}) \\ \hat{a}_2 - a_{11}a_{22} + a_{21}a_{12} \\ \hat{a}_3 \end{bmatrix} \tag{7D.3}$$

The block-diagram representation of the observer in this case is given in Fig. 7.9.

Case B Full-order observer with two measured variables The equations for the observer in this case are

$$\begin{aligned} \dot{\hat{v}}_1 &= -a_{11}\hat{v}_1 + a_{12}\hat{v}_2 + e_1\hat{x}_d + b_1u + k_{11}(y_1 - \hat{x}_1) + k_{12}(y_2 - \hat{x}_2) \\ \dot{\hat{v}}_2 &= a_{21}\hat{v}_1 - a_{22}\hat{v}_2 + e_2\hat{x}_d + b_2u + k_{21}(y_1 - \hat{x}_1) + k_{22}(y_2 - \hat{x}_2) \\ \dot{\hat{x}}_0 &= k_{31}(y_1 - \hat{x}_1) + k_{32}(y_2 - \hat{x}_2) \end{aligned} \tag{7D.4}$$

The observer gain matrix in this case is

$$K = \begin{bmatrix} k_{11} & k_{12} \\ k_{21} & k_{22} \\ k_{31} & k_{32} \end{bmatrix}$$

There are six gains to be selected: twice as many as are needed to place all the closed-loop poles. From the structure (see Fig. 6.5) it would seem reasonable to use y_1 for

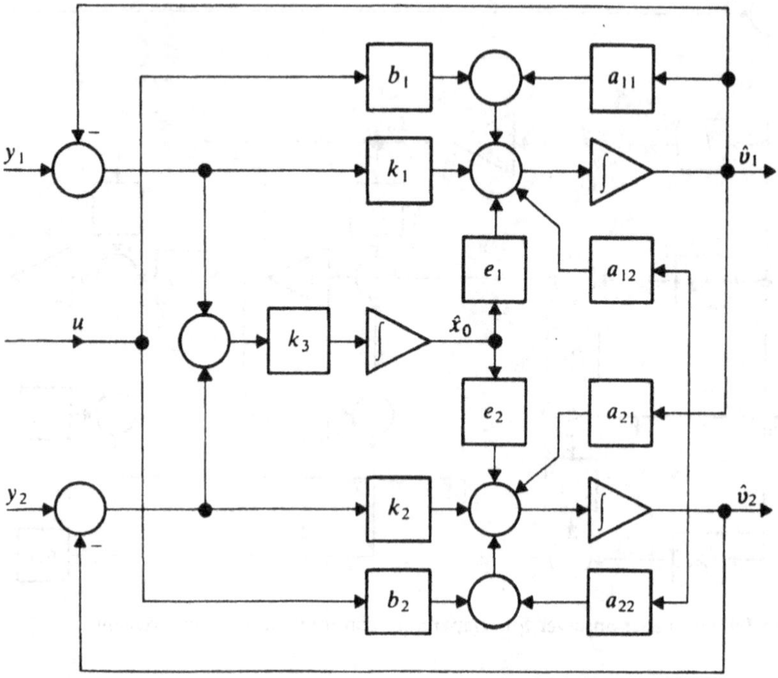

Figure 7.10 Full-order observer for temperature control with two measurements.

estimating x_1 and y_2 for estimating x_2. For estimating x_0 we might consider using the sum of $y_1 - \hat{x}_1$ and $y_2 - \hat{x}_2$, which would happen when $k_{31} = k_{32} = k_3$. Thus

$$K = \begin{bmatrix} k_1 & 0 \\ 0 & k_2 \\ k_3 & k_3 \end{bmatrix} \tag{7D.5}$$

The observer has the block-diagram representation shown in Fig. 7.10. Determination of the three gains k_1, k_2, and k_3 needed to place the eigenvalues of $\hat{A} = A - KC$ is a straightforward problem in algebra.

Case C Reduced-order observer with one measurement This case fits the theory developed in (7.43) through (7.52). Since $y_1 = v_1$ is directly measured we have

$$\hat{v}_1 = y$$

and

$$\hat{v}_2 = l_1 y + z_1$$

$$\hat{x}_d = l_2 y + z_2$$

with

$$\dot{z}_1 = f_{11}\hat{v}_2 + f_{12}\hat{x}_d + \bar{g}_1 y + h_1 u$$

$$\dot{z}_2 = f_{21}\hat{v}_2 + f_{22}\hat{x}_d + \bar{g}_2 y + h_2 u$$

These relations lead to the block-diagram of Fig. 7.11. The matrices† F, $\bar{G}$, and H are given by

$$F = A_{22} - LA_{12} = \begin{bmatrix} -a_{22} & e_2 \\ 0 & 0 \end{bmatrix}\begin{bmatrix} f_2 \\ 0 \end{bmatrix} - \begin{bmatrix} l_1 \\ l_2 \end{bmatrix}[a_{12} \quad f_1]$$

† Don't confuse the forcing terms f_1 and f_2, with the f_{ij}'s of the F matrix.

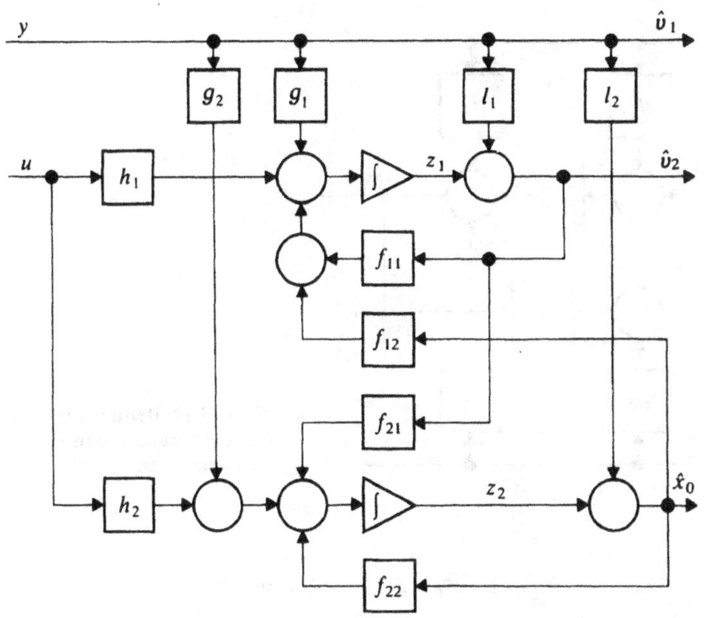

Figure 7.11 Reduced-order observer for temperature control with one measurement.

$$\bar{G} = A_{21} - LA_{11} = \begin{bmatrix} a_{21} \\ 0 \end{bmatrix} - \begin{bmatrix} l_1 \\ l_2 \end{bmatrix} a_{11} = \begin{bmatrix} a_{21} - l_1 a_{11} \\ -l_2 a_{11} \end{bmatrix} = \begin{bmatrix} \bar{g}_1 \\ \bar{g}_2 \end{bmatrix}$$

$$H = B_2 - LB_1 = \begin{bmatrix} b_2 \\ 0 \end{bmatrix} - \begin{bmatrix} l_1 \\ l_2 \end{bmatrix} b_1 = \begin{bmatrix} b_2 - l_1 b_1 \\ -l_2 b_2 \end{bmatrix} = \begin{bmatrix} h_1 \\ h_2 \end{bmatrix}$$

The characteristic equation for F is

$$s^2 - (f_{11} - f_{22})s + f_{11}f_{22} - f_{12}f_{21} = s^2 + \hat{a}_1 s + \hat{a}_2 = 0$$

where

$$\hat{a}_1 = -a_{22} - l_1 a_{12} - l_2 f_1$$

$$\hat{a}_2 = (a_{22}f_1 + a_{12}f_2)l_2$$

Thus

$$l_2 = \frac{\hat{a}_2}{a_{22}f_1 + a_{12}f_2}$$

$$l_1 = \frac{1}{a_{12}}\left[\hat{a}_1 + a_{22} - \frac{\hat{a}_2 f_1}{a_{22}f_1 + a_{12}f_2}\right]$$

Case D Reduced-order observer with two measurements This case also fits the theory developed in (7.43)-(7.52). In this case the observer is particularly simple:

$$\hat{v}_1 = y_1$$

$$\hat{v}_2 = y_2$$

$$\hat{x}_0 = l_1 y_1 + l_2 y_2 + z$$

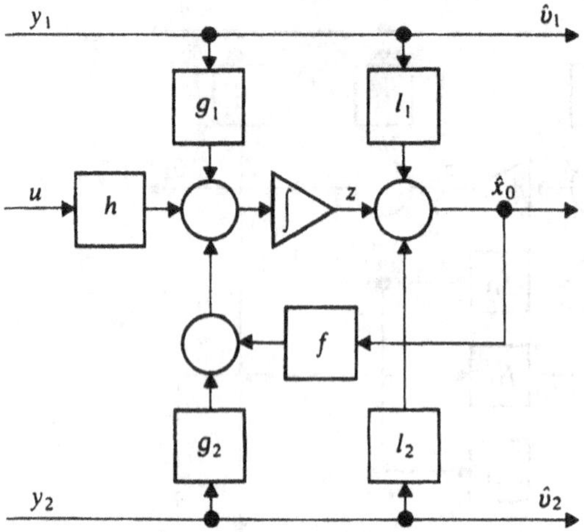

Figure 7.12 Reduced-order observer for temperature control with two measurements.

with

$$\dot{z} = f\hat{x}_0 + \bar{g}_1 y_1 + \bar{g}_2 y_2 + hu$$

as shown in Fig. 7.12. The matrices are given by

$$F = f = A_{22} - LA_{12} = 0 - [l_1 \quad l_2]\begin{bmatrix} f_1 \\ f_2 \end{bmatrix} = -l_1 f_1 - l_2 f_2$$

$$\bar{G} = A_{21} - LA_{11} = [0 \quad 0] - [l_1 \quad l_2]\begin{bmatrix} -a_{11} & a_{12} \\ a_{21} & -a_{22} \end{bmatrix}$$

$$= [l_1 a_{11} - l_2 a_{21}, -l_1 a_{12} + l_2 a_{22}]$$

$$H = B_2 - LB_1 = [0] - [l_1 \quad l_2]\begin{bmatrix} b_1 \\ b_2 \end{bmatrix} = -l_1 b_1 - l_2 b_2 = h$$

Since f_1 and f_2 are positive (see Example 6E) any positive pair of gains l_1 and l_2 will stabilize the first-order observer.

The case in which the state variables can be grouped in a manner that permits solving for a subvector x_1 as in (7.41) occurs quite frequently in practice. But there are situations in which this is not possible, and a more general reduced-order observer is needed. We assume that the matrix C is of "full-rank" l, in other words that the l rows of C are linearly independent. Then we can define a nonsingular transformation matrix

$$T = \overbrace{\begin{bmatrix} C \\ \hdashline M \end{bmatrix}}^{k \text{ columns}} \begin{matrix} \updownarrow \ l \text{ rows} \\ \updownarrow \ k - l \text{ rows} \end{matrix} \tag{7.55}$$

where M is a $(k - l) \times k$ matrix having rows which are linearly independent of each other and of the rows of C. The matrix M is of rank $(k - l)$ and the transformation matrix T is a $k \times k$ matrix of rank k and is thus nonsingular. We denote its inverse by

$$T^{-1} = [\underbrace{\quad P \quad}_{l \text{ columns}} \;\vdots\; \underbrace{\quad Q \quad}_{(k-l) \text{ columns}}] \updownarrow k \text{ rows} \tag{7.56}$$

We define the vector w by

$$w = \begin{bmatrix} w_1 \\ w_2 \end{bmatrix} = Tx \tag{7.57}$$

which by (7.55) becomes

$$w_1 = Cx \tag{7.58}$$

$$w_2 = Mx \tag{7.59}$$

and

$$x = T^{-1}w = [P \,\vdots\, Q]\begin{bmatrix} w_1 \\ \hdashline w_2 \end{bmatrix} = Pw_1 + Qw_2 \tag{7.60}$$

The reason behind the choice of the transformation (7.57) is that the observation y is a direct measurement of w_1, which will permit us to make use of the results derived earlier in this section. First we will obtain estimates $\hat{w}_1$ and $\hat{w}_2$, and then obtain $\hat{x}$ using (7.60), i.e.,

$$\hat{x} = P\hat{w}_1 + Q\hat{w}_2 \tag{7.61}$$

To obtain the observer for $\hat{w}_1$ and $\hat{w}_2$, we note that

$$w_1 = y$$

using (7.58) thus

$$\hat{w}_1 = y \tag{7.62}$$

and using the results of the previous analysis (7.43) and (7.54), we obtain the estimate

$$\hat{w}_2 = Ly + z \tag{7.63}$$

where

$$\dot{z} = \bar{F}\hat{w}_2 + (\bar{A}_{21} - L\bar{A}_{11})y + \bar{H}u \tag{7.64}$$

$$\bar{F} = \bar{A}_{22} - L\bar{A}_{12}$$

$$H = \bar{B}_2 - L\bar{B}_1$$

The overbars on $\bar{A}_{ij}$ and $\bar{B}_i$ denote that these matrices come from the transformed system

$$\dot{w} = \bar{A}w + \bar{B}u$$

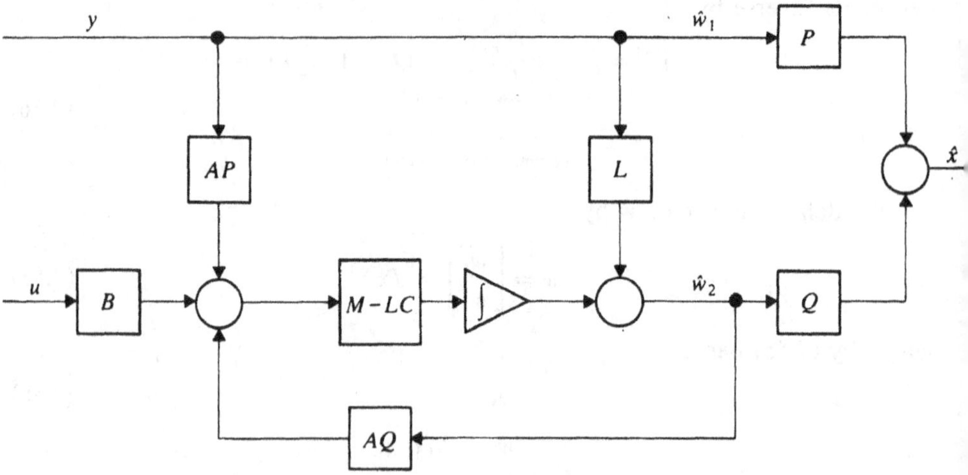

Figure 7.13 Reduced-order observer for $y = Cx$.

where

$$\bar{A} = \begin{bmatrix} \bar{A}_{11} & \bar{A}_{12} \\ \bar{A}_{21} & \bar{A}_{22} \end{bmatrix} = TAT^{-1} = \begin{bmatrix} C \\ M \end{bmatrix} A[P \quad Q] = \begin{bmatrix} CAP & CAQ \\ MAP & MAQ \end{bmatrix}$$

$$\bar{B} = \begin{bmatrix} \bar{B}_1 \\ \bar{B}_2 \end{bmatrix} = TB = \begin{bmatrix} C \\ M \end{bmatrix} B = \begin{bmatrix} CB \\ MB \end{bmatrix}$$

Thus

$$\bar{F} = MAQ - LCAQ = (M - LC)AQ \tag{7.65}$$

$$\bar{H} = MB - LCB = (M - LC)B \tag{7.66}$$

$$\bar{A}_{21} - L\bar{A}_{11} = MAP - LCAP = (M - LC)AP \tag{7.67}$$

The general form of the reduced-order observer, as given by (7.61)-(7.64) with the matrices satisfying (7.65), has the structure shown in Fig. 7.13.

Since $TT^{-1} = T^{-1}T = I$, the following relations must hold between C, M, P, and Q

$$\begin{bmatrix} C \\ M \end{bmatrix} [P \quad Q] = \begin{bmatrix} CP & CQ \\ MP & MQ \end{bmatrix} = \begin{bmatrix} I_l & 0 \\ 0 & I_{k-l} \end{bmatrix}$$

and

$$[P \quad Q] \begin{bmatrix} C \\ M \end{bmatrix} = PC + QM = I_k$$

where I_ν is a $\nu \times \nu$ identity matrix.

PROBLEMS

Problem 7.1 Two-car train

Consider the two-car train of Prob. 3.9 et seq. Design a reduced-order observer for each of the following conditions of measurement:

(*a*) Position of car 1 measurable;

(*b*) Positions of cars 1 and 2 measurable;

(*c*) Position and velocity of car 1 measurable;

(*d*) Position of car 1 and velocity of both cars measurable.

In each case draw the block diagram of the observer and discuss criteria that might be used to establish the gains.

Problem 7.2 Inverted pendulum on cart: full-order observer

An observer for the inverted pendulum on a motor-driven cart is to be designed using the measurement of the displacement of the cart ($y = x_1$). Determine the observer gain matrix for which the observer poles lie in a fourth-order Butterworth pattern of radius 5, i.e., the characteristic equation is to be

$$\left(\frac{s}{5}\right)^4 + 2.613\left(\frac{s}{5}\right)^3 + (2+\sqrt{2})\left(\frac{s}{5}\right)^2 + 2.613\left(\frac{s}{5}\right) + 1 = 0$$

Problem 7.3 Inverted pendulum on cart: reduced-order observer

A reduced-order observer is to be designed for the inverted pendulum on a motor-driven cart of Prob. 7.2. Given the observation of the cart displacement $y = x_1$, design a third-order observer with poles in a third-order Butterworth configuration of radius 5, i.e., the characteristic equation is to be

$$\left(\frac{s}{5}\right)^3 + 2\left(\frac{s}{5}\right)^2 + 2\left(\frac{s}{5}\right) + 1 = 0$$

Problem 7.4 Hydraulically actuated gun turret: full-order observer

Only the tracking error $e = \theta - \theta_0$ can be measured in the gun-turret control of Prob. 6.2. A full-order observer is to be designed to estimate all the state and exogenous variables.

(*a*) Assume that the reference angle θ_0 and the disturbances are all zero. Determine the gains of the observer such that the observer poles lie in a fourth-order Butterworth configuration with radius $\omega = 20$.

(*b*) Supplement the observer design of part *a* to estimate the exogenous variables of the system. Draw the block diagram for the system and select the supplementary gain matrix K_0 to place the remaining poles in a third-order Butterworth configuration with radius $\omega = 30$.

Problem 7.5 Two-axis gyro: observer design

In the two-axis gyro control problem considered in Prob. 6.9 it was assumed that all the state and exogenous variables are measurable. But in reality only the angular displacements δ_x and δ_y are measurable by means of pick-offs. Thus we need an observer to estimate the unmeasurable variables from the outputs of the pick-offs. A reduced-order observer, of the form

$$\hat{\delta} = \begin{bmatrix} \hat{\delta}_x \\ \hat{\delta}_y \end{bmatrix} = \begin{bmatrix} \delta_x \\ \delta_y \end{bmatrix} \qquad \text{(measured quantities)}$$

$$\begin{bmatrix} \hat{\omega}_{xB} \\ \hat{\omega}_{yB} \\ \hat{\omega}_{xE} \\ \hat{\omega}_{yE} \end{bmatrix} = L\begin{bmatrix} \delta_x \\ \delta_y \end{bmatrix} + z$$

where L is a 4×2 matrix and z is a four-component vector given by

$$\dot{z} = Fz + \bar{G}y + Hu$$

where the relations between F, $\bar{G}$, H, and L are given by (7.50)-(7.52) with $C_1 = I$. For reasons of symmetry (as already seen in Prob. 6.9) we would require L to be of the form

$$L = \begin{bmatrix} l_1 & l_2 \\ -l_2 & l_1 \\ l_3 & l_4 \\ -l_4 & l_3 \end{bmatrix} \tag{P7.5}$$

Hence there are only four different elements f_1, f_2, f_3, f_4 in the observer dynamics matrix $F = A_{22} - LC_1A_{12}$. (What are A_{22}, A_{12}, and C_1 in this case?)

(*a*) Express the matrices F, $\bar{G}$, and H in terms of the elements of L as given by (P7.5) and the elements of A, B, and E as defined by (3.9*b*) and (3.9*c*).

(*b*) Find the characteristic equation for the observer, that is, $|sI - F|$ where $F = A_{22} - LC_1A_{12}$ with L given by (P7.5).

(*c*) For the numerical values given in Prob. 6.9, find the observer gains $l_1, \ldots, l_4$ which place the real parts of the closed-loop poles at $\sigma = -1000$.

Problem 7.6 Three-capacitance thermal system: full-state observer

Only the temperature v_3 is measurable in the three-capacitance temperature control system of Prob. 6.8.

(*a*) Design an observer to estimate x_1, x_2, $e = v_3 - \bar{v}$ from the measurement $y = x_3$, assuming the external temperature v_0 and the reference temperature $\bar{v}$ are known quantities. The observer poles are to be located in a third-order Butterworth configuration on a circle of radius $\omega_0 = 3$.

(*b*) Modify the observer design to estimate not only x_1, x_2, and e, but also v_0 and $\bar{v}$ (assuming them to be constant) using the three steps outlined in Sec. 7.4. Draw the block diagram of the observer and locate the poles of the observer for the exogenous state at $s = -5 \pm j5$.

(*c*) Combine the states of parts *a* and *b* into a metastate $x = [x_1, x_2, e, v_0, \bar{v}]$ and design an observer for the metastate having its poles in the same locations as the system of part *b*. (See Prob. 7.1.)

Problem 7.7 Distillation column

The gains for the observer having the configuration described in Example 7C are to be determined, using the numerical data given in Example 2G.

(*a*) Find the observer gains which place its poles at $s = -36$, $s = -18$, and $s = -9(1 \pm j\sqrt{3}/2)$.

(*b*) Calculate the matrix V.

(*c*) Draw a block diagram of the observer.

NOTES

Note 7.1 Background of Luenberger observers

The use of a dynamic system to provide estimates of the unmeasurable states goes back to the earliest work with state-space methods. By the early 1960s[7, 8] this notion was well established although not expressed in its most general form. The basic idea of the separation principle, as discussed in Chap. 8, was also known by 1961. The fundamental papers on optimal filtering by Kalman and Bucy[4, 5] also appeared in the early 1960s. Luenberger's work of the mid-1960s[1, 2, 3] systematized and generalized much of the earlier results. It was Luenberger who first used the term *observer*, and who introduced the idea of a reduced-order observer to estimate only those states of a system that are not accessible to direct measurement. Since the optimum filter of Kalman is the same order as the process under observation, it is a "full-order" observer. The reduced-order observer of Luenberger was originally regarded as being distinct from a Kalman filter and hence could not be an "optimum" observer. In 1971, however, Friedland[8] considered the behavior of the Kalman filter used as an observer when the observation noise vanishes, and showed that the Kalman filter portion of the compensator is in fact a reduced-order Luenberger observer. (A discussion of reduced-order Kalman filters, following a somewhat different approach than was used in [8], is given in Sec. 11.6 of this book.)

Note 7.2 Linear matrix equations

Linear matrix equations of the form $VA + BV = C$ are treated in many books on matrix theory. (See Chap. VIII of Gantmacher[9] for example.) Although the equation is linear in the unknown matrix V, the solution cannot generally be expressed in terms of the known matrices A, B, and C. But the elements of V can be determined by solving a system of linear equations the unknowns of which are the elements of the matrix V. (Exhibiting these equations is rather messy, but it is easy to write these equations in any specific instance.) It should be noted that none of the matrices in the linear equation is required to be square, as long as the sums and products are all of proper dimension.

REFERENCES

1. Luenberger, D. G., "Observing the State of a Linear System," *IEEE Trans. on Military Electronics*, vol. MIL-8, April 1964, pp. 74–80.
2. Luenberger, D. G., "Observers for Multivariable Systems," *IEEE Trans. on Automatic Control*, vol. AC-11, no. 2, April 1966, pp. 190–197.
3. Luenberger, D. G., "An Introduction to Observers," *IEEE Trans. on Automatic Control*, vol. AC-16, no. 6, December 1971, pp. 596–602.
4. Kalman, R. E., "A New Approach to Linear Filtering and Prediction Problems," *Trans. ASME (J. Basic Engineering)*, vol. 82D, no. 1, March 1960, pp. 35–45.
5. Kalman, R. E., and Bucy, R. S., "New Results in Linear Filtering and Prediction Theory," *Trans. ASME (J. Basic Engineering)*, vol. 83D, no. 1, March 1961, pp. 95–108.
6. Gilles, E. D., and Retzbach, B., "Reduced Models and Control Distillation Columns with Sharp Temperature Profiles," *IEEE Trans. on Automatic Control*, vol. AC-28, no. 5, May 1983, pp. 628–630.
7. Joseph, P. D., and Tou, J. T., "On Linear Control Theory," *Trans. AIEE, Pt. II*, vol. 80, no. 11, September 1960, pp. 193–196.
8. Friedland, B., "Limiting Forms of Optimum Stochastic Linear Regulators," *Trans. ASME (J. Dynamics Systems, Measurement & Control)*, vol. 93G, no. 3, September 1971, pp. 134–141.
9. Gantmacher, F. R., *The Theory of Matrices*, Chelsea Publishing Company, New York, 1959.

CHAPTER

EIGHT

COMPENSATOR DESIGN BY THE SEPARATION PRINCIPLE

8.1 THE SEPARATION PRINCIPLE

In Chap. 6 we studied the design of control algorithms for processes in which the state variables are all accessible for measurement. We promised to overcome the difficulty of not being able to measure all the state variables by the use of an observer to estimate those state variables that cannot be measured. Then in Chap. 7 we studied the design of observers for processes with known inputs, but not when the state estimate is used for the purpose of control. We are now ready to redeem the promise of Chap. 6, namely to combine the control law for full-state feedback, the subject of Chap. 6, with the observer of Chap. 7 to obtain a general control law for linear processes in which not all the state variables can be measured.

The separation principle is so disarmingly simple that it almost comes as a surprise: it is hard to imagine that the observer designed for a known input can serve to estimate the state of the process for the purpose of generating the control input. But, as we shall see in this chapter, it *does* work.

The separation principle was first stated for discrete-time systems by Joseph and Tou[1] and later was generalized for continuous-time systems. (See Note 8.1.) In the present chapter we shall show only one property of systems designed by the separation principle, namely that the closed-loop system comprising the process under control and the compensator combines the dynamics of the closed-loop process designed for full-state feedback and the observer. In particular, the poles of the overall system occur at the locations selected for the process with full-state feedback and at the locations selected for the observer. Hence, in concept, each may be designed without regard to the other.

The separation principle goes beyond this result, however. When we consider optimum control in the presence of noisy observations, in Chaps. 9–11, we shall see that the separation principle leads to a statistically *optimum* design: if the process is excited by white, gaussian noise and if the noise on the observations is also white and gaussian, then the separation principle becomes the "separation theorem" of optimum stochastic control, and asserts that the optimum control law in the presence of such noise is obtained by combining the optimum deterministic control with the optimum observer (also called the *Kalman filter*).

Underlying the separation principle is a critical assumption, namely that the observer includes an exact dynamic model of the "plant": the process under control. This assumption is almost never valid in reality. In practical systems, the precise dynamic model is rarely known. Even that which is known about the real process dynamics is often too complicated to include in the observer. Thus the observer must in practice be configured to use only an approximate model of the plant. This encounter with the real world does not vitiate the separation principle, but it means that the effect of an inaccurate plant model must be considered. If the design achieved through use of the separation principle is *robust*, it will be able to tolerate uncertainty of the plant dynamics. The robustness tests discussed in Chap. 4 (gain and phase margins, singular value analysis, etc.) can be used to assess robustness of control laws designed by use of the separation principle. These are issues to be addressed in the present chapter.

8.2 COMPENSATORS DESIGNED USING FULL-ORDER OBSERVERS

Consider the "standard" dynamic process

$$\dot{x} = Ax + Bu \tag{8.1}$$

with observations given by

$$y = Cx \tag{8.2}$$

Suppose we have designed a "full-state feedback" control law

$$u = -Gx \tag{8.3}$$

using the methods of Chap. 6. And also suppose we have designed an observer

$$\dot{\hat{x}} = A\hat{x} + Bu + K(y - C\hat{x}) \tag{8.4}$$

using the methods of Chap. 7. On the basis of the separation principle, the control law we should use, if the full state *cannot* be measured, is

$$u = -G\hat{x} \tag{8.5}$$

where $\hat{x}$ is the state estimate given by (8.4).

The control system based on combining (8.4) and (8.5) has the configuration shown in Fig. 8.1. On the right-hand side is the "plant"—or process under control—with the control input u and observed output y. At the left is the "compensator," the input to which is the observed output y of the plant, and the output from which is the control input u. Since the observer (8.4) contains a model of the plant, that model is part of the compensator. Note also that the number of state variables in the compensator is equal to the order of the embedded observer and hence is equal to the order of the plant. Thus the overall closed-loop system, when a full-order observer is used in the compensator, is $2k$ for a plant of order k.

We are interested in the dynamic behavior of the $2k$th-order system comprising the plant and the compensator. With the control law (8.5) used, the plant dynamics become

$$\dot{x} = Ax - BG\hat{x} \tag{8.6}$$

and the observer (8.4) becomes

$$\dot{\hat{x}} = A\hat{x} - BG\hat{x} + K(Cx - C\hat{x}) \tag{8.7}$$

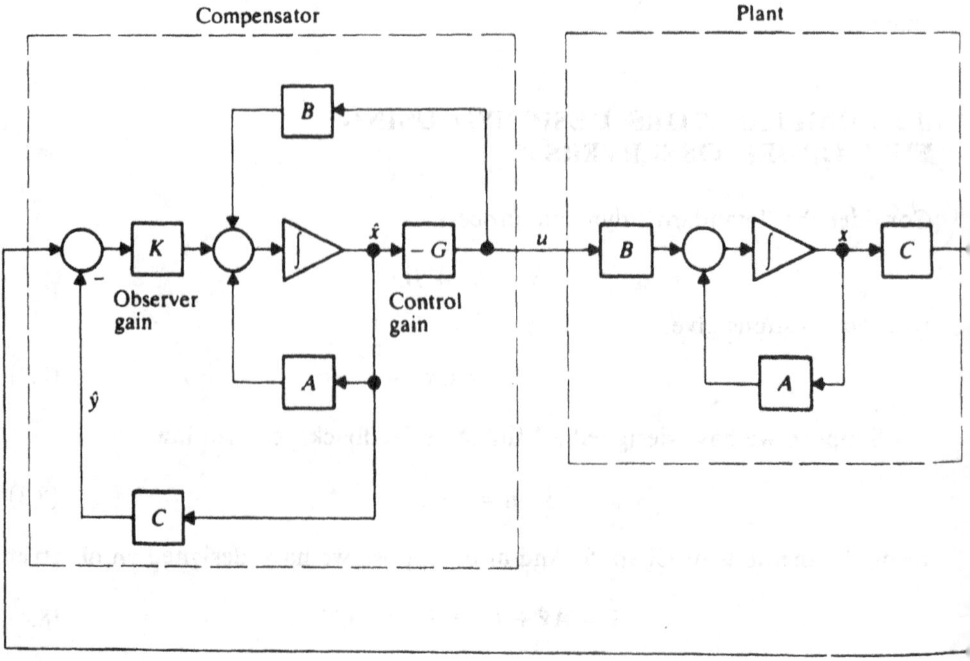

Figure 8.1 Control system using observer in compensator.

upon use of (8.2). As was done in Chap. 7, we define the observer error

$$e = x - \hat{x}$$

Then, (8.6) becomes

$$\dot{x} = Ax - BG(x - e) = (A - BG)x + BGe \tag{8.8}$$

and, after subtracting (8.7) from (8.6),

$$\dot{e} = (A - KC)e \tag{8.9}$$

The dynamics of the closed-loop system are thus given by (8.8) and (8.9), a block diagram representation of which is shown in Fig. 8.2. It is seen that when the state x and the error e are used for the state of the $2k$th-order system, the block diagram consists of two subsystems in tandem, the system on the left generates the estimation error and the estimation error, in turn, forces the evolution of the state. It thus follows that the dynamics of the overall system is that of two systems in tandem and hence that the closed-loop eigenvalues of the overall system comprise the eigenvalues of $A_c = A - BG$ and the eigenvalues of $\hat{A} = A - KC$. But A_c is the closed-loop A matrix of the system using full-state feedback, and $\hat{A}$ is the closed-loop matrix of the observer. We thus conclude that when an observer-based compensator is used to generate the control, the eigenvalues (poles) of the closed-loop system are simply the eigenvalues of the observer and the eigenvalues of the full-state feedback system. If the observer and the full-state feedback system are designed to have "favorable" dynamics, and the observer contains an accurate model of the plant, then the closed-loop dynamics of the overall system will also be "favorable."

We can study the system behavior in greater detail by taking the Laplace transform of (8.8) and (8.9):

$$(sI - A_c)\mathsf{x}(s) = BG\mathsf{e}(s) + x_0 \tag{8.10}$$

$$(sI - \hat{A})\mathsf{e}(s) = e_0 \tag{8.11}$$

where $\mathsf{x}(s)$ and $\mathsf{e}(s)$ are the Laplace transforms of $x(t)$ and $e(t)$, respectively; and x_0 and e_0 are the initial state and observation error, respectively. From (8.11)

$$\mathsf{e}(s) = (sI - \hat{A})^{-1}e_0$$

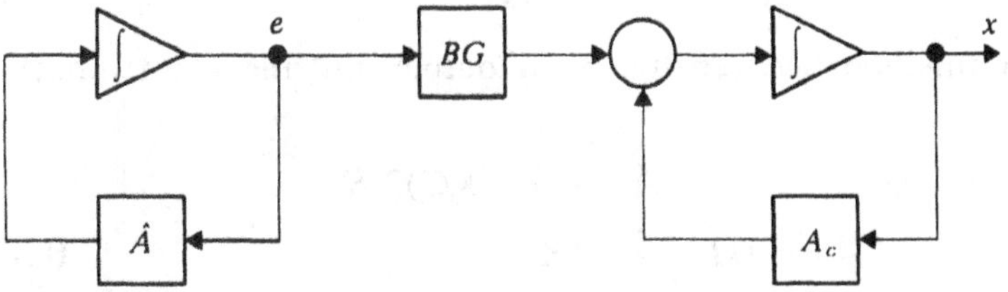

Figure 8.2 Block-diagram representation of state and error in system with compensator designed by separation principle.

and, from (8.10)

$$\begin{aligned} \mathsf{x}(s) &= (sI - A_c)^{-1}BG\mathsf{e}(s) + (sI - A_c)^{-1}x_0 \\ &= (sI - A_c)^{-1}BG(sI - \hat{A})^{-1}e_0 + (sI - A_c)^{-1}x_0 \end{aligned} \tag{8.12}$$

From (8.12) it is seen that the transient response of the state $x(t)$, which would be obtained by taking the inverse Laplace transform of (8.12) consists of two terms: The first term

$$(sI - A_c)^{-1}BG(sI - \hat{A})^{-1}e_0 \tag{8.13}$$

depends on the initial estimation error $e_0 = x_0 - \hat{x}_0$, and is the result of not being able to measure all the state variables and having to use an observer. The second term is due to the initial state x_0 and is present whether or not an observer is present.

The matrix that multiplies the initial estimation error can be written as follows:

$$(sI - A_c)^{-1}BG(sI - \hat{A})^{-1} = \frac{\operatorname{adj}(sI - A_c)BG\operatorname{adj}(sI - \hat{A})}{|sI - A_c||sI - \hat{A}|} \tag{8.14}$$

The matrix in the numerator of (8.14) contains polynomials of degree at most $2(k-2)$; and the denominator of (8.14), which are the poles of the system with respect to the initial estimation error, are the zeros of $|sI - A_c|$ and the zeros of $|sI - \hat{A}|$. Since the poles with respect to the initial state are the zeros of $|sI - A_c|$ it is again seen that the only poles of the closed-loop system are the zeros of $|sI - \hat{A}|$ and the zeros of $|sI - A_c|$.

The transfer function of the compensator is obtained by the use of (8.4) and (8.5). The former can be written

$$\dot{\hat{x}} = (A - BG - KC)\hat{x} + Ky$$

or

$$\hat{\mathsf{x}}(s) = (sI - A + BG + KC)^{-1}K\mathsf{y}(s)$$

Thus, from (8.5)

$$\mathsf{u}(s) = -G\hat{\mathsf{x}}(s) = -G(sI - A + BG + KC)^{-1}K\mathsf{y}(s) \tag{8.15}$$

The transfer function $\mathsf{D}(s)$ of the compensator, defined by

$$\mathsf{u}(s) = -\mathsf{D}(s)\mathsf{y}(s) \tag{8.16}$$

(i.e., the transfer function between the plant output and the plant input) is given by

$$\begin{aligned} \mathsf{D}(s) &= G(sI - A + BG + KC)^{-1}K \\ &= G(sI - \hat{A}_c)^{-1}K \end{aligned} \tag{8.17}$$

where

$$\hat{A}_c = A - BG - KC = \hat{A} - BG = A_c - KC$$

It should be quite apparent that the poles of the compensator, in general, do not occur at the poles of the open-loop plant (i.e., the zeros of $|sI - A|$) nor at poles of the closed-loop full-state feedback system (i.e., the zeros of $|sI - A_c|$), nor at the poles of the observer (i.e., the zeros of $|sI - \hat{A}|$). If $\hat{A}$ and A_c are chosen independently, it may even happen that $\hat{A}_c$ has one or more eigenvalues in the right half-plane! The compensator, in other words, could turn out to be unstable! But yet the closed-loop system, if so designed, would be stable. Since no requirement has been imposed on the stability of the compensator, one should not be too surprised to learn that the compensator might turn out unstable. After all, the loop transmission is the product of the compensator transfer function and the plant transfer function. If it is possible for a compensator to stabilize an unstable plant, it ought to be possible for the plant to stabilize an unstable compensator.

The two situations, however, are not the same. The instability of the plant is the problem that the control system is supposed to cure, and the compensator is the remedy. A remedy that is worse than the problem it cures may justifiably be frowned upon by the patient—the ultimate user of the control system. One of the consequences of an unstable compensator is that the closed-loop system is only conditionally stable. The open-loop poles of the system are the poles of the plant and the poles of the compensator. If the latter are in the right half-plane, then the closed-loop poles will also be in the right half when the loop gain becomes too small. These considerations are addressed in Sec. 8.4 that deals with robustness.

Postponing the question of robustness for the present, we summarize the steps of the compensator design using observers.

Step 1. Design the control law under the assumption that all state variables in the process can be measured.

Step 2. Design an observer to estimate the state of the process for which the control law of step 1 was designed.

Step 3. Combine the full-state control law design of step 1 with the observer design of step 2 to obtain the compensator design.

Example 8A Compensator for stabilizing inverted pendulum In Example 6B we considered the design of a control law for an inverted pendulum, and in Example 7B we considered the design of an observer under the assumption that a constant disturbance x_0 is present. Since the design of Example 6B did not account for the presence of the constant disturbance, however, we should amend that design accordingly. Thus this example will provide another illustration of the design of control systems for processes with constant disturbances as well as an illustration of the use of observers in the realization of the dynamic compensator.

Step 1. Full-state feedback design

The dynamics, including the disturbance x_0 are given by

$$\dot{x} = \begin{bmatrix} \dot{\theta} \\ \dot{\omega} \end{bmatrix} = \begin{bmatrix} 0 & 1 \\ \Omega^2 & -\alpha \end{bmatrix} \begin{bmatrix} \theta \\ \omega \end{bmatrix} + \begin{bmatrix} 0 \\ \beta \end{bmatrix} u + \begin{bmatrix} 0 \\ 1 \end{bmatrix} x_0 \qquad (8A.1)$$

The control law for this process is

$$u = -Gx - G_0 x_0 \tag{8A.2}$$

The gain matrix G was obtained in Example 6B using the Bass-Gura formula:

$$G = g' = \begin{bmatrix} (\bar{a}_2 + \Omega^2)/\beta \\ (\bar{a}_1 - \alpha)/\beta \end{bmatrix}'$$

In addition we need the disturbance gain g_0 which we compute using (6.49)

$$g_0 = B^{\#} E \tag{8A.3}$$

where E is the matrix that multiplies the disturbance, in this case

$$E = \begin{bmatrix} 0 \\ 1 \end{bmatrix}$$

and, from (6.47)

$$B^{\#} = (CA_c^{-1}B)^{-1}CA_c^{-1} \tag{8A.4}$$

where

$$A_c = A - BG = \begin{bmatrix} 0 & 1 \\ \Omega^2 & -\alpha \end{bmatrix} - \begin{bmatrix} 0 \\ \beta \end{bmatrix}[(\bar{a}_2 + \Omega^2)/\beta, (\bar{a}_1 - \alpha)/\beta]$$

$$= \begin{bmatrix} 0 & 1 \\ -\bar{a}_2 & -\bar{a}_1 \end{bmatrix}$$

The observation matrix C is needed for the computation of $B^{\#}$. The reason for the need of C is that we cannot expect to force both components of the state x to zero in the presence of a constant disturbance, if only one can be measured. Thus, for this example we assume that our sole measurement is of the pendulum angular position θ, that is,

$$y = Cx$$

with

$$C = [1 \quad 0] \tag{8A.5}$$

Thus

$$CA_c^{-1} = [1 \quad 0]\begin{bmatrix} -\bar{a}_1 & -1 \\ \bar{a}_2 & 0 \end{bmatrix}\frac{1}{\bar{a}_2} = [-\bar{a}_1/\bar{a}_2, -1/\bar{a}_2]$$

Hence, by (8A.4) with

$$B = \begin{bmatrix} 0 \\ \beta \end{bmatrix}$$

$$B^{\#} = -\begin{bmatrix} \dfrac{\bar{a}_2}{\beta} \end{bmatrix}[-\bar{a}_1/\bar{a}_2, -1/\bar{a}_2] = [\bar{a}_1/\beta, 1/\beta]$$

and, by (8A.3)

$$g_0 = \frac{1}{\beta} \tag{8A.6}$$

(This result does not come as a surprise, does it?) Thus the full-state feedback control law by (8A.2) is

$$u = -\left(\frac{\bar{a}_2 + \Omega^2}{\beta}\right)\theta - \left(\frac{\bar{a}_1 - \alpha}{\beta}\right)\omega - \frac{1}{\beta}x_0 \tag{8A.7}$$

Step 2. Observer design with known control

The observer, designed under the assumption that the control is known, was found in Example 7B to be given by

$$\begin{aligned}\dot{\hat{\theta}} &= \hat{\omega} + k_1(y - \hat{\theta})\\ \dot{\hat{\omega}} &= \Omega^2\hat{\theta} - \alpha\hat{\omega} + \beta u + \hat{x}_0 + k_2(y - \hat{\theta})\\ \dot{\hat{x}}_0 &= k_3(y - \hat{\theta})\end{aligned} \tag{8A.8}$$

with the observer gain matrix given by

$$K = \begin{bmatrix} k_1 \\ k_2 \\ k_3 \end{bmatrix} = \begin{bmatrix} \hat{a}_1 - \alpha \\ \hat{a}_2 + \Omega^2 - \alpha(\hat{a}_1 - \alpha) \\ \hat{a}_3 \end{bmatrix}$$

where $\hat{a}_1$, $\hat{a}_2$, $\hat{a}_3$ are the coefficients of the observer characteristic polynomial.

Step 3. Compensator design

The compensator dynamic equations are obtained by using the estimated state $\hat{x} = [\hat{\theta}, \hat{\omega}, \hat{x}_0]'$ in (8A.7), i.e.,

$$u = -\left(\frac{\bar{a}_2 + \Omega^2}{\beta}\right)\hat{\theta} - \left(\frac{\bar{a}_1 - \alpha}{\beta}\right)\hat{\omega} - \frac{1}{\beta}\hat{x}_0 \tag{8A.9}$$

and also using this control in (8A.8).

A block-diagram representation of (8A.8) and (8A.9) is shown in Fig. 8.3, which is the same as the block diagram (Fig. 7.6) for the observer with known input, but with the input u given by (8A.9).

Although the structure of Fig. 8.3 explicitly exhibits the estimates of the state variables, it is not necessary that the compensator be implemented by that structure. As long as the transfer function between the measured state $y = \theta$ and the control output u is the same as the transfer function between y and u in Fig. 8.3, the closed-loop system will have the same behavior.

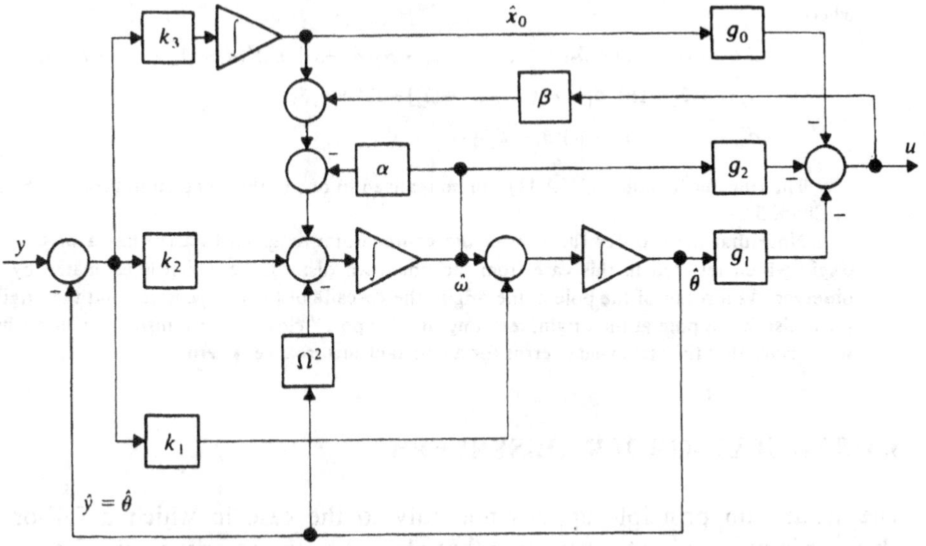

Figure 8.3 Compensator for inverted pendulum control with disturbance estimation.

Using (8.17), or working directly with (8A.8) and (8A.9), we find that the compensator transfer function is

$$D(s) = G(sI - \hat{A}_c)^{-1}K$$

where

$$\hat{A}_c = \hat{A} - BG = \begin{bmatrix} -k_1 & 1 & 0 \\ -k_2 + \Omega^2 & -\alpha & 1 \\ -k_3 & 0 & 0 \end{bmatrix} - \begin{bmatrix} 0 \\ \beta \\ 0 \end{bmatrix} [g, g_0]$$

$$= \begin{bmatrix} -\hat{a}_1 + \alpha & 1 & 0 \\ -\hat{a}_2 + \alpha(\hat{a}_1 - \alpha) - \bar{a}_2 - \Omega^2 & -\bar{a}_1 & 0 \\ -\hat{a}_3 & 0 & 0 \end{bmatrix}$$

The resolvent for $\hat{A}_c$ is given by

$$(sI - \hat{A}_c)^{-1}$$

$$= \frac{1}{\Delta} \begin{bmatrix} s(s - \bar{a}_1) & s & 0 \\ -s[\bar{a}_2 + \hat{a}_2 + \Omega^2 \alpha(\hat{a}_1 - \alpha)] & s(s + \hat{a}_1 - \alpha) & 0 \\ -\hat{a}_3(s + \hat{a}_1) & -\hat{a}_3 & s^2 + (\hat{a}_1 + \bar{a}_1 - \alpha)s + (\hat{a}_1 - \alpha)(\bar{a}_1 - \alpha) + \hat{a}_2 + \bar{a}_2 + \Omega^2 \end{bmatrix}$$

where

$$\Delta = s[s^2 + (\hat{a}_1 + \bar{a}_1 - \alpha)s + (\hat{a}_1 - \alpha)(\bar{a}_1 - \alpha) + \bar{a}_2 + \hat{a}_2 + \Omega^2] \qquad (8A.10)$$

After some calculation the transfer function of the compensator is determined to be

$$D(s) = \frac{d_1 s^2 + d_2 s + d_3}{\beta s[s^2 + (\hat{a}_1 + \bar{a}_1 - \alpha)s + (\hat{a}_1 - \alpha)(\bar{a} - \alpha) + \bar{a}_2 + \hat{a}_2 + \Omega^2]} \qquad (8A.11)$$

where

$$d_1 = \bar{a}_2(\hat{a}_1 - \alpha) + \hat{a}_2(\bar{a}_1 - \alpha) - \alpha(\bar{a}_1 - \alpha)(\hat{a}_1 - \alpha) + \Omega^2(\hat{a}_1 + \bar{a}_1 - 2\alpha) + \hat{a}_3$$

$$d_2 = \bar{a}_2\hat{a}_2 + \Omega^2[(\bar{a}_1 - \alpha)(\hat{a}_1 - \alpha) + \hat{a}_2] + \Omega^4 + \hat{a}_1\hat{a}_3$$

$$d_3 = \hat{a}_3[\bar{a}_2 + (\hat{a}_1 - \alpha)(\bar{a}_1 - \hat{a}_1)]$$

The transfer function of (8A.11) can be realized in one of the companion forms discussed in Chap. 3.

Note that the transfer function of the compensator as given by $D(s)$ has a pole at the origin which resulted in this case from the unknown disturbance which is estimated by the observer. As a result of the pole at the origin, the cascade of the compensator and the original plant also has a pole at the origin, resulting in a "type 1" closed-loop transfer function which will ensure that the steady-state error for a constant disturbance is zero.

8.3 REDUCED-ORDER OBSERVERS

The separation principle applies not only to the case in which a full-order observer is used for state estimation, but also when the compensator is based on a reduced-order observer. In particular, for the special case in which the observation can be used to solve for a substate:

$$y = C_1 x_1$$

with C_1 being a nonsingular matrix

$$\hat{x}_1 = x_1 = C_1^{-1} y \tag{8.18}$$

Then in accordance with the theory developed in Chap. 7, we take

$$\hat{x}_2 = Ly + z \tag{8.19}$$

where

$$\dot{z} = Fz + \bar{\bar{G}}y + Hu \tag{8.20}$$

where

$$\begin{aligned} F &= A_{22} - LC_1 A_{12} \\ \bar{\bar{G}} &= A_{21} - LC_1 A_{11} \\ H &= B_2 - LC_1 B_1 \end{aligned} \tag{8.21}$$

To implement the compensator we use the control law

$$u = -G\hat{x} = -[G_1, G_2]\begin{bmatrix} \hat{x}_1 \\ \hat{x}_2 \end{bmatrix} \tag{8.22}$$

Hence the compensator has the configuration shown in Fig. 8.4.

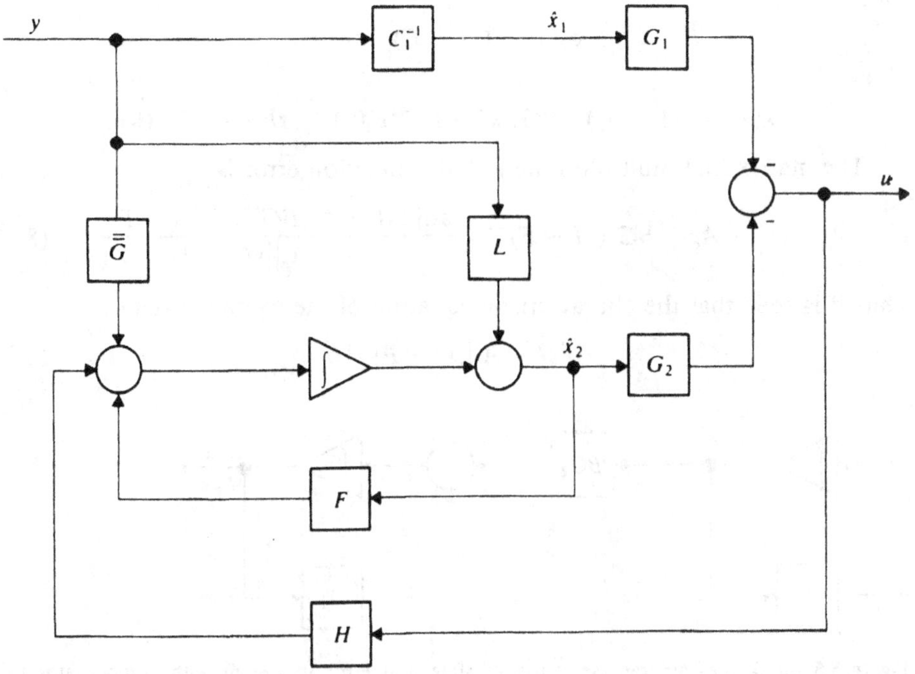

Figure 8.4 Block diagram of compensator using reduced-order observer.

As in the case of a full-order observer the closed-loop dynamics are expressed in terms of the state x of the plant and the estimation error

$$e = x - \hat{x} = \begin{bmatrix} x_1 - \hat{x}_1 \\ \hline x_2 - \hat{x}_2 \end{bmatrix}$$

The dynamics of the plant, using the control law (8.22) are described by

$$\dot{x} = Ax - BG(x - e) = (A - BG)x + B(G_1 e_1 + G_2 e_2) \tag{8.23}$$

By (8.18) $\quad \hat{x}_1 = x_1$

So $\quad e_1 \equiv 0 \quad (8.24)$

and, as shown in Chap. 7,

$$\dot{e}_2 = F e_2 \tag{8.25}$$

for any u when F, $\bar{G}$, and H are chosen to satisfy (8.21). Thus the closed-loop dynamics are given by (8.23) with $e_1 = 0$ and (8.25) as shown in Fig. 8.5. The structure is the same as shown in Fig. 8.1 for the full-order observer except that only the error e_2, in estimating the unmeasured substate x_2, produces a driving term to the state dynamics.

Following the development of Sec. 8.2 we can obtain the Laplace transform of the response due to nonzero initial conditions:

$$\mathrm{x}(s) = (sI - A_c)^{-1} BG_2 \mathrm{e}_2(s) + (sI - A_c)^{-1} x(0)$$

with

$$\mathrm{e}_2(s) = (sI - F)^{-1} e_2(0)$$

Thus

$$\mathrm{x}(s) = (sI - A_c)^{-1} BG_2 (sI - F)^{-1} e_2(0) + (sI - A_c)^{-1} x(0)$$

The matrix that multiplies the initial estimation error is

$$(sI - A_c)^{-1} BG_2 (sI - F)^{-1} = \frac{\operatorname{adj}(sI - A_c) BG_2 \operatorname{adj}(sI - F)}{|sI - A_c||sI - F|} \tag{8.26}$$

Thus it is seen that the characteristic equation of the overall system is

$$|sI - A_c||sI - F| = 0 \tag{8.27}$$

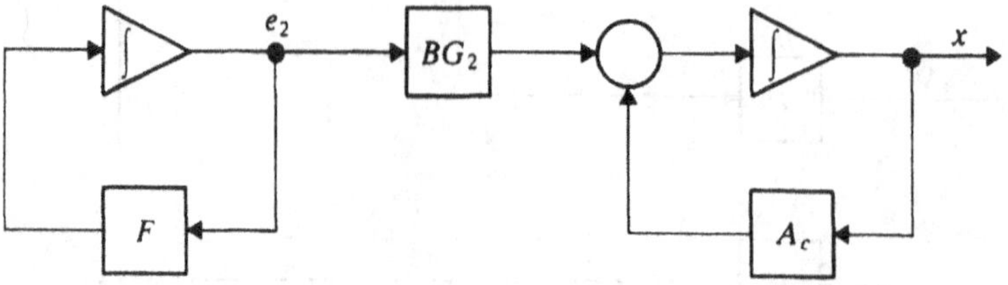

Figure 8.5 Block diagram representation of state and error in system with compensator using reduced-order observer.

Hence the poles of the closed-loop system, when a reduced-order observer is used in the compensator, are the zeros of $|sI - A_c|$ and the zeros of $|sI - F|$. The number of zeros in $|sI - A_c|$ is exactly equal to the order of the plant and the number of zeros of $|sI - F|$ is equal to the order of the compensator. Thus (8.27) accounts for all the eigenvalues of the closed-loop system.

The same method of analysis that was used in the above development for the special case in which a substate x_1 can be determined from the output by (8.18) can also be used to obtain an equation for the closed-loop dynamics in the more general case represented by (7.55)-(7.67). See Prob. 8.2.

8.4 ROBUSTNESS: EFFECTS OF MODELING ERRORS

In the foregoing analysis we learned that the closed-loop poles of a system using a compensator designed by the separation principle has its poles at the poles of the observer and at the poles of the full-state feedback control system. Since both sets of poles have putatively been selected by the designer, we can assume that their location is favorable to overall system operation. In demonstrating that the pole locations are exactly as specified, however, we made use of the fact that the observer design includes an *exact* dynamic model of the plant. There are several reasons why it is impractical to assume that the dynamic model of the plant is exact:

The physics of the plant may be understood only approximately, or
The exact dynamics may be known but too complicated to include in the control system design, or
The plant may change slowly with time owing to aging of components.

The reader no doubt can supply additional reasons.

The best that the designer can do is to design a compensator on the basis of a *nominal* plant model, i.e., a plant model defined for purpose of design. Since the true plant will (almost) never be the same as the nominal plant, the closed-loop poles will (almost) never be located in the exact locations intended. If their actual locations are not far from their intended locations when the actual plant does not differ greatly from the nominal plant, the nominal design will probably be satisfactory. If, on the other hand, a small change in the plant causes a large change in the closed-loop pole locations—perhaps going so far as to move them into the right half-plane—then the nominal design will surely be unsatisfactory.

To analyze the effect of "mismatches" between the nominal plant and the actual plant, suppose that the observer is designed on the basis of a nominal plant

$$\dot{\hat{x}} = \bar{A}\hat{x} + \bar{B}u + K(y - \bar{C}\hat{x}) \tag{8.28}$$

where

$$A = \bar{A} + \delta A$$
$$B = \bar{B} + \delta B \qquad (8.29)$$
$$C = \bar{C} + \delta C$$

with δA, δB, and δC representing changes in the plant dynamics, control, and observation matrices from their respective nominal values $\bar{A}$, $\bar{B}$, and $\bar{C}$.

a. Variations in Plant and Observation Matrices

Consider the effect of a variation in the plant matrix A and in the observation matrix C. The plant dynamics are then given by

$$\dot{x} = (\bar{A} + \delta A)x + Bu \qquad (8.30)$$

Subtract (8.28) from (8.30) to obtain

$$\dot{e} = \dot{x} - \dot{\hat{x}} = \bar{A}(x - \hat{x}) + \delta A x - K[(\bar{C} + \delta C)x - \bar{C}\hat{x}]$$
$$= (\bar{A} - K\bar{C})e + (\delta A - K\delta C)x \qquad (8.31)$$

Now

$$u = -G\hat{x} = -G(x - e)$$

Then (8.30) becomes

$$\dot{x} = (A - BG)x + BGe \qquad (8.32)$$

We can rewrite (8.32) and (8.31) as

$$\dot{x} = A_c x + BGe$$
$$\dot{e} = \hat{A}e + \Delta_{AC}x \qquad (8.33)$$

where

$$\hat{A} = \bar{A} - K\bar{C} = \text{nominal observer dynamics matrix} \qquad (8.34)$$

$$\Delta_{AC} = \delta A - K\delta C \qquad (8.35)$$

The characteristic equation of (8.33) is

$$\begin{vmatrix} sI - A_c & -BG \\ -\Delta_{AC} & sI - \hat{A} \end{vmatrix} = 0 \qquad (8.36)$$

Because of the presence of Δ_{AC} in the determinant of (8.36), the closed-loop poles are no longer at the zeros of $|sI - A_c|$ and at the zeros of $|sI - \hat{A}|$, but at other locations that depend on Δ_{AC}. In a practical application, we can usually estimate the size of Δ_{AC} in the worst circumstances. If we can assure that the closed-loop poles do not move too far from their nominal locations for the largest possible values of Δ_{AC}, then we have verified that the nominal design is sufficiently *robust* for the intended application. Otherwise it may be necessary to reconsider the nominal design.

One can readily appreciate that direct evaluation of (8.36) for all possible values of Δ_{AC}, although feasible, is a formidable task, and we seek a more palatable alternative. To accomplish this objective we take the Laplace transform of (8.33) and obtain

$$(sI - A_c)\mathbf{x}(s) = BG\mathbf{e}(s) + x(0)$$

and

$$(sI - \hat{A})\mathbf{e}(s) = \Delta_{AC}\mathbf{x}(s) + e(0)$$

or

$$\begin{aligned} \mathbf{x}(s) &= (sI - A_c)^{-1}(BG\mathbf{e}(s) + x(0)) \\ \mathbf{e}(s) &= (sI - \hat{A})^{-1}(\Delta_{AC}\mathbf{x}(s) + e(0)) \end{aligned} \tag{8.37}$$

A block-diagram representation of (8.37), in terms of transfer functions, is shown in Fig. 8.6. The block diagram is in the form of a closed-loop system, the loop being closed through Δ_{AC}. If Δ_{AC} is a scalar, it would be appropriate to study its effect by a root-locus-, Nyquist-, or Bode-diagram analysis. But Δ_{AC} is a $k \times k$ matrix, not a scalar, and the effect of Δ_{AC} on the closed-loop system must be studied by more general means, such as the singular-value analysis methods described in Chap. 4.

One special case is particularly easy to analyze, namely the case in which there is only a single output

$$y = Cx = c'x \qquad c' = [c_1, c_2, \ldots, c_k]$$

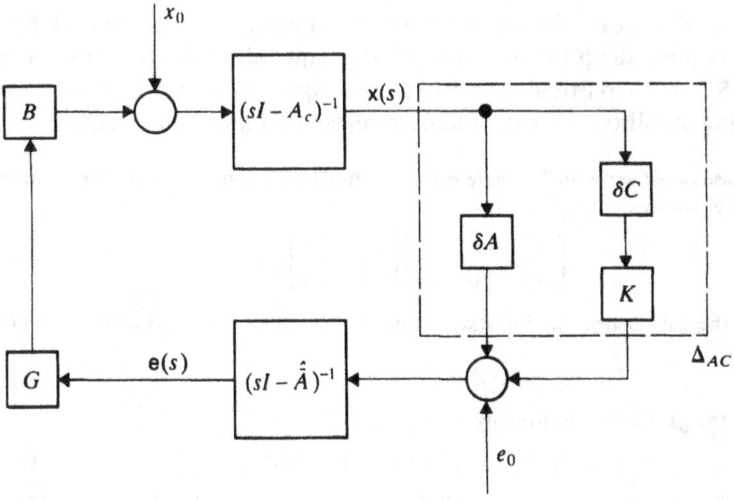

Figure 8.6 Block diagram to represent closed-loop dynamics due to change in plant and observation matrices.

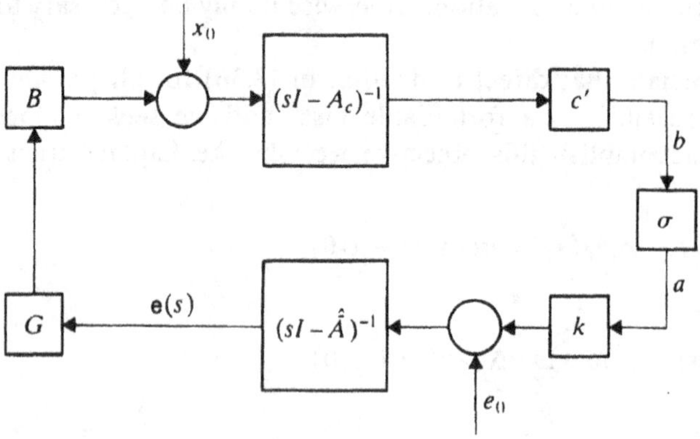

Figure 8.7 Special case of Fig. 8.6 when $\Delta_{AC} = \sigma c'$.

and where the only mismatch is due to a scale factor error, i.e.,

$$y = (1 + \sigma)c'x \tag{8.38}$$

where σ is the scale factor error. In this case

$$\delta A = 0 \quad \text{and} \quad \delta C = \sigma c'$$

The block diagram of Fig. 8.6 reduces to that shown in Fig. 8.7. (Since there is only a single output in this case, the observer gain matrix K becomes the column vector k.) If σ is regarded as a scalar gain, it is seen that the 1×1 transfer function from the output of σ (point a in the diagram) back to its input (point b) is

$$\mathsf{F}(s) = c'(sI - A_c)^{-1}BG(sI - \hat{A})^{-1}k \tag{8.39}$$

Hence the effect of a scale factor error in the output can be studied by a root-locus or Nyquist diagram analysis of the equivalent loop transmission $\mathsf{F}(s)$ given by (8.39). In particular, we can investigate how much σ can vary without degrading stability; we can calculate phase and gain margins, etc.

Example 8B Instrument servo In Example 6A, we considered a control system for an instrument servo governed by

$$\begin{bmatrix} \dot{e} \\ \dot{\omega} \end{bmatrix} = \begin{bmatrix} 0 & 1 \\ 0 & -\alpha \end{bmatrix} \begin{bmatrix} e \\ \omega \end{bmatrix} + \begin{bmatrix} 0 \\ \beta \end{bmatrix} u \tag{8B.1}$$

and designed the gain matrix for full-state feedback. For a closed-loop characteristic polynomial

$$\mathsf{D}_c(s) = s^2 + \bar{a}_1 s + \bar{a}_2$$

we found that the gain matrix is given by

$$G = g' = [g_1 \quad g_2] = [\bar{a}_2/\beta, (\bar{a}_1 - \alpha)/\beta] \tag{8B.2}$$

The observer design for the same example was found in Example 7A. For an observer characteristic polynomial

$$\mathsf{D}_0(s) = s^2 + \hat{a}_1 s + \hat{a}_2$$

we found the observer gain matrix

$$k = \begin{bmatrix} k_1 \\ k_2 \end{bmatrix} = \begin{bmatrix} \hat{a}_1 - \alpha \\ \hat{a}_2 - \alpha(\hat{a}_1 - \alpha) \end{bmatrix} \tag{8B.3}$$

This gain matrix was for an observation matrix

$$c' = [1 \quad 0] \tag{8B.4}$$

We now have all the matrices to evaluate the equivalent loop transmission $\mathsf{F}(s)$ given by (8.39) and are ready to assess the effect of an erroneous scale factor in the measurement of the system error e.

Note that $\mathsf{F}(s)$ conveniently factors into two scalar transfer functions:

$$\mathsf{F}(s) = \mathsf{F}_c(s)\mathsf{F}_0(s) \tag{8B.5}$$

where

$$\mathsf{F}_c(s) = c'(sI - A_c)^{-1}b, \qquad (B = b)$$

$$\mathsf{F}_0(s) = g'(sI - \hat{A})^{-1}k$$

where

$$A_c = A - bg' = \begin{bmatrix} 0 & 1 \\ 0 & -\alpha \end{bmatrix} - \begin{bmatrix} 0 \\ \beta \end{bmatrix}\begin{bmatrix} \dfrac{\bar{a}_2}{\beta}, & \dfrac{\bar{a}_1 - \alpha}{\beta} \end{bmatrix} = \begin{bmatrix} 0 & 1 \\ -\bar{a}_2 & -\bar{a}_1 \end{bmatrix}$$

$$\hat{A} = A - kc' = \begin{bmatrix} 0 & 1 \\ 0 & -\alpha \end{bmatrix} - \begin{bmatrix} \hat{a}_1 - \alpha \\ \hat{a}_2 - \alpha(\hat{a}_1 - \alpha) \end{bmatrix}[1, 0] = \begin{bmatrix} -\hat{a}_1 + \alpha & 1 \\ -\hat{a}_2 + \alpha(\hat{a}_1 - \alpha) & -\alpha \end{bmatrix}$$

Thus

$$\mathsf{F}_c(s) = [1, 0]\begin{bmatrix} s + \bar{a}_1 & 1 \\ -\bar{a}_2 & s \end{bmatrix}\begin{bmatrix} 0 \\ \beta \end{bmatrix}\frac{1}{s^2 + \bar{a}_1 s + \bar{a}_2} = \frac{\beta}{s^2 + \bar{a}_1 s + \bar{a}_2}$$

$$\mathsf{F}_0(s) = \begin{bmatrix} \dfrac{\bar{a}_2}{\beta}, & \dfrac{\bar{a}_1 - \alpha}{\beta} \end{bmatrix}\begin{bmatrix} s + \alpha & 1 \\ -\hat{a}_2 + \alpha(\hat{a}_1 - \alpha) & s + \hat{a}_1 - \alpha \end{bmatrix}\begin{bmatrix} \hat{a}_1 - \alpha \\ \hat{a}_2 - \alpha(\hat{a}_1 - \alpha) \end{bmatrix}\frac{1}{s^2 + \hat{a}_1 s + \hat{a}_2}$$

$$= \frac{\bar{a}_2[(\hat{a}_1 - \alpha)s + \hat{a}_2] + (\bar{a}_1 - \alpha)[\hat{a}_2 - \alpha(\hat{a}_1 - \alpha)]s}{\beta(s^2 + \hat{a}_1 s + \hat{a}_2)}$$

Thus we find that

$$\mathsf{F}(s) = \frac{s[\bar{a}_2(\hat{a}_1 - \alpha) + \hat{a}_2(\bar{a}_1 - \alpha) - \alpha(\hat{a}_1 - \alpha)(\bar{a}_1 - \alpha)] + \bar{a}_2\hat{a}_2}{(s^2 + \bar{a}_1 s + \bar{a}_2)(s^2 + \hat{a}_1 s + \hat{a}_2)} \tag{8B.6}$$

We also compute the transfer function $\mathsf{D}(s)$ of the compensator using (8.17):

$$\mathsf{D}(s) = G(sI - \hat{A}_c)^{-1}K = g'(sI - A + kc' + bg')^{-1}k$$

Now, using (8B.2), (8B.4), and (8B.3), we find

$$\hat{A}_c = A - kc' - bg' = \begin{bmatrix} -\hat{a}_1 + \alpha & 1 \\ -\bar{a}_2 - \hat{a}_2 + \alpha(\hat{a}_1 - \alpha) & -\bar{a}_1 \end{bmatrix}$$

Thus

$$\mathsf{D}(s) = \frac{1}{\beta}[\bar{a}_2, \bar{a}_1 - \alpha]\begin{bmatrix} s + \bar{a}_1 & 1 \\ -\bar{a}_2 - \hat{a}_2 + \alpha(\hat{a}_1 - \alpha) & s + \hat{a}_1 - \alpha \end{bmatrix}^{-1}\begin{bmatrix} \hat{a}_1 - \alpha \\ \hat{a}_2 - \alpha(\hat{a}_1 - \alpha) \end{bmatrix}$$

After a bit of messy algebra we find that

$$\mathsf{D}(s) = \frac{s[\bar{a}_2(\hat{a}_1 - \alpha) + \hat{a}_2(\bar{a}_1 - \alpha) - \alpha(\hat{a}_1 - \alpha)(\bar{a}_1 - \alpha)] + \bar{a}_2\hat{a}_2}{\beta[s^2 + (\bar{a}_1 + \hat{a}_1 - \alpha)s + \hat{a}_2 + \bar{a}_2 + (\hat{a}_1 - \alpha)(\bar{a}_1 - \alpha)]} \tag{8B.7}$$

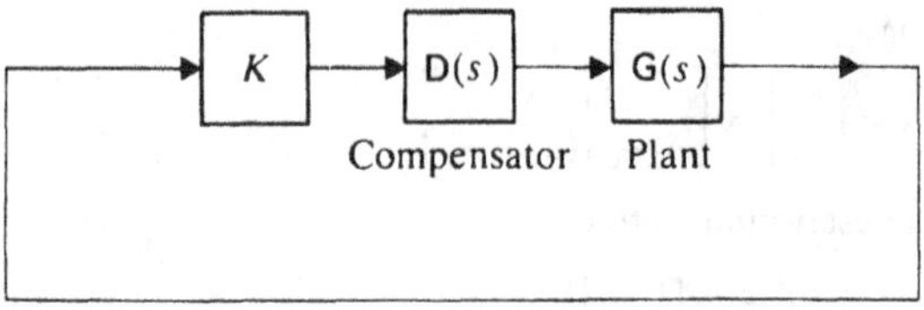

Figure 8.8 Block diagram for root-locus plot.

Note that the numerator of D(s) is exactly the same as the loop transmission F(s) at the nominal operating point. This is no accident. In this case the plant is a single-input, single-output system having the transfer function

$$\mathsf{G}(s) = c'(sI - A)^{-1}b = [1 \quad 0]\begin{bmatrix} s & -1 \\ 0 & s+\alpha \end{bmatrix}^{-1}\begin{bmatrix} 0 \\ \beta \end{bmatrix} = \frac{\beta}{s(s+\alpha)}$$

And the transmission of the open-loop is the product of D(s) and G(s). Thus the open-loop transmission zeros are the zeros of the compensator and the zeros of the plant. In this case the compensator has one zero and the plant has none. The open-loop poles of the system comprise the poles of the plant and the poles of the compensator. The former are at $s = 0$ and at $s = -\alpha$, and the latter are at the zeros of the denominator. When the loop is closed through a variable gain, as shown in Fig. 8.8, the poles of course will move, but *the zeros remain fixed.* As the gain is raised to unity the poles will move to the values for which the system was designed, namely the poles of the observer (at the zeros of $s^2 + \hat{a}_1 s + \hat{a}_2$) and at the poles of the full-state feedback system (i.e., at the zeros of $s^2 + \bar{a}_1 s + \bar{a}_2$). Further increase of the gain, beyond unity, will drive the poles away from these values and toward the open-loop zeros or to minus infinity in accordance with the excess of poles over zeros. In this example the excess is 3 and thus three branches of the root locus will tend to minus infinity, as discussed in Chap. 4. We shall consider a root-locus plot later in this example.

Although the calculations were somewhat tedious, we left the parameters α, $\bar{a}_i$, and $\hat{a}_i$ as variables rather than substituting numerical values. Now we can reap the reward.

First we note that the compensator transfer function is symmetric in the parameters of the observer and the full-state feedback controller: if the overbars and carets were interchanged in (8B.7) the compensator would have the same transfer function. (This may not be the case in other systems.)

Next we note that the compensator may turn out unstable. In particular we must have

$$\bar{a}_1 + \hat{a}_1 > \alpha \tag{8B.8}$$

Since $\hat{a}_1$ and $\bar{a}_1$ are proportional to the damping factors ($a_1 = 2\zeta\omega$) selected for the observer and the full-state feedback control, (8B.8) implies that the damping factors cannot be made too small without creating an unstable compensator. This is a fairly easy requirement to achieve. To avoid a compensator zero in the right half-plane, there is a requirement on the coefficient of s in the numerator of D(s). This requirement is also easy to achieve.

To proceed farther with the example without getting mired in the algebra, we finally must use some numbers. In particular, suppose

$$\alpha = 1$$

which means that an open-loop pole occurs at $s = -1$. Using only a gain and no compensator, we can see that the closed-loop root locus has the appearance shown in Fig. 8.9. Since the root locus lies in the left half-plane for all positive values of gain the uncompensated system has an infinite gain margin. But the system bandwidth cannot be much greater than 1 rad/s without having a very oscillatory response. For a damping factor of $\zeta = \sqrt{2}/2$, in fact, the natural frequency is only 0.5 rad/s as seen from Fig. 8.9.

The compensator (8B.7) is obviously more complex than a simple gain. What does the complexity of the compensator achieve? At least it should provide a larger bandwidth. So

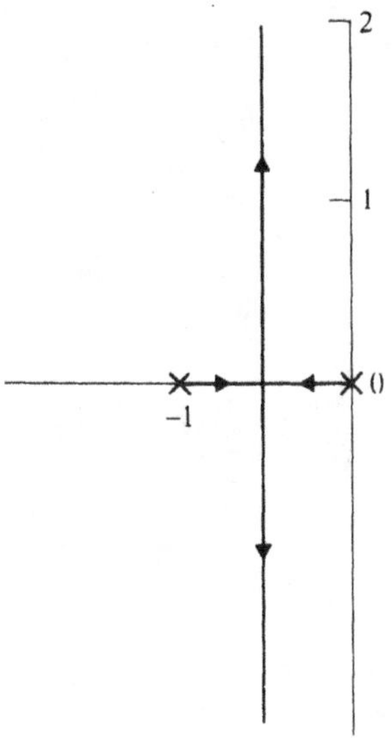

Figure 8.9 Root locus for uncompensated plant ($D(s) = 1$).

suppose we select the closed-loop poles of the full-state feedback controller and of the observer such that the natural frequency is 2 rad/s with a damping factor of $\sqrt{2}/2$. This means that

$$s^2 + \bar{a}_1 s + \bar{a}_2 = s^2 + \hat{a}_1 s + \hat{a}_2 = (s+2)^2 + 2^2 = s^2 + 4s + 8$$

i.e.,

$$\bar{a}_1 = \hat{a}_1 = 4$$

$$\bar{a}_2 = \hat{a}_2 = 8$$

Substituting these into (8B.7) gives a compensator transfer function

$$D(s) = \frac{39s + 64}{\beta(s^2 + 7s + 25)}$$

The zero of $D(s)$ occurs at

$$s = -\tfrac{64}{39} = -1.64$$

and the poles occur at

$$s = \tfrac{1}{2}(-7 \pm j\sqrt{51}) = -3.50 \pm j3.57$$

The root-locus plot is shown in Fig. 8.10. It is seen that the two branches of the root locus intersect at the complex frequencies $s = -2 \pm j2$ which are the frequencies selected for the normal operation of the compensator. The root loci cross the imaginary axis at $\omega = 5.80$ and this occurs for a gain K of 3.63, which is the gain margin of the system. Thus, the price we pay to raise bandwidth by a factor of 4 (and still maintain a damping factor of $\sqrt{2}/2$) is a fairly complicated compensator and a reduced gain margin. Whether this is an acceptable price depends of course upon the application.

We note that the finite gain margin is the result of the fact that the compensator has two poles and one zero, i.e., that the excess of poles over zeros in the loop transmission has been increased by one. A reduced-order observer would introduce only one pole and hence not increase the excess in the loop transmission. Hence a reduced-order observer (see Prob. 8.3) might be preferable to the full-order observer considered here.

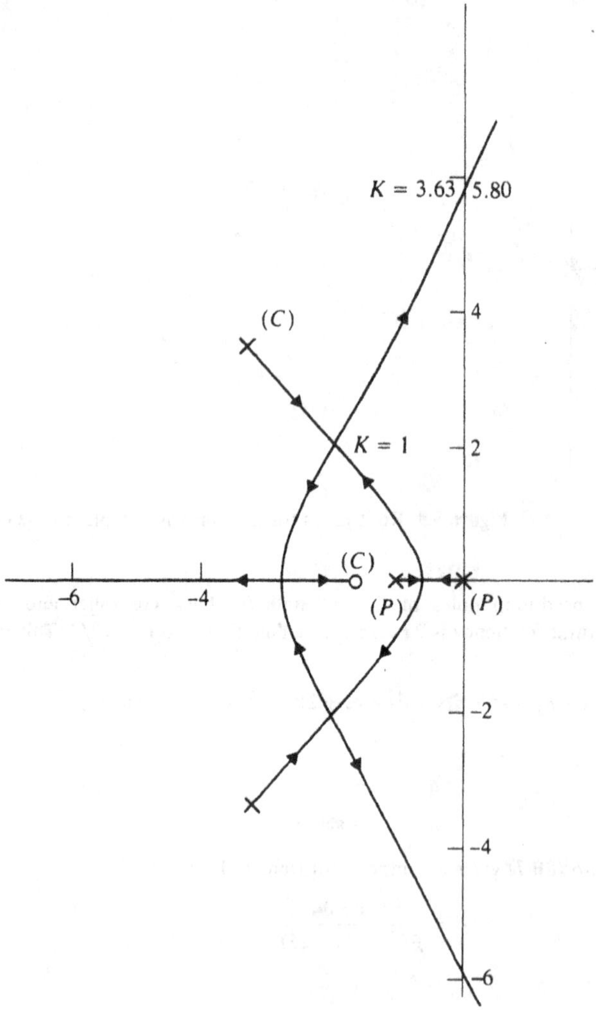

Figure 8.10 Root locus of compensated instrument servo. (C) denotes pole or zero due to compensator; (P) denotes plant pole.

It is hardly worth the effort of using the apparatus of the separation principle for a plant as simple as this. The advantage of the separation principle is in applications such as systems of high order, or with multiple inputs and outputs.

b. Variations in Control Matrix

A slightly different analysis is appropriate for the case in which the control matrix B differs from its nominal value. In this case the closed loop is analyzed using $\hat{x}$ and $e = x - \hat{x}$ as the state variables. We use

$$A = \bar{A} \qquad \text{and} \qquad C = \bar{C}$$

but

$$B = \bar{B} + \delta B \tag{8.40}$$

The plant dynamics can be written

$$\dot{\hat{x}} + \dot{e} = A(\hat{x} + e) + (\bar{B} + \delta B)G\hat{x} = (A - \bar{B}G)\hat{x} + Ae - \delta BG\hat{x} \tag{8.41}$$

and the observer is given by

$$\dot{\hat{x}} = (A - \bar{B}G)\hat{x} + KCe \tag{8.42}$$

Subtract (8.42) from (8.41) to obtain the differential equations for the error:

$$\dot{e} = (A - KC)e - \delta BG\hat{x} \tag{8.43}$$

The characteristic equation of the $2k$th-order system comprising (8.42) and (8.43) is

$$\begin{vmatrix} sI - \bar{A}_c & KC \\ -\delta BG & sI - \hat{A} \end{vmatrix} = 0 \tag{8.44}$$

where

$$\bar{A}_c = A - \bar{B}G \qquad \hat{A} = A - KC$$

Note that the determinant of (8.44) is triangular for $\delta B = 0$ and hence that the closed-loop poles are the zeros of $|sI - \bar{A}_c|$ and $|sI - \hat{A}|$.

A Laplace transform analysis of (8.42) and (8.43) is still more revealing: Proceeding as in the development of (8.37) we find that

$$\begin{aligned} \hat{\mathrm{x}}(s) &= (sI - \bar{A}_c)^{-1}(KC\mathrm{e}(s) + \hat{x}(0)) \\ \hat{\mathrm{e}}(s) &= (sI - \hat{A})^{-1}(-\delta BG\hat{\mathrm{x}}(s) + e(0)) \end{aligned} \tag{8.45}$$

which has the block-diagram representation shown in Fig. 8.11. Note that the loop is closed through the perturbation δB in the control matrix.

The case in which there is only a single input, that is u is a scalar, and the perturbation in the control matrix is only a scale factor change, is also particularly easy to analyze. In this case

$$B = b \qquad G = g'$$

and

$$\delta B = \sigma b \tag{8.46}$$

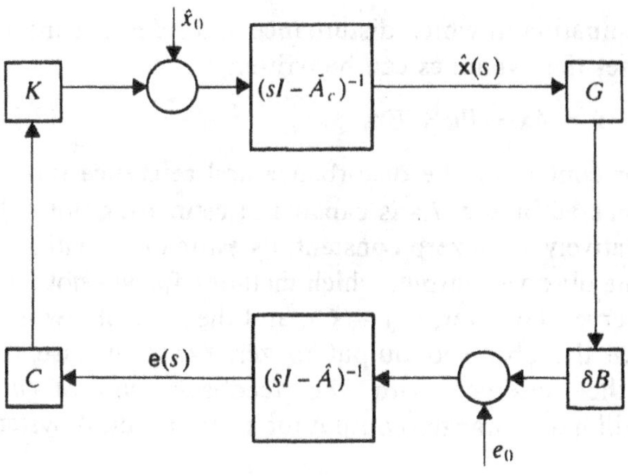

Figure 8.11 Block diagram to represent closed-loop dynamics due to change in control matrix.

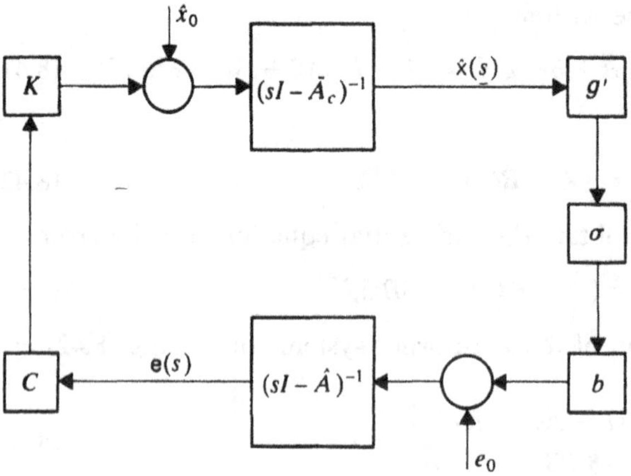

Figure 8.12 Special case of Fig. 8.11 when $\delta B = \sigma b$.

In this special case Fig. 8.11 reduces to Fig. 8.12, and it is seen that the loop transmission is given by the scalar transfer function

$$F(s) = g'(sI - \bar{A}_c)^{-1}KC(sI - \hat{A})^{-1}b \tag{8.47}$$

Note the similarity (actually duality) of (8.47) and (8.39).

When the plant has both a single input *and* a single output both (8.39) and (8.47) factor into two-scalar transfer functions:

$$F_s = F_c F_0 \tag{8.48}$$

where

$$F_c = g'(sI - \bar{A}_c)^{-1}k \tag{8.49}$$

and

$$F_0 = c'(sI - \hat{A})^{-1}b \tag{8.50}$$

8.5 DISTURBANCES AND TRACKING SYSTEMS: EXOGENOUS VARIABLES

In Sec. 7.4 we discussed the situation in which disturbance and reference inputs may be present. For such cases the dynamics can be written

$$\dot{x} = Ax + Bu + Ex_0 \tag{8.51}$$

where x_0 is a constant vector containing the disturbance and reference inputs. An observer designed as discussed in Sec. 7.4 is capable of estimating not only x but also x_0. Since x_0 is putatively a nonzero constant, its estimate $\hat{x}_0$ will tend to the same constant. Thus the observer output, which includes $\hat{x}_0$, will not tend to zero. The input to the observer, however, is $y = Cx$, and the control law may have been designed to force the observed output to zero. In that case the compensator, comprising the observer with the feedback control $u = -G\hat{x} - G_0\hat{x}_0$ implemented, will have a nonzero output for a zero input. A system can have a constant nonzero output for an input of zero only if it has a pole at the origin, i.e., if it contains a pure integrator. Thus the compensator affords the possibility of providing "integral control."

In our discussion in Sec. 6.4 of systems designed for tracking and with disturbances, we noted that it is not possible, in general, to achieve a zero error unless the number of components in the control vector u is at least equal to the number of components in the vector $y = Cx$ that we want to reduce to zero (in the steady state). In many applications, however, it may be possible to measure a larger number of variables than the number that can be reduced to zero. In that case, the observation vector y may not go to zero in the steady state and the estimate $\hat{x}_0$ of the constant vector x_0 may not require pure integral control, since a nonzero steady state signal will be present in the observation y.

Since the observer provides an estimate of x_0 in all cases, we can justify the claim that it provides "quasi-integral" control. It provides true integral control, however, only when the observed output $y = Cx$ is designed to go to zero in the steady state.

The quasi-integral control provided by the compensator can be present when either a full-order observer or a reduced-order observer is used. In particular, suppose that the observation vector consists of the state x itself

$$y = x = [I \quad 0]\begin{bmatrix} x \\ x_0 \end{bmatrix} = \mathbf{C}\mathbf{x} \tag{8.52}$$

Then, in the metastate representation,

$$\mathbf{C} = [I \quad 0] = [C_1 \quad C_2]$$

and

$$\mathbf{A} = \begin{bmatrix} A & E \\ 0 & 0 \end{bmatrix} = \begin{bmatrix} A_{11} & A_{12} \\ A_{21} & A_{22} \end{bmatrix}$$

$$\mathbf{B} = \begin{bmatrix} B \\ 0 \end{bmatrix} = \begin{bmatrix} B_1 \\ B_2 \end{bmatrix}$$

The reduced-order observer, defined in Sec. 7.5, for the components of the metastate $\mathbf{x}$, namely

$$x_1 = x$$

$$x_2 = x_0$$

is

$$\hat{x}_1 = \hat{x} = y \tag{8.53}$$

$$\hat{x}_2 = \hat{x}_0 = Ly + z \tag{8.54}$$

with

$$\dot{z} = Fz + \bar{G}y + Hu \tag{8.55}$$

where F is chosen as desired, and, in accordance with (7.50)–(7.52)

$$LE = -F \qquad LB = -H \qquad \bar{G} = -LA + FL \tag{8.56}$$

Using (8.56) we obtain for (8.55)

$$\begin{aligned}\dot{z} &= -LEz - (LA + LEL)y - LBu \\ &= -L(E\hat{x}_0 + Ay + Bu)\end{aligned} \tag{8.57}$$

(The last line of (8.57) is obtained using (8.54).)

In accordance with the separation principle, we are to generate the control u in (8.57) using the control law of Chap. 6 (Sec. 6.4) with the estimated states. Thus

$$u = -Gy - G_0\hat{x}_0 \tag{8.58}$$

A block-diagram representation of the compensator using the observer of (8.57) and (8.58) is shown in Fig. 8.13(a). An alternate form of the compensator is obtained by substituting (8.58) into (8.57) to yield

$$\dot{z} = -L[(E - BG_0)\hat{x}_0 + A_c y] \tag{8.59}$$

where

$$A_c = A - BG$$

The block-diagram representation of (8.54), (8.58), and (8.59) is shown in Fig. 8.13(b). The block diagram shows a direct path from the measured state $y = x$ through the gain matrix G to the control, and a parallel path, through the dynamic system, whose state is z, which contributes to generating the exogenous estimate $\hat{x}_0$ which is multiplied by G_0 to produce the component of control due to $\hat{x}_0$.

In what circumstances will the compensator defined by (8.54), (8.58), and (8.59), i.e., Fig. 8.13(b), provide integral control? The transfer function of the compensator, i.e., from y to the control u, can be obtained by taking the Laplace transform of (8.59)

$$sz = -L[(E - BG_0)\hat{\mathrm{x}}_0 + A_c\mathrm{y}]$$

where, by (8.54),

$$s\hat{\mathrm{x}}_0 = Lsy + sz = L[(sI - A_c)\mathrm{y} - (E - BG_0)\hat{\mathrm{x}}_0]$$

Thus

$$[sI - L(E - BG_0)]\hat{\mathrm{x}}_0 = L(sI - A_c)\mathrm{y}$$

and

$$\hat{\mathrm{x}}_0 = [sI - L(E - BG_0)]^{-1}L(sI - A_c)\mathrm{y} \tag{8.60}$$

The characteristic equation of the compensator is

$$|sI - L(E - BG_0)| = 0 \tag{8.61}$$

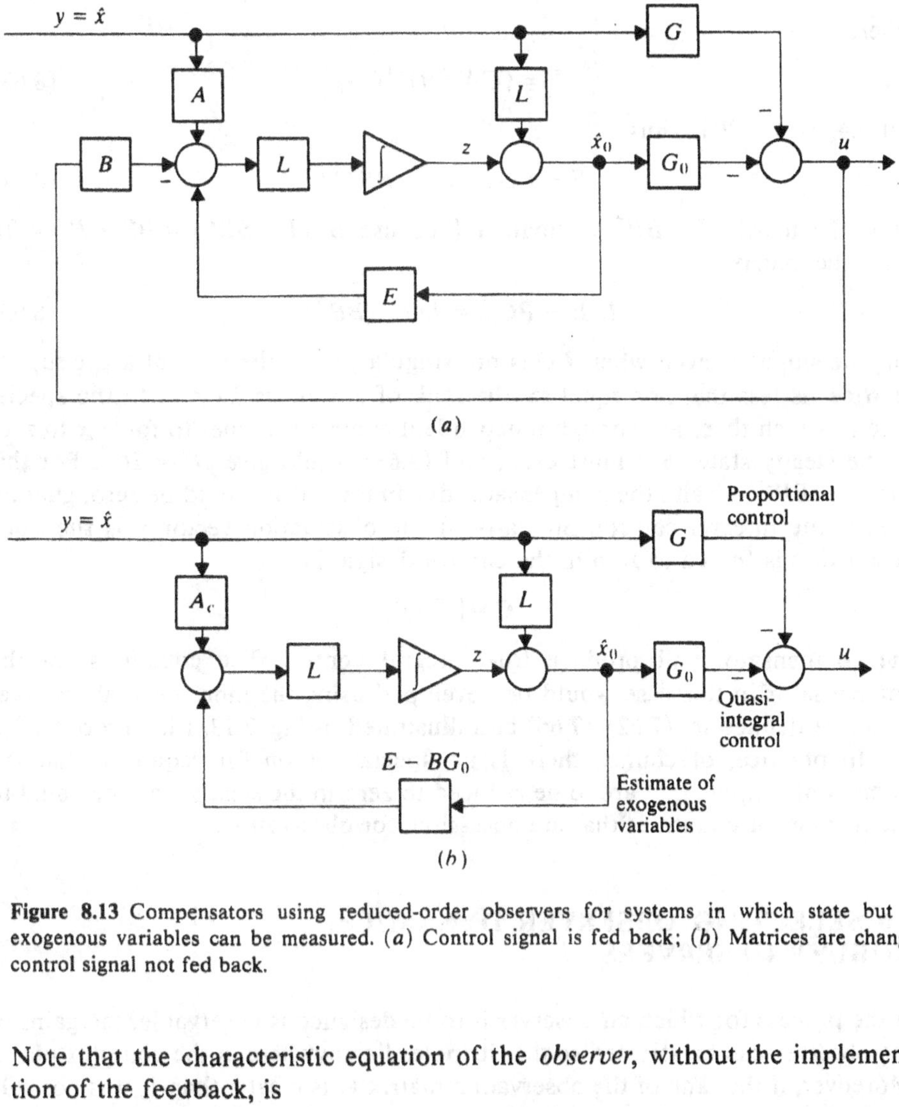

Figure 8.13 Compensators using reduced-order observers for systems in which state but not exogenous variables can be measured. (*a*) Control signal is fed back; (*b*) Matrices are changed, control signal not fed back.

Note that the characteristic equation of the *observer*, without the implementation of the feedback, is

$$|sI - LE| = 0 \tag{8.62}$$

The poles of the latter are closed-loop poles of the overall system consisting of the compensator *and* the plant. By design, we select the L matrix so that the observer dynamics matrix LE is a stability matrix, which implies that LE is not singular. In order for the compensator to provide true (and not quasi) integral control, however, the characteristic equation (8.61) of the compensator must have at least one zero at the origin. This means that $L(E - BG_0)$ must be singular. Is it possible for $L(E - BG_0)$ to be singular when LE is not? To learn the answer to this question, recall from Sec. 6.4 that

$$G_0 = B^{\#} E$$

where

$$B^{\#} = (CA_c^{-1}B)^{-1}CA_c^{-1} \tag{8.63}$$

with $A_c = A - BG$. Thus

$$E - BG_0 = (I - BB^{\#})E \tag{8.64}$$

Now the matrix $I - BB^{\#}$ is singular (because $B^{\#}(I - BB^{\#}) = B^{\#} - B^{\#} = 0$). Thus the matrix

$$L(E - BG_0) = L(I - BB^{\#})E \tag{8.65}$$

may be singular, even when LE is not singular, since the rank of a product of matrices is less than, or equal to, the rank of any of its factors. In the special case in which there are enough independent control variables to force x to zero in the steady state, B^{-1} must exist, and (8.63) would give $B^{\#} = B^{-1}$. For this case $I - BB^{\#} = 0$ and the compensator dynamics matrix would be zero, guaranteeing true integral control. Similarly, if the observation vector y is the same vector that is forced to zero in the control design, i.e.,

$$\mathbf{C} = [C \quad 0]$$

the compensator will produce true integral control. The equations for the compensator in this case would be developed using the more general observer structure defined by (7.62)-(7.66) and illustrated in Fig. 7.13. (See Prob. 8.2.)

In practice, of course, there is no special reason for requiring that the number of outputs that are to be reduced to zero in the steady state be equal to the number of quantities that are accessible for observation.

8.6 SELECTING OBSERVER DYNAMICS: ROBUST OBSERVERS

If the process for which an observer is to be designed is observable, the gains of the observer can be selected so that its poles lie anywhere in the complex plane. Moreover, if the rank of the observation matrix C is greater than 1, it is possible to accomplish more than simply placing the poles. Thus the question arises, as it did in the case of the design of the feedback control law $u = -Gx$, of the choice of an appropriate dynamics matrix for the observer.

One of the considerations in the design of a gain matrix G in the full-state feedback control law is that the resulting control signal u not be too large to be realized by the existing actuator (power source). Since the function of the observer is only to process data, however, there is no practical limitation on the size of the observer gain matrix K: The observer can be made to respond as rapidly as one might desire without concern that the signals will be larger than physically realizable. (Nowadays, it is all but certain that the entire compensator would be realized by a digital computer. With floating-point numerics, a digital computer would be capable of handling variables of any reasonable dynamic range.) Though the hardware may impose no limitation on the observer dynamics, it may nevertheless be desirable to limit the observer speed-of-response (bandwidth). Remember that real sensors are noisy. And much of the noise occurs at relatively high frequencies. By limiting the bandwidth of the observer, we can thus attenuate and smooth the noise contribution to the compensator output—which is the control input. This is a method for reducing the undesired response of the dynamic process to the sensor noise. But reducing the bandwidth of the closed-loop process in response to sensor noise also reduces the speed of response to reference inputs. The benefits of one cannot be achieved without some sacrifice of the benefits of the other. It is possible, at least in principle, to trade off the noise attenuation for dynamic response "optimally" using the methods of Chaps. 10 and 11. There we learn how to formulate the optimum observer design problem in terms of the intensity (i.e., spectral density) of the noise at the sensors. The word "optimally" is enclosed in quotes to emphasize that the optimum trade-off requires a precise knowledge of the sensor noise characteristics. In practice, these characteristics are rarely known with sufficient accuracy to justify the claim of true optimality. The methods of Chaps. 10 and 11 are readily amenable to machine calculation, however, and can be used as effective design tools, even when only rough estimates of the sensor noise characteristics are available, and hence when the appelation of optimal is scarcely justified.

One of the problems that has been found to arise when observers are designed strictly on the basis of the sensor noise parameters, is a possible lack of robustness, i.e., stability margins, when parameters of the process are different from those used in the design. We already discussed this problem in Sec. 8.4 and studied how the effect of parameter changes might be studied by analyzing the return difference for variations in the plant parameters.

One of the considerations that should be addressed in the design of an observer is the robustness of the closed-loop dynamic process. When the observer dynamics are perfectly matched to the process, then the error $e = x - \hat{x}$ between the estimated state $\hat{x}$ and the true state x goes to zero in the steady state. This should mean that the transfer function from the control input u to the error e is zero, or that the transfer functions from the control input u to x and to $\hat{x}$ are both the same. Does it? To determine the answer to this question consider the two block-diagrams of the closed-loop process as drawn in Fig. 8.14. The first block diagram shows the full-state feedback control system and an extraneous input u_0. Also

$$\bar{u} = u_0 - Gx$$

The transfer function from the extraneous input u_0 to the state x is calculated as follows

$$x(s) = \Phi B\bar{u}(s) \tag{8.66}$$

where Φ is the "resolvent" of the plant, i.e.,

$$\Phi = (sI - A)^{-1} \tag{8.67}$$

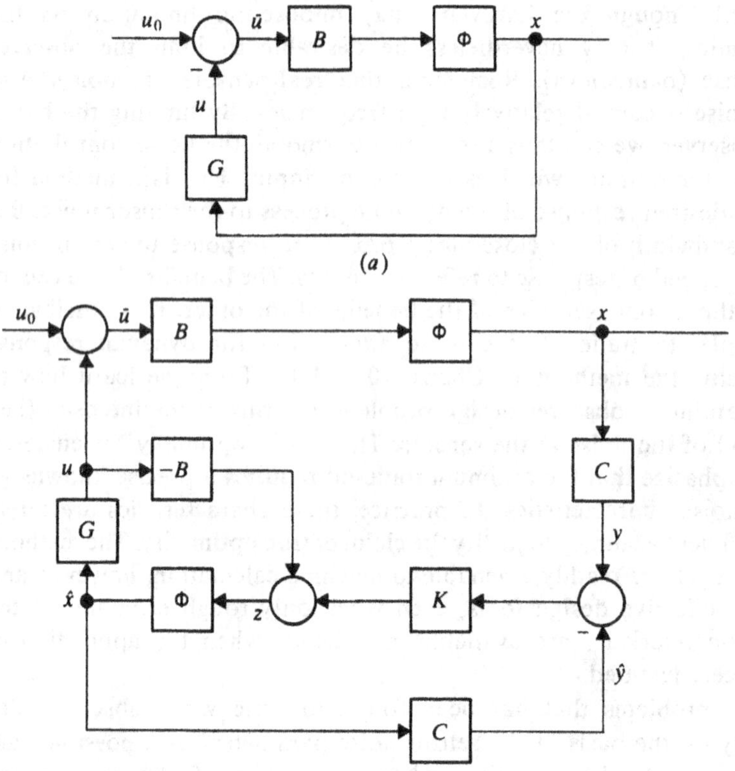

Figure 8.14 Development of robust observers. (*a*) Full-state feedback; (*b*) Observer used to estimate state.

and

$$\bar{\mathrm{u}}(s) = \mathrm{u}_0 - G\mathrm{x}(s) \tag{8.68}$$

Thus, by (8.68) and (8.66),

$$\mathrm{x}(s) = (I + \Phi BG)^{-1}\Phi B \mathrm{u}_0 \tag{8.69}$$

Thus, the transfer function from u_0 to the state x, using full-state feedback, is

$$\begin{aligned} \mathrm{H}_0(s) &= (I + \Phi BG)^{-1}\Phi B \\ &= [I + (sI - A)^{-1}BG]^{-1}(sI - A)^{-1}B \end{aligned} \tag{8.70}$$

The form of the transfer function given by (8.70) is more complicated than the more usual form

$$\mathrm{H}_0(s) = (sI - A + BG)^{-1}B = (\Phi^{-1} + BG)^{-1}B \tag{8.71}$$

but is particularly suited to our present development. (See Note 8.2 and Prob. 8.1.)

Now suppose that we realize the control law by the separation principle, using an observer as shown in Fig. 8.14(b). Again we seek the transfer function from the extraneous input u_0 to the state x. It is crucial to recognize that the extraneous input u_0 is presumed unknown and hence not available for feeding into the observer.

The transfer function from the extraneous input u_0 to the state x is still determined from (8.66); but now

$$\bar{\mathrm{u}}(s) = \mathrm{u}_0 - G\hat{\mathrm{x}}(s) \tag{8.72}$$

For an arbitrary control gain matrix G the transfer function from u_0 to x in Fig. 8.14(b) will not be the same as that of Fig. 8.14(a) unless the transfer function from $\bar{u}$ to $\hat{x}$ in Fig. 8.14(b) is the same as that from $\bar{u}$ to x in Fig. 8.14(a). The transfer function from $\bar{u}$ to x in Fig. 8.14(a) is given by (8.66). In Fig. 8.14(b)

$$\hat{\mathrm{x}}(s) = \Phi[B\mathrm{u} - KC(\hat{\mathrm{x}}(s) - \mathrm{x}(s))]$$

or

$$(\Phi^{-1} + KC)\hat{\mathrm{x}}(s) = B\mathrm{u}(s) + KC\Phi B\bar{\mathrm{u}}(s)$$

or

$$\hat{\mathrm{x}}(s) = (\Phi^{-1} + KC)^{-1}[B\mathrm{u}(s) + KC\Phi B\bar{\mathrm{u}}(s)] \tag{8.73}$$

The transfer function from $\bar{\mathrm{u}}(s)$ to $\hat{\mathrm{x}}(s)$ given by (8.73) is generally not the same as that given by (8.66). As first shown by Doyle and Stein,[2] however, they are equal when

$$K(I + C\Phi K)^{-1} = B(C\Phi B)^{-1} \tag{8.74}$$

This is shown with the aid of the Schur matrix identity (see Appendix (A.47) and Note 8.3)

$$(\Phi^{-1} + KC)^{-1} = \Phi - \Phi K(I + C\Phi K)^{-1}C\Phi \tag{8.75}$$

By (8.75) we find that (8.73) becomes

$$\hat{\mathrm{x}}(s) = [\Phi - \Phi K(I + C\Phi K)^{-1}C\Phi]B\mathrm{u}(s)$$
$$+ [\Phi - \Phi K(I + C\Phi K)^{-1}C\Phi]KC\Phi B\bar{\mathrm{u}}(s) \tag{8.76}$$

On use of the Doyle-Stein condition (8.74), the matrix multiplying $\mathrm{u}(s)$ becomes zero and the matrix multiplying $\bar{\mathrm{u}}(s)$ becomes B. Thus when (8.74) holds

$$\hat{\mathrm{x}}(s) = \Phi B\bar{\mathrm{u}}(s)$$

which is the same form as (8.66).

It is noted that the Doyle-Stein condition depends only on the open-loop characteristics of the observer; it is independent of the control gain. When the Doyle-Stein condition holds, the transfer function from an extraneous input u_0 to the state x is given by (8.69), independent of the observer: The dynamics of the observer do not influence this transfer function.

Another property of a "Doyle-Stein observer," i.e., an observer satisfying the Doyle-Stein condition, is obtained by computing the transfer function from the observable output y to the state estimate $\hat{x}$. With reference to Fig. 8.14(b) we see that

$$\Phi^{-1}\hat{x} = z = K(y - C\hat{x}) - BG\hat{x}$$

or
$$(\Phi^{-1} + KC)\hat{x} = Ky - BG\hat{x}$$

or
$$\hat{x} = (\Phi^{-1} + KC)^{-1}Ky - (\Phi^{-1} + KC)^{-1}BG\hat{x} \tag{8.77}$$

But, by the Doyle-Stein condition,

$$(\Phi^{-1} + KC)^{-1}B = 0 \tag{8.78}$$

This means that the transfer function from y to the estimated state $\hat{x}$ does not entail the feedback of the control signal u. The path from u to z may be omitted. Thus, if K can be selected to satisfy the Doyle-Stein condition (8.74), the

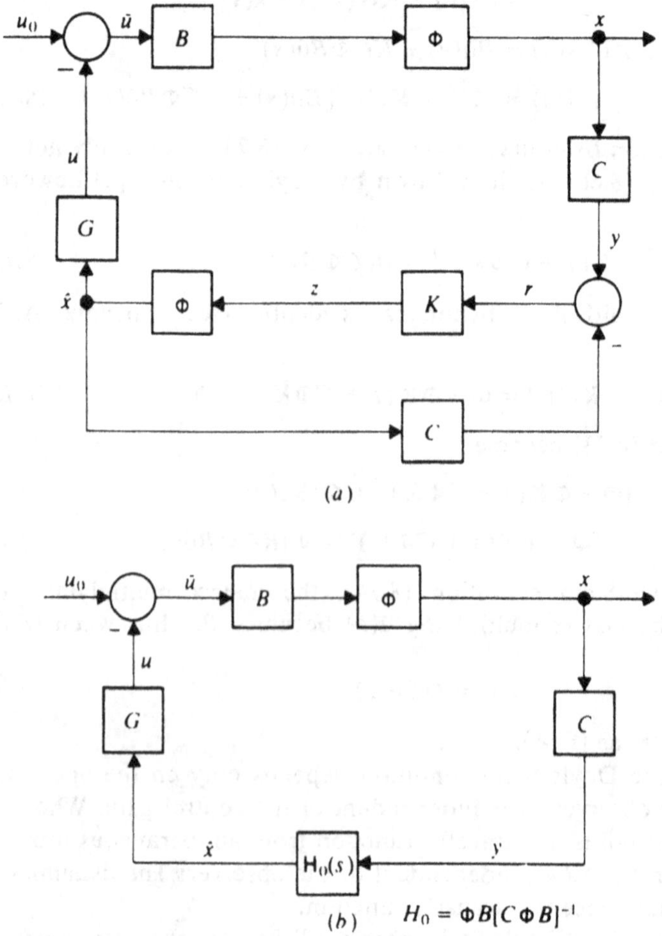

Figure 8.15 Representations of control systems with Doyle-Stein observers. (a) Feedback path from u to z may be omitted in Doyle-Stein observer; (b) Alternate representation of closed-loop system with Doyle-Stein observer.

closed-loop system of Fig. 8.14(b) can be represented by Fig. 8.15(a). Since there is no feedback from the control u to the observer through the control distribution matrix B, the observer transfer function

$$\mathsf{H}_0(s) = (\Phi^{-1} + KC)^{-1}K = (sI - A + KC)^{-1}K \tag{8.79}$$

is the same as it would be for the unforced system $\dot{x} = Ax$ with observation $y = Cx$. When a general observer is used and the control matrix B changes by an amount δB, it would be necessary to change the control distribution matrix B in the observer in order to preserve the separation property. Otherwise the closed-loop poles would change in the manner discussed in Sec. 8.4. When a Doyle-Stein observer is used, however, the observer poles remain unchanged, and only the poles due to the change of B in the full-state feedback control law change. The invariance of those closed-loop poles of the overall system that are attributed to the observer imparts a considerable measure of robustness to the closed-loop system. This has motivated Rynaski[3] to call observers that satisfy the Doyle-Stein condition "robust observers."

The Doyle-Stein condition (8.74) has another interesting interpretation. Note that the left-hand side of (8.74) can be written

$$K(I + C\Phi K)^{-1} = K[I + C(sI - A)^{-1}K]^{-1} = (sI - A)(sI - A + KC)^{-1}K \tag{8.80}$$

The first expression in (8.80) is obtained by finding the transfer function from y to $z = Kr$ (with $r = y - Cx$) in Fig. 8.15(a). The right-hand side of (8.80) is obtained by noting that

$$\mathsf{z} = \Phi^{-1}\hat{\mathsf{x}} = (sI - A)\mathsf{H}_0(s)\mathsf{y} \tag{8.81}$$

with $\mathsf{H}_0(s)$ given by (8.79). Thus, by (8.80), the Doyle-Stein condition (8.74) can be written

$$(sI - A)\mathsf{H}_0(s) = B[C(sI - A)^{-1}B]^{-1}$$

Thus the transfer function of a Doyle-Stein observer is

$$\mathsf{H}_0(s) = (sI - A)^{-1}B[C(sI - A)^{-1}B]^{-1} \tag{8.82}$$

and the closed-loop system can be depicted as shown in Fig. 8.15(b). It is readily established, using Fig. 8.15(b), that the transfer function from u_0 to x is

$$\mathsf{H}_c(s) = \Phi B(I + G\Phi B)^{-1} \tag{8.83}$$

which is the transfer function of the closed-loop system when full-state feedback is used. From Fig. 8.14(a)

$$\mathsf{x} = \Phi B\bar{\mathsf{u}}$$

and

$$\bar{\mathsf{u}} = \mathsf{u}_0 - G\Phi B\bar{\mathsf{u}}$$

or

$$\bar{\mathsf{u}} = (I + G\Phi B)^{-1}\mathsf{u}_0$$

Thus

$$\mathsf{x} = \Phi B(I + G\Phi B)^{-1}\mathsf{u}_0$$

which implies that the transfer function from u_0 to x is given by (8.83).

The foregoing analysis shows that, with regard to the transfer function from u_0 to x, the Doyle-Stein observer acts as if it were the inverse of the observation matrix C. (If C *is* a nonsingular matrix, then there is of course no need for an observer in the first place.)

The transfer function in the presence of the closed-loop process, with a Doyle-Stein observer in place, is the same as it would be for full-state feedback. This means that the transfer function has the same order as the open-loop system. But the true order of the closed-loop system is equal to the order k of the open-loop plant plus the order of the observer (k, for a full-order observer). Whenever a transfer function between some input and some output of a system is of lower order than the true dynamic order of the system, either the condition of unobservability or of uncontrollability obtains. In this case it is unobservability: The observer states are not observable in the plant output.

From (8.82) it is seen that

$$C\mathsf{H}_0(s) = I \tag{8.84}$$

Thus a Doyle-Stein observer is a "right inverse" of C. There are many matrices that satisfy (8.84), but the transfer function from u_0 to x does not entail $C\mathsf{H}_0$ but rather $\mathsf{H}_0 C$ and, unless H_0 satisfies other requirements, e.g., the Doyle-Stein conditions, the transfer function from u_0 to x will not be given by (8.83).

In order for a Doyle-Stein observer (8.81) to exist it is necessary that the open-loop system be "square," i.e., that there are exactly as many outputs as inputs. Otherwise the open-loop matrix $C\Phi B = C(sI - A)^{-1}B$ would not be a square matrix and its inverse, needed in (8.82), would not be defined. While many systems are square, many more are not. For the latter, the Doyle-Stein conditions cannot be satisfied. (But see Note 8.4.)

Even for square systems, it may not be possible to achieve a stable Doyle-Stein observer. Note that the resolvent $\Phi(s)$ can be written

$$\Phi(s) = (sI - A)^{-1} = \frac{\operatorname{adj}(sI - A)}{|sI - A|}$$

where $\operatorname{adj}(sI - A)$ is the adjoint matrix of $sI - A$. (Recall Chap. 4.) Thus

$$[C(sI - A)^{-1}B]^{-1} = [C \operatorname{adj}(sI - A)B]^{-1}|sI - A|$$

and, by (8.82) the transfer function of the Doyle-Stein observer is

$$\mathsf{H}_0(s) = \operatorname{adj}(sI - A)B\frac{\operatorname{adj}[C \operatorname{adj}(sI - A)B]}{|C \operatorname{adj}(sI - A)B|} \tag{8.85}$$

The denominator of $\mathsf{H}_0(s)$ is thus the determinant of the numerator of the transfer matrix of the open-loop plant. As we noted in our discussion of multivariable systems in Chap. 4, the zeros of the determinant are the so-called "transmission zeros" of the plant. This means that the poles of the Doyle-Stein observer are transmission zeros of the plant. Consequently, if the open-loop plant has one or more transmission zeros in the right half of the s plane (i.e., nonminimum phase zeros) then a *stable* Doyle-Stein observer does not exist. An unstable Doyle-Stein observer can (perhaps) be realized, but its use in a closed-loop control system would be disastrous, because the unstable poles are unobservable, being cancelled by the transmission zeros of the open-loop plant. Any noise in the observer, or parameter mismatch, would initiate an unstable oscillation that could destroy the system.

Although it may not be possible to realize an observer having all the properties of a Doyle-Stein observer, it may be possible to design an observer that has some of its properties:

Makes the closed-loop transfer function from u_0 to x the same as it is for full-state feedback.
Has its poles at the transmission zeros of the open-loop plant.
Does not require feedback of the control signal and thus has a constant transfer function independent of the control gain.

An observer having some, if not all, of these properties might be called a robust observer, in a more restricted sense than a Doyle-Stein observer. Rynaski[3] considered not only full-order observers which we have been addressing up to now, but also reduced-order observers. In the latter it may be possible to select the observer "gain matrix" L to eliminate feedback of the control signal even when the open-loop system is not "square." There still may be enough freedom left with the L matrix to place the observer poles at locations desired, including at some of the transmission zeros of the open-loop transfer functions. In particular, if the state x can be partitioned into two substates, one of which can be observed directly in the output and the other cannot, i.e.,

$$y = C_1 x_1$$

with C_1 being a nonsingular matrix, the reduced-order observer is given by (8.18)-(8.21). To eliminate feedback of the control matrix, H in (8.20) must be zero. By (8.21) this requires that L be selected in order to satisfy

$$B_2 - LC_1B_1 = 0 \tag{8.86}$$

The number of free parameters in the gain matrix L and the dimensions of B_1 and B_2 determine whether (8.86) can be satisfied.

Example 8C Aircraft stability augmentation In Example 4C we considered the open-loop dynamics of a typical aerodynamically unstable aircraft. In particular, using data for the AFTI-16 aircraft, we found that the open-loop poles are located at

$$\begin{aligned} s &= 0.724 \\ s &= -1.705 \\ s &= -0.039 \pm j0.200 \end{aligned} \tag{8C.1}$$

Because of the pole at $s = 0.724$ in the right half-plane, it is mandatory for the aircraft to have a stability augmentation system (SAS): it cannot safely be flown without one. (And since safety is a prime consideration, the control system hardware would most likely be designed to use redundant sensors and feedback loops.)

The objective of this example is to design a suitable stability augmentation compensator and to assess its performance.

Based on the data given in Example 4C, the matrices of the standard dynamic description are

$$A = \begin{bmatrix} -0.0507 & -3.861 & 0 & -32.17 \\ -0.00117 & -0.5164 & 1 & 0 \\ -0.000129 & 1.4168 & -0.4932 & 0 \\ 0 & 0 & 1 & 0 \end{bmatrix} \qquad B = \begin{bmatrix} 0 \\ -0.0717 \\ -1.645 \\ 0 \end{bmatrix} \tag{8C.2}$$

The sensor that we will consider for use in the stability augmentation system is a rate gyro which measures the pitch rate $q = x_3$. Thus

$$y = x_3$$

and the observation matrix is

$$C = c' = [0 \quad 0 \quad 1 \quad 0] \tag{8C.3}$$

A. Full state feedback control law According to Rynaski, who has studied the dynamics of this aircraft in considerable detail,[3] suitable pole locations for the closed-loop system are as given in Table 8C.1.

This results in a closed loop characteristic equation

$$[s^2 + 2(0.5)(2.5)s + (2.5)^2][s^2 + 2(0.1)(0.1) + (0.1)^2]$$
$$= s^4 + 2.52s^3 + 6.31s^2 + 0.150s + 0.0625 = 0 \tag{8C.4}$$

The open-loop characteristic equation, as found in Example 4C is

$$s^4 + 1.0603s^3 - 1.115s^2 - 0.0565s - 0.0512 = 0 \tag{8C.5}$$

Thus the vectors of coefficients needed for the control system design are:

$$\hat{a} = \begin{bmatrix} 2.52 \\ 6.31 \\ 0.150 \\ 0.0625 \end{bmatrix} \qquad a = \begin{bmatrix} 1.0603 \\ -1.115 \\ -0.0565 \\ -0.0512 \end{bmatrix}$$

Numerical solution for the gain matrix, using the Bass-Gura formula (6.35), yields

$$G = g' = [-0.00429, -3.872, -0.7186, -0.09875] \tag{8C.6}$$

Note that all the terms in the gain matrix G are negative. This happens in the present application because the elements of the control matrix B are negative. (See Example 4C.)

Table 8C.1 Desirable closed-loop poles

Mode	Frequency	Damping
Short period	$\omega_{sp} = 2.5$ rad/s	$\zeta = 0.5$
Phugoid	$\omega_{ph} = 0.1$ rad/s	$\zeta = 0.1$

B. Robust observer design It is readily established that the system is observable using the rate gyro, i.e., with the observation matrix given by (8C.3), so the observer poles can be placed wherever desired. Following Rynaski's robust observer approach, the observer poles are to coincide with the plant zeros that are in the left half-plane. The plant transfer function from the input (control surface deflection) to the rate-gyro output is

$$\mathsf{H}(s) = C(sI - A)^{-1}B \tag{8C.7}$$

Since C is a row vector and B is a column vector the plant transfer function is a 1×1 matrix, the numerator of which is found to be

$$\begin{aligned}\mathsf{N}(s) &= -1.645s^3 - 1.0345s^2 - 0.040\,75s \\ &= -1.645s(s^2 + 0.6289s + 0.024\,77) \\ &= -1.645s(s + 0.0421)(s + 0.587)\end{aligned}$$

Thus the open-loop plant, with the pitch rate defined as the output, has its zeros at $s = 0$, $s = -0.0421$, and $s = -0.587$. Since none of the open-loop zeros are in the right half-plane, we can use these as poles of the observer; the zero at the origin will result in the observer having a pole at the origin. Since the full-order observer is 4th-order, one more pole remains to be placed. An appropriate location of this pole would be at $s = -1$. Thus the observer characteristic polynomial is

$$s(s^2 + 0.6289s + 0.024\,77)(s + 1) = s^4 + 1.6289s^3 + 0.6534s^2 + 0.024\,77s$$

The corresponding coefficient vector for the observer design is thus

$$\hat{a} = \begin{bmatrix} 1.6289 \\ 0.6534 \\ 0.02477 \\ 0 \end{bmatrix} \tag{8C.8}$$

The Bass-Gura formula (7.15) for the observer yields the observer gain matrix:

$$G = g = \begin{bmatrix} -0.1728 \\ 1.02 \\ 0.5686 \\ 1.0 \end{bmatrix} \tag{8C.9}$$

C. Compensator and closed loop The transfer function of the compensator is found to be

$$\begin{aligned}\mathsf{D}(s) &= G(sI - A + BG + KC)^{-1}K = G(sI - \hat{A}_c)^{-1}K \\ &= -\frac{4.46s^3 + 5.63s^2 + 0.245s + 0.115}{s(s^3 + 3.09s^2 + 1.572s + 0.0609)} \\ &= -\frac{4.46(s + 1.234)[(s + 0.0138)^2 + (0.1436)^2]}{s(s + 0.0421)(s + 0.587)(s + 2.459)}\end{aligned} \tag{8C.10}$$

Note that two of the compensator poles (those at $s = -0.0421$ and at $s = -0.587$) are precisely at the same locations as selected for the observer, as is to be expected in view of our discussion in Sec. 8.5.

The compensator is seen to have three zeros on the negative real axis and a pair of zeros rather close to the imaginary axis. The presence of these zeros produces an antiresonance (notch) in the compensator frequency response characteristic, which is shown in Fig. 8.16.

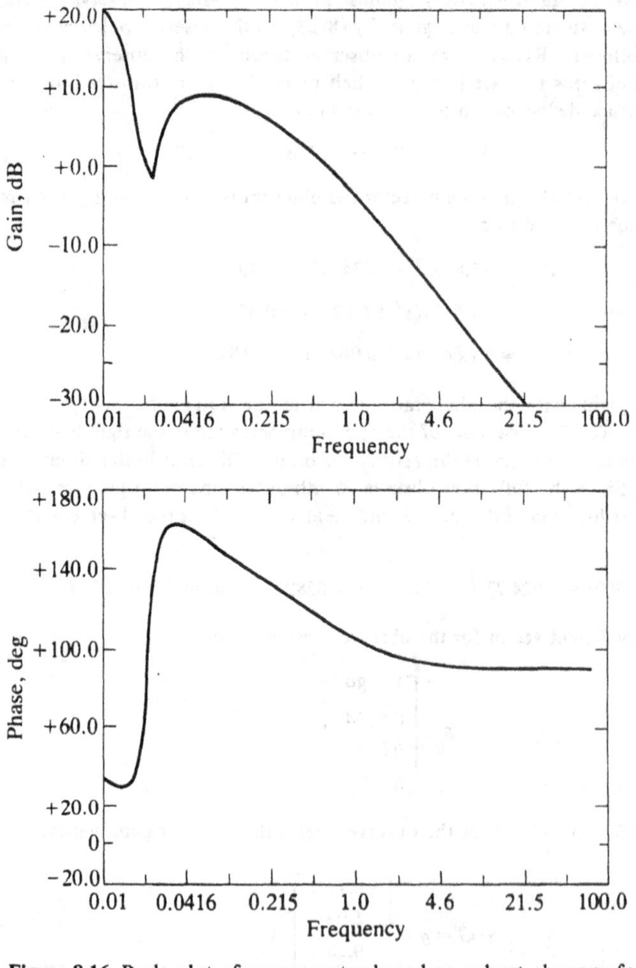

Figure 8.16 Bode plot of compensator based on robust observer for aerodynamically unstable aircraft.

To assess the performance of the closed-loop system as the gain is varied, we note that the return difference is

$$1 + KD(s)G(s)$$

$$= 1 + K\frac{-4.46(s + 1.234)[(s + 0.0138)^2]}{s(s + 0.0421)(s + 0.587)(s + 2.459)}\frac{-1.645s(s + 0.0421)(s + 0.587)}{(s - 0.724)(s + 1.705)[(s + 0.138)^2 + (0.151)^2]}$$

$$= 1 + \bar{K}\frac{(s + 1.234)[(s + 0.0138)^2 + (0.1436)^2]}{(s + 2.459)(s + 1.705)(s - 0.724)[(s + 0.0138)^2 + (0.151)^2]}$$

where

$$\bar{K} = (4.46)(1.645)K = 7.34K$$

The short-period dynamics are determined by the locations of the poles and zeros on the real axis; the long-period dynamics of the pair by poles and the pair of zeros near the imaginary axis. The latter set of poles and zeros are relatively close and will result in root locus

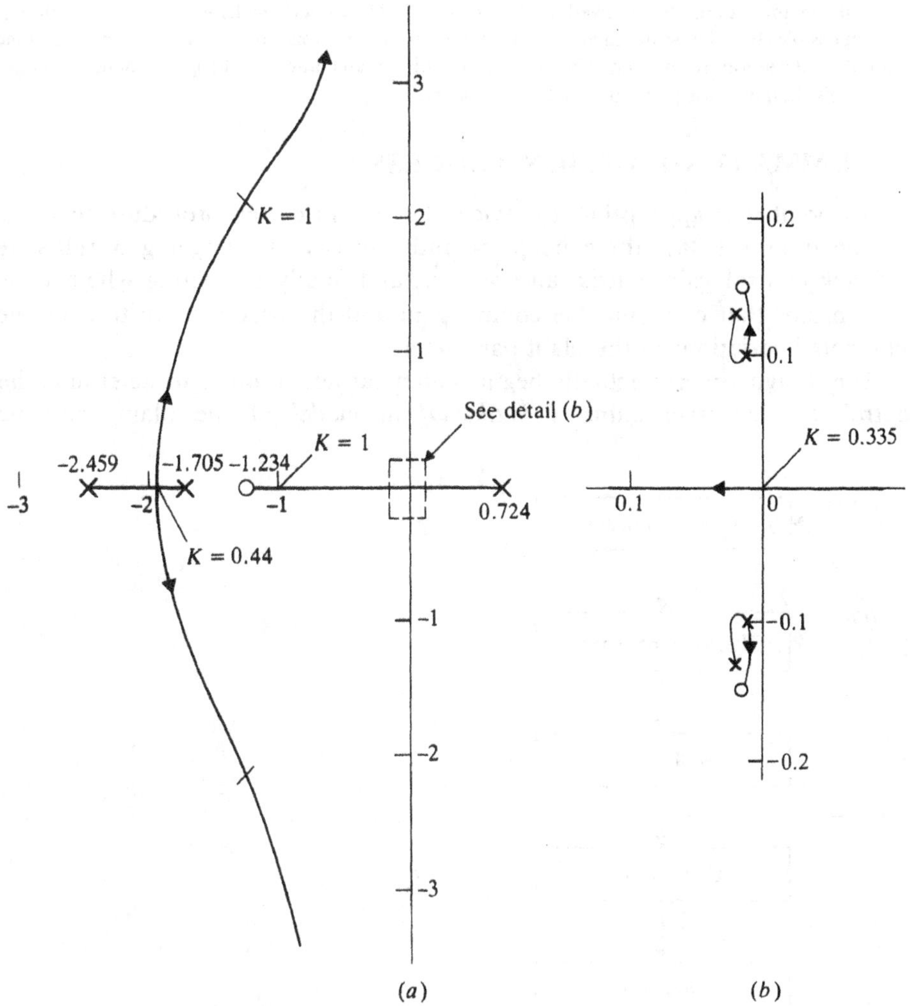

Figure 8.17 Root-locus plots for stability-augmented aerodynamically unstable aircraft. (*a*) Short-period poles; (*b*) Phugoid poles.

branches which pass through the desired phugoid-mode pole defined in Table 8C.1 when the gain loop K is unity. (See Fig. 8.17.) The root-locus plot has two more branches corresponding to the short-period dynamics. One branch goes from the right half-plane pole at $s = 0.724$ to the zero at $s = -1.234$, crossing the imaginary axis for a loop-gain $K = 0.335$. The other branch moves along the real axis from the poles at $s = -2.459$ and $s = -1.705$ to the breakaway point at $s = -1.92$ (at a gain $K = 0.44$) and thence upward, through the desired short-period pole at $s = -1.25(1 \pm j\sqrt{3})$ and then asymptotic to the imaginary axis. The phugoid mode variation with gain is shown in the detail of Fig. 8.17.

From this root locus we see that the gain margin is infinite: the gain can be raised arbitrarily high without compromising stability. Since the aircraft is aerodynamically unstable, the gain cannot of course be reduced to zero. In fact, it cannot be reduced below 0.335. Thus a nominal gain of unity gives a gain reduction margin of 2.98. One of the possible physical causes for a reduction in loop gain would be a decrease in the gyro scale factor. Thus the gyro scale factor cannot be allowed to fall below 0.335. In any realistic system it is all but impossible for the scale factor of a rate gyro to fall this low without triggering some failure-detection mechanism. Thus it is reasonable to conclude that the gain-reduction margin of this design is adequate for practical purposes.

8.7 SUMMARY OF DESIGN PROCESS

At this point it is appropriate to review the system design procedure that was developed in the last three chapters and consists of designing a full-state feedback control gain matrix, an observer, and finally evaluating whether the compensator that combines the control gain and the observer can tolerate the anticipated variations in the plant parameters.

The design process actually begins much earlier. In order to determine the control and observer gains a *mathematical model* of the plant must be

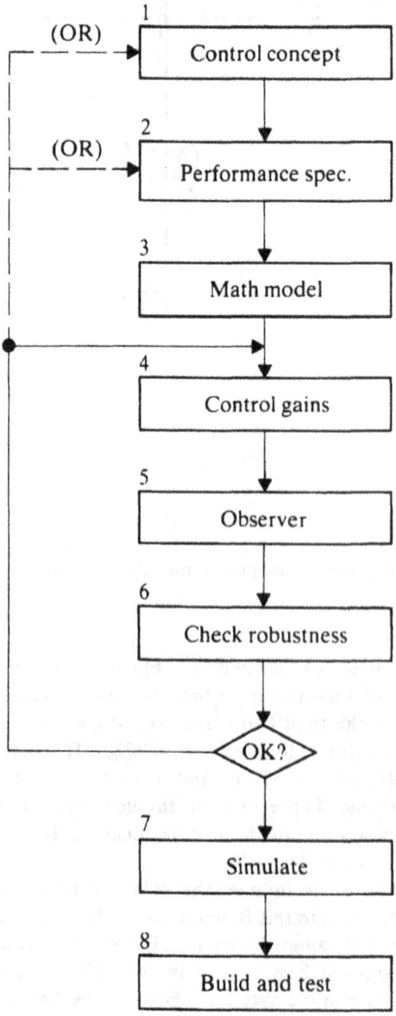

Figure 8.18 Sequence of steps in control system design.

developed. And even before that is done, an overall system concept must be developed in which it is decided by what means the control is to be accomplished and what variables are to be measured. About this time it is also necessary to determine performance specifications: characteristics of the desired system behavior and the quality of components available that might be able to achieve the desired performance.

The design steps, moreover, do not end with the specification of the compensator. In almost every case the process is nonlinear; there are usually limits on the control signals; the process may be of higher order than that used in the model; if the control system is implemented by a digital computer there are effects of sampling and amplitude quantization to be studied. And so forth. The only way that these issues can be resolved is by a very thorough simulation of all the effects that the system designer thinks might conceivably influence system behavior. If the results of a comprehensive simulation are favorable, the system would finally be fabricated and tested. If the steps leading to this last step were done skillfully, and if fortune smiles, the system will work as expected.

The design process described above is illustrated by the flow chart of Fig. 8.18. The emphasis of the last three chapters has been on steps 4, 5, and 6. These are the easy steps. The hard steps—the tedious steps—are those that come before and after.

Example 8D Compensator for missile autopilot The missile autopilot design given in Example 6F assumes measurement of the pitch rate q, the acceleration error $e = a_{NC} - a_N$, and the commanded acceleration a_{NC}. Since these quantities can all be measured, the autopilot can be implemented directly as shown in Fig. 6.9. But the feedforward gain G_D to eliminate the steady state error must be determined accurately. Any difference between the true aerodynamic parameters and those used in the computation of G_D will result in the presence of a steady state error. Since the missile and its autopilot are only the inner loop of an overall guidance system, a small steady state error may well be acceptable in practice. Nevertheless, the "performance spec" for the autopilot may require negligible steady state errors. A possible way of achieving this is by using an observer to estimate the commanded acceleration and thus to eliminate the feedforward signal. It is noted that one of the measured quantities, pitch rate, does not go to zero when the steady state error in acceleration goes to zero. Thus we should not expect the compensator to necessarily produce pure integral control. We would certainly not want a control signal proportional to the integral of pitch rate to be present in the output of the compensator. Although sensors to measure the control surface deflection are readily available, the deployment of such a sensor adds an additional item of cost which the designer may desire to avoid. In Example 6F, we went to great lengths to design a full-state feedback control law that avoids feeding back the control surface deflection. We would not want to reintroduce this measurement in the design of the observer. Hence the observer, and ultimately the compensator, should be designed to use at most two measurable variables, namely, the acceleration error $e = a_{NC} - a_N$ and the pitch rate q.

Thus we consider the design of a reduced-order observer using the measurement vector

$$y = \begin{bmatrix} e \\ q \end{bmatrix} = [I \mid 0] \begin{bmatrix} e \\ q \\ \hline \delta \\ a_{NC} \end{bmatrix}$$

where

$$\mathbf{x} = [e, q, \delta, a_N]'$$

is the metastate of the process. The matrices that define the (metastate) dynamics are:

$$\mathbf{A} = \left[\begin{array}{cc|cc} Z_\alpha/V & -Z_\alpha & Z_\delta/\tau & -Z_\alpha/V \\ -M_\alpha/Z_\alpha & M_q & \bar{M}_\delta & M_\alpha/Z_\alpha \\ \hline 0 & 0 & -1/\tau & 0 \\ 0 & 0 & 0 & 0 \end{array}\right] = \left[\begin{array}{c|c} A_{11} & A_{12} \\ \hline A_{21} & A_{22} \end{array}\right]$$

$$\mathbf{B} = \left[\begin{array}{c} -Z_\delta/\tau \\ 0 \\ \hline 1/\tau \\ 0 \end{array}\right] = \left[\begin{array}{c} B_1 \\ \hline B_2 \end{array}\right] \tag{8D.1}$$

$$\mathbf{C} = \left[\begin{array}{cc|cc} 1 & 0 & 0 & 0 \\ 0 & 1 & 0 & 0 \end{array}\right] = [C_1 \mid 0] \qquad (C_1 = I)$$

The matrices in (8D.1) are shown partitioned into submatrices corresponding to the two measured states e and q and the two states that need to be estimated δ and a_{NC}. (Although the control law has been designed not to require knowledge of δ, its estimate is needed to provide the proper observer for estimating a_{NC}. If we had initially decided upon use of an observer, there would have been no benefit in choosing a control law in which δ is not used.)

In accordance with the theory developed in Sec. 7.5, the reduced-order observer for this application is given by

$$\hat{x}_1 = \begin{bmatrix} \hat{e} \\ \hat{q} \end{bmatrix} = \begin{bmatrix} e \\ q \end{bmatrix} = y \tag{8D.2}$$

and

$$\hat{x}_2 = \begin{bmatrix} \hat{\delta} \\ \hat{a}_{NC} \end{bmatrix} = Ly + z \tag{8D.3}$$

where

$$\dot{z} = F\hat{x}_2 + (A_{21} - LA_{11})y + (B_2 - LB_1)u \tag{8D.4}$$

with

$$F = A_{22} - LA_{12} \tag{8D.5}$$

The matrix C_1 used in Sec. 7.5 is the identity matrix here, and the submatrices $A_{11}, \ldots, A_{22}$, B_1, and B_2 are the matrices defined by (8D.1).

Selecting the observer gain matrix The only matrix that is not specified is the observer "gain" matrix L which in this case is a 2×2 matrix. The four elements in L are the parameters in the observer that are available for adjustment. Since the observer dynamics matrix is also 2×2, only two parameters are needed to adjust the observer poles. Two other things can be accomplished by proper adjustment of the remaining parameters. We might also wish to adjust the poles of the compensator, the dynamics of which are obtained by using the control law

$$u = -G_1 y - G_2 \hat{x}_2$$

where

$$G_1 = [g_e, g_q] \qquad G_2 = [0, G_D] \tag{8D.6}$$

into (8D.4). The result is

$$\dot{z} = [(A_{22} - B_2G_2) - L(A_{12} - B_1G_2)]\hat{x}_2 + [(A_{21} - B_2G_1) - L(A_{11} - B_1G_1)]y \tag{8D.7}$$

Using the numerical data from Example 6F we find:

$$A_{11} = \begin{bmatrix} -3.328 & 4170. \\ -0.05947 & 0. \end{bmatrix} \qquad A_{12} = \begin{bmatrix} -111\,500. & 3.328 \\ -595.68 & 0.05947 \end{bmatrix} \qquad A_{22} = \begin{bmatrix} -100. & 0. \\ 0. & 0. \end{bmatrix}$$

and

$$A_{11} - B_1G_1 = \begin{bmatrix} 3.770 & 8551. \\ -0.05947 & 0. \end{bmatrix} \qquad A_{12} - B_1G_2 = \begin{bmatrix} -111\,500. & 17.97 \\ -595.68 & 0.05947 \end{bmatrix}$$

$$A_{21} - B_2G_1 = \begin{bmatrix} 0.006366 & 3.929 \\ 0. & 0. \end{bmatrix} \qquad A_{22} - B_2G_2 = \begin{bmatrix} -100. & 0.001313 \\ 0. & 0. \end{bmatrix}$$

A diagonal matrix

$$L = \begin{bmatrix} l_1 & 0 \\ 0 & l_2 \end{bmatrix}$$

is one possible choice for the observer gain matrix. In this case the observer dynamics matrix is

$$F = A_{22} - LA_{12} = \begin{bmatrix} -100 + 111\,500 l_1 & -3.328 l_1 \\ 595.68 l_2 & -0.05947 l_2 \end{bmatrix} \tag{8D.8}$$

and the compensator dynamics matrix is

$$F_c = A_{22} - B_2G_2 - L(A_{12} - B_1G_2) = \begin{bmatrix} -100 + 111\,500 l_1 & 0.001313 - 17.97 l_1 \\ 595.68 l_2 & -0.05947 l_2 \end{bmatrix} \tag{8D.9}$$

True integral control would be desirable if it could be achieved using the structure selected. We check to see whether this is possible. To achieve integral control, the compensator must have a pole at the origin. This means that $|F_c|$ must be zero. By (8D.9)

$$\begin{aligned} |F_c| &= l_2[-0.05947(-100 + 111\,500 l_1) - 595.68(0.001313 - 17.97 l_1)] \\ &= l_2(-1.874 + 4073.5 l_1) \end{aligned}$$

In order to have $|F_c| = 0$ either l_2 or the other factor must be zero.

If we set $l_2 = 0$ this would also make the observer determinant $|F| = 0$. But this is undesirable because the observer (and hence the closed-loop system) would then have a pole at the origin and not be asymptotically stable. Thus our only feasible choice is

$$-1.874 + 4073.5 l_1$$

or

$$l_1 = 4.600 \times 10^{-4}$$

With this choice of l_1 we obtain

$$|F| = \begin{vmatrix} -48.70 & -1.531 \times 10^{-3} \\ 595.68 l_2 & -0.059\,47 l_2 \end{vmatrix} = 3.808 l_2$$

Thus, by setting l_2 to a positive number we can achieve a stable observer and a compensator having a pole at the origin. The characteristic equation of the observer is

$$|sI - F| = s^2 + (48.7 + 0.05947 l_2)s + 3.808 l_2 = 0 \tag{8D.10}$$

The compensator transfer function can be obtained with the aid of (8D.7) and (8D.3):

$$s(\mathrm{x}_2 - L\mathrm{y}) = F_c\mathrm{x}_2 + [(A_{21} - B_1G_2) - L(A_{11} - B_1G_1)]\mathrm{y}$$

or

$$\mathrm{x}_2 = (sI - F_c)^{-1}[sL + (A_{21} - B_1G_2) - L(A_{11} - B_1G_1)]\mathrm{y}$$

and

$$\mathrm{u}(s) = -G_1\mathrm{y} - G_2\mathrm{x}_2 = -\mathrm{D}(s)\mathrm{y} = -\mathrm{D}(s)\begin{bmatrix} \mathrm{e}(s) \\ \mathrm{q}(s) \end{bmatrix}$$

where

$$\mathsf{D}(s) = [\mathsf{D}_E(s), \mathsf{D}_Q(s)] = G_1 + G_2(sI - F_c)^{-1}[sL + (A_{21} - B_1G_2) - L(A_{11} - B_1G_1)] \quad (8D.11)$$

is the compensator transfer function. From (8D.9)

$$F_c = \begin{bmatrix} -48.70 & -4.864 \times 10^{-3} \\ -595.68l_2 & -0.05947l_2 \end{bmatrix}$$

Thus

$$(sI - F_c)^{-1} = \frac{1}{s^2 + (48.7 + 0.05947l_2)s} \begin{bmatrix} s + 0.05947l_2 & 4.864 \times 10^{-3} \\ 595.68l_2 & s + 48.70 \end{bmatrix}$$

Also

$$sL + A_{21} - B_1G_2 - L(A_{11} - B_1G_1) = \begin{bmatrix} 4.600 \times 10^{-4}s + 4.63 \times 10^{-3} & 0 \\ 0.05947l_2 & sl_2 \end{bmatrix}$$

Thus, using (8D.11), the transfer functions of the compensator are

$$\begin{aligned} \mathsf{D}_E(s) &= -0.6366 \times 10^{-4} - 1.313 \times 10^{-4}l_2 \frac{0.3334s + 5.654}{s(s + 48.7 + 0.05947l_2)} \\ \mathsf{D}_Q(s) &= -0.3929 \times 10^{-1} - 1.313 \times 10^{-4}l_2 \frac{1}{s + 48.7 + 0.05947l_2} \end{aligned} \quad (8D.12)$$

The transfer function $\mathsf{D}_E(s)$ is the transfer function between the measured error $e = a_{NC} - a_N$ and the control input to the actuator. Note that the denominator contains the factor s and hence true integral control is provided, as we intended. Note, however, that the transfer function $\mathsf{D}_Q(s)$ from the pitch rate q to the control surface deflection *does not* have a pole at the origin. This is as expected, since the steady state pitch rate does not go to zero.

The compensator parameter l_2 may be selected in any manner desired, provided it is positive. One possible method of choosing l_2 would be to enhance the closed-loop stability margins. For this purpose, the closed-loop block diagram of the system is drawn as shown in Fig. 8.19. The output δ of the actuator is shown driving two outputs: the normal acceleration a_N and the pitch rate q. The transfer functions $\mathsf{G}_A(s)$ and $\mathsf{G}_Q(s)$ can readily be calculated (see

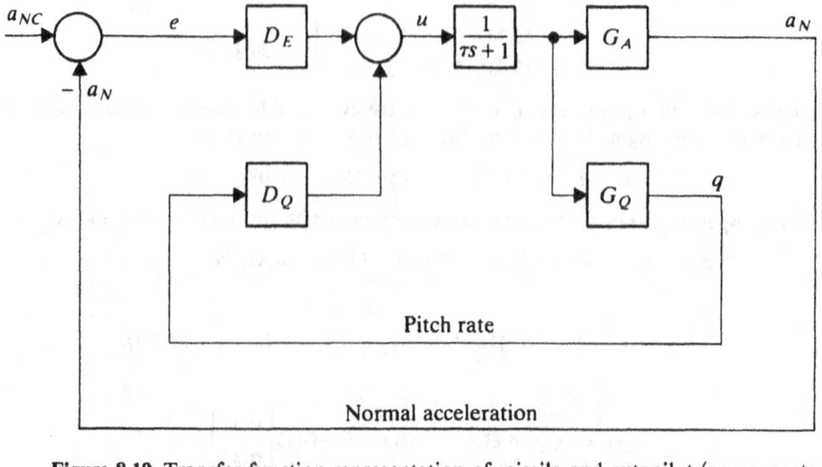

Figure 8.19 Transfer function representation of missile and autopilot (compensator).

Example 3F):

$$\mathsf{G}_A(s) = \frac{Z_\delta s^2 + Z_\alpha M_\delta - M_\alpha Z_\delta}{s^2 + \dfrac{Z_\alpha}{V}s - M_\alpha} = \frac{-1115s^2 + 2.485 \times 10^6}{s^2 + 3.328s + 248}$$

$$\mathsf{G}_Q(s) = \frac{M_\delta s - \dfrac{Z_\alpha M_\delta - M_\alpha Z_\delta}{V}}{s^2 + \dfrac{Z_\alpha}{V}s - M_\alpha} = \frac{-662s - 1983}{s^2 + 3.328s + 248} \tag{8D.13}$$

By examining Fig. 8.19 we see that the return difference of the system is

$$1 + \frac{1}{\tau s + 1}[\mathsf{G}_A(s)\mathsf{D}_E(s) + \mathsf{G}_Q(s)\mathsf{D}_Q(s)] \tag{8D.14}$$

The loop transmission is $[\mathsf{G}_A(s)\mathsf{D}_E(s) + \mathsf{G}_Q(s)\mathsf{D}_Q(s)]/(\tau s + 1)$. The closed-loop poles are located at the zeros of $(s^2 + 3.328s + 248)$, at $s = -100$, and at $s = -(48.7 + 0.05947l_2)$. The loop zeros are located at the zeros of $\mathsf{G}_A(s)\mathsf{D}_E(s) + \mathsf{G}_Q(s)\mathsf{D}_Q(s)$ which are functions of l_2 that can be obtained from (8D.12) and (8D.13). It would be appropriate to select l_2 so that these zeros give a favorable root locus.

Robust observer Another way of selecting the observer gain matrix L is to satisfy some of the requirements of a robust observer. Since there are two observables (e and q) and only one control input u, the system is not "square" so the Doyle-Stein condition (8.73) cannot be met. But in this case it is possible, with the reduced-order observer, to eliminate feedback of the control u to the compensator dynamics, since (8.86) can be satisfied. In particular, since C_1 is the 2×2 identity matrix, (8.86) becomes, in this application,

$$B_2 - LB_1 = \begin{bmatrix} 1/\tau \\ 0 \end{bmatrix} - \begin{bmatrix} l_{11} & l_{12} \\ l_{21} & l_{22} \end{bmatrix} \begin{bmatrix} -Z_\delta/\tau \\ 0 \end{bmatrix}$$

which is satisfied when

$$\frac{1}{\tau} - l_{11}\frac{Z_\delta}{\tau} = 0$$

$$-l_{21}\frac{Z_\delta}{\tau} = 0$$

which requires that

$$l_{11} = \frac{1}{Z_\delta}$$

$$l_{21} = 0$$

whence

$$L = \begin{bmatrix} 1/Z_\delta & l_{12} \\ 0 & l_{22} \end{bmatrix} \tag{8D.15}$$

The second column of the observer gain matrix L can be chosen to satisfy any other requirements.

Since $B_2 - LB_1$ has been made zero by this choice of gain matrix L the dynamics matrix of the observer F and of the compensator F_c are identical. Thus we cannot place a pole of the compensator at the origin without also having an observer pole at the origin. Since the observer poles are among the closed-loop poles, it would not do to have an observer pole at the origin. Thus, in the interest of asymptotic stability, we must give up the desire that the observer provide true integral control.

The observer dynamics matrix (which in this case is also the compensator dynamics matrix) is

$$F = A_{22} - LA_{12} = \begin{bmatrix} -1/\tau & 0 \\ 0 & 0 \end{bmatrix} - \begin{bmatrix} 1/Z_\delta & l_{12} \\ 0 & l_{22} \end{bmatrix} \begin{bmatrix} Z_\delta/\tau & -Z_\alpha/V \\ \bar{M}_\delta & M_\alpha/Z_\alpha \end{bmatrix}$$

$$= \begin{bmatrix} -l_{12}\bar{M}_\delta & \dfrac{1}{Z_\delta}\dfrac{Z_\alpha}{V} - l_{12}\dfrac{M_\alpha}{Z_\alpha} \\ -l_{22}\bar{M}_\delta & -l_{22}\dfrac{M_\alpha}{Z_\alpha} \end{bmatrix}$$

The characteristic equation corresponding to F is

$$|sI - F| = s^2 + \left(l_{12}\bar{M}_\delta + l_{22}\frac{M_\alpha}{Z_\alpha} \right) s + l_{22}\frac{\bar{M}_\delta}{Z_\delta}\frac{Z_\alpha}{V} = 0$$

$$= s^2 + (-595.68 l_{12} + 0.05947 l_{22})s - 1.778 l_{22} \tag{8D.16}$$

In keeping with the spirit of the robust observer concept of Rynaski, we can select the poles of the observer to occur at the zeros of the plant. Since the plant is not square, the zeros of $|C(sI - A)B|$ are not defined. But $G_A(s)$ and $G_Q(s)$ each have one zero in the left half-plane and these can be chosen as the poles of the compensator. The zeros of $G_A(s)$ occur at

$$s^2 - \frac{2.485 \times 10^6}{1115} = 0$$

i.e., at

$$s = \pm 47.21$$

and the zero of $G_Q(s)$ occurs at

$$662s + 1983 = 0$$

i.e., at

$$s = -2.995$$

Thus, as the observer characteristic equation we use

$$(s + 47.21)(s + 2.995) = s^2 + 50.21s + 141.39 = 0 \tag{8D.17}$$

Equating the corresponding coefficients of s in (8D.16) and (8D.17) gives

$$l_{12} = -0.09229$$
$$l_{22} = -79.52 \tag{8D.18}$$

The dynamics of the observer are given by

$$\dot{z} = F\hat{x}_2 - LA_{11}y$$
$$\hat{x}_2 = Ly + z$$

PROBLEMS

Problem 8.1 Closed-loop system

Show that the resolvent of the closed-loop system can be expressed by

$$\mathbf{\Phi}_c(s) = (sI - A + BG)^{-1} = [\mathbf{\Phi}^{-1}(s) + BG]^{-1} = \mathbf{\Phi} - \mathbf{\Phi}B(I + G\mathbf{\Phi}B)^{-1}G\mathbf{\Phi}$$

where

$$\Phi(s) = (sI - A)^{-1}$$

is the resolvent of the open-loop system.

Hint: Use block-diagram manipulations.

Problem 8.2 Characteristic equation of system with general reduced-order observer

Consider the general reduced-order observer where the observation is given by

$$y = Cx$$

and, as explained in Chap. 7, (7.61)-(7.64)

$$\hat{x} = Py + Q(Ly + z)$$

with

$$\dot{z} = \bar{F}(Ly + z) + (A_{21} - LA_{11})y + \bar{H}u$$

Derive the characteristic equation of the closed-loop system following the steps used to derive (8.27).

Problem 8.3 Instrument servo: reduced-order observer

A compensator based on a reduced-order observer is to be designed for the instrument servo of Example 8B.

(*a*) Under the assumption that the output position is measured, design the reduced-order observer to estimate the angular velocity.

(*b*) Find the transfer function of the compensator using the regulator gains used in Example 8B.

(*c*) For the numerical data used in Example 8B (i.e., $\alpha = 1$) find the range of variation of the loop gain. Draw the root locus and the Nyquist plot.

Problem 8.4 Inverted pendulum on cart: compensator design

A compensator based on a full-order observer is to be designed for the inverted pendulum on the motor-driven cart of Prob. 3.6 et seq.

(*a*) Using the regulator gains of Prob. 6.1 and the observer gains of Prob. 7.2, determine the transfer function $\mathsf{D}(s)$ of the compensator.

(*b*) Assume a gain variation at the control input so that the return difference is

$$\mathsf{T}(s) = 1 + K\mathsf{D}(s)\mathsf{H}(s)$$

where $\mathsf{H}(s)$ is the transfer function of the plant (i.e., the cart and pendulum). Find the range of K for which the closed-loop system is stable.

(*c*) In view of these results, discuss the circumstances under which the compensator design is satisfactory.

Problem 8.5 Inverted pendulum on cart: reduced-order compensator

Repeat Prob. 8.4 except in this case, use the reduced-order observer design of Prob. 7.3.

Problem 8.6 Three-capacitance thermal system

A compensator for the three-capacitance thermal system of Prob. 3.7 et seq. is to be designed by combining the full-state control law of Prob. 6.8 with the observer designs of Prob. 7.6.

(*a*) Draw the block diagram of the compensator using the observer structure of Prob. 7.6 part *a*. Find the transfer function of the compensator. Does the compensator provide true integral control?

(*b*) Repeat part *a*, except use the observer structure of part *b* of Prob. 7.6.

(*c*) For the compensator design of part *a* draw the root locus with the loop gain as a parameter. Find the gain and phase margins.

Problem 8.7 Two-car train—one-drive motor

Consider the compensator obtained by combining the full-state feedback control law of Prob. 6.3 with the observer of Prob. 7.1 part *a*.

(*a*) Design a robust observer in the sense defined by Rynaski (Sec. 8.6). Choose a Butterworth pattern for adjustable observer poles.

(*b*) Draw the root locus.

(*c*) Determine the gain and phase margins.

Problem 8.8 Two-car train—separated sensor and driver

Again consider the two-car train of Prob. 6.3 except in this case the position of car 2 is measured, but the motor drives car 1. (This is an example of a system in which the sensor and actuator are said to be "noncolocated." When might such a condition arise in a realistic application?)

(*a*) Design a robust observer as in Prob. 8.7.

(*b*) Draw the root locus.

(*c*) Determine the gain and phase margins.

Discuss the similarities and differences between this control system and that of Prob. 8.7.

Problem 8.9 Compensator for aircraft lateral dynamics

A compensator for the lateral dynamics of the aircraft of Prob. 4.4 is to be designed. The compensator is to use a reduced-order (two-state) to estimate β, e from measurements of p, r, and ϕ_0.

(*a*) Arrange the dynamic equations to display the submatrices A_{11}, A_{12}, A_{21}, A_{22}, B_1, and B_2 (with $C_1 = I$) that are needed to design the reduced-order observer.

(*b*) Draw the block diagram of the observer.

(*c*) Is it possible to design an observer to satisfy (8.86)? If so is the resulting observer asymptotically stable? Are the poles in a suitable location?

(*d*) Since there are four observer gains, is it possible to design a robust observer having its zeros at the transmission zeros of the process?

Problem 8.10 Two-axis gyro

We continue the design of the control law for the two-axis gyro. The state-variable feedback gains have been determined in Prob. 6.9 and the reduced-order observer structure has been established in Prob. 7.5. Now we put these together to obtain the compensator.

Because of symmetry there are only two different compensator transfer functions

$$\frac{\mathsf{u}_x(s)}{\delta_x(s)} = \frac{\mathsf{u}_y(s)}{\delta_y(s)} = \mathsf{H}_1(s)$$

and

$$\frac{\mathsf{u}_x(s)}{\delta_y(s)} = -\frac{\mathsf{u}_y(s)}{\delta_x(s)} = \mathsf{H}_2(s)$$

(*a*) Find the transfer functions $\mathsf{H}_1(s)$ and $\mathsf{H}_2(s)$ in terms of the control gains $g_1, \ldots, g_4$ and the observer parameters $l_1, \ldots, l_4$.

(*b*) Draw the block diagram of the compensator.

Problem 8.11 Hydraulically actuated gun turret

A compensator for the gun turret of Probs. 6.2 and 7.4 is to be designed by combining the full-state feedback design and the observer design of these problems, respectively.

(*a*) Find the transfer function of the compensator using the gains of part *a* of Prob. 6.2 and the observer design of Prob. 7.4.

(*b*) Draw the root locus of the closed-loop system.

(*c*) Find the gain and phase margin.

Problem 8.12 Full-order compensator for constant-altitude autopilot

The only instrument available for implementation of the constant-altitude autopilot of Prob. 6.6 is an altimeter. Hence an observer is needed to estimate the other state variables.

(*a*) Assuming that a full-order (i.e., fourth-order) observer is to be used, find the gains that place the observer poles at

$$s = 4(-\tfrac{1}{2} \pm j\sqrt{3}/2) \qquad s = 4(-\sqrt{3}/2 \pm j\tfrac{1}{2})$$

(*b*) Determine the transfer function of the compensator.

(*c*) Plot the closed-loop root locus and find the range of gain variation for which the closed-loop system is stable.

Problem 8.13 Reduced-order compensator for constant altitude autopilot

(*a*) Repeat Prob. 8.12 except use a reduced-order (i.e., third-order) compensator with observer poles at $s = -4$, $s = -2\sqrt{2} \pm j2\sqrt{2}$.

(*b*) Compare the advantages and disadvantages of the reduced-order compensator over the full-order compensator.

NOTES

Note 8.1 Separation principle and separation theorem

In this book we make a distinction between the general *separation principle* and the more specialized *separation theorem*. The former is a statement of the fact that the regulator gains can be calculated as if all the states were accessible to measurement, and the observer gains can be designed without regard to the process input, and then the two can be combined to obtain a compensator that guarantees stability of the closed-loop system. The separation principle, as we use it here, thus makes no assertion about the *optimality* of the resulting system. That assertion is made by the separation *theorem*, as we use the term in this book. The latter, considered in Sec. 11.7, asserts that when the gains of the regulator are designed to optimize a *deterministic* quadratic performance criterion, and the observer is optimized for the noise present in the system, then the resulting compensator is the optimum *stochastic* controller.

Additional background of the separation theorem is given in Note 11.5.

Note 8.2 Block diagrams for transfer-function calculations

In deriving the relationships needed for the study of robustness, a number of identities between transfer functions are developed. Although these identities may seem surprising from the algebraic viewpoint, most are derived with the aid of a block diagram of the system and represent transfer functions from one variable to another obtained by "going around the loops" by a different sequence of steps. Many identities can be obtained by this process.

Note 8.3 The Schur identity

The Schur identity, equation (A.47), has found many applications in system and control theory. One application, totally unrelated to anything in this book, deals with the propagation of the covariance matrix in discrete-time Kalman filtering.

Because of the utility of this formula, it is sometimes referred to as "*the* matrix inversion lemma."

Note 8.4 Nonsquare systems

If the open-loop system for which a compensator is being designed is not square, having fewer inputs than outputs, then Doyle and Stein[2] suggest "squaring it up" by adding one or more fictitious inputs with corresponding columns of the B matrix. Since these inputs are fictitious the full-state feedback control law is then designed so that the gains from the state variables to these inputs are all zero. Alternatively, the Doyle-Stein condition can be generalized to nonsquare systems. This has been done recently.[4]

REFERENCES

1. Joseph, P. D., and Tou, J., "On Linear Control Theory," *Trans. AIEE, Pt. II*, vol. 80, no. 11, September 1961, pp. 193–196.
2. Doyle, J. C., and Stein, G., "Robustness with Observers," *IEEE Trans. On Automatic Control*, vol. AC-24, no. 4, August 1979, pp. 607–611.
3. Rynaski, E. G., "Flight Control Synthesis Using Robust Output Observers," *Proc. AIAA Guidance and Control Conference, San Diego, CA*, August 1982, pp. 825–831.
4. Madiwale, A. N., and Williams, D. E., "Some Extensions of Loop Transfer Recovery," *Proc. American Control Conference, Boston, MA*, June 1985, pp. 790–795.

CHAPTER

NINE

LINEAR, QUADRATIC OPTIMUM CONTROL

9.1 WHY OPTIMUM CONTROL?

In the previous chapters we learned how to design a compensator for a single-input, single-output process which places the closed-loop poles wherever we want them to be (assuming the process is controllable and observable). Since the closed-loop poles determine the speed (bandwidth) and damping of the response, isn't this enough? Why should we want to go any farther? There are several good reasons.

The first reason for seeking an optimum controller is that in a multiple-input or multiple-output system, the pole-placement technique described in the earlier chapters does not completely specify the controller or compensator parameters (gains). Consider, for example, a kth-order plant with m inputs and the entire state vector accessible for feedback. A nondynamic controller has km parameters to be determined, but only k possible closed-loop pole locations. Thus we have to set m times as many parameters as there are poles; there are infinitely many ways by which the same closed-loop poles can be attained. Which way is best? What algorithm can be used to determine the feedback gains? From a practical standpoint, of course, the availability of more adjustable parameters than the minimum number needed to achieve the desired closed-loop pole location is a great benefit because other things can be accomplished besides placing the closed-loop poles. But the absence of a definitive algorithm for determining a unique control law is a detriment to the system designer who does not know how to handle this "embarrassment of riches." By choosing a control law to optimize performance (in the precise sense to be defined shortly) this embarrassment is avoided.

A more cogent reason for seeking an optimum controller is that the designer may not really know the desirable closed-loop pole locations. Choosing pole locations far from the origin may give very fast dynamic response but require control signals that are too large to be produced with the available power source. Use of gains that would be able to produce these signals, in the absence of power limitations, could cause the control signals to exceed physical limits (i.e., to "saturate"). In such cases the closed-loop dynamic behavior will not be as predicted by the linear analysis, and may even be unstable. To avoid these problems it often is necessary to limit the speed of response to that which can be achieved without saturation. Another reason for limiting speed of response is a desire to avoid problems of noise that typically accompany high-gain systems. The engineer who has acquired extensive experience with a particular type of process generally has an intuitive "feel" about the proper closed-loop pole locations. But, faced with an unfamiliar process to control and a lack of time to acquire the necessary insight, the engineer will appreciate a design method that can provide an initial design while insight is developed. The optimization theory, to be developed in this chapter, can serve this purpose.

Still another reason for using optimum control theory for design is that the process to be controlled may not be controllable, in the sense defined in Chap. 5. There may be some subspace of the process state-space in which the state vector cannot be moved around by application of suitable control signals. The dynamic behavior in that subspace is not subject to control and hence not all the poles of the closed-loop system can be placed at will. Hence design by pole placement will not work. But, by use of optimum control theory, and not demanding impossible behavior, it is possible to design a control system to control as much as can be controlled. If the behavior of the uncontrollable part is stable, the overall system will behave in an acceptable manner.

9.2 FORMULATION OF THE OPTIMUM CONTROL PROBLEM

The dynamic process considered here as elsewhere in this text is, as usual, characterized by the vector-matrix differential equation

$$\dot{x} = Ax + Bu \tag{9.1}$$

where x is the process state, u is the control input, and A and B are known matrices. Again, as before, we seek a linear control law

$$u(t) = -Gx(t) \tag{9.2}$$

where G is a suitable gain matrix. Here, however, instead of seeking a gain matrix to achieve specified closed-loop pole locations, we now seek a gain to minimize a specified performance criterion V (or "cost function") expressed as the integral of a quadratic form in the state x plus a second quadratic form in the control u; i.e.,

$$V = \int_t^T [x'(\tau)Q(\tau)x(\tau) + u'(\tau)Ru(\tau)]\, d\tau \tag{9.3}$$

where Q and R are symmetric matrices.

Some explanatory remarks about this performance criterion are in order before we attempt to find the optimum gain matrix G.

First, we note that minimization of V also minimizes ρV where ρ is any positive constant. So the problem is not altered by multiplying V by any positive constant. Often the constant 1/2 is used in front of V to simplify expressions resulting in other developments. (See Note 9.1.)

Second, regarding the limits on the integral, the lower limit t is identified as the *present* time, and the upper limit T is the *terminal* time, or *final* time. The time difference $T - t$ is the control interval, or "time-to-go." If the terminal time T is finite and fixed, the time-to-go keeps decreasing to zero, at which time the control process ends. This situation is characteristic of missile guidance problems, as will be discussed in an example below. The more customary case, however, is that in which the terminal time is infinite. In this case we are interested in the behavior of the process "from now on," including the steady state. This is precisely the case addressed by pole placement, and is the case that will receive the major portion of our attention subsequently.

Finally, consider the weighting matrices Q and R. These are often called the *state weighting* matrix and *control weighting* matrix, respectively. We are about to derive a "recipe" for finding the control gain matrix G in terms of these weighting matrices. In other words, we can plug the matrices Q and R—along with the matrices A and B that define the dynamic process—into a computer program and direct it to find G. If the process is controllable and Q and R are suitable, the computer will not fail to find G. (This is not to say that the calculation is a numerically trivial problem—far from it—but only that the problem of determining G once A, B, Q, and R are given, is not a control system design problem but a problem in numerical analysis.)

The question of concern to the control system designer is the selection of the weighting matrices Q and R. In candor one must admit that minimization of a quadratic integral of the form of (9.3) is rarely the true design objective. The problem, however, is that the true design objective often cannot be expressed in mathematical terms. And even in those instances when the design objective is amenable to mathematical expression, it is usually all but impossible to solve for the optimum control law. Expression of the design objective in the form of a quadratic integral is a practical compromise between formulating the real problem that cannot be solved, and formulating a somewhat artificial problem that can be solved easily. The need for such compromises arises in many contexts, and the control system designer should not feel guilty about being acquiescent to the need.

In the performance or cost function defined by (9.3) two terms contribute to the integrated cost of control: the quadratic form $x'Qx$ which represents a penalty on the deviation of the state x from the origin and the term $u'Ru$ which represents the "cost of control." This means, of course, that the desired state is the origin, not some other state. (In Chap. 5 we studied how it is possible to formulate a problem with a nonzero desired state in the form of a regulator problem. This discussion will be resumed in Sec. 9.6.) The weighting matrix Q specifies the importance of the various components of the state vector relative to each other. For example, suppose that x_1 represents the system error, and that $x_2, \ldots, x_k$ represent successive derivatives, i.e.,

$$\begin{aligned} \dot{x}_2 &= \dot{x} \\ \dot{x}_3 &= \ddot{x} \\ &\cdots\cdots \\ \dot{x}_k &= x^{(k-1)} \end{aligned}$$

If only the error and none of its derivatives are of concern, then we might select a state weighting matrix

$$Q = \begin{bmatrix} 1 & 0 & \cdots & 0 \\ 0 & 0 & \cdots & 0 \\ \cdots & \cdots & \cdots & \cdots \\ 0 & 0 & \cdots & 0 \end{bmatrix} \tag{9.4}$$

which will yield the quadratic form

$$x'Qx = x_1^2$$

But the choice of (9.4) as a state weighting matrix may lead to a control system in which the velocity $x_2 = \dot{x}$ is larger than desired. To limit the velocity, the performance integral might include a velocity penalty, i.e.,

$$x'Qx = x_1^2 + c^2 x_2^2$$

which would result from a state weighting matrix

$$Q = \begin{bmatrix} 1 & 0 & \cdots & 0 \\ 0 & c^2 & \cdots & 0 \\ \cdots & \cdots & \cdots & \cdots \\ 0 & 0 & \cdots & 0 \end{bmatrix}$$

Another possible situation is one in which we are interested in the state only through its influence on the system output

$$y = Cx$$

For example, for a system with a single output

$$y = c'x$$

a suitable performance criterion might well be

$$y^2 = x'cc'x$$

So in this case

$$Q = cc'$$

It should by now be obvious that the choice of the state weighting matrix Q depends on what the system designer is trying to achieve.

The considerations alluded to above with regard to Q apply as well to the control weighting matrix R. The term $u'Ru$ in the performance index (9.3) is included in an attempt to limit the magnitude of the control signal u. Unless a "cost" is imposed for use of control, the design that emerges is liable to generate control signals that cannot be achieved by the actuator—the physical device that produces the control signal—and the result will be that the control signal will saturate at the maximum signal that can be produced. This is often exactly what the designer desires. In most cases, saturation of the control will produce the fastest possible response. But when saturation occurs, the closed-loop system behavior that was predicted on the basis that saturation will not occur, may be very different from the actual system behavior. A system that a linear design predicts to be stable may even be unstable when the control signal is saturated. Thus in a desire to avoid saturation and its consequences, the control signal weighting matrix is selected large enough to avoid saturation of the control signal under normal conditions of operation.

The relationship between the weighting matrices Q and R and the dynamic behavior of the closed-loop system depend of course on the matrices A and B and are quite complex. It is impractical to predict the effect on closed-loop behavior of a given pair of weighting matrices. A suitable approach for the designer would be to solve for the gain matrices G that result from a range of weighting matrices Q and R, and calculate (or simulate) the corresponding closed-loop response. The gain matrix G that produces the response closest to meeting the design objectives is the ultimate selection. With the software that is now widely available, it is a simple matter to solve for G given A, B, Q, and R. In a few hours time, the gain matrices and transient response that result for a dozen or more combinations of Q and R can be determined, and a suitable selection of G can be made.

Further comments relating to the selection of the weighting matrices will be given after the general theory is developed and illustrated by a few examples.

9.3 QUADRATIC INTEGRALS AND MATRIX DIFFERENTIAL EQUATIONS

When the control law (9.2) is used to control the dynamic process (9.1), the closed-loop dynamic behavior is given by

$$\dot{x} = Ax - BGx = A_c x \tag{9.5}$$

where

$$A_c = A - BG \tag{9.6}$$

is the "closed-loop" dynamics matrix. In most cases considered in this text, we are interested in the case in which A, B, and G are constant matrices, but there is really no need to restrict them to be constant; in fact, the theoretical development is much easier if we do *not* assume that they are constant. Thus, we permit the closed-loop matrix A_c to vary with time. Since A_c may be time-varying we cannot write the solution to (9.5) as a matrix exponential. But the solution to (9.5) can be written in terms of the general state transition matrix introduced in Chap. 3:

$$x(\tau) = \Phi_c(\tau, t)x(t) \tag{9.7}$$

where Φ_c is the state-transition matrix corresponding to A_c. Equation (9.7) merely states that the state at any time τ depends linearly on the state at any other time t. In what follows there will be no need to have an expression for Φ_c; this is fortunate, because in general no simple expression is available.

Using (9.7), the performance index (9.3) can be expressed as a quadratic form in the initial state $x(t)$. In particular

$$\begin{aligned} V &= \int_t^T [x'(\tau)Qx(\tau) + x'(\tau)G'RGx(\tau)]\,d\tau \\ &= \int_t^T x'(t)\Phi_c'(\tau, t)\{Q + G'RG\}\Phi_c(\tau, t)x(t)\,d\tau \end{aligned} \tag{9.8}$$

The initial state $x(t)$ can be moved outside the integral to yield

$$V = x'(t)M(t, T)x(t) \tag{9.9}$$

where

$$M(t, T) = \int_t^T \Phi_c'(\tau, t)\{Q + G'RG\}\Phi_c(\tau, t)\,d\tau \tag{9.10}$$

(Note that M is a symmetric matrix.)

For purposes of determining the optimum gain, i.e., the matrix G which results in the closed-loop dynamics matrix $A_c = A - BG$ which minimizes the resulting integral (9.10), it is convenient to find a differential equation satisfied by (9.10). For this purpose, we note that V in (9.8) and (9.9) is a function of the initial time t. Thus we can write (9.8) as

$$V(t) = \int_t^T x'(\tau)L(\tau)x(\tau)\,d\tau \tag{9.11}$$

where

$$L = Q + G'RG \tag{9.12}$$

(Note that L is not restricted to be constant.) Thus, by the definition of an integral

$$\frac{dV}{dt} = -x'(\tau)Lx(\tau)\Big|_{\tau=t} = -x'(t)Lx(t) \tag{9.13}$$

But, from (9.9)

$$\frac{dV}{dt} = \dot{x}'(t)M(t, T)x(t) + x'(t)\dot{M}(t, T)x(t) + x'(t)M(t, T)\dot{x}(t) \tag{9.14}$$

(The dot over M in (9.14) denotes differentiation with respect to t, that is,

$$\dot{M}(t, T) = \partial M(t, T)/\partial t)$$

On using the closed-loop differential equation (9.5) we obtain from (9.14)

$$\frac{dV}{dt} = x'(t)[A_c'(t)M(t, T) + \dot{M}(t, T) + M(t, T)A_c(t)]x(t) \tag{9.15}$$

We thus have two expressions for dV/dt: one given by (9.13) and one given by (9.15). Both are quadratic forms in the initial state $x(t)$, which is arbitrary. The only way two quadratic forms in x can be equal for any (arbitrary) x is if the matrices underlying the forms are equal. Thus we have found that the matrix M satisfies the differential equation

$$-L = A_c'M + \dot{M} + MA_c$$

or, in the more customary form

$$-\dot{M} = MA_c + A_c'M + L \tag{9.16}$$

This is an important differential equation. It appears in many forms in control theory and estimation. To make it look neater, the arguments have been omitted in (9.16). But one should not forget that

$$M = M(t, T) \qquad A_c = A_c(t) \qquad L = L(t)$$

We have already determined the solution to (9.16) which, using (9.10), is

$$M(t, T) = \int_t^T \Phi_c'(\tau, t)L(t)\Phi_c(\tau, t)\, d\tau \tag{9.17}$$

Equation (9.16) is a first-order matrix differential equation and thus requires a single "initial condition" to pin it down completely. This condition is obtained from the integral (9.17). Clearly

$$M(T, T) = 0 \tag{9.18}$$

is the required condition.

9.4 THE OPTIMUM GAIN MATRIX

When any gain matrix G is chosen to close the loop, the corresponding closed-loop performance has been shown to be given by

$$V(t) = x'(t)M(t, T)x(t)$$

where $M(t, T)$ is the solution to (9.16), which, in terms of the matrices A, B, G,

Q, and R becomes

$$-\dot{M} = M(A - BG) + (A' - G'B')M + Q + G'RG \tag{9.19}$$

Our task now is to find the matrix G which makes the solution to (9.19) as small as possible. What does it mean for one matrix to be smaller than another? We are really interested in the quadratic forms resulting from these matrices, and thus we are seeking the matrix $\hat{M}$ for which the quadratic form

$$\hat{V} = x'\hat{M}x < x'Mx$$

for any arbitrary initial state $x(t)$ and any matrix $M \neq \hat{M}$.

The problem of finding an optimum gain matrix can be approached by a number of avenues. (See Note 9.1.) The approach we adopt here is to assume that a minimizing gain $G = \hat{G}$ exists and results in an optimum (i.e., minimum $M = \hat{M}$). We will then find a matrix differential equation that $\hat{M}$ must satisfy in order for it to result in a smaller value of V than results from any other matrix.

Now the minimizing matrix $\hat{M}$ that results from the minimizing gain $\hat{G}$ must of course satisfy (9.19), i.e.,

$$-\dot{\hat{M}} = \hat{M}(A - B\hat{G}) + (A' - \hat{G}'B')\hat{M} + Q + \hat{G}'R\hat{G} \tag{9.20}$$

Any nonoptimum gain matrix G and the corresponding matrix M can be expressed in terms of these matrices:

$$M = \hat{M} + N$$

$$G = \hat{G} + Z$$

Thus (9.19) becomes

$$-(\dot{\hat{M}} + \dot{N}) = (\hat{M} + N)[A - B(\hat{G} + Z)] + [A' - (\hat{G}' + Z')B'](\hat{M} + N)$$
$$+ Q + (\hat{G}' + Z')R(\hat{G} + Z) \tag{9.21}$$

On subtracting (9.20) from (9.21) we obtain the following differential equation for N

$$-\dot{N} = NA_c + A_c'N + (\hat{G}'R - \hat{M}B)Z + Z'(R\hat{G} - B'\hat{M}) + Z'RZ \tag{9.22}$$

where $A_c = A - BG = A - B(\hat{G} + Z)$.

The differential equation (9.22) is exactly in the form of (9.16) with L in the latter being given by

$$L = (\hat{G}'R - \hat{M}B)Z + Z'(\hat{G}'R - \hat{M}B)' + Z'RZ \tag{9.23}$$

Using (9.17) we see that the solution to (9.22) is of the form

$$N(t, T) = \int_t^T \Phi_c'(\tau, t)L\Phi_c(\tau, t)\, d\tau \tag{9.24}$$

Now if $\hat{V}$ is minimum, then we must have

$$x'\hat{M}x \leq x'(\hat{M} + N)x = x'\hat{M}x + x'Nx$$

which implies that the quadratic form $x'Nx$ must be positive-definite, or at least positive semidefinite. Now look at L as given by (9.23). If Z is sufficiently small the linear terms dominate the quadratic term $Z'RZ$. Thus, one can readily find values of Z which make L negative-definite, unless the linear term in (9.23) is absent altogether! Thus we conclude that for the control law $\hat{G}$ to be optimum, we must have

$$R\hat{G} - B'\hat{M} = 0 \tag{9.25}$$

or, on the assumption that the control weighting matrix R is nonsingular

$$\hat{G} = R^{-1}B'\hat{M} \tag{9.26}$$

This gives the optimum gain matrix in terms of the solution to the differential equation (9.20) that determines $\hat{M}$. When (9.26) is substituted into (9.20) the following differential equation results for $\hat{M}$:

$$-\dot{\hat{M}} = \hat{M}A + A'\hat{M} - \hat{M}BR^{-1}B'\hat{M} + Q \tag{9.27}$$

This matrix differential equation, one of the most famous in the literature of modern control theory, gives the matrix $\hat{M}$ which, using (9.26), gives the optimum gain matrix $\hat{G}$. (A historical discussion of the background of this equation is given in Note 9.2.)

It is noted that in addition to the linear terms $\hat{M}A$ and $A'\hat{M}$ in (9.27) there is also present the quadratic term $-\hat{M}BR^{-1}B'\hat{M}$. A scalar first-order differential equation with a linear term and a quadratic term (as well as a constant term) is known as a *Riccati equation* in the mathematical literature and the terminology was extended by Kalman[1] to the matrix case. Nowadays (9.27) is identified in the literature of optimum control as *the* Riccati equation.

Because of the presence of the quadratic term, no general formula for the solution to (9.27), analogous to the integral (9.17) for the linear equation (9.16), can be given. There are, of course, special cases—one of which is contained in Example 9E below—in which $\hat{M}(t, T)$ can be determined analytically. But in most practical cases of interest, it is necessary to solve for $\hat{M}(t, T)$ by some appropriate numerical method.

One obvious method of solving is the numerical integration of (9.27). Since $\hat{M}$ is symmetric, there are $k(k+1)/2$ coupled, scalar equations to be integrated. It should be noted that these equations are integrated *backward* in time, because the condition that must be satisfied is

$$\hat{M}(T, T) = 0$$

and we are interested in $\hat{M}(t, T)$ for $t < T$.

9.5 THE STEADY STATE SOLUTION

In an application in which the control interval is finite, the gain matrix G will generally be time-varying even when the matrices A, B, Q, and R are all constant, because the solution matrix $\hat{M}(t, T)$ of the matrix Riccati equation will not be constant. But suppose the control interval is infinite. We want a control gain G which minimizes the performance integral

$$V_\infty = \int_t^\infty (x'Qx + u'Ru)\, d\tau \tag{9.28}$$

In this case the terminal time T is infinite, so the integration (backward in time) of (9.28) will either converge to a constant matrix $\bar{M}$ or grow without limit. If it converges to a limit, the derivative $\hat{M}$ tends to zero. Hence for an infinite terminal time

$$V_\infty = x'\bar{M}x$$

where $\bar{M}$ satisfies the algebraic quadratic equation (sometimes called the *algebraic Riccati equation* or ARE)

$$0 = \bar{M}A + A'\bar{M} - \bar{M}BR^{-1}B'\bar{M} + Q \tag{9.29}$$

and the optimum gain in the steady state is given by

$$\bar{G} = R^{-1}B'\bar{M} \tag{9.30}$$

The single matrix equation of (9.29) represents a set of $k(k+1)/2$ coupled scalar quadratic equations. Each quadratic equation in the set in general has two solutions, so we may reasonably expect that there are $2[k(k+1)/2] = k(k+1)$ different (symmetric) solutions to (9.29). Are all the solutions correct? Is only one solution correct? Are there perhaps no correct solutions?

The answers to these questions are, as one might imagine, connected with the issues of stability and controllability, although from a strictly mathematical standpoint they depend on the three matrices A, $BR^{-1}B$, and Q and their relationships with each other. Kalman[1, 2] and others after him have addressed the issues. A complete discussion of this subject entails not only controllability, but also observability, and the more subtle concepts of *stabilizability*, *reconstructability*, and *detectability*, and is well beyond the scope of this book. (See Note 5.2.)

For most design applications the following facts about the solution of (9.29) will suffice:

(*a*) If the system is asymptotically stable, or

(*b*) If the system defined by the matrices (A, B) is controllable, and the system defined by (A, C) where $C'C = Q$, is observable,

Then the algebraic Riccati equation (ARE) has a *unique*, *positive definite* solution $\bar{M}$ which minimizes V_∞ when the control law $u = -R^{-1}B'\bar{M}x$ is used.

It should be understood that the *total* number of symmetric solutions (counting those with complex elements) is still $k(k+1)$. The assertion of the last paragraph is that one of these solutions (and not more than one) is positive-definite. Since the integral (9.28) is clearly positive-definite, that solution is the correct one. Let us examine the other possibilities.

It may happen that (9.29) may have no positive definite solutions. In this case, there is no control law which minimizes V_∞. This must mean that V_∞ becomes infinite for any possible control law, and helps to explain why asymptotic stability guarantees that the ARE has a unique positive-definite solution: the control law $u = 0$ will result in a finite value of V_∞ and we would suppose that other control laws exist which can reduce V_∞ still further. If the system $\dot{x} = Ax + Bu$ is not asymptotically stable, however, the control law $u = 0$ does not yield a finite value of V_∞ for any initial condition $x = x(t)$, and it will be necessary to actively intervene with a nonzero control. This is how the idea of controllability arises, since if the system is controllable, a control law can be found which produces a closed-loop dynamics matrix $A_c = A - BG$ with eigenvalues at arbitrary locations. Even if the system is not controllable, but merely "stabilizable," i.e., a control law can be found which can move the unstable eigenvalues to the left half-plane, a finite value of V_∞ can be achieved.

How does observability enter the picture? The system defined by A and B may be uncontrollable and even unstabilizable, but the matrix Q may be chosen so that those state variables corresponding to the unstabilizable portion of $\dot{x} = Ax + Bu$ are not weighted in $x'Qx$. In this case, there is reason to expect that V_∞ can be made finite. It might at first seem strange and impractical to consider a control law which does not stabilize a system, but there are many situations in which this is entirely reasonable. The most common instance is when the state is really a "metastate" comprising both the dynamic state x and the exogenous state x_0. By hypothesis the latter cannot be controlled by the input, and it may not be asymptotically stable.

In addition to the possibility that the ARE does not have even one positive-definite solution, it is also conceivable that the ARE has more than one. Since the total number of possible solutions is finite, obviously the one we are looking for to minimize V_∞ is the one that yields the smallest value of $x'\bar{M}x$. If we could find *all* the positive-definite solutions we should surely find the proper one.

If we could only find all the solutions to (9.29) there would be no difficulty in establishing which, if any, of the solutions is the correct one. The great difficulty arises because in most practical cases, the ARE (9.29) must be solved numerically and the numerical problem is far from being easy. (See Note 9.3.) If a computer program that embodies the algorithm for solving the ARE is set to work crunching out a solution that does not exist, we should not be surprised to find it grinding away forever. So it is important to be able to find out whether the sought-after solution exists before the crunching starts.

Example 9A Inverted pendulum It is recalled from Example 3D that the state variables are $x_1 = \theta$ (angular position) and $x_2 = \dot{\theta}$ (angular velocity). The matrices defining the dynamics, as determined earlier are

$$A = \begin{bmatrix} 0 & 1 \\ \Omega^2 & 0 \end{bmatrix} \qquad B = \begin{bmatrix} 0 \\ 1 \end{bmatrix} \tag{9A.1}$$

A control law is sought to minimize the performance index

$$V = \int_t^\infty \left(\theta^2 + \frac{u^2}{c^2}\right) d\tau \tag{9A.2}$$

where u is the angular acceleration. For this performance criterion, the weighting matrices are seen to be

$$Q = \begin{bmatrix} 1 & 0 \\ 0 & 0 \end{bmatrix} \qquad R = \frac{1}{c^2}$$

Let the performance matrix $\hat{M}$ be given by

$$\hat{M} = \begin{bmatrix} m_1 & m_2 \\ m_2 & m_3 \end{bmatrix}$$

The gain matrix, in terms of the elements of $\hat{M}$ is

$$\hat{G} = R^{-1}B'\hat{M} = c^2[0, 1]\begin{bmatrix} m_1 & m_2 \\ m_2 & m_3 \end{bmatrix} = [c^2m_2, c^2m_3] \tag{9A.3}$$

The terms needed for the matrix quadratic (9.29) are

$$\hat{M}A = \begin{bmatrix} m_1 & m_2 \\ m_2 & m_3 \end{bmatrix}\begin{bmatrix} 0 & 1 \\ \Omega^2 & 0 \end{bmatrix} = \begin{bmatrix} m_2\Omega^2 & m_1 \\ m_3\Omega^2 & m_2 \end{bmatrix}$$

$$A'\hat{M} = (\hat{M}A)' = \begin{bmatrix} m_2\Omega^2 & m_3\Omega^2 \\ m_1 & m_2 \end{bmatrix}$$

$$\hat{M}BR^{-1}B'\hat{M} = \begin{bmatrix} c^2m_2^2 & c^2m_2m_3 \\ c^2m_2m_3 & c^2m_3^2 \end{bmatrix}$$

$$Q = \begin{bmatrix} 1 & 0 \\ 0 & 0 \end{bmatrix}$$

Thus, the individual terms of (9.29) are

$$\begin{aligned} 0 &= 2m_2\Omega^2 - c^2m_2^2 + 1 \\ 0 &= m_1 + m_3\Omega^2 - c^2m_2m_3 \\ 0 &= 2m_2 - c^2m_3^2 \end{aligned} \tag{9A.4}$$

In this instance (9A.4) are simple enough to solve algebraically. In particular, the first equation of (9A.4) has the solution

$$m_2 = \frac{\Omega^2 \pm \sqrt{\Omega^4 + c^2}}{c^2}$$

We do not yet know which sign on the radical in m_2 is correct, but we will find out shortly. From the third equation in (9A.4)

$$m_3 = \frac{1}{c}\sqrt{2m_2}$$

We note that if the lower (−) sign is used then m_2 would be negative and this would make m_3 imaginary. Since the matrix $\hat{M}$ of the quadratic form $x'(t)\hat{M}x(t)$ must be real, an imaginary number for one of its elements is unacceptable. Thus we conclude that the upper (+) sign must be used. This gives

$$m_2 = \frac{\Omega^2 + \sqrt{\Omega^4 + c^2}}{c^2} \qquad m_3 = \frac{\sqrt{2}}{c^2}[\Omega^2 + \sqrt{\Omega^4 + c^2}]^{1/2}$$

And hence the gain matrix (9A.3) has elements

$$g_1 = \Omega^2 + \sqrt{\Omega^4 + c^2} \qquad g_2 = \sqrt{2}\,[\Omega^2 + \sqrt{\Omega^4 + c^2}]^{1/2}$$

The remaining term m_1 in $\hat{M}$ is obtained from the second equation in (9A.4), but it is not needed in the control law.

The closed-loop behavior of the system is of interest. The matrix of the closed-loop system is

$$A_c = A - BG = \begin{bmatrix} 0 & 1 \\ \Omega^2 & 0 \end{bmatrix} - \begin{bmatrix} 0 \\ 1 \end{bmatrix} [g_1 \quad g_2]$$

$$= \begin{bmatrix} 0 & 1 \\ -\sqrt{\Omega^4 + c^2} & -\sqrt{2}(\Omega^2 + \sqrt{\Omega^4 + c^2})^{1/2} \end{bmatrix}$$

And the characteristic equation is

$$s^2 + \sqrt{2}(\Omega^2 + \sqrt{\Omega^4 + c^2})^{1/2} s + \sqrt{\Omega^4 + c^2} = 0$$

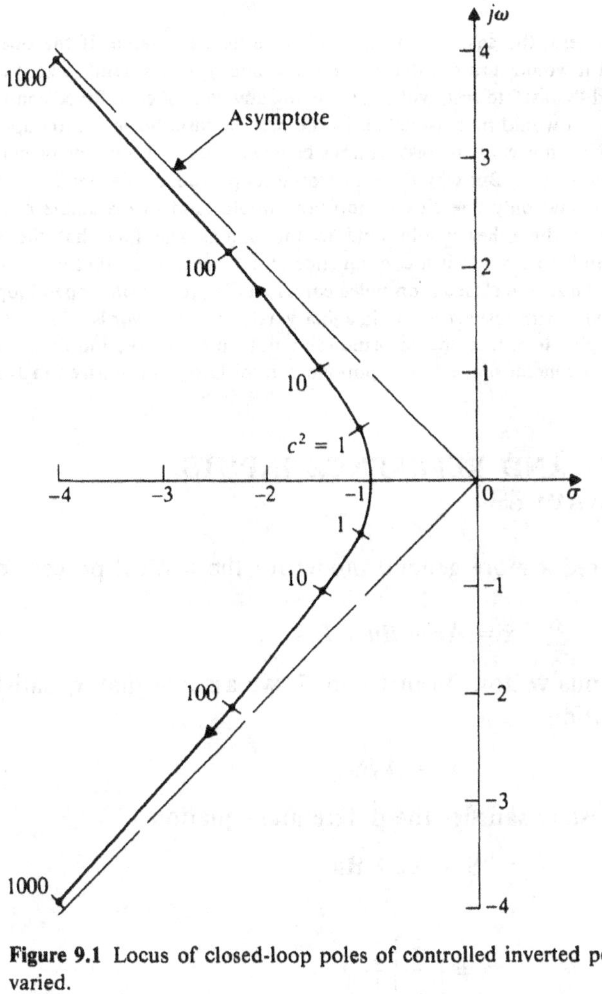

Figure 9.1 Locus of closed-loop poles of controlled inverted pendulum as weighting factor is varied.

the roots of which are

$$s_1, s_2 = -\frac{\sqrt{2}}{2}[(\bar{\Omega}^2 + \Omega^2)^{1/2} \pm j(\bar{\Omega}^2 - \Omega^2)^{1/2}] \tag{9A.5}$$

where

$$\bar{\Omega}^2 = \sqrt{\Omega^4 + c^2}$$

The locus of closed-loop poles as the weighting factor c is varied from ∞ to 0 as shown in Fig. 9.1. The following characteristics of the locus are noteworthy:

(*a*) As c increases, the closed-loop roots tend to asymptotes at 45° to the real axis, and move out to ∞ along these asymptotes. This implies that the response time tends to zero and the damping factor tends to $\zeta = \sqrt{2}/2 = 0.707$. That the response time tends to zero is not surprising, since increasing c decreases the cost of control and hence makes it desirable to have a rapid response time. The asymptotic damping factor of 0.707 is entirely reasonable, since this entails good response without overshoot. But why $\zeta \to \sqrt{2}/2$ exactly and not some other value may seem astonishing. It turns out that the root loci of second-order systems under very general conditions tend to have a damping factor of $\zeta = \sqrt{2}/2$. A discussion of this feature is given below.

(*b*) As c tends to zero, the cost of control tends to a nonzero value. If the open-loop system were stable, and it would turn out that the gains g_1 and g_2 would tend to zero and the open-loop system would "coast" to rest, without incurring any control cost. Since control cost is paramount, this solution would be reasonable. In the present case, however, the open-loop system is unstable, and cannot coast to rest without control. A certain amount of control is necessary to stabilize the system. But why do both closed-loop system poles tend to $s = -\Omega$? One might have thought that only one closed-loop pole would tend to the stable open-loop pole at $s = -\Omega$ and that the other would tend to the origin. The fact that the second closed-loop pole also tends to $s = -\Omega$ is a consequence of a general result that as the control weighting becomes very large, the closed loop poles corresponding to *unstable* open loop poles tend to their *mirror images* with respect to the imaginary axis. In other words, if $s_i = +\alpha + j\beta$ ($\alpha \geqq 0$) in the open-loop system, then the corresponding pole in the closed-loop system tends to $\bar{s}_i = -\alpha + j\beta$. This is a general property of optimum control laws, as discussed in Note 9.4.

9.6 DISTURBANCES AND REFERENCE INPUTS: EXOGENOUS VARIABLES

In Chap. 7 we considered a more general model for the control process of the form

$$\dot{x} = Ax + Bu + Ex_0 \tag{9.31}$$

where x_0 is the exogenous vector. As in Chap. 7, we assume that x_0 satisfies a known differential equation

$$\dot{x}_0 = A_0 x_0 \tag{9.32}$$

Hence the entire (meta)state satisfies the differential equation

$$\dot{\mathbf{x}} = \mathbf{A}\mathbf{x} + \mathbf{B}u$$

where

$$\mathbf{x} = \begin{bmatrix} x \\ \hline x_0 \end{bmatrix} \begin{matrix} \updownarrow k \\ \updownarrow l \end{matrix}$$

$$\mathbf{A} = \left[\begin{array}{c|c} A & E \\ \hline 0 & A_0 \end{array}\right] \begin{array}{l} \updownarrow k \\ \updownarrow l \end{array} \qquad \mathbf{B} = \left[\begin{array}{c} B \\ \hline 0 \end{array}\right] \tag{9.33}$$

$$\underset{k}{\longleftrightarrow}\ \underset{l}{\longleftrightarrow}$$

Obviously, the exogenous state x_0 is not controllable; hence an appropriate performance integral would be

$$V = \int_t^T (x'Qx + u'Ru)\, d\tau \tag{9.34}$$

Thus, the weighting matrix for the metastate is of the form

$$\mathbf{Q} = \left[\begin{array}{c|c} Q & 0 \\ \hline 0 & 0 \end{array}\right]$$

The upper limit on the integral in (9.34) is intentionally not made infinite as one might at first be tempted to do. Why? Suppose that the exogenous state does not tend to zero. It may not be possible to achieve a steady state error of zero with a control u that also goes to zero; it usually isn't possible to do so. In that case, either $x'Qx$ doesn't go to zero, or $u'Ru$ doesn't go to zero. In either case, the integral in (9.34) will become infinite as $T \to \infty$. One way of approaching this problem is to find a control $\bar{u}$ which satisfies the requirements of zero steady state error. For $x = \dot{x} = 0$, the required steady state control $\bar{u}$ must satisfy

$$B\bar{u} + Ex_0 = 0 \tag{9.35}$$

and then express the total control u as the sum of the steady state control and a "corrective" control v:

$$u = \bar{u} + v$$

In this case (9.31) becomes

$$\dot{x} = Ax + Bv$$

Now the corrective control v does tend to zero and it is proper to minimize

$$\bar{V} = \int_t^\infty (x'Qx + v'Rv)\, d\tau \tag{9.36}$$

There are several problems with this approach. First, there may not be a control $\bar{u}$ which satisfies (9.35); in other words, it may not be possible to achieve zero steady state error, but it would still be possible to minimize (9.34) for any finite time. The control *law* that is approached by the solution to (9.34) as $T \to \infty$, even if the limiting integral does not exist, may be just fine. Second, the control which achieves (9.35) may not be unique, hence determination of a unique v by minimizing (9.36) does not pin down $u = \bar{u} + v$. And finally, minimizing the quadratic form

$$V = \int_t^T (x'Qx + v'Rv)\, d\tau$$

is not the same as minimizing

$$\bar{V} = \int_t^T (x'Qx + u'Ru)\, d\tau = \int_t^T [x'Qx + (\bar{u} + v)'R(\bar{u} + v)]\, d\tau$$

for a fixed $\bar{u}$, because of the presence of the cross terms $v'R\bar{u}' + \bar{u}'Rv$. If we really want to minimize V we don't want to minimize $\bar{V}$.

The finite time duration problem (9.3) can be solved without theoretical difficulty. Partition the performance matrix $\hat{\mathbf{M}}$ for the metasystem correspondingly

$$\hat{\mathbf{M}} = \left[\begin{array}{c|c} \hat{M}_1 & \hat{M}_2 \\ \hline \hat{M}_2' & \hat{M}_3 \end{array}\right] \tag{9.37}$$

The gain matrix $\hat{\mathbf{G}}$ for the metasystem is given by

$$\hat{\mathbf{G}} = R^{-1}[B' \mid 0]\left[\begin{array}{c|c} \hat{M}_1 & \hat{M}_2 \\ \hline \hat{M}_2' & \hat{M}_3 \end{array}\right] = [R^{-1}B'\hat{M}_1 \mid R^{-1}B'\hat{M}_2] \tag{9.38}$$

Note that the submatrix $\hat{M}_3$ is not needed. This is a welcome fact, as we shall soon see.

Performing the matrix multiplications required by (9.27) we obtain the differential equations for the submatrices in (9.37):

$$-\dot{\hat{M}}_1 = \hat{M}_1 A + A'\hat{M}_1 - \hat{M}_1 BR^{-1}B'\hat{M}_1 + Q \tag{9.39}$$

$$-\dot{\hat{M}}_2 = \hat{M}_1 E + \hat{M}_2 A_0 + (A' - \hat{M}_1 BR^{-1}B')\hat{M}_2 \tag{9.40}$$

$$-\dot{\hat{M}}_3 = \hat{M}_3 A_0 + A_0'\hat{M}_3 + \hat{M}_2'E + E'\hat{M}_2 - \hat{M}_2'BR^{-1}B'\hat{M}_2 \tag{9.41}$$

Owing to the special structure of **A**, **B**, and **Q**, the following facts about the submatrices of $\hat{\mathbf{M}}$ emerge:

(*a*) The solution for $\hat{M}_1$, and hence the corresponding gain $R^{-1}B'\hat{M}_1$, is the same as it would have been with x_0 absent from the problem, i.e., if we were designing the control law for the simple regulator problem. A steady state solution for $\hat{M}_1$ can be obtained if the pair (A, B) is controllable, as explained above.

(*b*) The differential equation for $\hat{M}_2$, from which the gain $R^{-1}B'\hat{M}_2$ is determined, does not depend on $\hat{M}_3$, and in fact is a linear equation, which can also be written

$$-\dot{\hat{M}}_2 = \hat{M}_1 E + \hat{M}_2 A_0 + A_c'\hat{M}_2 \tag{9.42}$$

where

$$A_c = A - BR^{-1}B'\hat{M}_1$$

is the closed-loop dynamics matrix of the regulator subsystem. A steady state solution to (9.42) generally also can be found. It must satisfy

$$0 = \bar{M}_1 E + \bar{M}_2 A_0 + A_c'\bar{M}_2 \tag{9.43}$$

We thus have the necessary gains to realize the control law

$$u = -R^{-1}B'\bar{M}_1 x - R^{-1}B'\bar{M}_2 x_0 \tag{9.44}$$

(*c*) The differential equation (9.41) for $\hat{M}_3$ is also linear. Whether it has a steady state solution depends on A_0. If $A_0 = 0$, then (9.41) does not have a steady state solution. But this doesn't matter because $\hat{M}_3$ is not used in the determination of the gain matrix.

The case of greatest interest is that in which the matrix A_0 is zero. In this case the exogenous subsystem produces signals that are constant. These are the most frequently used reference signals. For this case the equations for $\hat{M}_2$ and $\hat{M}_3$, as given by (9.40) and (9.41), become

$$-\dot{\hat{M}}_2 = \hat{M}_1 E + A_c'\hat{M}_2 \tag{9.45}$$

and

$$-\dot{\hat{M}}_3 = \hat{M}_2'E + E'\hat{M}_2 - \hat{M}_2'BR^{-1}B'\hat{M}_2 \tag{9.46}$$

Note that the right-hand side of (9.46) does not contain $\hat{M}_3$ and hence

$$M_3(t) = M_3(T) + \int_t^T (\hat{M}_2'E + E'\hat{M}_2 - \hat{M}_2'BR^{-1}B'\hat{M}_2)\,d\tau \tag{9.47}$$

In general the integral in (9.47) goes to infinity as $T \to \infty$; in other words a steady state solution to (9.46) does not exist, and any attempt to obtain such a solution, by setting $\dot{\hat{M}}_3$ to zero, will generally be erroneous. The correct relationship for $\bar{M}_2$ is given by the solution to (9.45) with $\dot{\hat{M}}_2 = 0$

$$\bar{M}_2 = -(A_c')^{-1}\bar{M}_1 E \tag{9.48}$$

where $\bar{M}_1$ is the steady state solution to (9.39), i.e., the control matrix for the regulator design. Thus the gain for the exogenous variables

$$G_0 = -R^{-1}B'(A_c')^{-1}\bar{M}_1 E = B^*E = G_0^*$$

where

$$B^* = -R^{-1}B'(A_c')^{-1}\bar{M}_1 \tag{9.49}$$

In Chap. 6 we considered the problem of reducing the steady state error to zero in the presence of exogenous variables. We found there that

$$G_0 = B^{\#}E$$

where

$$B^{\#} = (CA_c^{-1}B)^{-1}CA_c^{-1} \tag{9.50}$$

By use of the optimization technique of this chapter we have also found gains for the exogenous variables. Since the matrix $B^{\#}$ is unique for a given regulator gain matrix G, it follows that the gains given by (9.48) will reduce the steady state error to zero (for arbitrary E) only when B^* given by (9.49) and $B^{\#}$ given by (9.50) are the same.

Example 9B Accelerometer proof mass "capture" We previously determined that the differential equations governing the displacement z of the proof mass in an accelerometer, shown in Fig. 9.2, is given by

$$\begin{aligned}\dot{x}_1 &= x_2 \\ \dot{x}_2 &= -\frac{K}{\bar{M}}x_1 - \frac{B}{\bar{M}}x_2 + a\end{aligned} \tag{9B.1}$$

where $K/\bar{M}$ is the acceleration due to the spring, $B/\bar{M}$ is the acceleration due to friction, and a is the specific force (nongravitational acceleration) acting on the body. Suppose that the spring and damping forces are both absent. Then of course the proof mass would strike the end wall of the instrument after a short time. To "capture" the proof mass, i.e., to keep it from striking the end walls, a control force is generated in the typical instrument. (This can be accomplished magnetically, for example. The means of generating the force is not germane.) So instead of (9B.1), the differential equations for the proof mass, with the acceleration due to the capture force denoted by u, are

$$\begin{aligned}\dot{x}_1 &= x_2 \\ \dot{x}_2 &= u + a\end{aligned} \tag{9B.2}$$

These are just the equations of a double-integrator with an external disturbance a and a control u. For a constant acceleration a, the control acceleration must tend to $-a$; otherwise the proof mass will surely hit the wall. Thus, by measuring the control acceleration u that is needed to keep the proof mass from moving toward the walls, we can determine the external acceleration a. In Example 11F, using the separation principle, we will develop the design for a complete control system to capture the proof mass and provide an estimate $\hat{a}$ of the input acceleration a. But for now, let us consider only the control problem when all the state variables, including the input acceleration a, treated as a state variable, are assumed to be measurable. (If the acceleration a could be measured, there would of course be no need for this accelerometer.)

First, consider the control problem of returning the proof mass to the origin ($x_1 = x_2 = 0$) in the absence of an input acceleration ($a = 0$). The matrices for the dynamics are

$$A = \begin{bmatrix} 0 & 1 \\ 0 & 0 \end{bmatrix} \qquad B = \begin{bmatrix} 0 \\ 1 \end{bmatrix} \tag{9B.3}$$

We use a performance criterion of the form

$$V = \int_t^{\infty} \left(x_1^2 + \frac{u^2}{c^2} \right) dt \tag{9B.4}$$

The gain matrix for this control design is

$$G = R^{-1}B'M = c^2[0 \quad 1]\begin{bmatrix} m_1 & m_2 \\ m_2 & m_3 \end{bmatrix} = [c^2 m_2 \quad c^2 m_3] \tag{9B.5}$$

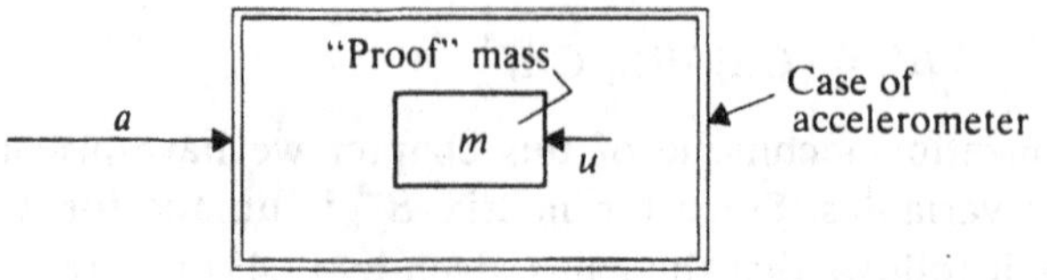

Figure 9.2 Force-rebalanced ("captured") accelerometer.

and the components of M are given by

$$\begin{aligned} 0 &= -c^2 m_2^2 + 1 \\ 0 &= m_1 - c^2 m_2 m_3 \\ 0 &= 2m_2 - c^2 m_3^2 \end{aligned} \tag{9B.6}$$

the solutions to which are

$$m_1 = 2^{1/2} c^{-1/2} \qquad m_2 = c^{-1} \qquad m_3 = 2^{1/2} c^{-3/2} \tag{9B.7}$$

(For details, see Example 9A for the inverted pendulum. The present example is a special case of Example 9A with $\Omega^2 = 0$.)

Using (9B.5) the gain matrix is obtained:

$$G = [c^2 m_2 \quad c^2 m_3] = [c \quad \sqrt{2c}] \tag{9B.8}$$

The dynamics matrix of the closed-loop system is given by

$$A_c = A - BG = \begin{bmatrix} 0 & 1 \\ 0 & 0 \end{bmatrix} - \begin{bmatrix} 0 \\ 1 \end{bmatrix} [c \quad \sqrt{2c}] = \begin{bmatrix} 0 & 1 \\ -c & -\sqrt{2c} \end{bmatrix} \tag{9B.9}$$

Hence the closed-loop poles are the roots of

$$|sI - A_c| = \begin{vmatrix} s & -1 \\ c & s + \sqrt{2c} \end{vmatrix} = s^2 + \sqrt{2c}\, s + c = 0$$

or

$$s_{1,2} = \frac{\sqrt{2}}{2} c(-1 \pm j)$$

The locus of the closed-loop poles are thus straight lines at 45 degrees to the coordinate axes and moving away from the origin as $c \to \infty$.

The case we really want to consider, of course, is a nonzero external acceleration. Any model for a can be used (e.g., a step, a ramp, etc.). Suppose that it is modeled as a step

$$\dot{a} = 0 \tag{9B.10}$$

Adjoining this to (9B.2) gives

$$\begin{bmatrix} \dot{x}_1 \\ \dot{x}_2 \\ \hline \dot{a} \end{bmatrix} = \left[\begin{array}{cc|c} 0 & 1 & 0 \\ 0 & 0 & 1 \\ \hline 0 & 0 & 0 \end{array}\right] \begin{bmatrix} x_1 \\ x_2 \\ \hline a \end{bmatrix} + \begin{bmatrix} 0 \\ 1 \\ \hline 0 \end{bmatrix} u \tag{9B.11}$$

The matrices are in the form of (9.33) with

$$E = \begin{bmatrix} 0 \\ 1 \end{bmatrix} \qquad A_0 = 0$$

Thus the theory developed below (9.33) applies. In particular, let

$$\hat{\mathbf{M}} = \left[\begin{array}{c|c} \hat{M}_1 & \hat{M}_2 \\ \hline \hat{M}_2' & \hat{M}_3 \end{array}\right] = \left[\begin{array}{cc|c} m_1 & m_2 & m_4 \\ m_2 & m_3 & m_5 \\ \hline m_4 & m_5 & m_6 \end{array}\right]$$

Then, as already found,

$$\hat{M}_1 = \hat{M} = \begin{bmatrix} 2^{1/2} c^{-1/2} & c^{-1} \\ c^{-1} & 2^{1/2} c^{-3/2} \end{bmatrix}$$

The submatrix $\bar{M}_2$ is found using (9.47). In this application (9.47) is

$$0 = \begin{bmatrix} 2^{1/2}c^{-1/2} & c^{-1} \\ c^{-1} & 2^{1/2}c^{-3/2} \end{bmatrix} \begin{bmatrix} 0 \\ 1 \end{bmatrix} + A_c' \begin{bmatrix} m_4 \\ m_5 \end{bmatrix}$$

or, upon use of A_c given by (9B.9),

$$\bar{M}_2 = \begin{bmatrix} m_4 \\ m_5 \end{bmatrix} = -(A_c')^{-1} \begin{bmatrix} c^{-1} \\ 2^{1/2}c^{-3/2} \end{bmatrix} = \begin{bmatrix} 0 \\ c^{-2} \end{bmatrix} \tag{9B.12}$$

Thus the gain matrix is given by

$$\bar{G} = [G \mid R^{-1}B'\bar{M}_2]$$

The part of the gain matrix $\bar{G}$ due to the state $[x_1, x_2]'$ was already found in (9B.8). The additional gain due to the forcing acceleration $a = x_3$ is

$$G_a = R^{-1}B'\bar{M}_2 = c^2[0 \quad 1] \begin{bmatrix} m_4 \\ m_5 \end{bmatrix} = 1$$

It is not in the least bit surprising that the gain for the external acceleration should turn out to be 1 exactly. In fact any other gain would be surprising: Obviously, when x_1 and x_2 are zero, the control acceleration u should be exactly equal in magnitude and opposite in sign to the external acceleration. Thus the control law

$$u = -g_1x_1 - g_2x_2 - 1 \cdot a \tag{9B.13}$$

is exactly what one would have expected to obtain.

Note that we never needed to determine the remaining term m_6 of $\bar{M}$. The differential equation for m_6 is a special case of (9.45). In particular,

$$-\dot{m}_6 = 2m_5 - c^2m_5^2 = 2c^{-2} - c^2(c^{-2})^2 = c^{-2}$$

Thus

$$m_6(t) = m_6(T) + c^{-2}(T - t)$$

which implies that a steady state solution for m_6 does not exist. This is not surprising, in view of the fact that a constant value of external acceleration demands a constant, nonzero control, and this cannot result in a finite value of the performance integral V over an infinite time interval. Nevertheless, the control law (9B.13) is eminently reasonable, provided an observer is used to estimate the unmeasured state variables $x_3 = a$ and possibly also $x_2 = \dot{x}_1$. (An optimum observer design is the subject of Examples 11A and 11F.)

Example 9C Temperature control The temperature control considered in Example 6E and 7D can also be designed by the method of this section. Suppose we have a set of capacitances and resistances for which the dynamic model of (6E.1) becomes

$$\begin{aligned} \dot{x}_1 &= -x_1 + x_2 + u \\ \dot{x}_2 &= x_1 - 3x_2 + 2x_0 \end{aligned} \tag{9C.1}$$

where x_1 and x_2 are internal temperatures, and where x_0 is the outside (ambient) temperature which may be assumed constant, i.e.,

$$\dot{x}_0 = 0 \tag{9C.2}$$

From (9C.1) and (9C.2) we obtain the metasystem

$$\dot{\mathbf{x}} = \begin{bmatrix} \dot{x}_1 \\ \dot{x}_2 \\ \hline \dot{x}_0 \end{bmatrix} = \left[\begin{array}{cc|c} -1 & 1 & 0 \\ 1 & -3 & 2 \\ \hline 0 & 0 & 0 \end{array}\right] \begin{bmatrix} x_1 \\ x_2 \\ \hline x_0 \end{bmatrix} + \begin{bmatrix} 1 \\ 0 \\ 0 \end{bmatrix} u$$

Thus the submatrices of **A** and **B** are

$$A = \begin{bmatrix} -1 & 1 \\ 1 & -3 \end{bmatrix} \qquad E = \begin{bmatrix} 0 \\ 2 \end{bmatrix} \qquad A_0 = 0 \qquad B = \begin{bmatrix} 1 \\ 0 \end{bmatrix} \tag{9C.3}$$

Before starting the calculations, take note of some features of the problem due to the physics of the process. First, it is recalled that the temperatures x_1 and x_2 are measured with respect to any arbitrary reference temperature. If the ambient temperature is also at the reference, i.e., $x_0 = 0$, then the steady state condition is that all temperatures are equal with no heat input ($u = 0$). The control problem in this case is to add heat ($u > 0$) or remove heat ($u < 0$) to bring the temperature to the ambient in an optimum manner. To visualize the control problem, one might imagine that x_1 and x_2 are temperatures in a building which has cooled down overnight. In the morning, the indoor temperatures, having reached the nighttime ambient temperature, are lower than the daytime ambient, which just happens to be the desired temperature. Thus our problem is to heat the building to raise its temperature to that of the ambient. A similar problem might be to cool a building that has reached a high daytime temperature to the ambient temperature of a pleasant summer evening.

In most climates, of course, the ambient temperature is either too hot or too cold, so that $x_0 \neq 0$. In the winter, heat must be added continuously ($u > 0$) to keep the temperature above the ambient; in the summer, heat must be removed continuously ($u < 0$) to keep the temperature below the ambient. Since our model of (9C.1) includes only one control variable (one heater or air conditioner) it is clear that it is not possible to keep both x_1 and x_2 at the reference temperature. We can control x_1 or x_2 or a weighted average of the two, but not both independently. The thermal model (9C.2) suggests that x_1 is the temperature of the area nearest the source of heat (in a residence, perhaps downstairs) and x_2 is the temperature of the area farthest from the heat (perhaps upstairs) and most prone to heat loss to the ambient environment. In the daytime we might wish to give more weight to the temperature x_1, and in the nighttime we might wish to give more weight to x_2. Thus a performance criterion of the form

$$V = \int_t^T [(c_1 x_1 + c_2 x_2)^2 + k^2 u^2]\, d\tau \tag{9C.4}$$

might be used, with $c_1 \gg c_2$ in the daytime and $c_2 \gg c_1$ in the nighttime. The state and control weighting matrices would thus be

$$Q = \begin{bmatrix} c_1^2 & c_1 c_2 \\ c_1 c_2 & c_2^2 \end{bmatrix} \qquad R = k^2 \tag{9C.5}$$

We are now prepared to perform the required calculations. First we find the gain matrix for the case in which $x_0 = 0$ using

$$\hat{G} = R^{-1} B' \hat{M}_1 = [g_1, g_2] \tag{9C.6}$$

with

$$\hat{M}_1 = \begin{bmatrix} m_1 & m_2 \\ m_2 & m_3 \end{bmatrix} \tag{9C.7}$$

satisfying (9.39). Using the data matrices of (9C.3) and (9C.5), we find that the steady state values of m_1, m_2, and m_3 satisfy

$$-2m_1 + 2m_2 - k^{-2} m_1^2 + c_1^2 = 0 \tag{9C.8}$$

$$m_1 - 4m_2 + m_3 - k^{-2} m_1 m_2 + c_1 c_2 = 0 \tag{9C.9}$$

$$2m_2 - 6m_3 - k^{-2} m_2^2 + c_2^2 = 0 \tag{9C.10}$$

These equations are too complicated to solve other than numerically. But the numerical values are easily obtained, and from these, using (9C.6), the gains are obtained:

$$g_1 = k^{-2} m_1 \qquad g_2 = k^{-2} m_2 \tag{9C.11}$$

Numerical values for g_1 and g_2 for several values of c_1 and c_2 are given in Table 9C.1.

Table 9C.1 Temperature control gains

	$c_1 = 1, c_2 = 0$		$c_1 = 1, c_2 = 1$		$c_1 = 0, c_2 = 1$	
k	g_1	g_2	g_1	g_2	g_1	g_2
1	0.4957	0.1185	0.685	0.419	0.0586	0.0603
10^{-1}	9.120	0.7067	9.860	8.470	1.692	3.122
10^{-2}	99.02	0.9628	99.981	98.06	10.562	66.338
10^{-3}	999.0	0.99615	1000.	998.0	40.855	875.44
10^{-4}	9999.0	0.99931	10000.	9998.0	137.46	9585.6

The variation of gains shown in Table 9C.1 seems reasonable: as the control weighting is decreased ($k \to 0$) the gains get higher—as the cost of energy decreases the temperature can be brought to the ambient more rapidly. Also note that the higher gain is associated with the state variable that is weighted more heavily in the performance index—also as expected. But the gains are unequal when the states are equally weighted because the heat (or cooling) input is not distributed to x_1 and x_2 in the same way.

Having found g_1 and g_2 we can now determine the gain for the ambient temperature:

$$g_0 = B^* E \tag{9C.12}$$

Where B^* is given by (9.49). Performing the required calculations we obtain the results shown in Table 9C.2.

It is of interest to compare the gain g_0 given by (9C.12) with the gain g_0 given by (6.48), needed for zero steady state error. For the weighting matrix

$$C = [c_1, c_2] \tag{9C.13}$$

we find from (6.48) that

$$g_0 = B^{\#} E = \frac{2c_1(1 - g_2) + 2c_2(1 + g_1)}{3c_1 + c_2} \tag{9C.14}$$

$$= \begin{cases} \frac{2}{3}(1 - g_2) & \text{for } c_2 = 0 \\ \dfrac{g_1 - g_2 + 2}{2} & \text{for } c_1 = c_2 \\ 2(1 + g_1) & \text{for } c_1 = 0 \end{cases}$$

The results, using (9C.14) with g_1 and g_2 as given in Table 9C.1, are also shown in Table 9C.2 for purpose of comparison. It is evident that except for the largest values ($k > 0.1$) of

Table 9C.2 Ambient temperature gains

	$c_1 = 1, c_2 = 0$		$c_1 = c_2 = 1$		$c_1 = 0, c_2 = 1$	
k	B^*E	$B^{\#}E$	B^*E	$B^{\#}E$	B^*E	$B^{\#}E$
1	0.2178	0.5873	0.685	1.133	0.3283	2.1172
10^{-1}	0.1511	0.1960	1.645	1.695	4.991	5.924
10^{-2}	0.02036	0.0250	1.954	1.96	23.084	23.124
10^{-3}	0.00216	0.00253	1.995	2.0	83.65	83.71
10^{-4}	~0	~0	1.9995	2.0	277.0	276.9

control weighting, the gains for the ambient temperature as given by B^*E are very close to the gains given by $B^{\#}E$ required to reduce the steady state error precisely to zero. The differences are largely academic, because the error due to measuring temperature with any sensor of realistic quality would be greater than the errors caused by the differences between B^*E and $B^{\#}E$.

Example 9D Missile autopilot In Chap. 6 (Example 6F) we obtained the design of a missile autopilot using a pole-placement technique. In this example we will obtain the design using the optimization methods of this chapter.

The state of the system is the difference e between the commanded and the achieved angular acceleration, the pitch rate q, and the control surface deflection δ

$$x = [e, q, \delta]' .$$

The dynamics are given by

$$\dot{x} = Ax + Bu + Ex_0$$

where x_0 is the commanded normal acceleration a_{NC}. The matrices A, B, and E are given in Example 6F.

To use the methods of this chapter it is appropriate to use a performance criterion which weights the error e and the control surface deflection δ

$$V = \int_t^\infty (e^2 + R\delta^2)\, d\tau$$

For this performance criterion

$$Q = \begin{bmatrix} 1 & 0 & 0 \\ 0 & 0 & 0 \\ 0 & 0 & 0 \end{bmatrix}$$

and R is a scalar.

The matrix quadratic equation is much too complicated to solve analytically, but it can readily be solved by a suitable numerical method. The numerical values of the elements of the gain matrix $\bar{G} = R^{-1}B'\bar{M}$ are tabulated for a range of control weightings in Table 9D.1. Table 9D.1 also shows the closed-loop poles and the matrices B^*E and $B^{\#}E$ which constitute the feedforward gain G_0 for the reference input.

A graphical representation of the closed-loop poles, as the control weighting R is varied, is shown in Fig. 9.3. It is seen that as R becomes very large, i.e., the control surface deflection is very heavily weighted, the closed-loop poles approach the open-loop poles, as one would expect. But as the weighting on the control surface is reduced (R is decreased), the complex

Table 9D.1 Missile autopilot design

R	G_e	G_q	G_δ	Closed-loop poles		B^*E	$B^{\#}E$
E5	2.086E-3	−.492	5.818	−360.0,	−46.0 ± j6.5	1.0755E-3	1.0786E-3
5E5	.873E-3	−.235	2.557	−176.4,	−42.6 ± j13.4	0.5363E-3	0.5433E-3
E6	.588E-3	−.173	1.795	−139.7,	−38.0 ± j16.9	0.4070E-3	0.4169E-3
5E6	.211E-3	−.085	0.784	−107.5,	−25.3 ± j20.4	0.2250E-3	0.2460E-3
E7	.127E-3	−.063	0.545	−103.7,	−20.0 ± j19.9	0.1746E-3	0.2047E-3
5E7	.274E-4	−.0296	0.222	−100.7,	−10.8 ± j17.5	0.0881E-3	0.1457E-3
E8	.105E-4	−.0205	0.146	−100.3,	−8.2 ± j16.8	0.0602E-3	0.1308E-3
E9	−.174E-5	−.004618	0.02883	−100.03,	−3.18 ± j15.8	0.0112E-3	0.1064E-3

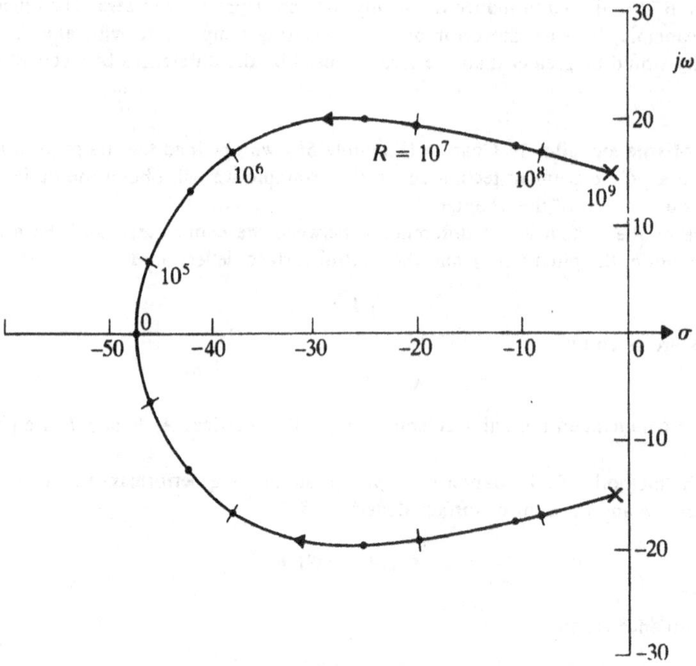

Figure 9.3 Locus of complex poles in missile autopilot design as weighting factor is varied.

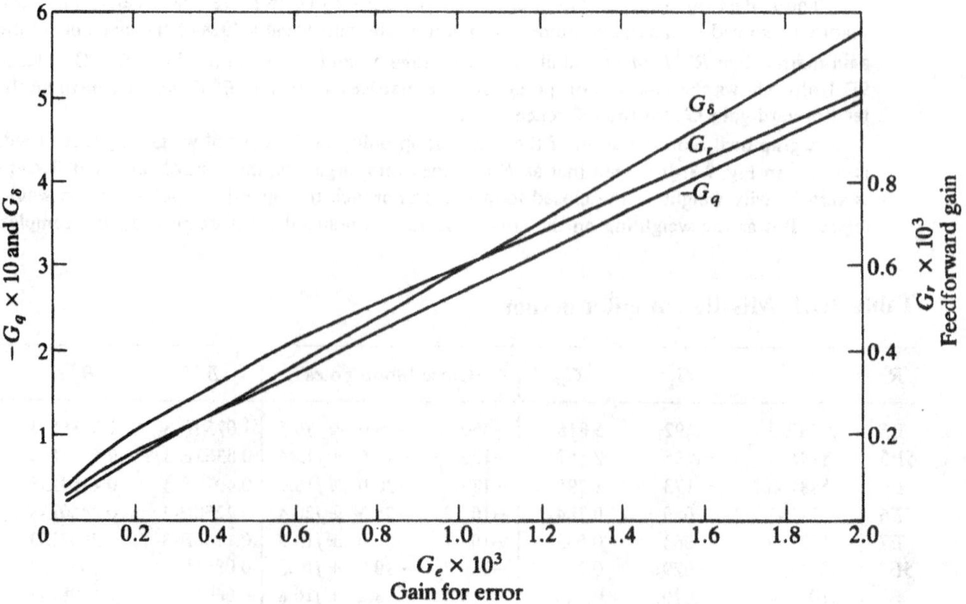

Figure 9.4 Variation of control gains for missile/autopilot design.

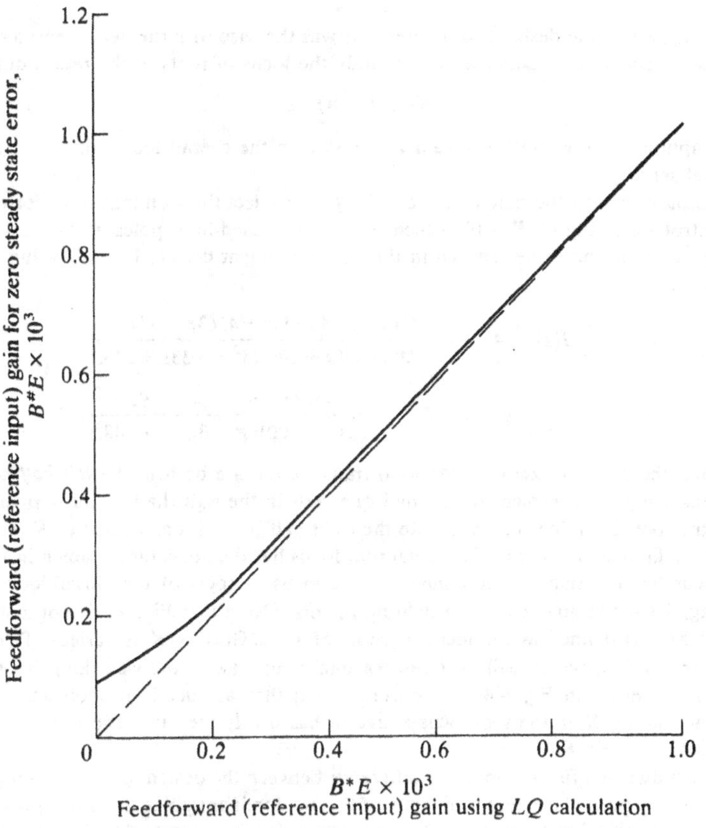

Figure 9.5 Comparison of feedforward gains.

poles move to the negative open-loop zero on the real axis, which is where we would expect it to go in view of our discussion in Note 9.4. The gain variations are illustrated graphically in Fig. 9.4.

Note that the gains B^*E and $B^{\#}E$ are not equal, although they converge as the control weighting tends to zero. (Fig. 9.5.) This again is as expected in view of our earlier discussion: If the control weighting is not zero, the cost of using control requires that its steady state value be reduced from that required to maintain a steady state error of zero. The discrepancy between the feedforward gains is largest when the control weighting is largest, as expected. Since the missile is stable, the feedback gains can be reduced to zero, which is what happens when the control weighting becomes infinite. But this also reduces the feedforward gain to zero and there is no connection between the reference input (the commanded acceleration) and missile: The achieved acceleration tends to zero leaving a steady state error equal to the commanded acceleration. But it is possible to track the input acceleration perfectly, even without feedback, by use of a feedforward gain given by $G_0 = (CA^{-1}B)^{-1}CA^{-1}E$ where A is the open-loop dynamics matrix. The numerical value of $G_0 = 0.1064 \times 10^{-3}$ is the feedforward gain that achieves this condition.

If zero steady state tracking error is a rigid requirement, then there is no reason for not using the gain $B^{\#}E$ as given in the last column of Table 9D.1 instead of the gain given by B^*E. Since these are feedforward gains, they have no effect on the stability of the system.

The robustness of the design is of interest. As was the case with the design based on pole placement, as considered in Example 6F, we study the locus of roots of the return difference

$$1 + KG(sI - A)^{-1}B$$

on the assumption that the effect of a gain variation is in the overall loop gain rather than in the individual sensors.

For comparison with the pole-placement design we select the gain matrix G corresponding to a control weighting of $R = 10^7$ which places the closed-loop poles at $s = -20 \pm j19.9$ which is very close to the values chosen in the pole-placement design. For this value of gain we find that

$$G_0(s) = G(sI - A)^{-1}B = \frac{N(s)}{D(s)} = \frac{40.35s^2 + 4363s + 57\,628}{(s + 100)(s^2 + 3.33s + 248.)}$$

$$= \frac{40.35(s + 15.4)(s + 92.7)}{(s + 100)(s^2 + 3.33s + 248.)}$$

Note that the apparent zeros of the loop transmission are both in the *left half* of the s plane whereas the pole-placement design had one zero in the right half of the s plane. This means that the root-locus does not cross into the right half-plane for any value of K. Thus this design has an infinite gain margin. The actual root locus has the appearance shown in Fig. 9.6. The root locus has the same general shape as the locus of roots of the closed-loop system shown in Fig. 9.3 for various control weighting factors. (Note that Fig. 9.3 is not a standard root locus, which is defined as the locus of roots of $1 + KG(s)$ as K is varied.) This might seem surprising at first, but it really is quite reasonable in view of the way the gains G_e, G_q, and G_δ vary, as shown in Fig. 9.4. It is observed that they are nearly proportional to each other, so that varying K in the root-locus equation has nearly the same effect as varying the control weighting matrix R.

It is worth dwelling further upon the difference between the design of this section and the design obtained by pole-placement in Example 6F. The dominant poles in both cases are very nearly in the same location ($s \approx -20 \pm j20$) so the transient responses of both systems would be just about the same. Yet the pole-placement design has a finite gain margin while the

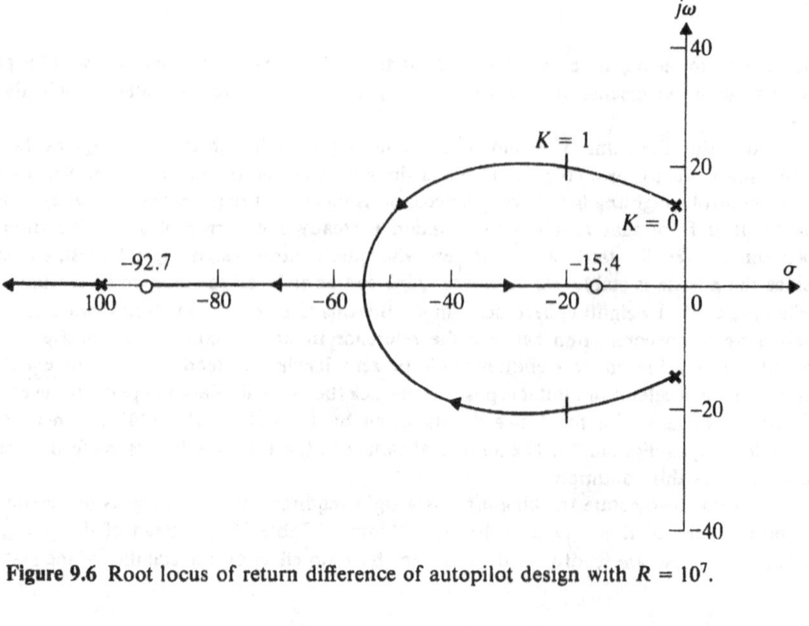

Figure 9.6 Root locus of return difference of autopilot design with $R = 10^7$.

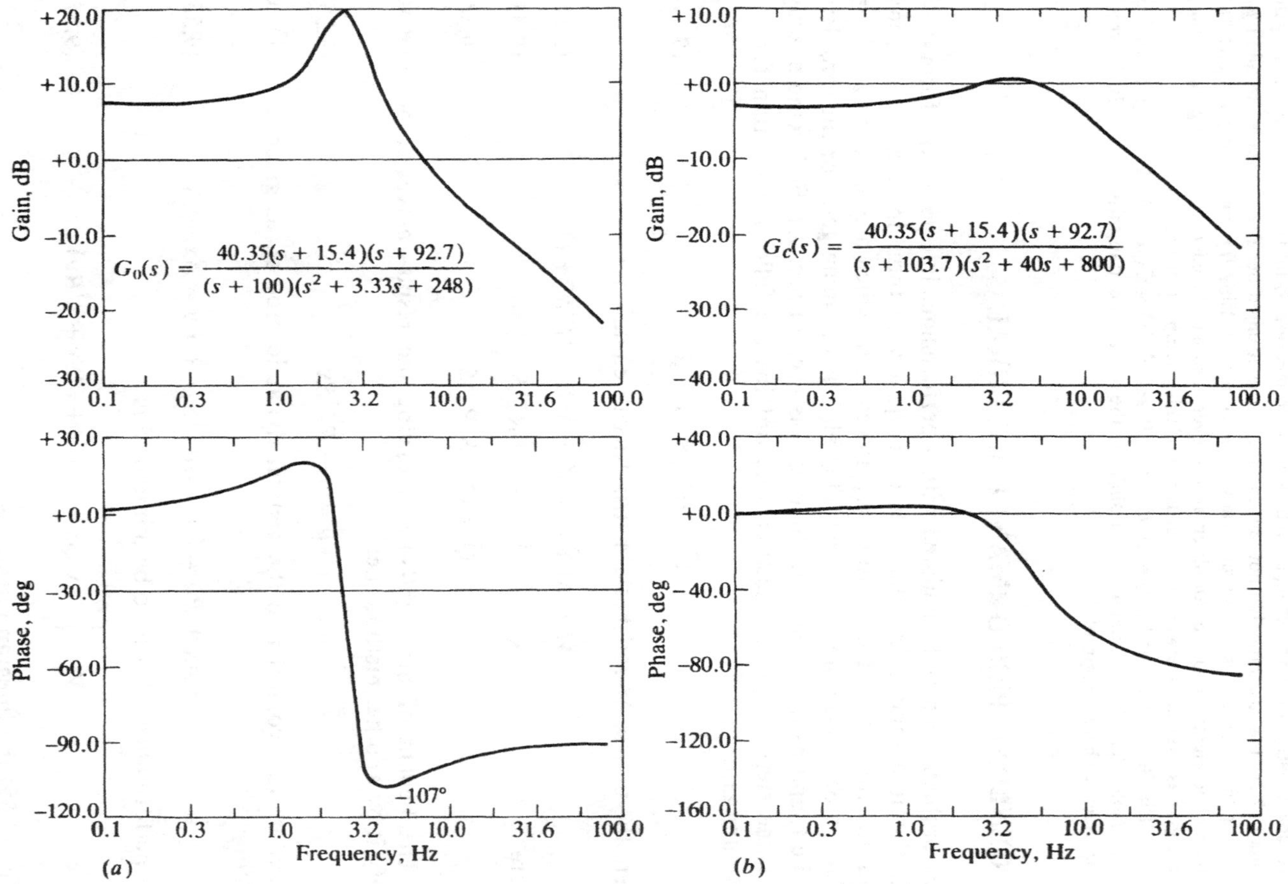

Figure 9.7 Bode plots for missile autopilot. (*a*) Open-loop transmission; (*b*) Closed-loop transmission.

linear-quadratic design of this example has an infinite gain margin. A gain margin of 14 is not at all bad, but a gain margin of infinity is better! On the other hand, the present design requires feedback of the actuator state δ. The pole-placement design intentionally eliminated this feedback path. Is it worth using an extra sensor (to measure δ) for the sake of raising the gain margin? In this case probably not, but in other cases it might be. The alternative to adding a sensor to measure δ is to use an observer to estimate δ using the measured pitch rate and normal acceleration. Use of an observer, however, also has the effect of reducing the stability margins as we shall see when our discussion of this example resumes.

The Bode plots for $G_0(s) = G(sI - A)^{-1}B$ and $G_c(s) = G(sI - A_c)B$ are given in Fig. 9.7. Note that the maximum phase shift of the open-loop transmission is $-107°$, which provides a phase margin of 73°.

9.7 GENERAL PERFORMANCE INTEGRAL

Most problems can be formulated with a performance integral of the form (9.3) with the integrand being the sum of a quadratic form in x and a second quadratic form in u. There are cases, however, in which a cross term $2x'S'u = x'S'u + u'Sx$ is also present in the integral. The optimum gain for this problem can be found using the same method as was used in Sec. 9.4. Following exactly the same steps as in that section, one obtains the following relation for the optimum gain

$$\hat{G} = R^{-1}(B'\hat{M} + S) \tag{9.51}$$

where the matrix $\hat{M}$ satisfies a matrix Riccati equation

$$-\dot{\hat{M}} = \hat{M}\bar{A} + \bar{A}'\hat{M} - \hat{M}BR^{-1}B'\hat{M} + \bar{Q} \tag{9.52}$$

where

$$\bar{A} = A - BR^{-1}S \tag{9.53}$$

$$\bar{Q} = Q - S'R^{-1}S \tag{9.54}$$

The benefit of hindsight—i.e., knowing the result—makes it possible to verify it by another method. Let

$$u = v - R^{-1}Sx \tag{9.55}$$

Substitute this control into the general dynamic process, as given by (9.1) to obtain

$$\dot{x} = Ax + Bu = (A - BR^{-1}S)x + Bv = \bar{A}x + Bv \tag{9.56}$$

The performance integral to be minimized is

$$V = \int_t^T (x'Qx + x'S'u + u'Sx + u'Ru)\, d\tau \tag{9.57}$$

Using (9.55) the integrand becomes

$$x'Qx + x'S'(v - R^{-1}Sx) + (v' - x'S'R^{-1})Sx + (v' - x'S'R^{-1})R(v - R^{-1}Sx)$$
$$= x'(Q - S'R^{-1}S)x + v'Rv \tag{9.58}$$

Thus minimization of (9.57) for the original process is equivalent to minimization of

$$V = \int_t^\infty (x'\bar{Q}x + v'Rv)\, d\tau \tag{9.59}$$

for the process

$$\dot{x} = \bar{A}x + Bv \tag{9.60}$$

Using the result of Sec. 9.4 the minimum value of V is obtained for

$$v = -\bar{G}x$$

where the gain for v is given by

$$\bar{G} = R^{-1}B'\hat{M}$$

and where $\hat{M}$ satisfies (9.52). Thus, finally, from (9.55)

$$u = -(R^{-1}B'\hat{M} + R^{-1}S)x = -\hat{G}x$$

where $\hat{G}$ is given by (9.51).

9.8 WEIGHTING OF PERFORMANCE AT TERMINAL TIME

In control processes of finite time duration, the terminal state $x(T)$ is often as important as, or more important than, the manner in which the state is reached. Thus a more general performance criterion is

$$V = \int_t^T [x'(\tau)Qx(\tau) + u'(\tau)Ru(\tau)]\, d\tau + x'(T)Zx(T) \tag{9.61}$$

The additional quadratic form $x'(T)Zx(T)$ may be called a *terminal penalty*—the cost of not getting to the origin at the terminal time.

The results of Secs. 9.3 and 9.4 are applicable to this problem except that the terminal condition to be used is

$$\hat{M}(T, T) = Z \tag{9.62}$$

instead of $\hat{M}(T, T) = 0$.

This is seen as follows. Since

$$x(T) = \Phi_c(T, t)x(t)$$

the quadratic form $x'(T)Zx(T)$ is also a quadratic form in the initial state:

$$x'(T)Zx(T) = x'(t)\Phi_c'(T, t)Z\Phi_c(T, t)x(t) \tag{9.63}$$

Thus

$$V = V(t, T) = x'(t)M(t, T)x(t) \tag{9.64}$$

where

$$M(t, T) = \int_t^T [x'(\tau)Qx(\tau) + u'(\tau)Ru(\tau)]\, d\tau + \Phi_c'(T, t)Z\Phi_c(T, t) \tag{9.65}$$

and clearly

$$M(T, T) = Z \tag{9.66}$$

since $\Phi_c(T, T) = I$ for any transition matrix. From (9.64)

$$\frac{\partial V}{\partial t} = \dot{x}'(t)M(t, T)x(t) + x'(t)\dot{M}(t, T)x(t) + x'(t)M(t, T)\dot{x}(t)$$

whereas from (9.61), with $x'(T)Zx(T)$ given by (9.63) and with $u = -Gx$,

$$\frac{\partial V}{\partial t} = x'(t)Lx(t) + \frac{\partial}{\partial t}[x'(t)\Phi_c'(T, t)Z\Phi_c(T, t)x(t)] \tag{9.67}$$

where $L = Q + G'RG$ as in (9.12).

The vector $x(T)$, when written

$$x(T) = \Phi_c(T, t)x(t)$$

seems to be a function of the initial time t, which appears in $\Phi_c(T, t)$ and $x(t)$. However

$$\frac{\partial x(T)}{\partial t} = \frac{\partial \Phi_c(T, t)}{\partial t} x(t) + \Phi_c(T, t)\dot{x}(t) \tag{9.68}$$

Recall from Chap. 2, however, that for any transition matrix

$$\Phi_c(t, T)\Phi_c(T, t) = I$$

Thus

$$\frac{\partial \Phi_c(t, T)}{\partial t}\Phi_c(T, t) + \Phi_c(t, T)\frac{\partial \Phi_c(T, t)}{\partial t} = 0 \tag{9.69}$$

But

$$\frac{\partial \Phi_c(t, T)}{\partial t} = A_c(t)\Phi_c(t, T) \tag{9.70}$$

for any transition matrix. Thus (9.69) becomes

$$A_c(t) + \Phi_c(t, T)\frac{\partial \Phi_c(T, t)}{\partial t} = 0$$

or

$$\frac{\partial \Phi_c(T, t)}{\partial t} = -\Phi_c(T, t)A_c(t) \tag{9.71}$$

Thus, finally, using (9.71) and the basic relation

$$\dot{x}(t) = A_c(t)x(t)$$

it is found that (9.68) becomes

$$\frac{\partial x(T)}{\partial t} = 0$$

Hence the second term in (9.67) vanishes and $M(t, T)$ satisfies the same differential equation as it did in Sec. 9.3, namely

$$-\dot{M} = MA_c + A_c'M + L$$

but subject to the condition (9.66). It follows that the optimum control law satisfies the usual matrix Riccati equation (9.27) with the terminal condition (9.61).

If the closed-loop system is asymptotically stable then $x(T) \to 0$ and the terminal penalty is zero. Thus a stabilizing steady state control designed by solving (9.28) is also the optimum control for the limit of V given by (9.61) with $T \to \infty$.

Example 9E Missile guidance In Example 3B we considered the approximate dynamic model of a missile which is controlled by the use of a control acceleration normal to the velocity vector:

$$\dot{z} = \bar{T}u \tag{9E.1}$$

where z is the projected miss distance if the relative velocity V between the missile and the target were unchanged (that is, $u = 0$), u is the normal acceleration, and $\bar{T} = T - t$ is the time-to-go, assumed to be a known quantity.

If z is brought to zero at any time, the missile will, in the absence of any further normal acceleration ($u = 0$), continue on a straight-line trajectory to intercept the target. Thus the control objective is to reduce z to zero. There are of course countlessly many ways that this can be accomplished. The only requirement is that

$$z(T) = z(t) + \int_t^T (T - \tau)u(\tau)\, d\tau = 0 \tag{9E.2}$$

Any control law satisfying (9E.2) will serve, provided the normal acceleration u is sufficiently large. And if u is not sufficiently large (i.e., the missile is not sufficiently maneuverable) no control law will do.

In order to formulate a suitable optimization problem we suppose that the control objective is to minimize

$$V = \int_t^T u^2(\tau)\, d\tau + k^2 z^2(T) \tag{9E.3}$$

The integral term in (9E.3) is a quadratic form in the normal acceleration; it penalizes large accelerations and hence is a way of limiting the acceleration requirement. The second term $k^2z^2(T)$ penalizes the terminal miss distance. The larger the value of k, the greater the cost attached to missing the target; as $k \to \infty$ the target must be hit at all costs.

In reality there is no reason to care about the integral of the square of the acceleration provided the missile hits the target, so we are not really interested in solving the optimization problem. But, as we shall see, the solution of the optimization problem provides a very reasonable guidance law and hence, even though the optimization problem is fishy, the result is good: the end justifies the means.

The matrices that define the problem are all scalars

$$A = 0 \qquad B(t) = T - t \qquad Q = 0 \qquad R = I \qquad Z = k^2 \tag{9E.4}$$

Thus the "optimum" guidance law is

$$u(t) = -R^{-1}B(t)M(t, T)z(t) = -(T-t)M(t, T)z(t) \tag{9E.5}$$

where $M(t, T)$ is a scalar satisfying the Riccati equation

$$-\dot{M} = -(T-t)^2M^2 \tag{9E.6}$$

subject to the terminal condition

$$M(T, T) = k^2 \tag{9E.7}$$

by (9E.4). To solve (9E.6) let

$$W(t) = 1/M(t, T) \tag{9E.8}$$

Then $\dot{M} = M^2\dot{W}$ and (9E.6) becomes

$$\dot{W} = (T-t)^2$$

which is integrated directly to give

$$W(T) = W(t) + \int_t^T (T-\tau)^2\, d\tau = W(t) + \frac{(T-t)^3}{3} \tag{9E.9}$$

But, by (9E.7) and (9E.8), $W(T) = 1/k^2$. Thus

$$W(t) = \frac{1}{k^2} + \frac{(T-t)^3}{3}$$

and hence, by (9E.8), we have the desired solution to the Riccati equation:

$$M(t, T) = \frac{3}{(T-t)^3 + 3/k^2} \tag{9E.10}$$

which, when substituted into (9E.5), gives the optimum guidance law. If we truly want the terminal miss to be zero, we must let k^2 be infinite. In this limiting case (9E.10) becomes

$$M(t, T) = \frac{3}{(T-t)^3} \tag{9E.11}$$

Note that $M(T, T)$ becomes infinite. This is a typical characteristic in control problems in which the state is to be brought to zero in a *finite* time. Using (9E.11) in (9E.5) gives the guidance law

$$u(t) = -\frac{3}{(T-t)^2}z(t) \tag{9E.12}$$

Note that the gain $G(t, T) = 3/(T-t)^2$ also becomes infinite as the time-to-go $T-t$ approaches zero. This is surely reasonable. As the time-to-go gets smaller, a larger acceleration is needed to reduce the projected miss distance to zero.

A form of (9E.12) more familiar in missile guidance is obtained by observing that

$$z(t) = V(T-t)^2\dot{\lambda}(t)$$

where $\dot{\lambda}$ is the inertial line-of-sight rate. (See Example 3B.) As a result (9E.12) becomes

$$u(t) = -3V\dot{\lambda}(t) \tag{9E.13}$$

This is a special case of the well-known "proportional navigation" law[3]

$$u(t) = -kV\dot{\lambda}(t) \tag{9E.14}$$

Proportional navigation has long been a popular guidance law, primarily for ease of implementation. In a typical missile, the seeker is gimbaled to keep it pointing to the target independent of the attitude of the missile. The output of a rate gyro mounted on the gimbal, in the absence of instrument errors, is precisely the inertial line-of-sight rate $\dot{\lambda}$. Thus (9E.14) is implemented by multiplying the measured line-of-sight rate by a constant (kV) and using that quantity as the reference acceleration to the autopilot which is designed to track this acceleration. (See Examples 8D, 9D.)

In this example we appear to have proved that proportional navigation, with a navigation constant of $k = 3$, precisely, is the optimum guidance law. What we have really done is to find a performance criterion which rationalizes proportional navigation (with $k = 3$). In fact, any value of $k > 2$ will guarantee interception; k may even be time-varying, so that there is no need to know the relative velocity V with great precision, since a change in V from the design value will have the same effect as a variation in the navigation constant. By including $z(t)$ under the integral sign in (9E.3), and giving it and $u(t)$ time-varying weighting, it is possible to rationalize almost any guidance law as being optimum. This should not be interpreted as diminishing the value of optimization theory but rather as evidence that optimization theory does not answer every goal of control system design.

PROBLEMS

Problem 9.1 Hamiltonian system equivalent to Riccati equation

If the optimum control law is derived using variational principles, the following "hamiltonian" $2k \times 2k$ system results

$$\begin{aligned} \dot{x} &= Ax - BR^{-1}B'p \\ \dot{p} &= -Qx - A'p \end{aligned} \tag{P9.1}$$

(a) By the substitution

$$p(t) = \hat{M}(t, T)x(t) \tag{P9.2}$$

show that (P9.1) is equivalent to the matrix Riccati equation (9.27).

(b) Let the transition matrix corresponding to the "hamiltonian matrix"

$$H = \begin{bmatrix} A & BR^{-1}B' \\ -Q & -A' \end{bmatrix} \tag{P9.3}$$

be

$$\Psi(t) = e^{Ht} = \begin{bmatrix} \Psi_{11}(t) & \Psi_{12}(t) \\ \Psi_{21}(t) & \Psi_{22}(t) \end{bmatrix} \tag{P9.4}$$

Using (P9.2) show that the solution to the matrix Riccati equation (9.27) can be written as

$$M(t, T) = [\Psi_{21}(t-T) + \Psi_{22}(t-T)M(T,T)][\Psi_{11}(t-T) + \Psi_{12}(t-T)M(T,T)]^{-1} \tag{P9.5}$$

or as

$$M(t, T) = [\Psi_{12}(T-t) - M(T,t)\Psi_{22}(T,t)]^{-1}[\Psi_{11}(T-t) - M(T,T)\Psi_{21}(T-t)] \tag{P9.6}$$

Problem 9.2 Transition matrix for hamiltonian system is symplectic

Using the two expressions (P9.5) and (P9.6) for $M(t, T)$ and the fact that

$$\begin{bmatrix} \Psi_{11}(T-t) & \Psi_{12}(T-t) \\ \Psi_{21}(T-t) & \Psi_{22}(T-t) \end{bmatrix} \begin{bmatrix} \Psi_{11}(t-T) & \Psi_{12}(t-T) \\ \Psi_{21}(t-T) & \Psi_{22}(t-T) \end{bmatrix} = \begin{bmatrix} I & 0 \\ 0 & I \end{bmatrix}$$

deduce that the transition matrix of a hamiltonian system is "symplectic," i.e.,

$$\Psi' J \Psi = J = \begin{bmatrix} 0 & I \\ -I & 0 \end{bmatrix}$$

Problem 9.3 Eigenvalues of hamiltonian system and steady state solution to Riccati equation

(*a*) Show that the $2k$ eigenvalues of the hamiltonian matrix occur in pairs that are the negatives of each other; i.e., if λ_i $(i = 1, \dots, k)$ is an eigenvalue of H then $-\lambda_i$ is an eigenvalue of H. Thus the eigenvalues in complex plane have the appearance shown in Fig. P9.3.

(*b*) Assume that the eigenvalues are distinct. Then H is similar to a diagonal matrix

$$H = \begin{bmatrix} T_{11} & T_{12} \\ T_{21} & T_{22} \end{bmatrix} \begin{bmatrix} \Lambda_1 & 0 \\ 0 & \Lambda_2 \end{bmatrix} \begin{bmatrix} U_{11} & U_{12} \\ U_{21} & U_{22} \end{bmatrix}$$

where

$$\begin{bmatrix} T_{11} & T_{12} \\ T_{21} & T_{22} \end{bmatrix} \begin{bmatrix} U_{11} & U_{12} \\ U_{21} & U_{22} \end{bmatrix} = \begin{bmatrix} I & 0 \\ 0 & I \end{bmatrix}$$

and where Λ_1 is the diagonal matrix corresponding to the left half-plane eigenvalues of H, and Λ_2 is the diagonal matrix corresponding to the right half-plane eigenvalues of H. Note that the columns of

$$T_1 = \begin{bmatrix} T_{11} \\ T_{21} \end{bmatrix} \quad \text{and} \quad T_2 = \begin{bmatrix} T_{12} \\ T_{22} \end{bmatrix}$$

are eigenvectors corresponding to the left half-plane and right half-plane eigenvalues, respectively.

Show that

$$e^{H\tau} = \begin{bmatrix} T_{11} & T_{12} \\ T_{21} & T_{22} \end{bmatrix} \begin{bmatrix} e^{\Lambda_1 \tau} & 0 \\ 0 & e^{\Lambda_2 \tau} \end{bmatrix} \begin{bmatrix} U_{11} & U_{12} \\ U_{21} & U_{22} \end{bmatrix}$$

Thus the solution to (P9.1) can be written

$$p(T) = (T_{11}\, e^{\Lambda_1 \tau} U_{11} + T_{12}\, e^{\Lambda_2 \tau} U_{21}) x(t) + (T_{11}\, e^{\Lambda_1 \tau} U_{12} + T_{12}\, e^{\Lambda_2 \tau} U_{22}) p(t)$$

$$x(T) = (T_{21}\, e^{\Lambda_1 \tau} U_{11} + T_{22}\, e^{\Lambda_2 \tau} U_{21}) x(t) + (T_{21}\, e^{\Lambda_1 \tau} U_{12} + T_{22}\, e^{\Lambda_2 \tau} U_{22}) p(t)$$

with $\tau = T - t$.

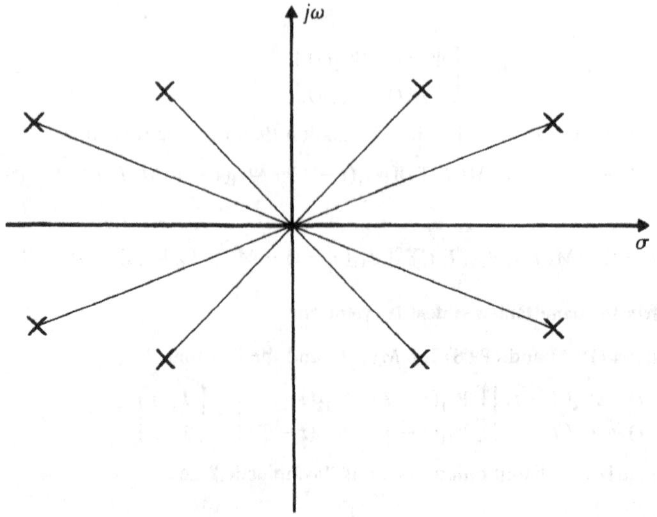

Figure P9.3 Eigenvalues of hamiltonian system are negatives of each other.

In order for $x(T)$ to go to zero in the steady state it is necessary for those terms containing increasing exponentials to vanish. This means that

$$T_{22} e^{\Lambda_2 T} U_{21} x(t) + T_{22} e^{\Lambda_2 T} U_{22} p(t) = 0$$

or

$$p(t) = U_{22}^{-1} U_{21} x(t)$$

It thus follows that the steady state solution $\bar{M}$ to the matrix Riccati equation is

$$\bar{M} = U_{22}^{-1} U_{21}$$

where U_{22} and U_{21} are submatrices formed from the eigenvectors of the hamiltonian system. Some numerical algorithms for determining $\bar{M}$ are based on this result.

Problem 9.4 Closed loop poles

Show that the closed-loop poles, i.e., the eigenvalues of $A_c = A - BR^{-1}B'\bar{M}$, of the optimum system are the eigenvalues of the hamiltonian system (P9.3) that lie in the left half-plane.

Problem 9.5 Closed loop return difference

(*a*) By block-diagram algebra show that the return-difference for the closed-loop system with full state feedback (that is, $u = -Gx$) is given by

$$T_c(s) = I + G\Phi(s)B \tag{P9.7}$$

where $\Phi(s) = (sI - A)^{-1}$ is the resolvent of the open-loop system.

(*b*) Let the control weighting matrix be of the form $R = r^2 I$ where r^2 is a scalar. Show that the closed-loop return difference $T_c(s)$ satisfies

$$T_c'(-s)T_c(s) = I + B'\Phi'(-s)Q\Phi(s)B/r^2 \tag{P9.8}$$

Hint: The matrix Riccati equation from which G is determined can be written

$$-MA - A'M + MBB'M/r^2 = Q \tag{P9.9}$$

Add and subtract sM to the left-hand side of (P9.9); then multiply the result by B' on the left and B on the right. Note that $MB = r^2 G$ to obtain (P9.8). This result and method of derivation was used by Kalman.[9]

(*c*) Use (P9.8) to show that the singular values of the closed-loop return differences are greater than unity for all $s = j\omega$.

Problem 9.6 Instrument servo

The gains for the instrument servo for which the gains were determined by pole placement in Example 6A are to be determined by minimization of a quadratic performance criterion. For numerical calculations use

$$\alpha = 1 \qquad \beta = 3$$

(*a*) For the performance criterion

$$V = \int_t^\infty (q_1^2 e^2 + u^2)\, d\tau$$

find and tabulate the control gains and corresponding closed-loop poles.

(*b*) Plot the transient response (e as a function of t) for an initial error of unity for several values of q_1^2.

(*c*) In addition to weighting the position error it is also desired to limit the velocity by using a performance criterion

$$V = \int_t^\infty (q_1^2 e^2 + q_2^2 \dot{e}^2 + u^2)\, d\tau$$

For several of the values of q_1^2 used in part *a* and for $q_2^2 = 0.1q_1^2$, q_1^2, and $10q_1^2$ find the control gains and corresponding closed-loop poles.

(*d*) Plot the transient response as in part *b* for a range of q_1^2 and q_2^2. Compare the results with those of part *b*. Are the results as expected?

Problem 9.7 Two-car train—one-drive motor

It is desired to rapidly bring to the origin the train of Prob. 3.9 et seq. driven by one motor, but without excessively stretching the spring. Explain why a suitable performance criterion is

$$V = \int_t^\infty [(x_1 + x_2)^2 + q^2(x_1 - x_2)^2 + r^2u^2]\, d\tau \tag{P9.10}$$

(*a*) What is the role of the weighting factor q^2?

(*b*) Specify the weighting matrices Q and R for the performance criterion (P9.10).

(*c*) Using the numerical data of Prob. 3.9 and for ranges of q^2 and r^2 chosen to cover the gamut of possible control laws, compute the gain matrix and the corresponding closed-loop poles.

(*d*) Discuss the choice of q^2 and r^2 in a realistic application. How does the strength of the spring enter into this selection? What other factors should be considered?

Problem 9.8 Two-car train—two-drive motors

Consider the two-car train of Prob. 9.7, except that both cars are motor-driven, so a suitable performance criterion in this case is

$$V = \int_t^\infty [(x_1 + x_2)^2 + q^2(x_1 - x_2)^2 + r^2(u_1^2 + u_2^2)]\, d\tau \tag{P9.11}$$

(*a*) Repeat the calculations of Prob. 9.7 to obtain the gain matrix and closed-loop poles.

(*b*) Is there any obvious advantage to using the two drive motors over using only one? It may be necessary to calculate transient responses to answer this question.

Problem 9.9 Two-axis gyro: optimum regulator gains

It is desired to choose the regulator gains for the two-axis gyro (Prob. 6.9 et seq.) control system to minimize a performance criterion of the form

$$V = \int_t^\infty [q^2(\delta_x^2 + \delta_y^2) + (u_x^2 + u_y^2)]\, d\tau$$

(*a*) Show that the symmetry (and antisymmetry) properties of the dynamics matrix make the regulator gain matrix be of the form

$$G = \left[\begin{array}{cc|cc} g_1 & g_2 & g_3 & 0 \\ -g_2 & g_1 & 0 & g_3 \end{array}\right]$$

Also the gain matrix for the exogenous variables is of the form

$$G_0 = \begin{bmatrix} g_4 & 0 \\ 0 & g_4 \end{bmatrix}$$

(*b*) Using the numerical data of Prob. 6.9 find and plot the control gains g_1, g_2, g_3, and the closed-loop poles of the regulator. Vary q^2 from 0 to the value for which the real parts of the poles lie at $\sigma = -5000$ rad/s.

(*c*) Find the matrices

$$G_0^* = B^*E \qquad \text{and} \qquad G_0^\# = B^\#E$$

Problem 9.10 Inverted pendulum on cart: optimum gains

The gains for stabilizing the inverted pendulum on a motor-driven cart of Probs. 2.1, 3.6 et seq. are to be optimized using a performance criterion of the form

$$V = \int_t^\infty (q_1^2 x_1^2 + q_3^2 x_3^2 + r^2 u^2)\, dt$$

A pendulum angle much greater than 1 degree = 0.017 rad would be precarious. Thus, a heavy weighting on $\theta = x_3$ is indicated: $q_3^2 = 1/(0.017)^2 \approx 3000$. For the physical dimensions of the system, a position error of the order of 10 cm = 0.1 m is not unreasonable. Hence $q_1^2 = 1/(0.1)^2 = 100$.

(*a*) Using these values of q_1^2 and q_3^2, determine and plot the gain matrices and corresponding closed-loop poles as a function of the control weighting parameter r^2 for $0.001 < r^2 < 50$.

(*b*) Repeat part *a* for a heavier weighting: $q_1^2 = 10^4$ on the cart displacement.

Problem 9.11 Hydraulically actuated gun turret: azimuth channel

The gains for the gun-turret considered in Example 2D et seq. are to be designed to minimize a quadratic performance criterion. According to Loh, Cheok, and Beck,[4] a suitable criterion would be

$$V = \int_t^\infty (q^2 x_1^2 + u^2)\, dt$$

where $10 \leq q^2 \leq 10\,000$.

(*a*) For the azimuth channel (see Table 2D.1 for numerical data) compute and tabulate or plot the gain matrix and corresponding closed-loop poles for q^2 in the range given above.

(*b*) It is desired to slew the gun to a constant angle θ_0 in the presence of disturbances. Let the state x_1 be $\theta - \theta_0$ and the exogenous vector be

$$x_0 = [\theta_0, d_T, d_p, d_q]'$$

be assumed constant. (See Prob. 6.2.) Find the feedforward gain matrices $G_0^* = B^* E$ and $G_0^\# = B^\# E$ for the range of q^2 studied in part *a*.

Problem 9.12 Hydraulically actuated gun turret: elevation channel

Repeat Prob. 9.11 but using the numerical data for the elevation channel as given in Table 2D.1.

Problem 9.13 Constant-altitude autopilot

The gains for the autopilot considered in Prob. 6.6 are to be determined by minimizing the performance criterion

$$J = \int_t^\infty [c_1^2 (h - h_0)^2 / V^2 + c_2^2 \alpha^2 + r^2 \delta_E^2]\, d\tau$$

(J is used to avoid confusion with the aircraft velocity V.) The weighting coefficient c_1^2 penalizes deviations from the desired altitude; c_2^2 penalizes excessive angle-of-attack which is desirable to avoid stalling.

(*a*) For $c_1^2 = c_2^2 = 1$, and $10^{-4} \leq r^2 \leq 100$, tabulate or plot the control gains and corresponding closed-loop poles.

(*b*) Repeat part *a* for $c_2^2 = 0.01 c_1^2$.

(*c*) Repeat part *a* for $c_2^2 = 100 c_1^2$.

Problem 9.14 Aircraft stability augmentation: optimum gains

The gains for the aircraft stability augmentation system of Example 8C are to be designed to minimize a quadratic performance criterion

$$V = \int_t^\infty (q_2^2 x_2^2 + q_3^2 x_3^2 + u^2)\, d\tau$$

(*a*) With $q_2^2 = 0$, find the value of q_3^2 which comes closest to achieving the desired pole locations as given in Example 8C.

(*b*) Determine how much closer the closed-loop poles can be made to the desired locations of part *a* by varying both q_2^2 and q_3^2.

Problem 9.15 Aircraft lateral dynamics: optimum control gains

For the aircraft lateral dynamics considered in Prob. 4.4 et seq. it is desired to achieve well-coordinated turns and reduce the roll angle error to zero rapidly. This suggests a performance criterion

$$V = \int_t^\infty [q_1^2 \beta^2 + q_2^2(\phi - \phi_0)^2 + (\delta_A^2 + \rho^2 \delta_R^2)]\, d\tau$$

where q_1^2 is the weighting on sideslip, q_2^2 is the weighting on the roll angle, ρ^2 is the weighting on the rudder deflection. (The weighting on the aileron deflection is normalized to unity. There are thus three parameters that can be used to design the control gains.)

(*a*) Find the feedback control ("regulator") gains for the case $q_1 = q_2 = \rho = 1$ and determine the corresponding pole locations.

(*b*) It is desired that approximate closed-loop pole locations be at $s = -1$, $s = -5$, $s = -1 \pm j3$. If the poles of part *a* are not located in the approximate vicinity of those specified, modify the weighting matrices to obtain more favorable locations.

(*c*) Determine the feedforward gains B^*E and $B^{\#}E$ for the control gains obtained in part *b*.

Problem 9.16 Three-capacitance thermal system

The gains for the three capacitance thermal system are to be optimized for the performance criterion

$$V = \int_t^\infty (x_3^2 + r^2 u^2)\, d\tau$$

(*a*) Find and tabulate (or plot) the regulator gains and pole locations as a function of the control weighting parameter r^2. Let r^2 range from 10^{-6} to 1.0.

(*b*) For each of the gains in part *a*, find the gain matrices $G_0^* = B^*E$ and $G_0^{\#} = B^{\#}E$ and compare the results.

Problem 9.17 Distillation column

The control gains from the distillation column considered in Example 6D et seq. are to be determined by minimization of a quadratic performance criterion. A suitable performance criterion would be

$$V = \int_t^\infty [100x_1^2 + 100x_2^2 + 100x_3^2 + 1500x_4^2 + r_1 u_1^2 + r_2 u_2^2]\, d\tau$$

(*a*) For this performance criterion, determine and tabulate or plot the gains and closed-loop poles as r_1 varies from 10^4 to 10^8 and r_2 varies from 10^{-2} to 1.

(*b*) The steady state control objective is to maintain $x_3 = \Delta z_1$ and $x_4 = \Delta z_2$ at zero in the presence of constant disturbances represented by the exogenous vector $x_0 = [\Delta x_{FA1}, \Delta F_A]'$ as described in Example 2G. Find the feedforward gain matrices $G_0^* = B^*E$ and $G_0^{\#} = B^{\#}E$ for the regulator gain matrices found in part *a*.

Problem 9.18 Double-effect evaporator

The control gains for the double-effect evaporator considered in Example 2E et seq. are to be designed by minimization of a quadratic performance criterion

$$V = \int_t^\infty [x_1^2 + 1.5x_4^2 + r^2(u_1^2 + u_2^2)]\, d\tau$$

without using u_3 (i.e., $u_3 = 0$).

Find and tabulate the gain matrix and the corresponding closed-loop poles as a function of r^2 for $0.001 \leq r^2 \leq 1$.

NOTES

Note 9.1 Optimization by calculus of variations

In this chapter we start with the assumption that the control law is linear: $u = -Gx$ and proceed to determine the optimum gain matrix $G = \hat{G}$. But how do we know that a linear control law is optimum in the first place? To show that a linear control law is optimum we must consider nonlinear control laws as well as linear ones and prove that better performance cannot be achieved by a nonlinear control law than by the optimum linear control law. This is a true statement, but only for a quadratic performance criterion. If we were to optimize a different performance criterion, e.g.,

$$V = \int_t^\infty (x^4 + u^2)\, d\tau$$

then in general a nonlinear control law would be optimal. Determination of the optimum control law for nonquadratic performance criteria, and possibly for nonlinear processes, must be approached by the methods of the calculus of variations or perhaps dynamic programming and is beyond the scope of this book. Readers interested in pursuing the variational approach should consult a standard textbook on optimum control theory, such as [5] or [6].

Note 9.2 The matrix Riccati equation

The matrix Riccati equation (9.27) may well be the most famous in all the literature of modern control theory. It arises not only in determining the optimum control gains but also, as we shall see in Chap. 11, in the determination of the gain matrix $\hat{K}$ of the optimum stochastic observer (i.e., Kalman filter).

Note 9.3 Numerical methods for the algebraic Riccati equation (ARE)

The importance of the ARE has spawned extensive research on numerical methods for its solution. There seem to be two basic types of algorithms: those that go after the solution directly by iteration and those that proceed indirectly, through eigenvalue and eigenvector expansions.

The iterative techniques are related to successive approximations and Newton-Raphson methods. The ARE is nothing more than a system of $k(k+1)/2$ coupled quadratic equations. Thus any method capable of solving systems of nonlinear equations should work for the ARE. The problem, however, is that even for a moderate value of k (say 10), such as may easily arise in practice, the number of equations in the system may be too high to assure reliable solutions. It was found early on that numerical methods that exploit the structure of the ARE are likely to be more efficient and better conditioned numerically. One method, due to Kleinman[7], is based on writing the ARE as

$$0 = \hat{M}\hat{A}_c + \hat{A}_c'\hat{M} = -(Q + \hat{M}BR^{-1}B'\hat{M})$$

where $\hat{A}_c = A - BR^{-1}B'\hat{M}$ = the closed-loop dynamics matrix. Start with an approximate solution

M_0 for $\hat{M}$ such that $A_0 = A - BR^{-1}B'M_0$ has its eigenvalues in the left half-plane. Then, by a stability theorem due to Liapunov[8], the solution M_1 to

$$0 = M_1A_0 + A_0'M_1 = -(Q + M_0BR^{-1}B'M_0)$$

is positive-definite, because the right-hand side is positive semidefinite. Using M_1 in place of M_0 we can solve for a second approximation M_2 and continue the process until there is no further change. At each step we must solve a linear matrix equation of the form

$$MA + A'M = -Q$$

which is known as the matrix Liapunov equation. Various standard algorithms are available for solving the Liapunov equation.

One problem with the iterative algorithms is that the numerical error due to solving the Liapunov equation with a finite precision computer may result in a matrix M_n which is not positive-definite. Once this happens, the subsequent iterations are likely to diverge very rapidly. Another problem with the iterative approach is finding a starting matrix M_0. (If the open-loop system is asymptotically stable, the starting value $M_0 = 0$ is satisfactory, but a starting value of zero cannot be used when the open-loop system is not asymptotically stable.)

A second class of numerical algorithms is based on the connection between the Riccati equation for a kth order system and a $2k \times 2k$ hamiltonian system as discussed in Prob. 9.1. Numerical algorithms are based on the eigenvector expansion described in Prob. 9.3. Reliable and efficient numerical methods for determining eigenvalues and eigenvectors, or equivalent quantities, are currently available even for very high order systems. Thus, this class of numerical algorithms is currently preferred over the iterative algorithms by most investigators.

Note 9.4 Asymptotic properties of optimum control law

The qualitative dependence of the closed-loop system upon the weighting matrices in the performance integral is frequently of interest. Consider the first-order system

$$\dot{x} = ax + bu$$

with

$$V = \int_t^\infty (x^2 + \rho^2u^2)\, d\tau$$

The gain (now a scalar) is given by

$$g = bm/\rho^2$$

where m is the positive root of $2am - b^2m^2/\rho^2 + 1 = 0$. The two roots of the quadratic are

$$m = \rho^2(a \pm \sqrt{a^2 + b^2/\rho^2})/b^2$$

Since the radical is always greater than $|a|$, it is clear that the top sign is the only possible choice for $m > 0$. Thus

$$g = (a + \sqrt{a^2 + b^2/\rho^2})/b$$

and

$$a_c = a - bg = -\sqrt{a^2 + b^2/\rho^2}$$

As the control weighting ρ^2 tends to zero, (i.e., the control becomes increasingly "cheap") it is seen that the closed-loop pole moves out to infinity along the negative real axis as b/ρ. The feedback gain becomes infinite as does the bandwidth of the system. This is entirely reasonable.

As the control weighting ρ^2 tends to ∞ (i.e., the control becomes increasingly "expensive") we see that the closed-loop pole tends to $-|a|$. If $a < 0$ (i.e., the open-loop system is asymptotically stable) then $a_c \to a$ as the control gain $g \to 0$. This again is reasonable. If control costs a great deal, don't use any of it and simply let the process coast to the origin. This strategy is obviously unsatisfactory when $a > 0$ (i.e., the open-loop system is unstable). In this case $a_c \to -a$ and the gain $g \to 2a$. This result is somewhat surprising. One might have thought that a gain $g \to a + \varepsilon$, (with ε arbitrarily small) might result in a lower cost of control. But, this is not the case, evidently because although the control signal in this case starts out only slightly more than half as large as the control with a gain of $2a$, it goes to zero much less rapidly than it does with a gain of $2a$.

To what extent do the results for a first-order system carry over to a general kth-order system? This question was first studied by Kalman[9] and subsequently by Kwakernaak[10] and others. The result may be summarized as follows:

(*a*) As the control weighting tends to ∞, the gains are such that the closed-loop poles tend to the open-loop poles when the latter are stable, or to their "mirror images" (with respect to the imaginary axis) when they are unstable.

(*b*) In a single-input system $x = Ax + bu$ with a performance integral

$$V = \int_t^\infty (x'c'cx + \rho^2 u^2)\, d\tau$$

the zeros of the "system" are defined by $c'(sI - A)^{-1}b = 0$. If there are r such zeros, then as $\rho^2 \to \infty$, r of the closed-loop poles approach these zeros and the remaining $k - r$ closed-loop poles radiate out toward infinity in configuration of a Butterworth polynomial of order $k - r$. In a multiple-input, multiple-output system the asymptotic behavior as $\rho^2 \to \infty$ is more luxuriant: The closed-loop poles form groups of Butterworth polynomials with the radius of each group tending to ∞ at a rate which may differ from the rates of the other groups.

Asymptotic properties of the optimum regulator can often be obtained by use of the return-difference equation (P9.8) which is a generalization of an equation first obtained by Kalman.[9]

REFERENCES

1. Kalman, R. E., "Contributions to the Theory of Optimal Control," *Proc. 1959 Mexico City Conf. on Differential Equations, Mexico City*, 1960, pp. 102–199.
2. Kalman, R. E., Ho, Y. C., and Narendra, K. S., "Controllability of Linear Dynamic Systems," *Contributions to Differential Equations*, vol. I, no. 2, 1963, pp. 189–213.
3. Locke, A. S. (ed.), *Guidance*, D. Van Nostrand, New York, 1955.
4. Loh, N. K., Cheok, K. C., and Beck, R. R., "Modern Control Design for a Gun-Turret Control," *Southcon/83 Convention Record, Atlanta GA*, Paper no. 1/5, 1983.
5. Athans, M., and Falb, P. L., *Optimal Control: An Introduction to the Theory and its Applications*, McGraw-Hill Book Co., New York, 1966.
6. Bryson, A. E., Jr., and Ho, Y.-C., *Applied Optimal Control*, Blaisdell Publishing Co., Waltham, MA, 1969.
7. Kleinman, D. L., "On an Iterative Technique for Riccati Equation Computations," *IEEE Trans. on Automatic Control*, vol. AC-13, no. 1, February 1968, pp. 114–115.
8. Gantmacher, F. R., *The Theory of Matrices*, vol. 2, Chelsea Publishing Company, New York, 1959.
9. Kalman, R. E., "When Is a Linear Control System Optimal?," *Trans. ASME (J. Basic Engineering)*, vol. 86D, no. 1, March 1964, pp. 51–60.
10. Kwakernaak, H., and Sivan, R., *Linear Optimal Control Systems*, Wiley-Interscience, New York, 1972.

CHAPTER

TEN

RANDOM PROCESSES

10.1 INTRODUCTION

One of the purposes of using feedback is to increase the immunity of the system to disturbance inputs. Up to this point we have been regarding disturbances as the responses of known dynamic systems with unspecified initial conditions. This is a convenient artifice, but an artifice nonetheless. The disturbances to a real system in almost every case are more complicated: they are *random processes.* Moreover, the sensors used in the measurement of the system output y are not perfect, but are also subject to errors, and these errors are also random processes. If a control system design is claimed to be optimum, it ought to account for the statistical nature of the random processes that act upon the system in the form of disturbances and that corrupt the sensor outputs.

Random processes are present everywhere in nature and technology. Here are some random processes with which everyone is familiar:

The environmental temperature at a given geographical location is a random process. The mean temperature and variance (both of which vary with season and geographic location) are known statistics, but knowledge of the temperature at any given time is not sufficient to predict the *precise* temperature in the future. Other environmental characteristics such as rainfall, wind velocity, etc., are other familiar random processes.

The output of a photodiode exposed to a source of light is a random process. The *mean* output is of course a function of the (mean) light intensity, but since each photon that impacts upon the photodiode contributes an electrical pulse to the output, the instantaneous output fluctuates about the mean value. The presence of these fluctuations makes the photodiode output a random process.

If one were to measure the voltage across an ordinary resistor not connected to anything at all with a very sensitive voltmeter one would find that the output is not always zero but actually a rapidly fluctuating signal of zero mean. This output is "thermal noise" and its statistical properties can be predicted using considerations of statistical mechanics.

The acceleration imparted to a vehicle when it travels over a road is a random process the statistical nature of which depends on the quality of the road.

The force acting upon a ship in rough seas is a random process, which may, however, have a strong periodic component.

The reader can no doubt supply many additional examples.

Our ultimate aim in these chapters is to design control systems when random processes such as these are disturbance inputs or noise on the sensors. Mathematically we wish to consider designing a control law for a system

$$\dot{x} = Ax + Bu + Fv$$

$$y = Cx + w$$

where v and w are random processes. Since the disturbances and sensor errors are random processes, the response of the system, either open-loop, or with the feedback control present, is also a random process. Hence we seek to optimize the behavior of the system not for a single disturbance, but for a suitably defined *average behavior.*

To accomplish this necessitates a study of some of the theory of random processes. A rigorous development of the theory of random processes is far beyond the scope of the present text. Other books, at every level of mathematical rigor, are available for this purpose (see Note 10.1). Our objective here is to represent known results in a manner that permits the reader to gain sufficient insight to be able to use them.

The reader with an elementary knowledge of the theory of probability should be able to follow the next two sections with little difficulty.

10.2 CONCEPTUAL MODELS FOR RANDOM PROCESSES

The essential feature of a random process, as distinguished from a deterministic process, which has occupied us until now, is that knowledge of the state of the process at any instant is *not* sufficient to predict the future evolution of the process in time. Conceptually, we can imagine a very large number of physical processes ("black boxes") each of which is indistinguishable from the others under reasonable conditions of measurement—like mass-produced radio receivers. Each black box has an output ("static," in the case of radio receivers tuned to no station) which is statistically similar to the outputs of the others, but because of very slight variations the outputs are all different. This "ensemble" of outputs can be regarded as a random process.

If we had access to the outputs of all the black boxes, we could determine empirical statistics of the ensemble. In particular let $x_i(t)$ denote the ith member (also called "sample function") of an ensemble of N members. Then we can define the following empirical statistics.

Mean:
$$\bar{x}(t) = \frac{1}{N}\sum_{i=1}^{N} x_i(t) \tag{10.1}$$

Mean square:
$$\overline{x^2(t)} = \frac{1}{N}\sum_{i=1}^{N} x_i^2(t) \tag{10.2}$$

Variance:
$$v(t) = \overline{x^2(t)} - [\bar{x}(t)]^2 \tag{10.3}$$

Correlation function:
$$r(t, \tau) = \frac{1}{N}\sum_{i=1}^{N} x_i(t)x_i(\tau) \tag{10.4}$$

and so forth.

Two issues are raised by this conceptual model: one philosophical and one practical. The philosophical issue is this: If we design a control system on the basis of the statistics of the random process, what assurance do we have that our statistics gathered over the past will continue to be valid? What about catastrophic storms, earthquakes, new phenomena? It takes a leap of faith to accept the hypothesis that what happened on the average in the past will continue into the future. Only if we are willing to take this leap of faith can we use statistical methods for system design. But what other choice is there?

The practical issue is simply the requirement to obtain the statistical parameters that are needed in order to use the methods. Gathering sufficient data to have confidence in the statistical parameters is often a time-consuming and costly operation. If we want to take statistical road characteristics into account in designing an active control system for an automotive suspension, for example, it is necessary to provide a test vehicle with an appropriate instrument (possibly an accelerometer) and go out and drive on roads having the physical characteristics for which our design is to be based. If we are designing an inertial guidance system for an aircraft, we need to test the instruments to be used in the system for extensive periods. We can test a hundred roads, a hundred gyros: perhaps a thousand. Is this enough testing? Who knows? The problem of performing enough tests, and the right tests, to get the required data is often overlooked. Moreover, it is often necessary to design a control system before any significant statistical data about the underlying random processes can be gathered. In such cases, the system designer has little alternative but to make an educated (or uneducated) guess about the required statistical parameters.

Because our knowledge of the required statistical parameters is often imperfect, with guesses substituted for real data, a claim that a given statistical design is optimum can be justified only on the basis that the calculations are sound. Thus the reservation that we made about the optimality of linear quadratic designs for deterministic systems holds even more forcefully in statistical designs. We often pick statistical parameters not because we are confident of their accuracy but because we need some numbers to put into the equations!

10.3 STATISTICAL CHARACTERISTICS OF RANDOM PROCESSES

First- and second-order statistics In theory, a random process is characterized by an infinite series of joint probability density functions

$$\begin{aligned} &\mathrm{pdf}[x;\,t] \\ &\mathrm{pdf}[x_1, x_2;\, t_1, t_2] \\ &\mathrm{pdf}[x_1, x_2, x_3;\, t_1, t_2, t_3] \\ &\quad\vdots \end{aligned} \tag{10.5}$$

where each density function describes the probability of finding x somewhere at some time. For example,

$$\mathrm{pdf}[x_1, x_2, x_3;\, t_1, t_2, t_3]\,\Delta x_1\,\Delta x_2\,\Delta x_3 = \mathrm{Prob}\,[x_1 < x(t_1) < x_1 + \Delta x_1,\, x_2 < x(t_2) < x_2 + \Delta x_2,\, x_3 < x(t_3) < x_3 + \Delta x_3]$$

Since there is no practical way of obtaining these probability density functions for a physical process, their main use is for mathematical development. In most cases we must be content with first- and second-order statistics:

Mean

$$\mu(t) = E\{x(t)\} = \int_{-\infty}^{\infty} x\,\mathrm{pdf}[x, t]\,dx \tag{10.6}$$

Mean square

$$E\{x^2(t)\} = \int_{-\infty}^{\infty} x^2\,\mathrm{pdf}[x, t]\,dx \tag{10.7}$$

Variance

$$\sigma^2(t) = E\{[x(t) - \mu(t)]^2\} = \int_{-\infty}^{\infty} [x - \mu(t)]^2\,\mathrm{pdf}[x, t]\,dt \tag{10.8}$$

Correlation function

$$\rho(t, \tau) = E\{x(t)x(\tau)\} = \int_{-\infty}^{\infty}\int_{-\infty}^{\infty} x_1 x_2\,\mathrm{pdf}[x_1, x_2;\, t, \tau]\,dx_1\,dx_2 \tag{10.9}$$

The first three of these statistical parameters are "first-order" statistics, since they entail only the first-order probability density function. The correlation function is a "second-order" statistic. The symbol $E\{\ \}$ denotes mathematical expectation, in other words the average computed in the probabilistic sense. The requisite probability functions are almost never known in practice. If we have gathered enough empirical data we may know the empirical averages (10.1)-(10.3) and, by our leap of faith, assume the requisite mathematical expectations (10.6)-(10.9) are the same as the former.

Note that when $\tau = t$ the correlation function (10.8) becomes

$$\rho(t, t) = \int_{-\infty}^{\infty}\int_{-\infty}^{\infty} x_1 x_2 \,\mathrm{pdf}[x_1, x_2; t, t]\, dx_1\, dx_2 \tag{10.10}$$

But

$$\mathrm{pdf}[x_1, x_2; t, t]\, dx_1\, dx_2 = \mathrm{Prob}\,[x_1 < x(t) < x_1 + dx_1, x_2 < x(t) < x_2 + dx_2]$$

Now at the same instant of time $x(t)$ cannot simultaneously be between x_1 and $x_1 + dx_1$ and between x_2 and $x_2 + dx_2$ unless $x_1 = x_2$. Thus

$$\mathrm{pdf}[x_1, x_2; t, t]\, dx_1\, dx_2 = \mathrm{pdf}[x_1, t]\delta(x_1 - x_2)\, dx_1\, dx_2 \tag{10.11}$$

where $\delta(x_1 - x_2)$ is a unit impulse at $x_1 - x_2 = 0$. Substitute (10.11) into (10.10) and integrate over the impulse to get

$$\rho(t, t) = \int_{-\infty}^{\infty} x_1^2 \,\mathrm{pdf}[x_1; t]\, dx_1 = E\{x^2(t)\} \tag{10.12}$$

which is not at all surprising, since $E\{x(t)x(\tau)\} = E\{x^2(t)\}$ as $\tau \to t$.

The first-order statistics (10.6)-(10.9) can be defined for vector processes as well as scalar processes. If

$$x(t) = \begin{bmatrix} x_1(t) \\ \vdots \\ x_n(t) \end{bmatrix}$$

then, for example,

$$\mu(t) = E\{x(t)\} = \begin{bmatrix} E\{x_1(t)\} \\ \vdots \\ E\{x_n(t)\} \end{bmatrix} \tag{10.13}$$

where $E\{x_i(t)\}$ is defined by an integral like (10.5).

A generalization of the correlation function is the correlation matrix

$$R(t, \tau) = E\{x(t)x'(\tau)\} = \begin{bmatrix} E\{x_1(t)x_1(\tau)\} & \cdots & E\{x_1(t)x_n(\tau)\} \\ \cdots\cdots & \cdots & \cdots\cdots \\ E\{x_n(t)x_1(\tau)\} & \cdots & E\{x_n(t)x_n(\tau)\} \end{bmatrix} \tag{10.14}$$

The diagonal entries in the correlation matrix are the *auto*correlation functions of the $x_i(t)$ with $x_i(\tau)$ as defined by (10.9). The off-diagonal terms are cross-correlations and are defined with the aid of suitable joint probability density functions.

It is obvious from (10.14) that

$$R(t, \tau) = R'(\tau, t) \tag{10.15}$$

and, as a special case

$$R(t, t) = E\{x(t)x'(t)\} = R'(t, t)$$

The matrix $R(t, t)$ is often called the *covariance matrix* for the vector process $x(t)$.

If the covariance matrix $R(t, t)$ is diagonal, i.e.,

$$R(t, \tau) = \begin{bmatrix} r_{11}(t, \tau) & 0 & 0 & \cdots & 0 \\ 0 & r_{22}(t, \tau) & 0 & \cdots & 0 \\ \cdots & \cdots & \cdots & \cdots & \cdots \\ 0 & 0 & 0 & \cdots & r_{nn}(t, \tau) \end{bmatrix}$$

this means

$$E\{x_i(t)x_j(\tau)\} = 0 \qquad \text{for } i \neq j$$

The components of the vector x are said to be *uncorrelated.* (This does *not* imply, however, that they are statistically independent.)

Stationary and ergodic processes The set of probability density functions in (10.5) are general functions of the time variables $t_1, t_2, t_3, \ldots$ for a general random process. If the functions are invariant to a translation of time, i.e.,

$$\text{pdf}[x; t + \tau] = \text{pdf}[x; t]$$

$$\text{pdf}[x_1, x_2; t_1 + \tau, t_2 + \tau] = \text{pdf}[x_1, x_2; t_1, t_2] \qquad \text{for all } \tau$$

$$\vdots$$

the process is called *stationary* (in the strict sense). If the process is stationary in the strict sense, all statistical parameters are invariant to a translation in time.

Since the statistics of the process of order higher than second are almost never known it is useful to deal with processes in which only the first two probability density functions are invariant to translation in time. These are called stationary "in the wide sense."

In all of the foregoing analysis, statistics were defined by averaging over the ensemble of members (sample functions) of the random process. Often it is not possible to get such statistics. If we wanted meteorological statistics for, say, New York we would have to conceive of an ensemble of infinitely many New Yorks and accumulate averages over this ensemble. But there is only one New York, so this experiment is inconceivable. Averages over time, however, can be obtained. We can get the average temperature of New York on January first at, say, 2 p.m. by measuring the temperature each January first at 2 p.m. for many years and calculating the average. How does an average of this type relate to an ensemble average? In general, there is no relationship. But, in a special subset of stationary processes, these time averages are the same as the ensemble averages. Such processes are called "ergodic." As with many other statistical properties, there is no practical way of verifying that a stationary process is ergodic. So, by another leap of faith, when it suits our needs, we assume that a process is ergodic. If a process is ergodic, a single sample function is representative of the ensemble. To get the ensemble statistics for the outputs for 10^6 radios tuned to no station, we can take time averages of the output of one radio!

For ergodic processes we can thus determine the ensemble statistics using any sample function in the ensemble:

$$\mu(t) = \lim_{T\to\infty} \frac{1}{T} \int_{-T/2}^{T/2} x(t)\, dt \tag{10.16}$$

$$\sigma^2(t) = \lim_{T\to\infty} \frac{1}{T} \int_{-T/2}^{T/2} [x(t) - \mu(t)]^2\, dt \tag{10.17}$$

$$\rho(\tau) = \lim_{T\to\infty} \frac{1}{T} \int_{-T/2}^{T/2} x(t)x(t+\tau)\, dt \tag{10.18}$$

Note that the correlation function in (10.18) is written as a function of a single argument τ, the time shift between $x(t)$ and $x(t+\tau)$. Since we assume the process is ergodic it must surely be stationary, and hence ρ is a function only of this time shift.

10.4 POWER SPECTRAL DENSITY FUNCTION

One of the most useful descriptions of a random process is its "power spectral density function" $S(\omega)$, also called the "power spectrum," which is defined as the Fourier transform of the correlation function, i.e.,

$$S(\omega) = \int_{-\infty}^{\infty} \rho(\tau)\, e^{-j\omega\tau}\, d\tau \tag{10.19}$$

Given the correlation functions $\rho(\tau)$, which can be computed from empirical data using a numerical approximation to (10.18), it is an easy task to compute the power spectral density function. Before the days of digital computers, however, computing a correlation function by multiplying $x(t)$ by $x(t+\tau)$ and integrating was not a simple task. Hence a different approach had to be used. This approach led to the name "spectral density." The idea was to connect the output of the process to a device known as a "spectrum analyzer" and measure the power contained in the random signal in different frequency bands. A spectrum analyzer is actually a sharply tuned (narrowband) filter with an adjustable center frequency. The signal is connected to the spectrum analyzer and the latter is tuned to a given frequency. The indicator is allowed to reach the steady state and the reading is recorded. The analyzer is then tuned to an adjacent frequency and the process is again repeated. This operation is continued until the entire range of frequencies of interest is scanned. By this means we have obtained a function of frequency approximated by:

$$|X_T(j\omega)|^2 = \left| \int_{-T/2}^{T/2} x(t)\, e^{-j\omega t}\, dt \right|^2 \qquad (10.20)$$

where T is the large but finite interval of operation at each frequency setting of the analyzer. We are assuming that the spectrum analyzer produces an approximation to the Fourier transform of the signal. Now

$$|X_T(j\omega)|^2 = X_T(j\omega)X_T(-j\omega) = \int_{-T/2}^{T/2} x(t)\, e^{-j\omega t}\, dt \cdot \int_{-T/2}^{T/2} x(\tau)\, e^{j\omega\tau}\, d\tau$$

$$= \int_{-T/2}^{T/2} \int_{-T/2}^{T/2} x(t)x(\tau)\, e^{-j\omega(t-\tau)}\, dt\, d\tau \qquad (10.21)$$

In this integral we make the change of variable $\lambda = t - \tau$ to obtain

$$\frac{1}{T}|X_T(j\omega)|^2 = \int_{-T/2}^{T/2} \left\{ \frac{1}{T} \int_{\lambda-T/2}^{\lambda+T/2} x(\tau+\lambda)x(\tau)\, d\tau \right\} e^{-j\omega\lambda}\, d\tau \qquad (10.22)$$

If the process is ergodic

$$\rho_T(\lambda) = \frac{1}{T} \int_{\lambda-T/2}^{\lambda+T/2} x(\tau+\lambda)x(\tau)\, dt \to \rho(\lambda) \qquad \text{as } T \to \infty \qquad (10.23)$$

(Actually a bit more care is needed in passing to the limit because λ will also go to ∞ in the second integral. See Papoulis.[1])

Thus

$$\lim_{T\to\infty} |X_T(j\omega)|^2 = \lim_{T\to\infty} \int_{-T/2}^{T/2} \rho_T(\lambda)\, e^{-j\omega\lambda}\, d\lambda = S(\omega) \qquad (10.24)$$

This relationship asserts that the power spectral density function is the limit of the magnitude square of the ordinary Fourier transform of the signal. The Fourier transform of a signal describes how its energy is distributed in frequency; division by T converts energy to power; hence we can properly call the result of the limiting operation the power spectral density of the signal or random process. The result expressed by (10.24) is sometimes called "Wiener-Khintchine" relation.

Returning to (10.19) we see that the correlation function is the inverse Fourier transform of the power spectral density

$$\rho(\tau) = \frac{1}{2\pi} \int_{-\infty}^{\infty} S(\omega)\, e^{j\omega\tau}\, d\omega \qquad (10.25)$$

for all values of τ including $\tau = 0$. For $\tau = 0$

$$\rho(0) = \frac{1}{2\pi} \int_{-\infty}^{\infty} S(\omega)\, d\omega \qquad (10.26)$$

But

$$\rho(0) = \lim_{T\to\infty} \int_{-T/2}^{T/2} x^2(t)\,dt = \text{mean square value of the signal } x$$

We thus conclude that the area under the spectral density function is 2π times the mean square value of the random process.

The unit of measurement of spectral density is related to the unit of measurement of the process $x(t)$. Suppose, for example, that the units of $x(t)$ are force: the units of $x^2(t)$ are $(\text{force})^2$. Since $S(\omega)$ is integrated over frequency to give $(\text{force})^2$, we infer that the units of $S(\omega)$ are $(\text{force})^2/(\text{rad/s})$ or $(\text{force})^2 \cdot \text{s}$. Remembering the units in which spectral density is measured often helps one to avoid reasoning errors in system design using statistical methods.

The spectral density is often expressed as a function of frequency $f = \omega/2\pi$. In that case the factor of 2π in (10.26) is absorbed into the integrand which can thus be written

$$\rho(0) = \int_{-\infty}^{\infty} \bar{S}(f)\,df = 2\int_0^{\infty} \bar{S}(f)\,df$$

where $\bar{S}(f)$ is the spectral density in $(\text{units})^2/\text{Hz}$. The factor of 2 in the last integral is sometimes also absorbed into the definition of $\bar{S}(f)$.

10.5 WHITE NOISE AND LINEAR SYSTEM RESPONSE

White noise is one of those theoretical abstractions which simplifies calculations, but grieves mathematicians. It is analogous to the Dirac δ function, used successfully by physicists for a number of years before being reduced to mathematical rigor in the 1930s. But while mathematicians now accept δ functions, they eschew white noise, preferring rather to deal with its (nonstationary) integral—the Wiener process, which we will encounter in Sec. 10.8.

White noise is simply a random process with an expected value (mean) of zero and with an absolutely flat power spectrum

$$S(\omega) = W = \text{constant for all } \omega \tag{10.27}$$

Since the inverse Fourier transform of a constant is a unit impulse, the correlation function of white noise is

$$\rho(\tau) = W\delta(\tau) \tag{10.28}$$

where $\delta(\tau)$ is a unit impulse at the origin.

Because the mean square value of any random process is the integral of the spectral density over all frequencies—see (10.26)—and since white noise has a constant spectral density for all frequencies, white noise has a theoretically infinite mean square value. This is also seen from the correlation function as given by (10.28)

$$\rho(0) = \lim_{\tau\to 0} \rho(\tau) = \infty$$

A physical process that has an infinite mean square value is inconceivable and we must conclude that white noise is a mathematical abstraction.

The bothersome feature of white noise is that its power spectral density does not decrease with frequency. The spectral density of known physical processes, on the other hand, always decreases and tends to zero as the frequency tends to infinity. When mystifying results are obtained for a calculation made under the assumption that some random process is white noise, the problem might be due to the theoretically infinite "bandwidth" of the white noise. If this should prove to be the case, it can usually be resolved by replacing the white noise by a random process which has a very large, but finite, bandwidth.

A vector random process is white noise if its correlation matrix is of the form

$$R(\tau) = W\delta(\tau) = E\{x(t)x'(t+\tau)\} \tag{10.29}$$

where W is a square matrix.

We can imagine white noise being the input to an amplifier having a time-varying gain. This will result in a random process with a time-varying spectral density, and so our white noise would have a correlation function which would be

$$\rho(t, \tau) = W(t)\rho(\tau)$$

Since this expression is not a function of only $t - \tau$, this white noise is nonstationary and the very meaning of the spectral density is problematical.

White noise is a convenient abstraction because it leads to relatively simple expressions for the correlation function and the power spectrum of the *output* of a linear system into which it is the input. To see this, consider a linear system, the input to which is the signal $u(t)$ and the output from which is $y(t)$. Whether or not the system is represented in state-space, its output can always be expressed as a superposition integral

$$y(t) = \int_0^t H(t, \lambda)u(\lambda)\, d\lambda \tag{10.30}$$

where $H(t, \lambda)$ is the "impulse response" matrix of the system (see [2]). It is assumed for (10.30) that the input starts at $t = 0$ which is the lower limit of the integral and that the system is causal, i.e., that $H(t, \lambda) = 0$ for $t < \lambda$. This explains why the upper limit on the integral is t, and not ∞. The correlation matrix for the output $y(t)$ is

$$\begin{aligned} R_y(t, \tau) = E\{y(t)y'(\tau)\} &= E\left\{\int_0^t H(t, \lambda)u(\lambda)\, d\lambda \cdot \int_0^\tau u'(\xi)H'(\tau, \xi)\, d\xi\right\} \\ &= E\left\{\int_0^t \int_0^\tau H(t, \lambda)u(\lambda)u'(\xi)H'(\tau, \xi)\, d\lambda\, d\xi\right\} \end{aligned} \tag{10.31}$$

(Writing the product of integrals as a double integral is a standard technique used often in analyses of this sort.)

Assume that the conditions hold that are required to permit interchanging expectation operator with the double integral and that the input $u(t)$ is white noise:

$$E\{u(\lambda)u'(\xi)\} = Q\delta(\lambda - \xi) \qquad Q = \text{const}$$

then (10.31) becomes

$$R_y(t, \tau) = \int_0^t \int_0^T H(\tau, \lambda)Q\delta(\lambda - \xi)H'(\tau, \xi)\, d\lambda\, d\xi \tag{10.32}$$

Using the fact that

$$\int_a^b f(\xi)\delta(\lambda - \xi)\, d\xi = f(\lambda)$$

(if $a \leqq \lambda \leqq b$) we obtain from (10.32)

$$R_y(t, \tau) = \int_0^t H(t, \lambda)QH'(\tau, \lambda)\, d\lambda \tag{10.33}$$

which is a single integral instead of a double integral.

If the system having the impulse response matrix $H(t, \tau)$ is time-varying, the output process is not stationary, in general. If the process is time-invariant, then

$$H(t, \tau) = H(t - \tau) \qquad \text{for all } t, \tau$$

and (10.33) becomes

$$R_y(t, \tau) = \int_0^t H(t - \tau)QH'(\tau - \lambda)\, d\lambda$$

or, replacing τ by $t + \tau$,

$$R_y(t, t + \tau) = \int_0^t H(t - \lambda)QH'(t - \lambda + \tau)\, d\lambda$$

$$= \int_0^t H(\xi)QH'(\xi + \tau)\, d\xi \tag{10.34}$$

upon making the change of variable $\xi = t - \lambda$. Because t is the upper limit of the integral, the correlation matrix (10.34) is not that of a stationary process. But, as $t \to \infty$, that is, as the process reaches a steady state

$$\lim_{t\to\infty} R_y(t, t + \tau) = \bar{R}_y(\tau) = \int_0^\infty H(\xi)QH'(\xi + \tau)\, d\xi \tag{10.35}$$

Of course (10.35) is valid only if the dynamic system *has* a steady state response. If the system is not asymptotically stable the integral (10.35) is not meaningful. If the integral is valid, however, the output has the correlation function of a stationary process.

We have not yet considered the expected value (mean) of the output. From (10.20)

$$E\{y(t)\} = E\left\{\int_0^t H(t, \lambda)u(\lambda)\,d\lambda\right\} = \int_0^t H(t, \lambda)E\{u(\lambda)\}\,d\lambda$$

If $u(t)$ is white noise or indeed any process with zero mean

$$E\{u(t)\} = 0$$

and hence

$$E\{y(t)\} = 0 \tag{10.36}$$

Thus, we have found that the response of a linear system to white noise in the steady state (if the system reaches a steady state) has a zero mean and has a correlation function (matrix) given by (10.35).

The power spectrum of the output y is obtained from (10.35) using the definition (10.19). (Note that if $y(t)$ is a vector, implying that $R_y(\tau)$ is a matrix, then the power spectrum is also a matrix.) Using (10.19)

$$S(\omega) = \int_{-\infty}^{\infty}\left[\int_0^{\infty} H(\xi)QH'(\xi+\tau)\,d\xi\right]e^{-j\omega\tau}\,d\tau \tag{10.37}$$

Invert the order of integration in (10.37) to obtain

$$S_y(\omega) = \int_0^{\infty} H(\xi)Q\left[\int_{-\infty}^{\infty} H'(\xi+\tau)\,e^{-j\omega\tau}\,d\tau\right]d\xi \tag{10.38}$$

The bracketed integral in (10.38) is

$$\int_{-\infty}^{\infty} H'(\xi+\tau)\,e^{-j\omega\tau}\,d\tau = \int_{-\infty}^{\infty} H'(\lambda)\,e^{-j\omega(\lambda-\xi)}\,d\lambda = e^{j\omega\xi}\mathsf{H}'(j\omega)$$

where

$$\mathsf{H}(j\omega) = \int_{-\infty}^{\infty} H(t)\,e^{-j\omega t}\,dt \tag{10.39}$$

is the transfer function of the linear system. Note that the transfer function is usually defined in terms of the Laplace transform

$$\mathsf{H}(s) = \int_0^{\infty} H(t)\,e^{-st}\,dt \tag{10.40}$$

Thus, in addition to the usual transfer function having complex frequency s instead of $j\omega$ as the frequency variable, the lower limit on the defining integral (10.40) is 0 instead of $-\infty$. But, in a causal system

$$H(t) = 0 \qquad \text{for } t < 0$$

so that there is no conflict between the two definitions. In other words

$$\mathsf{H}(j\omega) = \mathsf{H}(s)\big|_{s=j\omega}$$

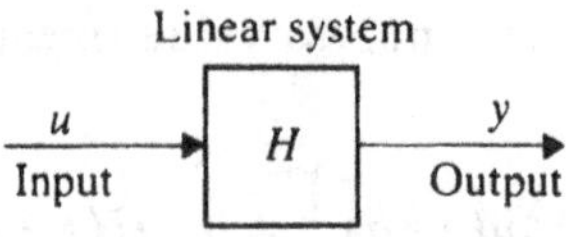

$H(t)$ = impulse response
$\mathsf{H}(s)$ = transfer function = $\mathscr{L}[H(t)]$

Domain	Deterministic inputs	White noise inputs
Time domain	$y(t) = \int_0^t H(t-\lambda)u(\lambda)\, d\lambda$	$R_y(\tau) = \int_0^\infty H(\xi)QH'(\xi+\tau)\, d\tau$
Frequency domain	$\mathsf{y}(s) = \mathsf{H}(s)\mathsf{u}(s)$	$S_y(\omega) = \mathsf{H}(-j\omega)Q\mathsf{H}'(j\omega)$

Figure 10.1 Input-output relation for linear system with deterministic input and excited by white noise.

Return now to (10.38). From (10.39)

$$S(\omega) = \int_0^\infty H(\xi)\, e^{j\omega\xi}\, d\xi \cdot Q\mathsf{H}'(j\omega) \tag{10.41}$$

By the causality argument just given above, the lower limit on the integral in (10.41) can be replaced by $-\infty$ so that

$$S_y(\omega) = \mathsf{H}(-j\omega)Q\mathsf{H}'(j\omega) \tag{10.42}$$

In words, the spectrum of the output y of a linear system excited by white noise is the product of the transfer function matrix at negative frequency with the spectral density matrix of the white noise, with the transfer function matrix, transposed.

The relationships in this section are summarized in Fig. 10.1.

Example 10A First-order Markov process The most common random process after white noise is the output of a first-order low pass filter having the transfer function

$$\mathsf{H}(s) = \frac{1}{s+\omega_0} \tag{10A.1}$$

The impulse response corresponding to $\mathsf{H}(s)$ is

$$h(t) = \begin{cases} e^{-\omega_0 t} & t \geqq 0 \\ 0 & t < 0 \end{cases}$$

The correlation function of this process, often known as a *first-order Markov process*, is given by (10.35) which, in this case, is

$$r(\tau) = Q\int_0^\infty e^{-\omega_0\xi}e^{-\omega_0(\xi+\tau)}\, d\xi = Q\, e^{-\omega_0\tau}\int_0^\infty e^{-2\omega_0\xi}\, d\xi$$

$$= \frac{Q}{2\omega_0}e^{-\omega_0\tau} \quad \text{for } \tau > 0 \tag{10A.2}$$

Since $h(\tau)$ is zero for $\tau < 0$, this expression is not valid for $\tau < 0$. To obtain $r(\tau)$ for negative τ we use the general relation $R(t, \tau) = R'(\tau, t)$, which in this case is

$$r(\tau) = r(-\tau)$$

by which (10A.2) becomes

$$r(\tau) = \frac{Q}{2\omega_0} e^{-\omega_0|\tau|} \tag{10A.3}$$

as shown in Fig. 10.2.

The power spectrum is obtained either as the Fourier transform of (10A.3) or using (10.42). The latter is easier.

$$S(\omega) = \frac{1}{-j\omega + \omega_0} \cdot Q \cdot \frac{1}{j\omega + \omega_0} = \frac{Q}{\omega^2 + \omega_0^2} \tag{10A.4}$$

In the physics literature a spectral density function of the form of (10A.4) is called *lorentzian.*

The mean square value of the output is given by

$$r(0) = \frac{Q}{2\omega_0}$$

which can also be obtained by evaluating the integral

$$r(0) = \frac{1}{2\pi} \int_{-\infty}^{\infty} \frac{Q}{\omega^2 + \omega_0^2} \, d\omega = \frac{Q}{\pi} \int_0^{\infty} \frac{d\omega}{\omega^2 + \omega_0^2}$$

by any well known method.

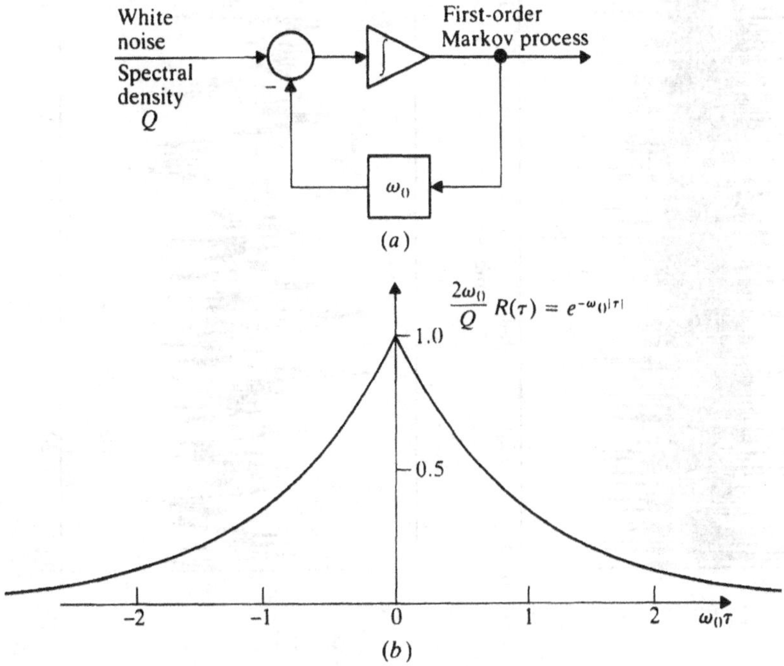

Figure 10.2 First-order Markov process. (*a*) Realization; (*b*) Correlation function.

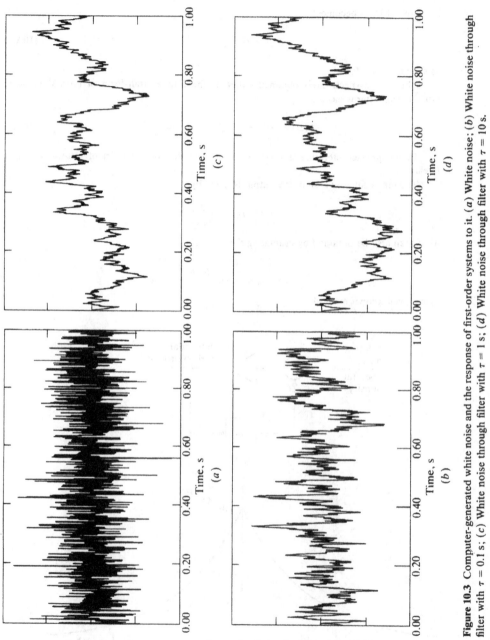

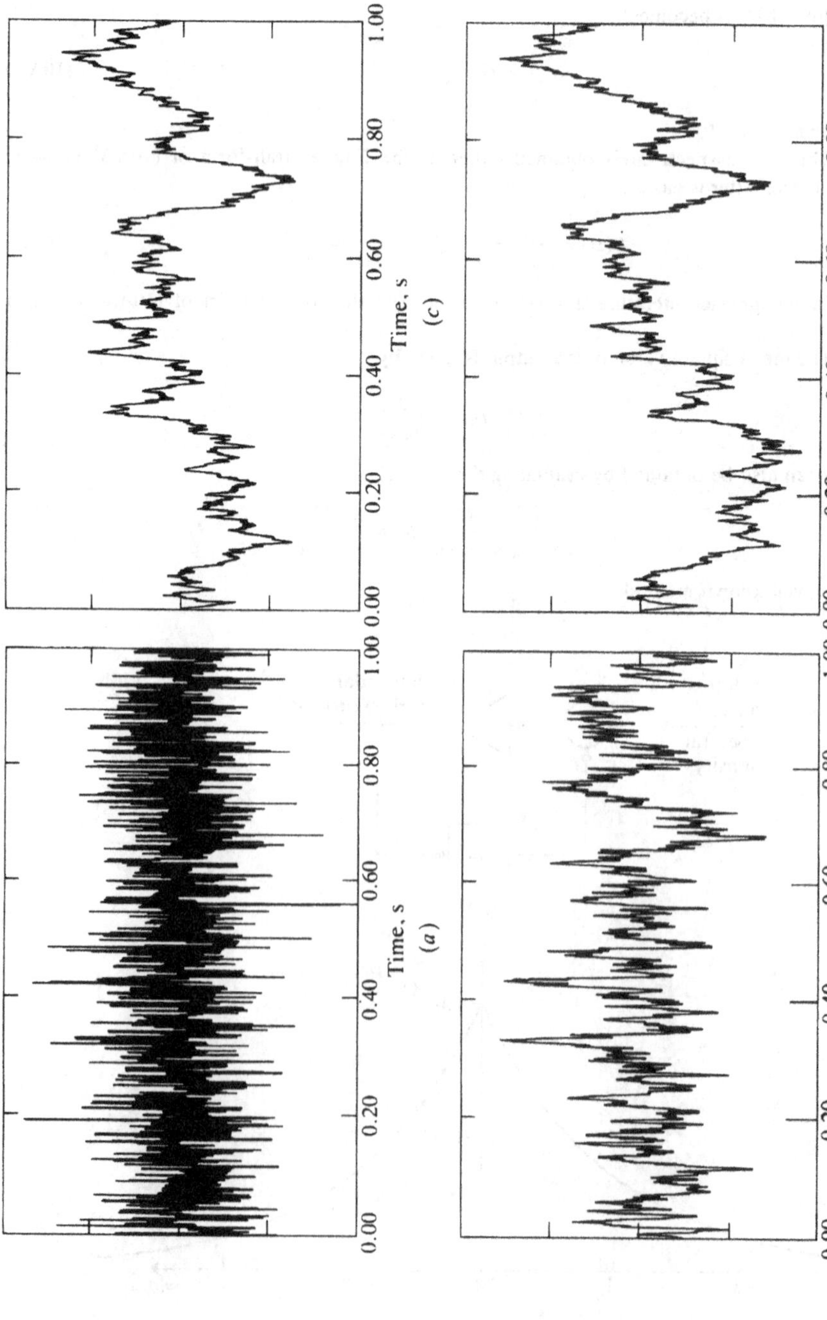

Figure 10.3 Computer-generated white noise and the response of first-order systems to it. (*a*) White noise; (*b*) White noise through filter with $\tau = 0.1$ s; (*c*) White noise through filter with $\tau = 1$ s; (*d*) White noise through filter with $\tau = 10$ s.

Note that $r(0)$, the mean square value of the signal, is $Q/2\omega_0$. Thus the spectral density of the white noise is

$$Q = 2\omega_0 \times (\text{mean square value of signal})$$

The units of the white noise spectral density Q are thus

$$(\text{units of the signal})^2 \times \text{sec}$$

which is in accordance with the discussion at the end of Sec. 10.4.

Examples of computer-generated first-order Markov processes for different corner frequencies ω_0 are shown in Fig. 10.3. It is noted that the output gets rougher as ω_0 tends to zero. In the limit ($\omega \to 0$) the result would be computer generated white noise. If the empirical correlation function were computed for this data it would resemble that shown in Fig. 10.2, but would not be nearly as smooth.

10.6 SPECTRAL FACTORIZATION

If statistical data needed for control system design is available at all, it is often in the form of empirical power spectral density curves, obtained either by means of a spectrum analyzer described earlier, or by a more modern instrument. (Some instruments sample the signal, compute its correlation function by numerical integration, and then determine the power spectral density by evaluating the Fourier transform using a *fast-Fourier-transform* (FFT) algorithm.)

The only systems that can be handled effectively by the linear quadratic design methods discussed in this book, or anywhere else for that matter, are those having a linear state-space representation and excited by white noise. The practical consequence of this is that the only random processes that can be accommodated within the framework of linear quadratic control theory are processes which can be represented as the response of linear systems to white noise.

If we know or have good reason to believe that a random disturbance is *not* the response of a linear system to white noise then two options are open to us: (1) we can approximate the actual process by one which has the required property or (2) we can develop the theory to deal with the actual process. The latter alternative is a fruitful source of research problems but it is not a practical course for the novice whose only real choice is thus the first.

If the only available data is based on a curve fit to empirical records it may be possible to change the form of the curve to meet the requirement that the random process be representable by a linear system excited by white noise. What is this requirement? The answer is given by (10.42).

First consider the case in which the output y is a scalar. In this case the transfer function of the system in response to white noise input is

$$\mathsf{H}(s) = \frac{\mathsf{y}(s)}{\mathsf{u}(s)} = \frac{\beta_1 s^{k-1} + \beta_2 s^{k-2} + \cdots + \beta_k}{s^k + \alpha_1 s^{k-1} + \cdots + \alpha_k} \tag{10.43}$$

(The process is assumed to be of order k and low pass, so that no $\beta_0 s^k$ term is present in (10.43).)

In factored form we can write

$$H(s) = c\frac{(s - z_1)\cdots(s - z_{k-1})}{(s - p_1)\cdots(s - p_k)} \tag{10.44}$$

where z_i and p_i are the zeros and poles, respectively, of the transfer function. Remember that the poles must lie in the left half-plane; otherwise the system will be unstable and will not have a stationary output. With $H(s)$ given by (10.44) the spectral density of y, using (10.42), is

$$\begin{aligned} S_y(\omega) &= c^2 Q\frac{(-j\omega - z_1)\cdots(-j\omega - z_{k-1})}{(-j\omega - p_1)\cdots(-j\omega - p_k)}\cdot\frac{(j\omega - z_1)\cdots(j\omega - z_{k-1})}{(j\omega - p_1)\cdots(j\omega - p_k)} \\ &= c^2 Q\frac{(\omega^2 - z_1^2)\cdots(\omega^2 - z_{k-1}^2)}{(\omega^2 - p_1^2)\cdots(\omega^2 - p_k^2)} \\ &= c^2 Q\frac{\omega^{2(k-1)} + a_1\omega^{2(k-1)} + \cdots + a_{k-1}}{\omega^{2k} + b_1\omega^{2(k-1)} + \cdots + b_k} \end{aligned} \tag{10.45}$$

In words, the spectrum of a (time-invariant) linear system is a rational function (i.e., a ratio of polynomials) in ω^2. The first-order Markov process considered in Example 10A has this property.

If a spectrum has the property of being a rational function of ω^2 as in (10.25) the process by which (10.45) was obtained from (10.43) can be reversed to obtain the rational transfer function $H(s)$. Then, one of the canonical realizations of the transfer function discussed in Chap. 3 can be used to obtain a state-space representation of the process.

The procedure of factoring an assumed rational function of ω^2 to obtain $H(s)$ is known as "spectral factorization." The calculations may be summarized as follows

Step 1. Write S_y as a rational function of $\lambda = \omega^2$

$$S_y = \gamma\frac{\lambda^{k-1} + a_1\lambda^{k-2} + \cdots + a_{k-1}}{\lambda^k + b_1\lambda^{k-1} + \cdots + b_k} = \frac{N(\lambda)}{D(\lambda)} \tag{10.46}$$

Step 2. Find the roots of $N(\lambda)$ and $D(\lambda)$

$$N(\lambda) = (\lambda - \alpha_1)\cdots(\lambda - \alpha_{k-1})$$
$$D(\lambda) = (\lambda - \beta_1)\cdots(\lambda - \beta_k)$$

Step 3. The poles and zeros of $H(s)$ are given by

$$z_i = \sqrt{\alpha_i} \qquad i = 1, 2, \ldots, k-1$$
$$p_i = \sqrt{\beta_i} \qquad i = 1, 2, \ldots, k$$

It is noted that there are two square roots of each β_i; unless β_i is negative real, one square root will have a positive real part and the other will have a negative real part. To have a stable transfer function the square root with positive real part is assigned to $\mathsf{H}(j\omega)$. The same can be done for the numerator roots and this will make $\mathsf{H}(s)$ have neither poles nor zeros in the right half-plane. It is not necessary to do so for purposes of stability, but zeros in the right half-plane are liable to cause other problems. Hence the spectral factorization that has neither poles nor zeros in the right half-plane (which is called *minimum phase*) is the factorization of choice.

Step 4. The spectral density Q of the white noise and the filter gain c are chosen to satisfy

$$\gamma = c^2 Q \tag{10.47}$$

The choice of c and Q are not unique.

Example 10B Dryden spectrum for air turbulence Based on experimental data (see Note 10.2) it has been found that the spectrum of the vertical component of random wind velocity in turbulent air has a "Dryden" spectrum

$$S(\omega) = \sigma_z^2 T \frac{1 + 3(\omega T)^2}{[1 + (\omega T)^2]^2} \tag{10B.1}$$

This spectral density has the appearance shown in Fig. 10.4. It has a maximum value of $1.125\sigma_z^2 T$ at $\omega T = 1/\sqrt{3}$.

To obtain a linear system which, when excited by white noise, has an output with a Dryden spectrum, write

$$S(\omega) = \sigma_z^2 T \frac{1 + j\sqrt{3}\omega T}{(1 + j\omega T)^2} \frac{1 - j\sqrt{3}\omega T}{(1 - j\omega T)^2} \tag{10B.2}$$

The factor of $S(\omega)$ that corresponds to a stable, minimum phase transfer function is identified with $\mathsf{H}(s)$. Thus

$$\mathsf{H}(s) = c \frac{1 + \sqrt{3}Ts}{(1 + Ts)^2} \tag{10B.3}$$

where c is any constant chosen such that

$$c^2 Q = \sigma_z^2 T$$

In particular, we can set $c = 1$ and $Q = \sigma_z^2 T$.

A companion form realization of the transfer function (10B.3) is shown in Fig. 10.5. This realization is not unique, of course.

Using (10.35) we can find the steady state correlation function. In the scalar case (10.35) is

$$R(\tau) = Q \int_0^\infty h(\xi) h(\xi + \tau)\, d\xi \tag{10B.4}$$

where $h(t)$ is the (scalar) impulse response corresponding to the transfer function given by (10B.3) with $c = 1$. This impulse response is the inverse Laplace transform of $\mathsf{H}(s)$ and is

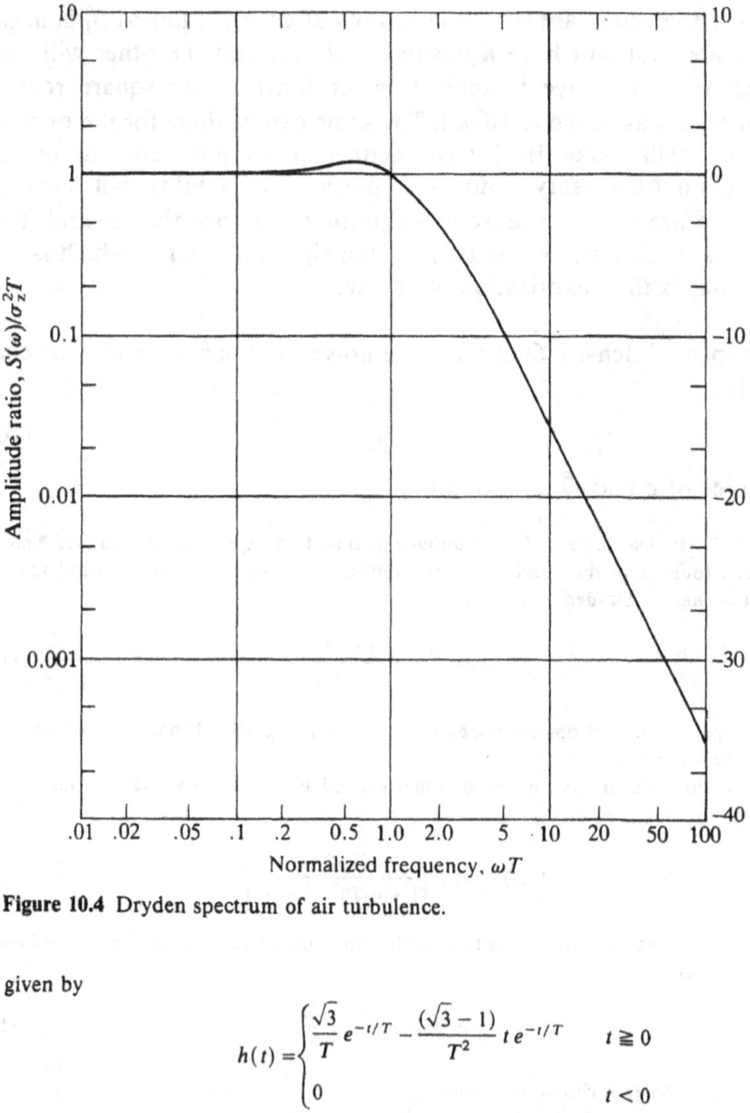

Figure 10.4 Dryden spectrum of air turbulence.

given by

$$h(t) = \begin{cases} \dfrac{\sqrt{3}}{T} e^{-t/T} - \dfrac{(\sqrt{3}-1)}{T^2} t\, e^{-t/T} & t \geqq 0 \\ 0 & t < 0 \end{cases} \tag{10B.5}$$

Substitute into (10B.4) and integrate (carefully) to obtain

$$R(\tau) = \sigma_z^2 \, e^{-|\tau|/T}\left(1 - \frac{1}{2}\frac{|\tau|}{T}\right) \tag{10B.6}$$

which has the shape illustrated in Fig. 10.6.

10.7 SYSTEMS WITH STATE-SPACE REPRESENTATION

We return now to the general linear system having the state-space representation with which this chapter started:

$$\dot{x} = Ax + Bu + Fv$$

$$y = Cx + w$$

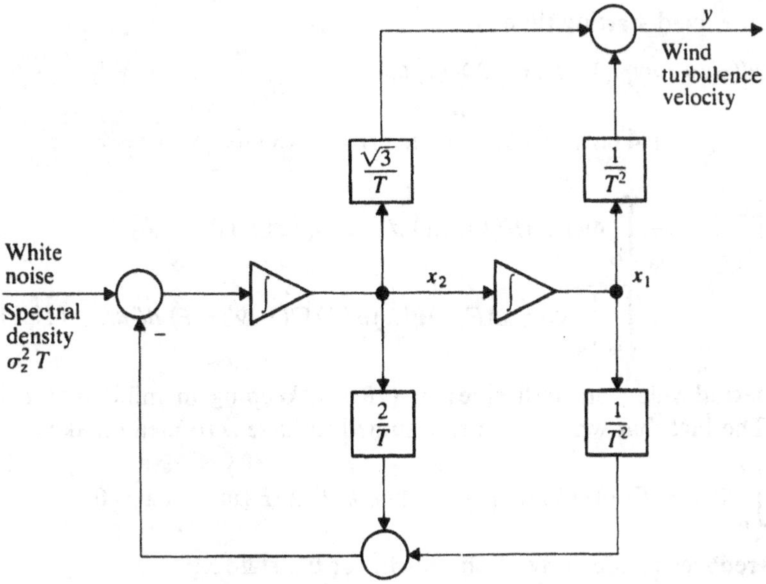

Figure 10.5 Realization of signal having Dryden spectrum.

where v and w are white-noise processes. For the present we ignore the "observation noise" w and focus on the effect on the output caused by the process excitation noise v. Also for the present we ignore the control u. This gives us

$$\dot{x} = Ax + Fv \tag{10.48}$$

$$y = Cx \tag{10.49}$$

In terms of the transition matrix, the solution to (10.48) is

$$x(t) = \Phi(t, t_0)x(t_0) + \int_{t_0}^{t} \Phi(t, \lambda)F(\lambda)v(\lambda)\, d\lambda \tag{10.50}$$

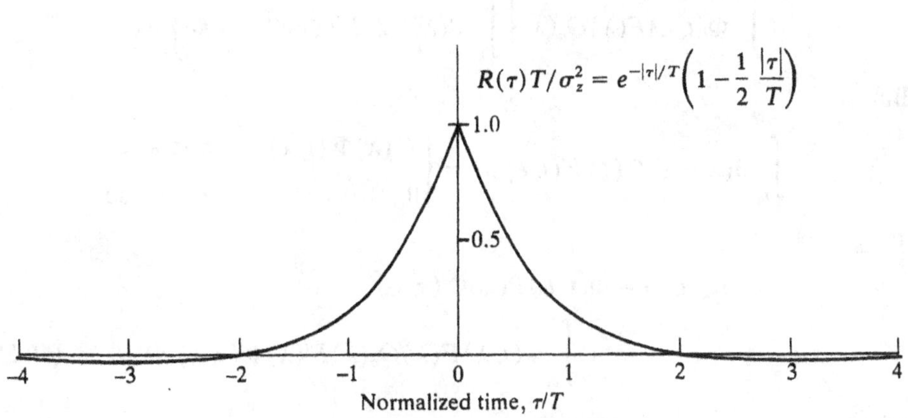

Figure 10.6 Correlation function for Dryden-spectrum wind turbulence.

where t_0 is some fixed starting time. Then

$$\begin{aligned} x(t)x'(\tau) &= \Phi(t, t_0)x(t_0)x'(t_0)\Phi'(\tau, t_0) \\ &\quad + \Phi(t, t_0)x(t_0)\left[\int_{t_0}^{t} \Phi(t, \lambda)F(\lambda)v(\lambda)\, d\lambda\right]' \\ &\quad + \int_{t_0}^{t} \Phi(t, \lambda)F(\lambda)v(\lambda)\, d\lambda \cdot x'(t_0)\Phi(\tau, t_0) \\ &\quad + \int_{t_0}^{t}\int_{t_0}^{\tau} \Phi(t, \lambda)F(\lambda)v(\lambda)v'(\xi)F'(\xi)\Phi'(\tau, \xi)\, d\xi\, d\lambda \end{aligned} \tag{10.51}$$

Take the expected value on both sides of (10.51), keeping in mind that v is white noise. The fact that white noise is assumed to have zero mean makes

$$E\left\{\int_{t_0}^{t} \Phi(t, \lambda)F(\lambda)v(\lambda)\, d\lambda\right\} = \int_{t_0}^{t} \Phi(t, \lambda)F(\lambda)E\{v(\lambda)\}\, d\lambda = 0$$

Thus (10.51) reduces to the correlation matrix for the state x:

$$\begin{aligned} R_x(t, \tau) &= \Phi(t, t_0)E\{x(t_0)x'(t_0)\}\Phi'(t, t_0) \\ &\quad + \int_{t_0}^{t}\int_{t_0}^{\tau} \Phi(t, \lambda)F(\lambda)E\{v(\lambda)v'(\xi)\}F'(\xi)\Phi'(\tau, \xi)\, d\xi\, d\lambda \end{aligned} \tag{10.52}$$

The starting state may be a random variable—not known precisely. Suppose that

$$E\{x(t_0)x'(t_0)\} = P(t_0) \tag{10.53}$$

The matrix $P(t_0)$ is known as the *covariance matrix* of $x(t_0)$.

Recall also that v is white noise

$$E\{v(\lambda)v'(\xi)\} = Q_v(\lambda)\delta(\lambda - \xi) \tag{10.54}$$

Thus the double integral in (10.52) becomes

$$\int_{t_0}^{t} \Phi(t, \lambda)F(\lambda)Q_v(\lambda)\left\{\int_{t_0}^{\tau} \delta(\lambda - \xi)F'(\xi)\Phi'(t, \xi)\, d\xi\right\} d\lambda$$

But

$$\int_{t_0}^{\tau} \delta(\lambda - \xi)F'(\xi)\Phi'(t, \xi)\, d\xi = \begin{cases} F'(\lambda)\Phi'(t, \lambda) & t_0 < \lambda < t \\ 0 & \text{otherwise} \end{cases}$$

Thus

$$\begin{aligned} R_x(t, \tau) &= \Phi(t, t_0)P(t_0)\Phi'(\tau, t_0) \\ &\quad + \int_{t_0}^{\bar{t}} \Phi(t, \lambda)F(\lambda)Q_v(\lambda)F'(\lambda)\Phi'(\tau, \lambda)\, d\lambda \end{aligned} \tag{10.55}$$

where $\qquad \bar{t} = \min(t, \tau)$

Recall that the state transition matrix $\Phi(t, \tau)$ has the property that

$$\Phi(t_1, t_3) = \Phi(t_1, t_2)\Phi(t_2, t_3) \qquad \text{for all } t_1, t_2, t_3$$

Thus

$$\Phi'(\tau, t_0) = [\Phi(\tau, t)\Phi(t, t_0)]' = \Phi'(t, t_0)\Phi'(\tau, t)$$

$$\Phi'(\tau, \lambda) = [\Phi(\tau, t)\Phi(t, \lambda)]' = \Phi'(t, \lambda)\Phi'(\tau, t)$$

Substitute these into (10.55) to obtain

$$R_x(t, \tau) = P(t)\Phi'(\tau, t) \qquad \text{for } \tau \geqq t \tag{10.56}$$

where

$$P(t) = R_x(t, t) = \Phi(t, t_0)P(t_0)\Phi'(t, t_0) + \int_{t_0}^{t} \Phi(t, \lambda)F(\lambda)Q_vF'(\lambda)\Phi'(t, \lambda)\, d\lambda \tag{10.57}$$

The matrix $P(t)$ is the *covariance matrix* of the state $x(t)$ at time t. Thus (10.57) describes how the covariance matrix of the state $x(t_0)$ at time t_0 evolves as time increases. Equation (10.56) describes how the correlation matrix $R_x(t, \tau)$, for τ not necessarily equal to t, is related to $P(t) = R_x(t, t)$. It is simply the product of $P(t)$ with the transpose of the transition matrix from τ to t.

It is important to note that (10.56) holds only for $\tau \geqq t$. The symmetry property (10.15) can be used to find $R(t, \tau)$ for $t \geqq \tau$ using $R(t, \tau)$ with $\tau \geqq t$.

An integral having the form of the integral in (10.57) already has been encountered in Chap. 9 in conjunction with deterministic control, namely the integral $M(t, T)$ defined by (9.17). Differences between (9.17) and the integral in (10.57) are that:

(*a*) $\Phi_c'(t, \tau)$ appears in (9.17) in place of $\Phi(t, \tau)$;
(*b*) $L(\tau)$ appears in (9.17) in place of $F(\tau)Q_vF'(\tau)$;
(*c*) The limits of integration in (9.17) are $[t, T]$ instead of $[t_0, t]$.

If the limits of integration were the same we could express the integral (10.55) as the solution to a differential equation of the form (9.16), but with the matrix A corresponding to Φ replacing A_c' which corresponds to Φ_c' and with FQ_vF' replacing L. This is almost correct. An additional change is needed also: in the sign of the derivative of P. In (9.16), the dot denoting the time derivative corresponds to the lower limit on the integral. In other words we are integrating backward from a known terminal state. The derivative on the differential equation for $P(t)$, however, implies time progressing in the forward direction, and hence the correct differential equation for P is

$$\dot{P} = AP + PA' + FQ_vF' \tag{10.58}$$

The term $\Phi(t, t_0)P(t_0)\Phi'(t_0)$ outside the integral in (10.57) is accounted for by the initial condition

$$P(t)|_{t=t_0} = P(t_0) \tag{10.59}$$

which is the covariance matrix defined by (10.53).

Equation (10.58), which is known as the *variance equation*, is very useful in the analysis of random processes excited by white noise, since it permits one to determine how the covariance propagates with the elapse of time, without the necessity of having to find the state-transition matrix. This is especially important in time-varying systems, in which state-transition matrices are all but impossible to determine, but it is also useful even in time-invariant systems, especially to evaluate steady state covariance matrices. In particular, if A is a constant matrix corresponding to a stable dynamic system, and F and Q_v are constant

$$P(t) \to \bar{P} = \text{const}$$

where $\bar{P}$ is the steady state covariance matrix which is obtained as the steady state solution to (10.56), namely

$$0 = A\bar{P} + \bar{P}A' + FQ_vF' \tag{10.60}$$

It might at first seem that the solution $\bar{P}$ to (10.60) would be negative. But recall that if A is stable it has eigenvalues with negative real parts. By a famous stability theorem of Liapunov (see Note 10.3), the location of the eigenvalues of A in (10.60) guarantee that (10.60) has a unique, positive-definite solution for $\bar{P}$.

The differential-equation form (10.60) of the integral (10.57) was justified by using the known results from Chap. 9. But it can be obtained directly by differentiating both sides of (10.57) with respect to t. In particular

$$\dot{P}(t) = \frac{\partial P}{\partial t} = \frac{\partial \Phi(t, t_0)}{\partial t} P(t_0)\Phi'(t, t_0) + \Phi(t, t_0)\frac{\partial \Phi'(t, t_0)}{\partial t}$$
$$+ \frac{\partial}{\partial t}\left[\int_0^t \Phi(t, \lambda)F(\lambda)Q(\lambda)F'(\lambda)\Phi'(t, \lambda)\, d\lambda\right]$$

The differential equation (10.58) results upon use of Leibnitz's rule for differentiating an integral

$$\frac{\partial}{\partial t}\int_0^t f(t, \lambda)\, d\lambda = f(t, t) + \int_0^t \frac{\partial f(t, \lambda)}{\partial t}\, d\lambda$$

and also that

$$\frac{\partial \Phi(t, \tau)}{\partial t} = A(t)\Phi(t, \tau) \qquad \text{for all } t, \tau$$

The calculation is quite straightforward.

The correlation matrix of the output y, as given by (10.39), is readily obtained from the correlation matrix for x:

$$y(t) = C(t)x(t)$$

Then

$$y(t)y'(\tau) = C(t)x(t)x'(\tau)C'(\tau)$$

Thus, on taking expectations on both sides,

$$R_y(t, \tau) = E\{y(t)y'(\tau)\} = C(t)E\{x(t)x'(\tau)\}C'(\tau) = C(t)R_x(t, \tau)C'(\tau) \qquad (10.61)$$

In particular, the covariance matrix of the output

$$P_y(t) = R_y(t, t) = C(t)P(t)C'(t) \qquad (10.62)$$

(The covariance matrix $P(t)$, corresponding to the state x is written without a subscript x for orthographic convenience. The state covariance matrix occurs so often that use of the subscript would make many equations unnecessarily messy.)

Example 10C First-order Markov process (continued) We previously found (Example 10A) that the correlation function of the stationary process obtained by passing white noise through a low-pass filter with the transfer function $H(s) = 1/(s + \omega_0)$ is $r(\tau) = Q e^{-\omega_0|\tau|}/2\omega_0$. We now obtain this result by the methods of Sec. 10.7.

First we note that the process having this transfer function $1/(s + \omega_0)$ has the state-space representation

$$\dot{x} = -\omega_0 x + v \qquad (10C.1)$$

Hence, for (10.60),

$$A = -\omega_0 \qquad F = 1 \qquad Q_v = Q$$

Thus (10.60) becomes

$$-2\omega_0\bar{P} + Q$$

or

$$\bar{P} = \frac{Q}{2\omega_0} = \text{steady state covariance}$$

The state transition matrix (1×1 in this case) is

$$\Phi(t, t + \tau) = e^{A\tau} = e^{-\omega_0\tau}$$

Thus, in accordance with (10.56), in the steady state

$$r(\tau) = \bar{P}e^{-\omega_0\tau} = \frac{Q}{2\omega_0}e^{-\omega_0\tau} \qquad \text{for } \tau \geqq 0$$

and, hence, by symmetry

$$r(\tau) = \frac{Q}{2\omega_0}e^{-\omega_0|\tau|} \qquad \text{for all } \tau$$

Note, however, that we can also find the covariance and correlation function for the nonstationary process, obtained by applying white noise v to the process defined by (10C.1) but not waiting until the steady state is reached. From (10.58)

$$\dot{P} = -2\omega_0 P + Q$$

which has the solution

$$P(t) = P(t_0)\, e^{-2\omega_0(t-t_0)} + \frac{Q}{2\omega_0}[1 - e^{-2\omega_0(t-t_0)}]$$

Thus

$$R(t, \tau) = P(t)\, e^{-\omega_0(\tau-t)}$$
$$= P(t_0)\, e^{-\omega_0(t+\tau-2t_0)} + \frac{Q}{2\omega_0}[e^{-\omega_0(\tau-t)} - e^{-\omega_0(t+\tau-2t_0)}]$$

Example 10D Air turbulence (continued) Using the state-variable representation illustrated in Fig. 10.5, the differential equations for the wind-turbulence process having a Dryden spectrum are

$$\dot{x}_1 = x_2$$
$$\dot{x}_2 = -\frac{1}{T^2}x_1 - \frac{2}{T}x_2 + v \tag{10D.1}$$

and the output is given by

$$y = \frac{1}{T^2}x_1 + \frac{\sqrt{3}}{T}x_2 \tag{10D.2}$$

The matrices representing (10D.1) and (10D.2) are

$$A = \begin{bmatrix} 0 & 1 \\ -\frac{1}{T^2} & -\frac{2}{T} \end{bmatrix} \qquad F = \begin{bmatrix} 0 \\ 1 \end{bmatrix} \qquad C = \begin{bmatrix} \frac{1}{T^2} & \frac{\sqrt{3}}{T} \end{bmatrix} \tag{10D.3}$$

Let the steady state covariance matrix be

$$\bar{P} = \begin{bmatrix} p_1 & p_2 \\ p_2 & p_3 \end{bmatrix}$$

Then, by (10.60), the elements of $\bar{P}$ are given by the solutions of

$$0 = 2p_2$$
$$0 = -\frac{p_1}{T^2} - \frac{2p_2}{T} + p_3$$
$$0 = 2\left(-\frac{p_2}{T^2} - \frac{2p_3}{T}\right) + \sigma_z^2 T$$

which yield

$$p_1 = \frac{\sigma_z^2 T^4}{4} \qquad p_2 = 0 \qquad p_3 = \frac{\sigma_z^2 T^2}{4}$$

Thus

$$\bar{P} = \frac{\sigma_z^2 T^2}{4}\begin{bmatrix} T^2 & 0 \\ 0 & 1 \end{bmatrix} \tag{10D.4}$$

The steady state covariance $\bar{P}_y$ of the output is obtained using (10.62):

$$\bar{P}_y = C\bar{P}C' = \begin{bmatrix} \frac{1}{T^2} & \frac{\sqrt{3}}{T} \end{bmatrix} \frac{\sigma_z^2 T^2}{4} \begin{bmatrix} T^2 & 0 \\ 0 & 1 \end{bmatrix} \begin{bmatrix} \frac{1}{T^2} \\ \frac{\sqrt{3}}{T} \end{bmatrix} = \sigma_z^2$$

as was obtained in Example 10B.

To obtain the correlation function of the output we need the state-transition matrix $e^{A\tau}$ for A given by (10D.4). By one of the methods discussed in Chap. 3 we obtain

$$e^{A\tau} = \begin{bmatrix} 1+\frac{\tau}{T} & \tau \\ -\frac{\tau}{T^2} & 1-\frac{\tau}{T} \end{bmatrix} e^{-\tau/T} \tag{10D.5}$$

Thus, by the steady state form of (10.56), namely

$$R_x(\tau) = \bar{P}e^{A'\tau} \qquad \tau > 0$$

we obtain, using (10D.4) and (10D.5)

$$R_x(\tau) = \frac{\sigma_z^2 T^2}{4} \begin{bmatrix} T^2 & 0 \\ 0 & 1 \end{bmatrix} \begin{bmatrix} 1+\frac{\tau}{T} & -\frac{\tau}{T^2} \\ \tau & 1-\frac{\tau}{T} \end{bmatrix} e^{-\tau/T} \tau > 0$$

$$= \frac{\sigma_z^2 T^2}{4} \begin{bmatrix} T^2\left(1+\frac{\tau}{T}\right) & -\tau \\ \tau & 1-\frac{\tau}{T} \end{bmatrix} e^{-\tau/T} \tau > 0 \tag{10D.6}$$

Note

$$[R_x(\tau)]_{12} = [R_x(-\tau)]_{21}$$

as is required by the steady state form of (10.15), which is

$$R_x(\tau) = \lim_{t\to\infty} R_x(t, t+\tau) = R_x'(-\tau) \tag{10D.7}$$

Finally, the output correlation function is obtained by use of (10.61):

$$R_y(\tau) = CR_x(\tau)C'$$

$$= \begin{bmatrix} \frac{1}{T^2} & \frac{\sqrt{3}}{T} \end{bmatrix} \cdot \frac{\sigma_z^2 T^2}{4} \left[\begin{array}{c|c} T^2\left(1+\frac{\tau}{T}\right) & -\tau \\ \hline \tau & 1-\frac{\tau}{T} \end{array}\right] e^{-\tau/T} \begin{bmatrix} \frac{1}{T^2} \\ \hline \frac{\sqrt{3}}{T} \end{bmatrix}$$

$$= \sigma_z^2\left(1-\frac{1}{2}\frac{\tau}{T}\right)e^{-\tau/T} \qquad \text{for } \tau > 0$$

And, by symmetry

$$R_y(\tau) = \sigma_z^2\left(1-\frac{1}{2}\frac{|\tau|}{T}\right)e^{-|\tau|/T} \tag{10D.8}$$

as obtained in Example 10B.

10.8 THE WIENER PROCESS AND OTHER INTEGRALS OF STATIONARY PROCESSES

If white noise is the input to an asymptotically stable time-invariant system, the state of the system evolves ultimately into a stationary process with a covariance matrix $\bar{P}$ which is the solution to (10.60). There are many important situations, however, in which the system is not asymptotically stable. In such situations the covariance matrix tends to grow without bounds, and, of course, the state of such a system is not a stationary process.

The simplest of these nonstationary processes is a Wiener process w, the integral of white noise:

$$\frac{dw}{dt} = v = \text{white noise}$$

Note that w can be a scalar, or more generally, a vector. From (10.58) the covariance matrix of a Wiener process is

$$P_w(t) = P_v(0) + Qt \tag{10.63}$$

Thus, a Wiener process has a covariance matrix which grows linearly with time, and ultimately becomes infinite. A computer-generated sample function of a Wiener process is shown in Fig. 10.7. It is observed that the signal itself appears to be trending downward, as one might expect from the growing variance. Yet, since white noise has zero mean, and an integrator is linear, the output Wiener process also has zero mean. The signal in Fig. 10.7 certainly doesn't look like it

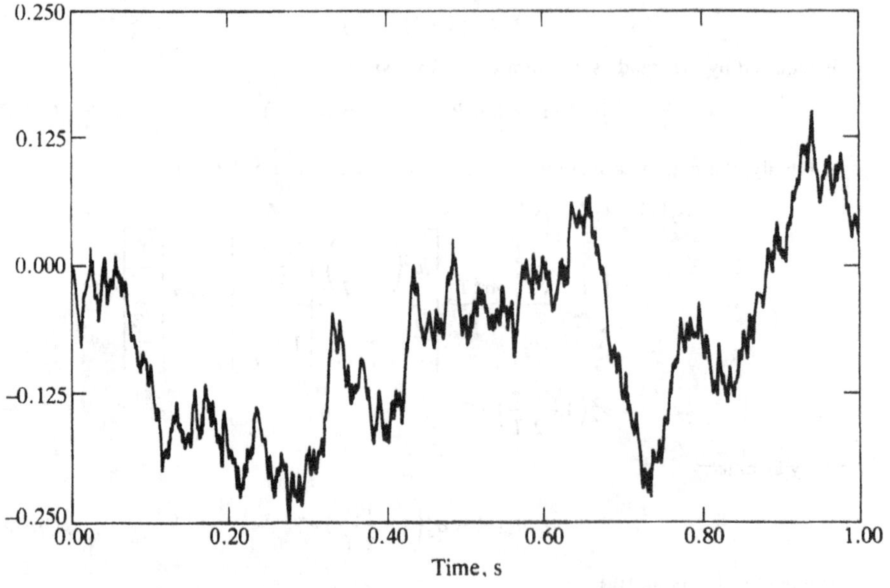

Figure 10.7 Computer-generated Wiener process.

has zero mean. But looks can be deceiving! Since the process is *not* stationary, it is not meaningful to consider time averages as equivalent to ensemble averages. The zero mean of the Wiener process refers to the *ensemble* average, and there is no reason to doubt that the ensemble has zero mean. Just imagine that for every signal $w(t)$ in the ensemble, its negative $-w(t)$ also is present and it is clear that the ensemble average is zero.

It is also of interest to determine the correlation matrix $R_w(t, \tau)$ for a Wiener process. This is accomplished through the use of (10.56). The state-transition matrix for a bank of integrators is

$$\Phi(\tau, t) = \mu_1(\tau - t)I$$

where $\mu_1(\tau - t)$ is a unit step starting at $\tau = t$.

Thus, assuming that $P_v(0) = 0$ in (10.61) we find from (10.56) that

$$R_x(t, \tau) = Qt\mu_1(\tau - t) = Qt \qquad \text{for } \tau \geqq t \tag{10.64}$$

Replacing τ by t in (10.64) gives

$$R_x(t, \tau) = Q\tau\mu_1(t - \tau) = Q\tau \qquad \text{for } t \geqq \tau \tag{10.65}$$

Thus, combining (10.64) and (10.65) we obtain

$$R_x(t, \tau) = Q \min(t, \tau) \tag{10.66}$$

A plot of constant values of $R_x(t, \tau)$ is shown in Fig. 10.8.

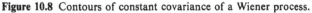

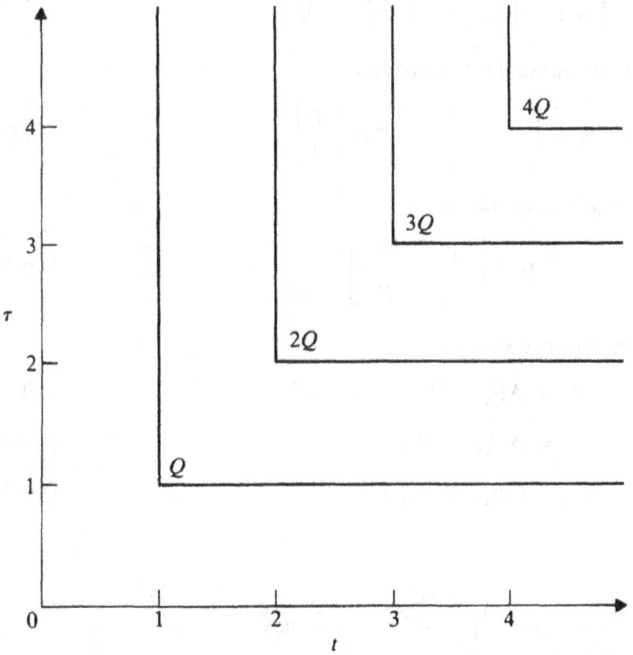

Figure 10.8 Contours of constant covariance of a Wiener process.

The Wiener process rather than white noise is the starting point for most rigorous treatments of random processes. In fact, the basic equation for a process excited by white noise is written

$$dx = Ax\,dt + dw \tag{10.67}$$

where w is a Wiener process; white noise $v = dw/dt$ is not used, because one of the theoretical properties of a Wiener process is that it is (almost) nowhere differentiable.

The Wiener process is frequently given other names, including:

Random walk (in continuous-time)
Brownian motion

and its properties are of considerable theoretical importance. See Note 10.1.

A more general class of nonstationary processes are the integrals of stationary processes. In particular, let w be the integral (Fig. 10.9) of the output $y = Cx$ of a stationary process having the usual state-space representation

$$\dot{x} = Ax + Fv$$

$$\dot{w} = Cx$$

Then we have the metastate equation

$$\begin{bmatrix} \dot{x} \\ \dot{w} \end{bmatrix} = \begin{bmatrix} A & 0 \\ C & 0 \end{bmatrix} \begin{bmatrix} x \\ w \end{bmatrix} + \begin{bmatrix} F \\ 0 \end{bmatrix} v \tag{10.68}$$

and (10.58) applies with the metastate matrices

$$\mathbf{A} = \begin{bmatrix} A & 0 \\ C & 0 \end{bmatrix} \qquad \mathbf{F} = \begin{bmatrix} F \\ 0 \end{bmatrix} \tag{10.69}$$

Thus, if the metastate covariance matrix is

$$\mathbf{P} = \begin{bmatrix} P_x & P_{xw} \\ P'_{xw} & P_w \end{bmatrix} \tag{10.70}$$

Then, by (10.58), its components satisfy

$$\dot{P}_x = AP_x + P_xA' + FQ_vF' \tag{10.71}$$

$$\dot{P}_{xw} = AP_{xw} + P_xC' \tag{10.72}$$

$$\dot{P}_w = CP_{xw} + P'_{xw}C' \tag{10.73}$$

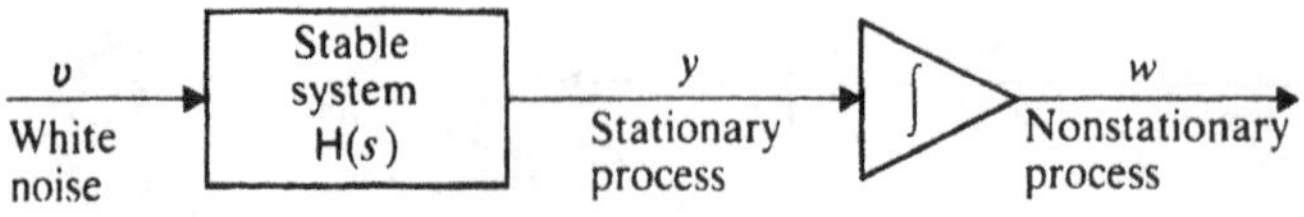

Figure 10.9 Integral of stationary process is nonstationary.

If A is the dynamics matrix of a stable system, P_x has a steady state solution $\bar{P}$ given by (10.60), as anticipated. Likewise the cross-covariance matrix P_{xw} has a steady state solution

$$\bar{P}_{xw} = -A^{-1}\bar{P}C' \tag{10.74}$$

But P_w does not have a steady state solution. Its "asymptotic growth rate" is obtained by substituting $\bar{P}_{xw}$ from (10.74) into (10.73)

$$\dot{\bar{P}}_w = -C[A^{-1}\bar{P} + \bar{P}(A')^{-1}]C' \tag{10.75}$$

Thus, if time is long enough for y to become stationary, then its integral behaves like a Wiener process having a white noise "driver" with a spectral density

$$\bar{Q} = -C[A^{-1}\bar{P} + \bar{P}(A')^{-1}]C' \tag{10.76}$$

Premultiply (10.60) by A^{-1} and postmultiply by $(A')^{-1}$ to obtain

$$-\bar{P}(A')^{-1} - (A^{-1})\bar{P} = A^{-1}FQ_vF'(A')^{-1}$$

Thus, the equivalent white-noise driver is

$$\bar{Q} = CA^{-1}FQ_vF'(A')^{-1}C' \tag{10.77}$$

This formula has a frequency domain interpretation. The spectral density matrix of y, the input to the (bank of) integrators is given by (10.42) with

$$\mathsf{H}(s) = C(sI - A)^{-1}F$$

Thus

$$S_y(\omega) = C(-j\omega I - A)^{-1}FQ_vF'(j\omega I - A)'C'$$

and hence

$$S_y(0) = CA^{-1}FQ_vF'(A')^{-1}C' \tag{10.78}$$

Compare (10.78) with (10.77) to find that

$$\bar{Q} = S_y(0) \tag{10.79}$$

This means that the white-noise equivalent of the stationary random process y that is integrated, is exactly the value of the spectral density of y at $\omega = 0$. Neither the mean square value, nor the spectral shape of the random process y, directly influence the asymptotic growth rate, as one might perhaps have believed.

PROBLEMS

Problem 10.1 Calculation of mean square value and spectral density

The input to the systems having the following transfer functions is white noise with a spectral density of unity

(a) $\mathsf{H}(s) = \dfrac{1}{(s+\alpha)(s+\beta)}$

$$(b)\ \mathsf{H}(s) = \frac{1}{(s+\alpha)(s+\beta)(s+\gamma)}$$

For the output of each system:

(i) Find the spectral density $S(\omega)$;
(ii) Find the mean square value by evaluating the integral (10.26);
(iii) Find the mean-square value by representing $\mathsf{H}(s)$ in some suitable canonical form, and then evaluating the steady state covariance matrix.

Problem 10.2 Realization of empirically measured power spectrum

The power spectral density of a random process was measured empirically and the data is approximated by the curve shown in Fig. P10.2. For purposes of control system design it is desired to approximate this process as the output of a linear system excited by white noise. The transfer function of the linear system is of the form

$$\mathsf{H}(s) = K\frac{s^2(s^2 + 2\zeta_3\omega_3 s + \omega_3^2)}{(s+\alpha)(s^2 + 2\zeta_1\omega_1 s + \omega_1^2)(s^2 + 2\zeta_2\omega_2 s + \omega_2^2)}$$

Find the values of K, α, ζ_i, ω_i such that the spectrum of the output of $\mathsf{H}(s)$ is a good approximation to the spectrum shown in Fig. P10.2.

Problem 10.3 Inertial navigation errors

In Example 3E the navigation errors due to constant (bias) errors on the gyro and the accelerometer of a single-axis inertial navigation system were computed. In addition to the constant errors, random instrument errors may also be present. Calculation of the covariance matrix of the navigation errors due to these instrument errors is the purpose of this problem. The dynamic model of Example 3E is used, except E_A and E_G are to be represented by independent white-noise sources of spectral densities W_A and W_G, respectively.

(*a*) Write the variance equation for the propagation covariance matrix of the navigation error, using

$$P = \begin{bmatrix} p_1 & p_2 & p_3 \\ p_2 & p_4 & p_5 \\ p_3 & p_5 & p_6 \end{bmatrix} = E\left\{\begin{bmatrix} \Delta x^2 & \Delta x\,\Delta v & \Delta x\,\Delta\phi \\ \Delta x\,\Delta v & \Delta v^2 & \Delta v\,\Delta\phi \\ \Delta x\,\Delta\phi & \Delta v\,\Delta\phi & \Delta\phi^2 \end{bmatrix}\right\}$$

(*b*) Solve these equations for the initial condition $P(0) = 0$ and plot $p_1(t)$, $p_4(t)$, and $p_6(t)$.

NOTES

Note 10.1 Mathematical theory of stochastic processes

The level of mathematical rigor with which stochastic processes can be treated ranges over an extremely wide spectrum. At one extreme is the treatment given in this and similar books such as Papoulis[1] and Schwarz and Friedland.[2] At the opposite extreme are the very rigorous treatments such as those of Doob[3] and Dynkin.[4] Between these extremes are a number of engineering and applied mathematics texts which emphasize different aspects of stochastic processes. From the standpoint of control engineering, the books by Åström[5] and Jazwinski[6] present very readable accounts of stochastic differential equations including the Ito and the Stratonovitch calculi and the differences between them. At a higher level of mathematical rigor is the book by Kushner.[7] A treatment of stochastic processes from the standpoint of communication theory is the encyclopedic text of Middleton.[8] And the viewpoint of the physicist and astronomer is represented by the compendium edited by Wax.[9]

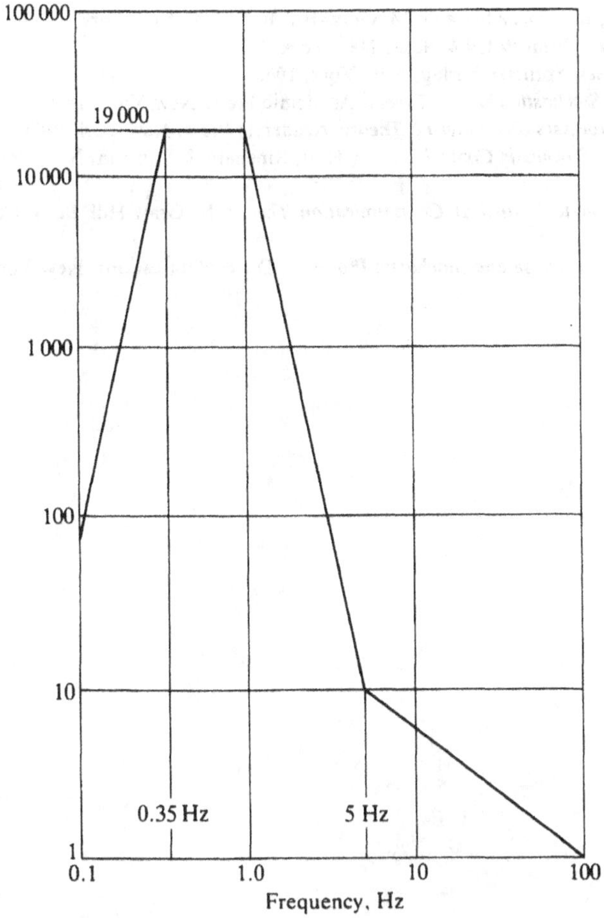

Figure P10.2 Approximate power spectrum of measured data.

Note 10.2 Dryden spectrum

The "Dryden" spectrum is named in honor of the aerodynamicist Hugh Dryden who spent most of his professional career with the National Advisory Committee for Aeronautics (NACA), the precursor of the National Aeronautics and Space Administration (NASA).

Note 10.3 Liapunov equation

The equation

$$AP + PA' = -Q$$

is often called the Liapunov equation, because it arises in the study of stability of linear systems by "Liapunov's Second Method". The theorem asserts that if A has its eigenvalues in the negative plane then, for any positive semidefinite matrix Q, the solution matrix P is positive-definite.

REFERENCES

1. Papoulis, A., *Probability, Random Variables, and Stochastic Processes,* John Wiley & Sons, New York, 1965.

2. Schwarz, R. J., and Friedland, B., *Linear Systems*, McGraw-Hill Book Co., New York, 1965.
3. Doob, J. L., *Stochastic Processes*, John Wiley & Sons, New York, 1953.
4. Dynkin, E. B., *Markov Processes*, Springer-Verlag, New York, 1965.
5. Åström, K. J., *Introduction to Stochastic Control Theory*, Academic Press, New York, 1970.
6. Jazwinski, A. H., *Stochastic Processes and Filtering Theory*, Academic Press, New York, 1970.
7. Kushner, H. J., *Introduction to Stochastic Control Theory*, Holt, Rinehart, & Winston, New York, 1971.
8. Middleton, D., *An Introduction to Statistical Communication Theory*, McGraw-Hill Book Co., New York, 1960.
9. Wax, N. (ed.), *Selected Papers on Noise and Stochastic Processes*, Dover Publications, New York, 1954.

CHAPTER

ELEVEN

KALMAN FILTERS: OPTIMUM OBSERVERS

11.1 BACKGROUND

The historic papers of Kalman and Bucy[1, 2] of the early 1960s are among the most important contributions to control and system science of the twentieth century. The impact of this work ranks it with the work of Nyquist and Bode of the 1920s and 1930s, and with the work of Wiener of the 1940s.

The contribution of Kalman and Bucy could not have come at a more auspicious time. The United States Apollo program was started only a year or two earlier. It was widely recognized that data processing algorithms based on the results of Kalman and Bucy would be of enormous benefit to this program and to other space projects.

State-space methods had only recently begun to receive a great deal of attention and the results of Kalman and Bucy were perfectly suited to these methods. A rather modest digital computer—the only kind available in the sixties—was appropriate for calculating the gains of the Kalman filter. The calculations required to implement other filtering algorithms, on the other hand, would have overwhelmed the digital computers of that period and were not even given serious consideration after the Kalman filter came upon the scene.

The Wiener filter, moreover, and its various extensions, which the Kalman filter supplanted, was limited inherently to linear systems because the results were expressed in terms of transfer functions or impulse responses, concepts that are meaningful only in the context of linear systems. There is no reasonable method for extending these results to nonlinear systems. The Kalman filter, however, is expressed in the form of differential equations (or difference equations), and methods of extending the original results to nonlinear systems, at least approximately, occurred to many investigators within a few years. (Most practical applications of Kalman filtering today are, in fact, applications to nonlinear systems.)

As often happens with results of fundamental importance, many potential users of the method were uncomfortable with their own understanding of the original presentation and they developed alternative derivations with which they felt more at ease. It is not a great exaggeration to say that in the mid-1960s every user had his own favorite derivation. Many of these were published and helped to illuminate the original results and make them accessible to a wider audience. (See Note 11.1 for additional historical background.)

11.2 THE KALMAN FILTER IS AN OBSERVER

The problem addressed by Kalman and Bucy is the following. Given a dynamic process†

$$\dot{x} = Ax + Bu + Fv \tag{11.1}$$

where u is a *known* input and having observations given by

$$y = Cx + w \tag{11.2}$$

where v and w are *white* noise processes, having known spectral density matrices. Find an *optimum* observer (or "state estimator") for the state x.

The solution of the problem, as given by Kalman and Bucy, is that the optimum state estimator is an observer, as shown in Fig. 11.1. In other words, it can be expressed by the differential equation

$$\dot{\hat{x}} = A\hat{x} + Bu + \hat{K}(y - C\hat{x}) \tag{11.3}$$

which defines an observer as given in Chap. 10, provided that the gain matrix $\hat{K}$ is optimally chosen. This might appear to be a routine and obvious result. It should be noted, however, that the very notion of an observer came only several years *after* the original work of Kalman and Bucy; the optimum observer came before the nonoptimum observer.

But what is meant by an *optimum* observer? How is optimum defined? A remarkable property of the Kalman filter (11.3) is that it is optimum under *any reasonable* performance criterion, provided the random processes are white and *gaussian.*

The gaussian requirement is a condition on the first-order probability density functions of w and v, e.g.,

$$\text{pdf}(w) = \frac{1}{(2\pi)^{n/2}|W|^{1/2}} \exp\left\{-\frac{1}{2}w'W^{-1}w\right\}$$

† In the original papers of Kalman and Bucy the control input u—which we assume here to be known—was not considered. But including u causes no problem, since we are assuming that the Kalman filter has the form of an observer.

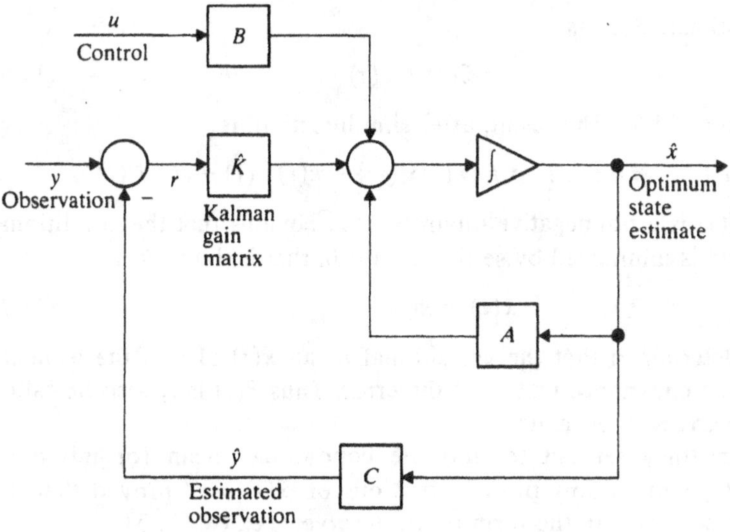

Figure 11.1 Kalman filter is an optimum observer.

which is a multidimensional gaussian probability density function. If v and w have such first-order density functions, then (as proved by Kalman and Bucy), (11.3) is the best of all possible filters: there is no other filter, linear or nonlinear, better than the linear Kalman filter of (11.3) when the dynamic process and observations are linear, and random processes v and w are gaussian white noise. (If the random processes are not gaussian, however, a nonlinear filter could be better.)

Kalman defined the state estimate $\hat{x}(t)$ as the conditional mean of $x(t)$, given the observation data $y(\tau)$ for $\tau \leqq t$. This is written as

$$\hat{x}(t) = E\{x(t)|y(\tau), \tau \leqq t\} \tag{11.4}$$

Let $\bar{x}(t)$ be any estimate of $x(t)$, and let the error be given by

$$e(t) = x(t) - \bar{x}(t)$$

Then

$$\begin{aligned} e(t)e'(t) &= [x(t) - \bar{x}(t)][x'(t) - \bar{x}'(t)] \\ &= x(t)x'(t) - \bar{x}(t)x'(t) - x(t)\bar{x}'(t) + \bar{x}(t)\bar{x}'(t) \end{aligned}$$

Now compute the conditional mean of $e(t)e'(t)$, that is, the conditional covariance matrix of e given the observation data $y(\tau)$ for $\tau \leqq t$. We have

$$\begin{aligned} E\{e(t)e'(t)|y(\tau), \tau \leqq t\} = E\{x(t)x'(t)\} &- \bar{x}(t)\hat{x}'(t) \\ &- \bar{x}'(t)\hat{x}(t) + \bar{x}(t)\bar{x}'(t) \end{aligned} \tag{11.5}$$

The middle terms in (11.5) are obtained by using the definition (11.4).

Write the estimate $\bar{x}(t)$ as

$$\bar{x}(t) = \hat{x}(t) + \zeta(t) \tag{11.6}$$

and substitute into (11.5). The result, after simplification, is

$$E\{e(t)e'(t)|y(\tau), \tau \leqq t\} = E\{x(t)x'(t)\} - \hat{x}(t)\hat{x}'(t) + \zeta(t)\zeta'(t)$$

Since $\zeta(t)\zeta'(t)$ is a nonnegative quantity, it is obvious that the conditional covariance matrix is minimized by setting $\zeta(t) = 0$, that is, by making

$$\bar{x}(t) = \hat{x}(t) \tag{11.7}$$

We have thus determined that the conditional mean $\hat{x}(t)$ also is the estimate that minimizes the covariance matrix of the error. Thus $\hat{x}(t)$ may also be called the "minimum variance" estimate.

Kalman and Bucy set out to find the conditional mean (or minimum variance estimate) without any prior restrictions on $\hat{x}(t)$, and proved that the estimate can be expressed in the form of an observer, i.e., by (11.3).

The derivation of the Kalman filter without making any assumptions about the form of the solution is beyond the scope of this text. But if it is accepted that the Kalman filter has the structure of an observer the problem is easier. All that is necessary is to find the optimum gain matrix $\hat{K}(t)$. As we shall see shortly, this problem is dual to the problem of finding the optimum gain for the linear quadratic control problem.

11.3 KALMAN FILTER GAIN AND VARIANCE EQUATIONS

Accepting that the Kalman filter has the structure of an observer, we now seek to find the gain matrix $K(t)$ which makes the covariance matrix of the error least. The optimum value of $K(t)$ will be designated by $\hat{K}(t)$.

The error now is defined by

$$e = x - \hat{x}$$

and, using (11.1), (11.2), and (11.3), the differential equation for the error is obtained:

$$\begin{aligned} \dot{e} = \dot{x} - \dot{\hat{x}} &= Ax + Fv - A\hat{x} - K(Cx + w - C\hat{x}) \\ &= (A - KC)e + Fv - Kw \end{aligned} \tag{11.8}$$

Since v and w are white noise processes, their weighted sum

$$\xi = Fv - Kw \tag{11.9}$$

is also white noise, with a covariance matrix Q_ξ. To find Q_ξ, observe that

$$\begin{aligned} E\{\xi(t)\xi'(\tau)\} = {} & F(t)E\{v(t)v'(\tau)\}F'(\tau) - K(t)E\{w(t)v'(\tau)\}F'(\tau) \\ & - F(t)E\{v(t)w'(\tau)\}K'(\tau) + K(t)E\{w(t)w'(\tau)\}K'(\tau) \end{aligned} \tag{11.10}$$

Assume that the expected values on the right-hand side of (11.10) are all those of white noise:

$$\begin{aligned} E\{v(t)v'(\tau)\} &= V(t)\delta(t-\tau) \\ E\{v(t)w'(\tau)\} &= X(t)\delta(t-\tau) \\ E\{w(t)w'(\tau)\} &= W(t)\delta(t-\tau) \end{aligned} \tag{11.10A}$$

Then

$$E\{\xi(t)\xi'(\tau)\} = Q_\xi(t)\delta(t-\tau)$$

where

$$\begin{aligned} Q_\xi(t) = F(t)V(t)F'(t) &- K(t)X'(t)F'(t) \\ &- F(t)X(t)K'(t) + K(t)W(t)K'(t) \end{aligned} \tag{11.11}$$

So (11.8) is the differential equation of a linear system excited by white noise ξ

$$\dot{e} = (A - KC)e + \xi \tag{11.12}$$

where ξ has the spectral density matrix given by (11.11).

Let P be the covariance matrix of the error. (The subscript e is omitted from P to keep the notation simpler.) Then, using the variance equation (10.58),

$$\begin{aligned} \dot{P} &= (A - KC)P + P(A' - C'K') + Q_\xi \\ &= (A - KC)P + P(A' - C'K') + FVF' - KX'F' - FXK' + KWK' \end{aligned} \tag{11.13}$$

If the cross-covariance X between the excitation noise v and the observation noise w were absent, (11.13) would have the same form as the optimum control equation (9.20) and we would be able to write the solution for the optimum gain matrix by use of the analysis of Chap. 9. This would give

$$\hat{K} = \hat{P}C'W^{-1} \tag{11.14}$$

where the optimizing covariance matrix is given by the matrix Riccati equation

$$\dot{\hat{P}} = A\hat{P} + \hat{P}A' - \hat{P}C'W^{-1}C\hat{P} + FVF' \tag{11.15}$$

These are the most familiar forms of the gain and optimum variance equations, respectively, and apply when v and w are uncorrelated ($X = 0$). More general relations hold when the cross-correlation matrix X is not zero, and these are derived by the same general method that was used in Chap. 9. In particular, we assume that

$$P = \hat{P} + U \tag{11.16}$$

$$K = \hat{K} + \Gamma \tag{11.17}$$

where $\hat{P}$ and $\hat{K}$ are the optimum covariance matrix and observer gain matrix, respectively. Substitute these into (11.13) to obtain

$$\begin{aligned} \dot{\hat{P}} + \dot{U} = (A - \hat{K}C - \Gamma C)(\hat{P} + U) &+ (\hat{P} + U)(A' - C'\hat{K}' - C'\Gamma') + FVF' \\ &- (\hat{K} + \Gamma)X'F' - FX(\hat{K}' + \Gamma') + (\hat{K} + \Gamma)W(\hat{K}' + \Gamma') \end{aligned} \tag{11.18}$$

But of course $\hat{P}$ also satisfies (11.13)

$$\dot{\hat{P}} = (A - \hat{K}C)\hat{P} + \hat{P}(A' - C'\hat{K}') + FVF' - \hat{K}X'F' - FX\hat{K}' + \hat{K}W\hat{K}' \tag{11.19}$$

Subtract (11.19) from (11.18). After combining terms, the result is

$$\begin{aligned}\dot{U} = (A - \hat{K}C - \Gamma C)U + U(A' - C'\hat{K}' - C'\Gamma') + \Gamma W \Gamma' \\ + \Gamma(W\hat{K}' - C\hat{P} - X'F') + (\hat{K}W - \hat{P}C' - FX)\Gamma'\end{aligned} \tag{11.20}$$

From this point onward, we reason as we did in Chap. 9: If $\hat{P}$ is the minimum covariance matrix P must be greater than $\hat{P}$ for any choice of Γ. This means U must be positive semidefinite. But the solution to (11.20) for U can be made negative definite by suitable choice of Γ unless the coefficient of Γ vanishes entirely, in which case U will be positive semidefinite. Thus, if $\hat{P}$ is optimum we must have

$$\hat{K}W = \hat{P}C' + FX$$

Assuming the observation noise spectral density matrix W to be nonsingular

$$\hat{K} = (\hat{P}C' + FX)W^{-1} \tag{11.21}$$

which is a generalization of (11.14).

To find the differential equation for $\hat{P}$, substitute (11.21) into (11.19). The result, after simplification, is the matrix Riccati equation

$$\dot{\hat{P}} = \tilde{A}\hat{P} + \hat{P}\tilde{A}' - \hat{P}C'W^{-1}C\hat{P} + F\tilde{V}F' \tag{11.22}$$

where

$$\tilde{A} = A - FXW^{-1}C \tag{11.23}$$

$$\tilde{V} = V - XW^{-1}X' \tag{11.24}$$

when the cross spectral density matrix X is zero then $\tilde{A} = A$ and $\tilde{V} = V$; then (11.22) reduces to the more familiar equation (11.15).

The effect of the cross spectral density is to modify the Kalman filter gain matrix by the addition of FX to $\hat{P}C'$ in (11.21) and to change the dynamics from A to $\tilde{A}$ and the process noise spectral density from V to $\tilde{V}$ in (11.22). It is of interest to compare (11.21) through (11.24) to the analogous equations of optimum control in which a cross term $2x'S'u$ is present in the performance integral. The latter are given by (9.51) through (9.54), and are in the same form as (11.21) through (11.24). The similarity of the Kalman filter equations derived in this chapter and the optimum control equations derived in Chap. 9 are emphasized by calling the filter problem and the control problem *duals* of each other.

Note that the "equivalent" excitation noise spectral density matrix $\tilde{V}$ is the difference of two positive semidefinite matrices, and we might think that if W is small enough $XW^{-1}X'$ would be larger than V and then $\tilde{V}$ would be negative. But this is not possible. Consider the joint process

$$z = \begin{bmatrix} v \\ w \end{bmatrix}$$

Then if v and w are white noise, z is white noise. Its spectral density is calculated as follows

$$E\{z(t)z'(\tau)\} = E\left\{\begin{bmatrix} v(t) \\ w(t) \end{bmatrix}[v'(\tau) \quad w'(\tau)]\right\}$$

$$= \begin{bmatrix} E\{v(t)v'(\tau)\} & E\{v(t)w'(\tau)\} \\ E\{w(t)v'(\tau)\} & E\{w(t)w'(\tau)\} \end{bmatrix}$$

$$= \begin{bmatrix} V & X \\ X' & W \end{bmatrix}\delta(t-\tau)$$

Thus the spectral density Z of the joint process z is given by

$$Z = \begin{bmatrix} V & X \\ X' & W \end{bmatrix} \tag{11.25}$$

In order for Z to be a spectral density matrix Z must be positive semidefinite. It can be shown that this implies that $\tilde{V} = V - XW^{-1}X'$ is also positive semidefinite. (See Note 11.2.)

11.4 STEADY-STATE KALMAN FILTER

The matrix Riccati equation (11.22)—or the special case (11.15)—are valid for any finite time interval. If time is allowed to become infinite, the solutions may tend to infinity or they may remain finite. If all the matrices on the right-hand side of (11.22) are constant, then a constant, steady state solution may exist, given by the solution of the matrix quadratic equation, also called the *algebraic Riccati equation*

$$0 = \tilde{A}\hat{P} + \hat{P}\tilde{A}' - \hat{P}C'W^{-1}C\hat{P} + F\tilde{V}F' \tag{11.26}$$

The algebraic Riccati equation (ARE) (11.26) for the steady state covariance matrix $\hat{P}$ is the same form as the ARE (9.29) for optimum deterministic control and hence the conditions under which (11.26) has a unique, positive definite solution are similar to those given in Chap. 9. In particular, the ARE (11.26) has a unique positive definite solution if either

(*a*) The system is asymptotically stable, or
(*b*) The system defined by the pair $[A, C]$ is observable and the system defined by the pair $[A, FV^{1/2}]$, (i.e., the system $\dot{x} = Ax + FV^{1/2}v$) is controllable.

The reason for observability is understandable. In line with the discussion of Sec. 9.5, the requirement of observability can be relaxed to detectability. (See Note 5.2.) The reason for what amounts to controllability by the noise (should we say "excitability"?) is similar to the reason for observability in the optimum control problem: in this instance we may have certain states (or "modes") which are not asymptotically stable but which cannot be excited by the noise.

The variance of these states would be zero and would cause the covariance matrix $\hat{P}$ to be singular.

Example 11A Inverted pendulum If it were possible to balance an inverted pendulum, it would not remain balanced without control owing to the inevitable presence of various types of disturbances, such as random air currents. Thus, if the accelerations due to the disturbances are represented by v, the differential equations for the pendulum are

$$\dot{\theta} = \omega$$
$$\dot{\omega} = \Omega^2\theta + u + v \tag{11A.1}$$

where u is the control acceleration and v is the disturbance acceleration. The matrices corresponding to (11A.1) are

$$A = \begin{bmatrix} 0 & 1 \\ \Omega^2 & 0 \end{bmatrix} \qquad B = F = \begin{bmatrix} 0 \\ 1 \end{bmatrix}$$

If the quantity observed is the position $x_1 = \theta$

$$y = [1 \quad 0]\begin{bmatrix} x_1 \\ x_2 \end{bmatrix} + w$$

Hence

$$C = [1 \quad 0]$$

Let the optimum covariance matrix be

$$\hat{P} = \begin{bmatrix} p_1 & p_2 \\ p_2 & p_3 \end{bmatrix}$$

Then, by (11.26) the elements of P satisfy

$$0 = 2p_2 - \frac{p_1^2}{W}$$

$$0 = p_3 + \Omega^2 p_1 - \frac{p_1 p_2}{W} \tag{11A.2}$$

$$0 = 2\Omega^2 p_2 - \frac{p_2^2}{W} + V$$

where V and W are the spectral density (1×1) matrices of the excitation noise and observation noise, respectively.

From the first equation in (11A.2) we have

$$p_1 = \pm\sqrt{2p_2 W} \tag{11A.3}$$

Since p_1 must be nonnegative, the top sign must be used and, moreover, p_2 must be positive. From the third equation in (11A.2)

$$p_2 = W[\Omega^2 \pm \sqrt{\Omega^4 + V/W}] \tag{11A.4}$$

Clearly, to make $p_2 > 0$ the top sign must be chosen.

Thus we have from (11A.3) and (11A.4)

$$p_1 = \Omega W\sqrt{2\gamma}$$
$$p_2 = \Omega^2 W\gamma$$

where

$$\gamma = 1 + \sqrt{1 + V/\Omega^4 W} \tag{11A.5}$$

Also, from the second equation in (11A.3)

$$p_3 = \Omega^3 W\sqrt{2\gamma}(\gamma - 1)$$

But p_3 is not needed to determine the Kalman filter gain matrix which is given by

$$\hat{K} = \hat{P}C'W^{-1} = \begin{bmatrix} p_1 & p_2 \\ p_2 & p_3 \end{bmatrix}\begin{bmatrix} 1 \\ 0 \end{bmatrix}\frac{1}{W} = \begin{bmatrix} \Omega\sqrt{2\gamma} \\ \Omega^2\gamma \end{bmatrix}$$

The closed-loop filter poles and transfer functions from the measured angle y to the estimated state $\hat{x}_1 = \hat{\theta}$ and $\hat{x}_2 = \hat{\omega}$ are of interest. Assuming that the input u is zero we have

$$s\hat{x}(s) = A\hat{x}(s) + \hat{K}[y(s) - C\hat{x}(s)] = (A - \hat{K}C)\hat{x}(s) + Ky(s)$$

or

$$\hat{x}(s) = (sI - A_0)^{-1}\hat{K}y(s) \tag{11A.6}$$

where $A_0 = A - \hat{K}C$ is the closed-loop observer dynamics matrix, in this example given by

$$A_0 = \begin{bmatrix} 0 & 1 \\ \Omega^2 & 0 \end{bmatrix} - \begin{bmatrix} \Omega\sqrt{2\gamma} \\ \Omega^2\gamma \end{bmatrix}[1 \quad 0] = \begin{bmatrix} -\Omega\sqrt{2\gamma} & 1 \\ \Omega^2(1-\gamma) & 0 \end{bmatrix}$$

Thus

$$(sI - A_0)^{-1} = \begin{bmatrix} s + \Omega\sqrt{\gamma} & -1 \\ \Omega^2(1-\gamma) & s \end{bmatrix}^{-1} = \frac{1}{\Delta(s)}\begin{bmatrix} s & 1 \\ \Omega^2(1-\gamma) & s + \Omega\sqrt{2\gamma} \end{bmatrix} \tag{11A.7}$$

where $\Delta(s)$ is the closed-loop characteristic polynomial, given by

$$\Delta(s) = s^2 + \Omega\sqrt{2\gamma}s + \Omega^2(\gamma - 1)$$

From (11A.5) and (11A.6)

$$\hat{x}(s) = \begin{bmatrix} \hat{\theta}(s) \\ \hat{\omega}(s) \end{bmatrix} = \frac{1}{\Delta(s)}\begin{bmatrix} \Omega(s\sqrt{2\gamma} + \Omega\gamma) \\ \Omega^2(\gamma s + \Omega\sqrt{2\gamma}) \end{bmatrix}$$

In particular

$$\begin{aligned} H_1(s) &= \frac{\hat{\theta}(s)}{y(s)} = \frac{\Omega(s\sqrt{2\gamma} + \Omega\gamma)}{s^2 + \Omega\sqrt{2\gamma}s + \Omega^2(\gamma - 1)} \\ H_2(s) &= \frac{\hat{\omega}(s)}{y(s)} = \frac{\Omega^2(s + \Omega\sqrt{2\gamma})}{s^2 + \Omega\sqrt{2\gamma}s + \Omega^2(\gamma - 1)} \end{aligned} \tag{11A.8}$$

The closed-loop poles of the filter are given by

$$s = -\Omega\left(\sqrt{\frac{\gamma}{2}} \pm j\sqrt{\frac{\gamma}{2} - 1}\right)$$

The zeros of the filters $H_1(s)$ and $H_2(s)$, respectively, lie at

$$\begin{aligned} s &= -\Omega\sqrt{\frac{\gamma}{2}} \qquad \text{for angular position} \\ s &= -\Omega\sqrt{\frac{2}{\gamma}} \qquad \text{for angular velocity} \end{aligned} \tag{11A.9}$$

As the excitation noise covariance matrix V tends to zero, γ, as given by (11A.5), approaches 2 and

$$H_1(s) \to \frac{\Omega(2s + 2\Omega)}{s^2 + 2\Omega s + \Omega^2} = \frac{2\Omega}{s + \Omega}$$

$$H_2(s) \to \frac{\Omega^2(2s + 2\Omega)}{s^2 + 2\Omega s + \Omega^2} = \frac{2\Omega^2}{s + \Omega}$$

In other words the zero of the numerator tends to one pole of the denominator, and both filters become first-order. Note also that the optimum estimate of the angular velocity is simply the natural frequency Ω times the angular position.

As the excitation noise covariance matrix tends to infinity (or as the observation noise covariance matrix tends to zero) γ tends to infinity and the closed-loop poles of the observer tend to

$$s = -\Omega\sqrt{\frac{\gamma}{2}}(1 \pm j)$$

which are lines at 45° angles from the real axis. The zero of the transfer function $H_1(s) = \hat{\theta}(s)/y(s)$ for angular position moves out to infinity along the negative real axis; but the zero of the transfer function $H_2(s) = \hat{\omega}(s)/y(s)$ for angular velocity tends to the origin. [See (11A.9).] This means that at low frequencies the transfer function for angular velocity behaves like a high-pass system, i.e., a "lead" network, whereas the transfer function for angular position behaves as low-pass ("lag") system. Loci of the poles of the filters, as γ varies from 2 to infinity, are shown in Fig. 11.2.

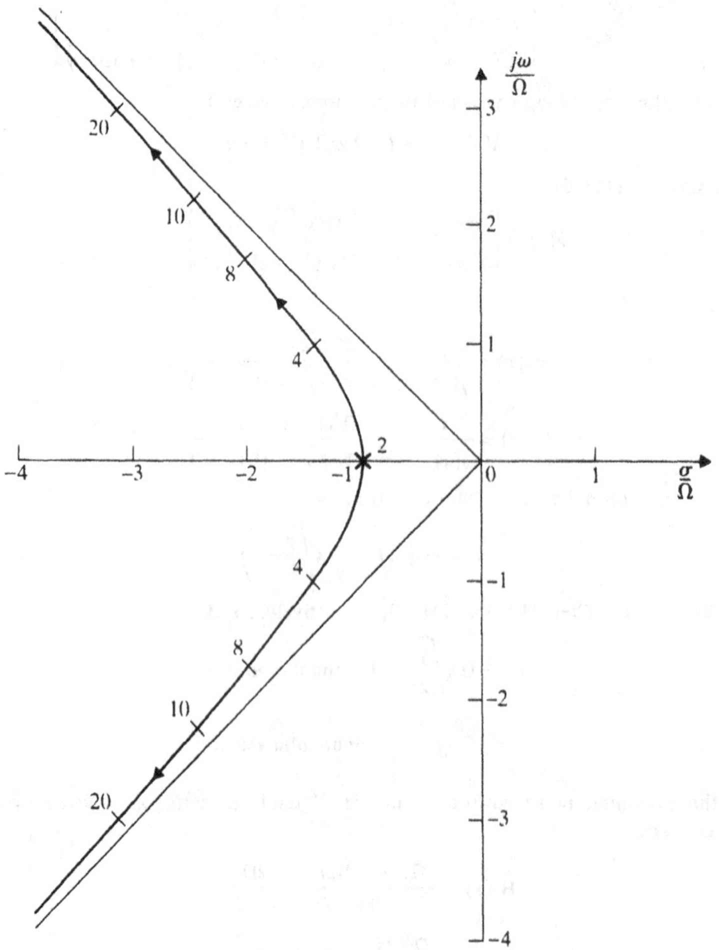

Figure 11.2 Poles of Kalman filter for inverted pendulum.

Example 11B Accelerometer pick-off The accelerometer considered in prior examples has the differential equations

$$\begin{aligned} \dot{x}_1 &= x_2 \\ \dot{x}_2 &= u + a \end{aligned} \tag{11B.1}$$

The position of the proof mass is determined by some sort of "pick-off"—perhaps magnetic or optical. The output of the pick-off is

$$y = x_1 + w$$

where w is the pick-off noise which we assume to be white.

We want to design a Kalman filter to estimate the acceleration a. (For the present, control of the proof mass position is not considered.) For this purpose it is necessary to model the unknown acceleration as a random process. If the spectral density of a were known, it would be appropriate to represent a as the response of the linear system which, when excited by white noise, produces an output with this known spectral density.

For the purpose of this illustration, however, we assume that the acceleration a is a Wiener process

$$\dot{a} = v \tag{11B.2}$$

where v is white noise with spectral density matrix V. If V were zero then (11B.2) would become $\dot{a} = 0$, that is, a would be an unknown constant. But as we will soon see, it is necessary to assume $V \neq 0$ in order to get a meaningful filter design.

Represent a by another state variable x_3 and adjoin (11B.2) to (11B.1):

$$\begin{aligned} \dot{x}_1 &= x_2 \\ \dot{x}_2 &= u + x_3 \\ \dot{x}_3 &= v \end{aligned}$$

For this system the defining matrices are

$$A = \begin{bmatrix} 0 & 1 & 0 \\ 0 & 0 & 1 \\ 0 & 0 & 0 \end{bmatrix} \qquad B = \begin{bmatrix} 0 \\ 1 \\ 0 \end{bmatrix} \qquad F = \begin{bmatrix} 0 \\ 0 \\ 1 \end{bmatrix}$$

$$C = [1 \quad 0 \quad 0]$$

Let the optimum covariance matrix be

$$\hat{P} = \begin{bmatrix} p_1 & p_2 & p_3 \\ p_2 & p_4 & p_5 \\ p_3 & p_5 & p_6 \end{bmatrix}$$

Then the components of $\hat{P}$ satisfy:

$$\dot{p}_1 = 0 = 2p_2 - \frac{p_1^2}{W}$$

$$\dot{p}_2 = 0 = p_3 + p_4 - \frac{p_1 p_2}{W}$$

$$\dot{p}_3 = 0 = p_5 - \frac{p_1 p_3}{W}$$

$$\dot{p}_4 = 0 = 2p_5 - \frac{p_2^2}{W}$$

$$\dot{p}_5 = 0 = p_6 - \frac{p_2 p_3}{W}$$

$$\dot{p}_6 = 0 = -\frac{p_3^2}{W} + V$$

These can be solved readily. The resulting solutions are

$$p_1 = 2V^{1/6}W^{5/6} \qquad p_2 = 2V^{1/3}W^{1/3} \qquad p_3 = V^{1/2}W^{1/2}$$
$$p_4 = 3V^{1/2}W^{1/2} \qquad p_5 = 2V^{2/3}W^{1/3}$$
$$p_6 = 2V^{5/6}W^{1/6}$$

Note that if the spectral density V of the noise v exciting the acceleration were allowed to go to zero, the steady state covariance matrix $\hat{P}$ would also tend to zero. From this one would infer that the acceleration (and other state variables x_1 and x_2) could be estimated without error, even with a noisy pick-off. This is correct, and the Kalman filter would have time-varying gains that would all tend to zero. To get a Kalman filter whose gains do not all tend to zero in the steady state, it is necessary to assume that V is not zero, which implies that the acceleration is not known, a priori, to be constant.

We can now compute the Kalman filter gain matrix

$$\hat{K} = \hat{P}C'W^{-1} = \begin{bmatrix} p_1/W \\ p_2/W \\ p_3/W \end{bmatrix} = \begin{bmatrix} 2(V/W)^{1/6} \\ 2(V/W)^{1/3} \\ (V/W)^{1/2} \end{bmatrix}$$

Since V is an indicator of the randomness of the acceleration that the instrument is trying to measure and W is an indicator of the random noise in making the measurement, the ratio V/W can be regarded as a "signal-to-noise ratio" and it is seen that the filter gains all increase with increasing signal-to-noise ratio, which seems reasonable.

The filter has the dynamics matrix

$$A_0 = A - KC = \begin{bmatrix} -2\bar{\Omega} & 1 & 0 \\ -2\bar{\Omega}^2 & 0 & 1 \\ -\bar{\Omega}^3 & 0 & 0 \end{bmatrix}$$

where

$$\bar{\Omega} = \left(\frac{V}{W}\right)^{1/6}$$

Thus the filter characteristic equation is

$$|sI - A_0| = \begin{vmatrix} s + 2\bar{\Omega} & -1 & 0 \\ 2\bar{\Omega}^2 & s & -1 \\ \bar{\Omega}^3 & 0 & s \end{vmatrix} = s^3 + 2\bar{\Omega}s^2 + 2\bar{\Omega}^2 s + \bar{\Omega}^3 = 0$$

The characteristic roots are

$$s_1 = -\bar{\Omega} \qquad s_2, s_3 = -\bar{\Omega}(\tfrac{1}{2} \pm j\sqrt{\tfrac{3}{2}})$$

Thus the roots lie upon a circle of radius $\bar{\Omega} = (V/W)^{1/6}$ on the negative real axis and at 60 degree angles, as shown in Fig. 6.7. This is a characteristic "Butterworth pattern" as discussed in more detail in Chap. 6.

Example 11C Velocity-aided inertial navigation The dynamics of the error propagation in an inertial navigation system (Example 3E) are not asymptotically stable, hence the effects of initial errors do not decay to zero, and the errors due to noisy instruments tend to grow with time (Example 3E and Prob. 10.3). To improve the performance of the system, external navigation aids, when available, may be used. TACAN, LORAN, Omega and navigation satellites (GPS) may provide position data; velocity data may be provided by a Doppler radar.

A number of studies have been made of the use of Kalman filtering to mix data from inertial and noninertial sources to optimize overall system accuracy. (See [3] for a detailed example.)

The purpose of this example, however, is not to explore the realm of aided-inertial navigation but rather to show an instance of the design of a Kalman filter for a system that is not observable. For this purpose, we assume that the only errors present in the inertial instruments are white noise in the gyro and in the accelerometer outputs. The dynamic model for this situation is

$$\begin{aligned} \Delta\dot{x} &= \Delta v \\ \Delta\dot{v} &= -g\Delta\psi + v_A \\ \Delta\dot{\psi} &= \frac{1}{R}\Delta v + v_G \end{aligned} \tag{11C.1}$$

where g is the acceleration of gravity, R is the earth's radius, and where v_A and v_G are the accelerometer and gyro noise processes, respectively, which are assumed to be independent, with spectral densities V_A and V_G. The relevant matrices are

$$A = \begin{bmatrix} 0 & 1 & 0 \\ 0 & 0 & -g \\ 0 & 1/R & 0 \end{bmatrix} \qquad F = \begin{bmatrix} 0 & 0 \\ 1 & 0 \\ 0 & 1 \end{bmatrix} \tag{11C.2}$$

Suppose that the velocity aid provides a (noisy) measurement of velocity v_m, independent of the velocity $\bar{v}$ indicated by the pure-inertial navigation system as shown in Fig. 11.3. The difference $y = \bar{v} - v_m$ is thus a (noisy) measurement of the velocity error, i.e.,

$$y = \Delta v + w \tag{11C.3}$$

where w is the white noise on the velocity measurement, with spectral density matrix W.

The Kalman filter will provide optimum estimates of the state:

$$\begin{aligned} \Delta\dot{\hat{x}} &= \Delta\hat{v} + k_1(y - \Delta\hat{v}) \\ \Delta\dot{\hat{v}} &= -g\Delta\hat{\psi} + k_2(y - \Delta\hat{v}) \\ \Delta\dot{\hat{\psi}} &= \frac{1}{R}\Delta\hat{v} + k_3(y - \Delta\hat{v}) \end{aligned} \tag{11C.4}$$

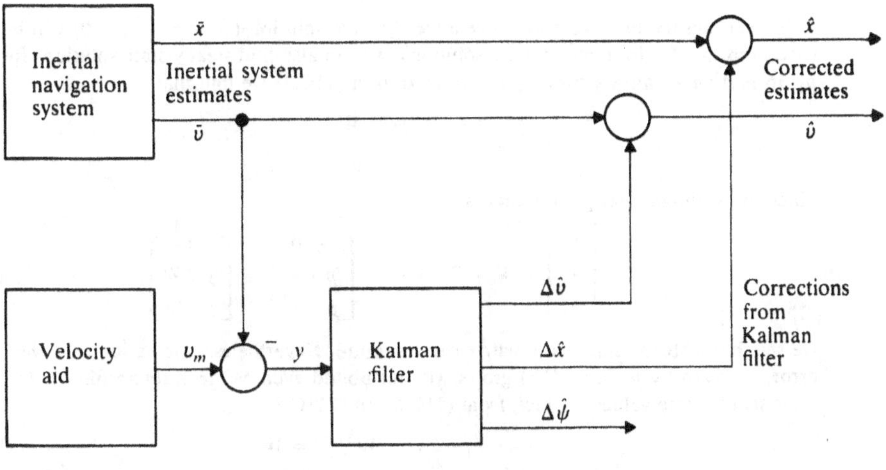

Figure 11.3 Use of Kalman filter for mixing inertial and noninertial navigation data.

These error estimates are then added to the outputs of the pure-inertial navigation system to produce (optimum) estimates $\hat{x}$ and $\hat{v}$ of the total position and velocity, respectively, as shown in Fig. 11.3.

It now remains to determine the Kalman filter gains k_1, k_2, and k_3. The observation matrix for corresponding to (11C.3) is

$$C = [0 \quad 1 \quad 0]$$

It is readily established that the pair $[A, C]$ in this instance is *not* observable. This is also obvious if a block-diagram for (11C.1) and (11C.3) is drawn; there is no path from Δx to y. Since the system is not observable and not asymptotically stable, we should expect problems with the variance equation. In fact, as we shall see, the steady state covariance matrix has one term, namely the variance of the position error, that grows without bound as $t \to \infty$. Hence the steady state solution to the variance equation does not exist. Nevertheless, the Kalman filter gains approach steady state values, thus permitting the use of a constant gain filter.

Let the desired covariance matrix be denoted by

$$\hat{P} = \begin{bmatrix} p_1 & p_2 & p_3 \\ p_2 & p_4 & p_5 \\ p_3 & p_5 & p_6 \end{bmatrix} \tag{11C.5}$$

The equations for the elements of $\hat{P}$, in accordance with (11.15), are

$$\begin{aligned} \dot{p}_1 &= 2p_2 - p_2^2/W \\ \dot{p}_2 &= p_4 - gp_3 - p_2 p_4/W \\ \dot{p}_3 &= p_5 + p_3/R - p_2 p_5/W \end{aligned} \tag{11C.6}$$

$$\begin{aligned} \dot{p}_4 &= -2gp_5 - p_4^2/W + V_1 \\ \dot{p}_5 &= -gp_6 + p_4/R - p_4 p_5/W \\ \dot{p}_6 &= 2p_5/R - p_5^2/W + V_2 \end{aligned} \tag{11C.7}$$

Note that the differential equations (11C.7) for p_4, p_5, p_6 do not depend on p_1, p_2, p_3, and correspond to the subsystem consisting of Δv and $\Delta\psi$ with the dynamics and observation matrices

$$A_s = \begin{bmatrix} 0 & -g \\ 1/R & 0 \end{bmatrix} \qquad C_s = [1 \quad 0]$$

This *is* an observable pair and hence a steady state solution ($\dot{p}_4 = \dot{p}_5 = \dot{p}_6 = 0$) can be found. (See Prob. 11.19.) In terms of these solutions, we can also find steady-state solutions for p_2 and p_3. In particular, after setting $\dot{p}_2$ and $\dot{p}_3$ to zero in (11C.7) we find that

$$\begin{aligned} p_2 &= W \\ p_3 &= 0 \end{aligned} \tag{11C.8}$$

Since the Kalman filter gain matrix is

$$\begin{bmatrix} k_1 \\ k_2 \\ k_3 \end{bmatrix} = K = PC'W^{-1} = \begin{bmatrix} p_2/W \\ p_4/W \\ p_5/W \end{bmatrix} = \begin{bmatrix} 1 \\ p_4/W \\ p_5/W \end{bmatrix} \tag{11C.9}$$

we see that a steady state gain matrix can be found. Nevertheless, the variance of the position error, as given by $\dot{p}_1$ in (11C.6) grows without bound even as the Kalman filter gains tend to their steady state values. In fact, from (11C.6) and (11C.8)

$$\dot{p}_1 = 2W - W^2/W = W \tag{11C.10}$$

Thus the steady state growth rate of p_1, which is the variance of the position error, is exactly the same as if the velocity aid alone were present. This means that (in the steady state) the position error of the aided-inertial navigation system is not less than if the inertial system were absent. In the transient case the inertial system, *when optimally used*, cannot but help the velocity-measuring system. Hence the solution for p_4 in (11C.6), as a function of time, should always be expected to be below $p(0) + Wt$, the solution to (11C.10). (See Prob. 11.19.)

Because $k_1 = p_2/W = 1$, the equation for the optimum velocity estimate (the first equation in (11C.4)) becomes

$$\Delta\dot{\hat{x}} = y \tag{11C.11}$$

i.e., the estimated velocity error $\Delta\hat{v}$ is not used in correcting the position estimate. This is the same estimation law that would be used if the inertial system were absent and the acceleration acting on the vehicle being navigated were purely random. (See Prob. 11.19.)

It is emphasized in concluding this example that a numerical algorithm for solving the algebraic Riccati equation will not work in this problem, since one element of $\hat{P}$ (namely p_1, upon which the gain matrix does not depend) does not have a steady state solution. In this example we found a way around this difficulty. In other situations, however, the detour might not be so obvious.

There is one device which often works when a numerical solution is sought for a system having poles only on the imaginary axis: Artificially stabilize the system by adding small negative numbers to some or all the diagonal elements of the A matrix. This brings the poles into the left half-plane making the resulting system asymptotically stable and thereby guaranteeing a unique positive definite solution to the ARE. The covariance matrix for the artificially stabilized system is of course not correct—it has all finite elements whereas the correct matrix will have at least one infinite element. Nevertheless, those elements of the covariance matrix that are needed in calculating the gains may differ only negligibly from their correct values, and hence may be good enough for all practical purposes. To minimize the possibility of errors in the filter gains, the artificial stabilization should be as small as can be coped with by the numerical algorithm for solving the ARE. This method cannot be expected to work if the system has poles to the right of the imaginary axis.

11.5 THE "INNOVATIONS" PROCESS

The difference between the actual observation y and the expected observation $\hat{y} = C\hat{x}$, that is,

$$r = y - \hat{y} \tag{11.27}$$

is a random process of considerable interest. This process is usually known as the "innovations process." (See Note 11.3.) The "residual process" or simply the "residuals" are other appellations for this process.

A remarkable property of the innovations process is that it is white noise. To see this we compute the correlation function of the innovations process, using the defining relation (11.27). Recall that the actual observation is given by

$$y = Cx + w$$

and the estimated observation is given by

$$\hat{y} = C\hat{x}$$

Thus the residual is given by

$$r = y - \hat{y} = C(x - \hat{x}) + w = Ce + w \tag{11.28}$$

where e is the estimation error $e = x - \hat{x}$.

The correlation matrix for the residual process is thus given by

$$\begin{aligned} R_r(t, \tau) &= E\{r(t)r'(\tau)\} \\ &= CE\{e(t)e'(\tau)\}C' + CE\{e(t)w'(\tau)\} \\ &\quad + E\{w(t)e'(\tau)\}C' + E\{w(t)w'(\tau)\} \end{aligned} \tag{11.29}$$

To compute the various terms in (11.29) we use the fact that the error e, as shown in (11.28), satisfies the differential equation

$$\dot{e} = A_c e + F(t)v - K(t)w \tag{11.30}$$

where

$$A_c = A - KC$$

is the closed-loop matrix of the observer.

In the first term on the right-hand side of (11.29) we have

$$E\{e(t)e'(\tau)\} = R_e(t, \tau) \tag{11.31}$$

which is the correlation matrix for the estimation error e. In accordance with the theory of Sec. 10.7 of the previous chapter

$$R_e(t, \tau) = P(t)\Phi_c'(\tau, t) \qquad \text{for} \qquad \tau \geqq t \tag{11.32}$$

where $P(t)$ is the covariance matrix for e, given by (11.13), and Φ_c is the state transition matrix corresponding to the closed loop dynamics matrix A_c in (11.30). Thus the first term in (11.29) is

$$CE\{e(t)e'(\tau)\}C' = CP(t)\Phi_c'(\tau, t)C' \qquad \text{for} \qquad \tau \geqq t. \tag{11.33}$$

In order to evaluate the second and third terms in (11.29) we recall that the solution to (11.30) is

$$e(t) = \Phi_c(t, t_0)e(t_0) + \int_{t_0}^{t} \Phi_c(t, \lambda)[F(\lambda)v(\lambda) - K(\lambda)w(\lambda)]\, d\lambda$$

Thus the second term in (11.29) is

$$\begin{aligned} CE\{e(t)w'(\tau)\} &= C\Phi_c(t, t_0)e(t_0)E\{w'(\tau)\} \\ &\quad + C\int_{t_0}^{t} \Phi_c(t, \lambda)[F(\lambda)E\{v(\lambda)w'(\tau)\} - K(\lambda)E\{w(\lambda)w'(\tau)\}]\, d\lambda \end{aligned} \tag{11.34}$$

Now, as given in (11.10A)

$$E\{v(\lambda)w'(\tau)\} = X(\lambda)\delta(\lambda - \tau)$$

$$E\{w(\lambda)w'(\tau)\} = W(\lambda)\delta(\lambda - \tau)$$

and

$$E\{w'(\tau)\} = 0$$

Thus (11.34) becomes

$$CE\{e(t)w'(\tau)\} = C\int_{t_0}^{t} \Phi_c(t,\lambda)[F(\lambda)X(\lambda) - K(\lambda)W(\lambda)]\delta(\lambda-\tau)\,d\lambda$$

$$= \begin{cases} C\Phi_c(t,\tau)[F(\tau)X(\tau) - K(\tau)W(\tau)] & \text{for} \quad t_0 < \tau \leqq t \\ 0 & \text{for} \quad \tau > t \end{cases} \tag{11.35}$$

The third term in (11.29) is given by

$$E\{w(t)e'(\tau)\}C'$$

$$= E\left\{w(t)\left[e'(t_0)\Phi_c(\tau,t_0) + \int_{t_0}^{\tau}[v'(\lambda)F'(\lambda) - w'(\lambda)K'(\lambda)]\Phi_c'(\tau,\lambda)\,d\lambda\right]\right\}$$

Reasoning as above we conclude that

$$E\{w(t)e'(\tau)\}C' = \begin{cases} [X'(t)F'(t) - W(t)K'(t)]\Phi_c'(\tau,t)C' & \text{for} \quad t_0 \leqq t \leqq \tau \\ 0 & \text{for} \quad t > \tau \end{cases} \tag{11.36}$$

Finally, the fourth term on the right-hand side of (11.29) is

$$E\{w(t)w'(\tau)\} = W(t)\delta(t-\tau) \tag{11.37}$$

Combine (11.33), (11.34), (11.35), and (11.36) to get $R_r(t, \tau)$ as given by (11.29). Note, however, that only one nonzero expression of (11.35) or (11.36) is used in the sum, depending on whether $\tau > t$ or $\tau \leqq t$. Suppose $\tau > t$. Then

$$R_r(t,\tau) = [CP(t) + X'(t)F'(t) - W(t)K'(t)]\Phi_c'(\tau,t)C' + W(t)\delta(t-\tau) \tag{11.38}$$

Thus, the correlation function for the innovations process has two terms: the second term is that of white noise with spectral density $W(t)$, and the first term is that of a linear system excited by white noise, and having the transition matrix $\Phi_c(t, \tau)$. The covariance matrix of that linear system is

$$CP(t) + X'(t)F'(t) - W(t)K'(t)$$

This term is zero, however, if the gain is chosen optimally, i.e., in accordance with (11.21). Hence we conclude that the innovations process in the optimum observer—the Kalman filter—is white noise having the spectral density W of the observation noise. Conversely, if the observer is not optimum, the spectral density of the innovations process is not white noise.

11.6 REDUCED-ORDER FILTERS AND CORRELATED NOISE

A crucial assumption used in the derivation of the Kalman filter is that the observation noise vector is white and has a nonsingular covariance matrix W, the inverse of which is used in the formula for the Kalman filter gain. In the discussion of white noise in Chap. 10, we made the point that white noise is a convenient abstraction, but it never exists in reality. We are thus faced with an annoying paradox: the entire edifice of Kalman filtering is based on an assumption which does not apply to the physical world! We should almost expect the whole structure to come tumbling down. Fortunately the structure is not at all precarious, because the white noise approximation is often quite good: the bandwidth of the noise, although finite, is much higher than the bandwidth of the dynamic process under consideration.

There are numerous cases, however, in which the white noise approximation is poor. The noise may have a bandwidth that is much lower than the dynamic process in question. In this case the white noise approximation is unjustified.

Another common problem is that some sensors may be of such high quality that the observation noise may be extremely small. If these sensors are combined in a system with others that are not of comparable quality, the resulting spectral density matrix may be very poorly conditioned. An observation noise matrix such as

$$W = \begin{bmatrix} 1 & 0 \\ 0 & 10^{-18} \end{bmatrix}$$

would wreak havoc with many numerical algorithms for computing the Kalman filter gains. It might be preferable to regard the small term as zero and develop a technique for dealing forthrightly with observation noise spectral density matrices that are singular.

Methods of dealing with a general singular observation spectral density matrix have been developed. (See Note 11.4.) The results, however, although not difficult are rather repulsive to look at and tedious. For this reason we confine our attention to a very special case, namely the case in which the spectral density matrix W is actually zero: there is no noise at all present in the observations. Any nonwhite noise (i.e., correlated or "colored" noise) present that can be modeled as the response of a linear system can be treated within this framework, as we shall show later on.

For now we consider the dynamic process

$$\dot{x} = Ax + Bu + Fv \tag{11.39}$$

with noise-free observations

$$y = Cx$$

A new observation can be obtained by taking the time derivative of the observation vector:

$$\dot{y} = C\dot{x} = C(Ax + Bu + Fv)$$

Consider the quantity

$$z = \dot{y} - CBu = CAx + CFv \tag{11.40}$$

This quantity is linear in the state x and contains white noise, so it is a candidate for use as the input to a Kalman filter. To obtain z from y, however, entails differentiating the latter which is an undesirable operation. It will turn out that the differentiation of y can be avoided, and as a byproduct, the order of the filter is reduced.

The Kalman filter for (11.39) with observations given by (11.40) is expressed by

$$\dot{\hat{x}} = A\hat{x} + Bu + \hat{K}(z - CA\hat{x}) \tag{11.41}$$

where $\hat{K}$ is the Kalman filter gain matrix.

In order to obtain $\hat{K}$ we note that the observation noise $w = CFv$ is obviously correlated with the excitation noise v. In particular if

$$E\{v(t)v'(\tau)\} = V\delta(t-\tau)$$

then

$$E\{v(t)w'(\tau)\} = VF'C'\delta(t-\tau)$$

$$E\{w(t)w'(\tau)\} = CFVF'C'\delta(t-\tau)$$

Thus the matrices needed to determine the Kalman filter gain, in addition to V, are

$$X = VF'C' \tag{11.42}$$

and

$$W = CFVF'C' \tag{11.43}$$

We now make the further assumption that W, as given by (11.43), is nonsingular. (The consequences of this assumption not being valid will be discussed later.) In this case, by (11.21), the Kalman filter gain matrix is given by

$$\hat{K} = (\hat{P}A' + FVF')C'(CFVF'C')^{-1} \tag{11.44}$$

where, by (11.22), the covariance matrix is given by

$$\dot{\hat{P}} = \tilde{A}\hat{P} + \hat{P}\tilde{A}' - \hat{P}A'C'(CFVF'C')^{-1}CA\hat{P} + F\tilde{V}F' \tag{11.45}$$

where

$$\begin{aligned}\tilde{A} &= A - FVF'C'(CFVF'C')^{-1}CA \\ &= [I - FVF'C'(CFVF'C')^{-1}C]A \end{aligned} \tag{11.46}$$

$$\tilde{V} = V - VF'C'(CFVF'C')^{-1}CFV \tag{11.47}$$

In the standard Kalman filtering problem, the covariance matrix $\hat{P}$ is usually positive definite. This is not the case here: $\hat{P}$ is a singular matrix. This can be seen by premultiplying both sides of (11.45) by C. This gives

$$C\dot{\hat{P}} = C\tilde{A}\hat{P} + C\hat{P}\tilde{A}' - C\hat{P}A'C'(CFVF'C')^{-1}CAP + CF\tilde{V}F' \tag{11.48}$$

But, from (11.46),

$$C\tilde{A} = CA - CFVF'C'(CFVF'C')^{-1}CA = 0$$

and, from (11.47),

$$CF\tilde{V} = CFV - CFVF'C'(CFVF'C')^{-1}CFV = 0$$

Hence (11.48) reduces to

$$C\dot{\hat{P}} = C\hat{P}\tilde{A} - C\hat{P}A'C'(CFVF'C')^{-1}CA\hat{P}$$

This is a homogeneous differential equation (i.e., there is no forcing term) and it has a solution

$$C\hat{P}(t) \equiv 0 \tag{11.49}$$

Thus, if the initial condition $C\hat{P}(0)$ is zero, then $C\hat{P}(t)$ remains zero thereafter. We assume that this condition holds, and hence that (11.49) is valid.

Now consider $C\dot{\hat{x}}$ where $\dot{\hat{x}}$ is given by (11.41)

$$C\dot{\hat{x}} = C(A\hat{x} + Bu) + C\hat{K}(z - CA\hat{x}) \tag{11.50}$$

But, using (11.42) and (11.49),

$$C\hat{K} = C(\hat{P}A' + FVF')C'(CFVF'C)^{-1} = I \tag{11.51}$$

and hence (11.50) becomes

$$C\dot{\hat{x}} = z + CBu = \dot{y} \tag{11.52}$$

using the definition of z as given by (11.40). Thus, upon integration of (11.52), we obtain

$$C\hat{x} = y \tag{11.53}$$

assuming the constant of integration to be zero. This assumption is valid because the derivative of y was defined for purposes of the theoretical development; it is not an actual observation. The actual observation is y.

The result given by (11.53) is not in the least surprising: it asserts that if a substate Cx of the state x can be observed without noise, then the optimum estimate of that substate is the observation itself. If C is a nonsingular matrix, then of course (11.53) gives $\hat{x} = C^{-1}y$, an obvious result. The important case, however, is when C is not a square matrix: when the number l of (independent) observed quantities is less than the number k of state variables. In this case (11.53) only partially specifies the state. The remaining $k - l$ relations are to be obtained from (11.41). Since only $k - l$ relations are needed, we should not be surprised to find that these can be obtained using a dynamic system of order $k - l$. This is exactly the way it turns out.

The key to the result we are seeking is in (11.51): $C\hat{K} = I$. If C were a nonsingular matrix $\hat{K}$ would equal C^{-1} and we could write (11.53) as

$$\hat{x} = C^{-1}y = \hat{K}y$$

When C is singular we seek a general estimate of the form

$$\hat{x} = \hat{K}y + L_2\gamma_2 \tag{11.54}$$

where γ_2 is the state of a system of order $k - l$. The route to (11.54) is circuitous. First consider (11.41), which can be written

$$\dot{\hat{x}} = (I - \hat{K}C)A\hat{x} + \hat{K}z + Bu \tag{11.55}$$

We have found earlier that $C\hat{K} = I$; but this doesn't mean that $\hat{K}C = I$, unless C^{-1} exists. The matrix $I - \hat{K}C$, however, has the property that

$$(I - \hat{K}C)^2 = (I - \hat{K}C)(I - \hat{K}C) = I - 2\hat{K}C + \hat{K}C = I - \hat{K}C \tag{11.56}$$

A matrix whose square is equal to itself is said to be *idempotent*.[4] All of the characteristic roots of an idempotent matrix are either zero or unity; the number of characteristic roots that are unity equals the rank of the matrix. For this case assume that the matrix C is of full rank l, that is, that the rows of C are linearly independent. Then it is readily shown that $I - \hat{K}C$ is also of rank l and hence, by the above property of idempotent matrices, there exists a similarity transformation T such that

$$I - \hat{K}C = T^{-1}E_lT \tag{11.57}$$

where

$$E_l = \begin{bmatrix} 0 & 0 \\ 0 & I_{k-l} \end{bmatrix} \begin{matrix} \updownarrow l \\ \updownarrow k-l \end{matrix}$$

(I_{k-l} is a $(k - l) \times (k - l)$ identity matrix.) Although a similarity transformation is not unique, one transformation satisfying (11.57) is

$$T = \begin{bmatrix} C \\ U_l(I - \hat{K}C) \end{bmatrix} \tag{11.58}$$

where $U_l(I - \hat{K}C)$ is a $(k - l) \times k$ matrix consisting of $(k - l)$ rows of $I - \hat{K}C$ linearly independent of the rows of C. (The matrix U_l is a $(k - l) \times k$ matrix, each row of which has all zero elements except a single 1 in the position of the row of $I - \hat{K}C$ to be selected. See the example below.)

To verify (11.57), we rewrite it as

$$T(I - \hat{K}C) = E_lT \tag{11.59}$$

Using (11.58), the left-hand side of (11.59) is

$$\begin{bmatrix} C \\ U_l(I - \hat{K}C) \end{bmatrix}(I - \hat{K}C) = \begin{bmatrix} 0 \\ U_l(I - \hat{K}C)^2 \end{bmatrix} = \begin{bmatrix} 0 \\ U_l(I - \hat{K}C) \end{bmatrix}$$

Note that the idempotency of $I - \hat{K}C$ was used to obtain the second relationship.

The right-hand side of (11.59) is

$$\begin{bmatrix} 0 & 0 \\ 0 & I_{k-l} \end{bmatrix}\begin{bmatrix} 0 \\ U_l(I - \hat{K}C) \end{bmatrix} = \begin{bmatrix} 0 \\ U_l(I - \hat{K}C) \end{bmatrix}$$

Thus (11.59) is established. Since C was assumed to be of full rank l and the rows of $U_l(I - \hat{K}C)$ are linearly independent of the rows of C, the rows of T are, by construction, linearly independent. Hence T is nonsingular and is a transformation matrix satisfying (11.57).

We return now to (11.55), which can be written

$$\begin{aligned}\dot{\hat{x}} &= (I - \hat{K}C)A\hat{x} + \hat{K}(\dot{y} - CBu) + Bu \\ &= (I - \hat{K}C)(A\hat{x} + Bu) + \hat{K}\dot{y}\end{aligned} \tag{11.60}$$

Let

$$\xi = \hat{x} - \hat{K}y \tag{11.61}$$

Then

$$\begin{aligned}\dot{\xi} = \dot{\hat{x}} - \hat{K}\dot{y} &= (I - \hat{K}C)[A(\xi + \hat{K}y) + Bu] \\ &= T^{-1}E_lT[A(\xi + \hat{K}y) + Bu]\end{aligned} \tag{11.62}$$

upon use of (11.57). Premultiplying both sides of (11.62) by T gives

$$T\dot{\xi} = E_lT[A\xi + A\hat{K}y + Bu] \tag{11.63}$$

Let

$$\gamma = T\xi \qquad \xi = T^{-1}\gamma \tag{11.64}$$

Then (11.63) becomes

$$\dot{\gamma} = E_l[TAT^{-1}(\gamma + T\hat{K}y) + TBu] \tag{11.65}$$

Partition γ and TAT^{-1} as follows:

$$\gamma = \left[\begin{array}{c}\gamma_1 \\ \hline \gamma_2\end{array}\right]\begin{array}{l}\updownarrow l \\ \updownarrow k-l\end{array} \qquad \underbrace{\left[\begin{array}{c|c}\bar{A}_{11} & \bar{A}_{12} \\ \hline \bar{A}_{21} & \bar{A}_{22}\end{array}\right]}_{\substack{\leftrightarrow \quad \leftrightarrow \\ l \quad\; k-l}} = TAT^{-1} \tag{11.66}$$

Note also that

$$T\hat{K} = \left[\begin{array}{c}C\hat{K} \\ \hline U_l(I - \hat{K}C)\hat{K}\end{array}\right] = \left[\begin{array}{c}I_l \\ \hline 0\end{array}\right]$$

Thus (11.65) becomes

$$\left[\begin{array}{c}\dot{\gamma}_1 \\ \hline \dot{\gamma}_2\end{array}\right] = \left[\begin{array}{c|c}0 & 0 \\ \hline 0 & I_{k-l}\end{array}\right]\left\{\left[\begin{array}{c|c}\bar{A}_{11} & \bar{A}_{12} \\ \hline \bar{A}_{21} & \bar{A}_{22}\end{array}\right]\left[\begin{array}{c}\gamma_1 + y \\ \hline \gamma_2\end{array}\right] + \left[\begin{array}{c}CBu \\ \hline U_l(I - \hat{K}C)Bu\end{array}\right]\right\}$$

Or, in component form,

$$\begin{aligned}\dot{\gamma}_1 &= 0 \\ \dot{\gamma}_2 &= \bar{A}_{21}(\gamma_1 + y) + \bar{A}_{22}\gamma_2 + U_l(I - \hat{K}C)Bu\end{aligned} \tag{11.67}$$

Note that

$$\gamma_1 + C\xi = C(\hat{x} - \hat{K}y) = C\hat{x} - y$$

which is assumed to be zero at $t = 0$. Thus, from (11.67), $\gamma_1 \equiv 0$ for $t \geqq 0$ and (11.67) becomes

$$\dot{\gamma}_2 = \bar{A}_{22}\gamma_2 + \bar{A}_{21}y + U_l(I - \hat{K}C)Bu \tag{11.68}$$

This is the differential equation of a system of order $k - l$ having inputs y and u.

The state estimate $\hat{x}$ is obtained using (11.63) and (11.64)

$$\hat{x} = \hat{K}y + \xi = \hat{K}y + T^{-1}\gamma \tag{11.69}$$

If T^{-1} is partitioned as

$$T^{-1} = [\,\underbrace{L_1}_{l} \mid \underbrace{L_2}_{k-l}\,]$$

then, since $\gamma_1 \equiv 0$, (11.69) reduces to the desired expression as given by (11.54), namely

$$\hat{x} = \hat{K}y + L_2\gamma_2 \tag{11.70}$$

A block-diagram representation of the state estimate expressed by (11.68) and (11.70) is shown in Fig. 11.4. Note that the structure of the Kalman filter with no observation noise is that of a reduced-order observer as developed in Chap. 7. It is readily established that the matrices appearing in Fig. 11.4 have the properties required of the general matrices of the reduced-order observer.

A special case of the reduced-order Kalman filter results when CF is a nonsingular matrix. In this case, the matrix $\tilde{V}$ that enters into the variance equation (11.45) is given by

$$\tilde{V} = V - V(CF)'(CF)^{-1}V^{-1}(CF)^{-1}CFV = 0$$

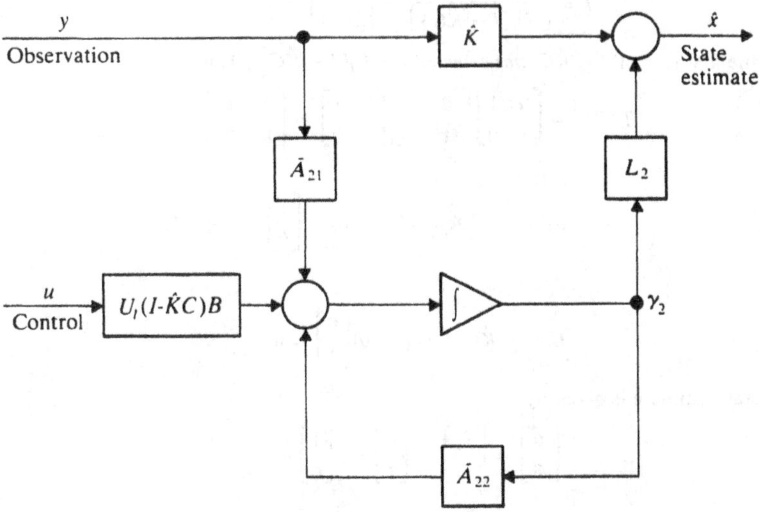

Figure 11.4 Kalman filter in absence of observation noise is a reduced-order observer.

This means that the forcing term $F\tilde{V}F'$ in (11.45) is zero. Then one of the solutions to (11.45) is $\hat{P} = 0$, which would imply that the state x can be estimated without steady state error. In this case

$$\hat{K} = FV(CF)'[(CF)']^{-1}V^{-1}(CF)^{-1} = F(CF)^{-1} \tag{11.71}$$

But $\hat{P} = 0$ may not be the only solution to (11.45) with $\dot{\hat{P}} = 0$. In that event, the correct solution is the one to which the solution of the differential equation converges as $t \to \infty$.

Example 11D Inverted pendulum with velocity sensor Consider the inverted pendulum of Example 11A, except with a velocity sensor instead of a position sensor. The observation equation is

$$y = \omega = x_2$$

Thus

$$C = [0 \quad 1]$$

Using

$$F = \begin{bmatrix} 0 \\ 1 \end{bmatrix}$$

as given in Example 11A, we obtain

$$CF = 1$$

Hence, by (11.71)

$$\hat{K} = F(CF)^{-1} = \begin{bmatrix} 0 \\ 1 \end{bmatrix}$$

$$I - \hat{K}C = \begin{bmatrix} 1 & 0 \\ 0 & 1 \end{bmatrix} - \begin{bmatrix} 0 \\ 1 \end{bmatrix}[0 \quad 1] = \begin{bmatrix} 1 & 0 \\ 0 & 0 \end{bmatrix}$$

Thus

$$T = \begin{bmatrix} C \\ \hline U_l(I - \hat{K}C) \end{bmatrix} = \begin{bmatrix} 0 & 1 \\ 1 & 0 \end{bmatrix} = T^{-1}$$

[Note that the first row of $I - \hat{K}C$ was selected for $U_l(I - \hat{K}C)$.] Thus

$$TAT^{-1} = \begin{bmatrix} 0 & 1 \\ 1 & 0 \end{bmatrix}\begin{bmatrix} 0 & 1 \\ \Omega^2 & 0 \end{bmatrix}\begin{bmatrix} 0 & 1 \\ 1 & 0 \end{bmatrix} = \begin{bmatrix} 0 & \Omega^2 \\ 1 & 0 \end{bmatrix}$$

Finally

$$\bar{A}_{21} = 1 \qquad \bar{A}_{22} = 0 \qquad L_2 = \begin{bmatrix} 1 \\ 0 \end{bmatrix}$$

and

$$U_l(I - \hat{K}C)B = [1 \quad 0]\begin{bmatrix} 0 \\ 1 \end{bmatrix} = 0$$

Hence the state estimate is given by

$$\begin{bmatrix} \hat{\theta} \\ \hat{\omega} \end{bmatrix} = \begin{bmatrix} \hat{x}_1 \\ \hat{x}_2 \end{bmatrix} = \begin{bmatrix} 0 \\ 1 \end{bmatrix} y + \begin{bmatrix} 1 \\ 0 \end{bmatrix} \gamma_2$$

with

$$\dot{\gamma}_2 = y$$

which reduces to

$$\hat{\theta} = \gamma_2 = \int_0^t y(\tau)\, d\tau$$
$$\hat{\omega} = y \tag{11D.1}$$

In other words, the angular velocity estimate is the measured angular velocity, and the estimated angular position is simply the integral of the measured angular velocity. The reduced-order Kalman filter has the very essence of simplicity.

It is of interest to compare the reduced-order filter with the full-order filter which would result when the noise in measuring the velocity is not zero. For the covariance matrix

$$\hat{P} = \begin{bmatrix} p_1 & p_2 \\ p_2 & p_3 \end{bmatrix}$$

The gain matrix is

$$\hat{K} = \begin{bmatrix} p_2 \\ p_3 \end{bmatrix} W^{-1} \tag{11D.2}$$

where W is the convariance matrix of the velocity observation noise (not the same W that appears in Example 11A).

The components of the variance equation are

$$0 = 2p_2 - \frac{p_2^2}{W} \tag{11D.3}$$

$$0 = p_3 + \Omega^2 p_1 - \frac{p_2 p_3}{W} \tag{11D.4}$$

$$0 = 2\Omega^2 p_2 - \frac{p_3^2}{W} + V \tag{11D.5}$$

From (11D.3) either $p_2 = 0$ or $p_2 = 2W$. But from (11D.4)

$$\Omega^2 p_1 = -p_3\left(1 - \frac{p_2}{W}\right)$$

Thus, if $p_2 = 0$ were correct, then $\Omega^2 p_1$ would equal $-p_3$ which is not acceptable since p_1 and p_3 must both be positive for a valid covariance matrix. We thus conclude that $p_2 = 2W$. In this case

$$p_1 = \frac{1}{\Omega^2}\sqrt{VW}$$

and, from (11D.5),

$$p_3 = \sqrt{VW + 4\Omega^2 W^2}$$

By (11D.2) the gain matrix is

$$\hat{K} = \begin{bmatrix} 2 \\ k_2 \end{bmatrix} \qquad k_2 = \sqrt{V/W + 4\Omega^2}$$

Thus the Kalman filter is given by

$$\dot{\hat{\theta}} = \hat{\omega} + 2(y - \hat{\omega})$$
$$\dot{\hat{\omega}} = \Omega^2 \hat{\theta} + k_2(y - \hat{\omega}) + u \tag{11D.6}$$

In the frequency domain, $\hat{\theta}$ and $\hat{\omega}$ are given by

$$\hat{\omega} = \frac{(k_2 s + 2\Omega^2)y + su}{s^2 + k_2 s + \Omega^2}$$

$$\hat{\theta} = \frac{(2s^2 + k_2)y + u}{s^2 + k_2 s + \Omega^2} \tag{11D.7}$$

If the sensor noise W tends to zero, then $k_2 = \sqrt{V/W + 4\Omega^2} \to \infty$ and (11D.7) reduces to

$$\hat{\omega} = y$$

$$\hat{\theta} = \frac{1}{s} y$$

which is the same as obtained in (11D.1) for the reduced-order filter. When the sensor noise is not zero, use of (11D.7) instead of (11D.1) provides the optimum smoothing of the output from the noisy sensor.

As mentioned in the introduction to this section, the application may arise in which the observation noise covariance matrix W is singular, but not zero. This case typically arises when some subset of observations are noise-free and another subset has the usual white noise:

$$y = \begin{bmatrix} y_1 \\ y_2 \end{bmatrix}$$

with

$$y_1 = C_1 x$$

$$y_2 = C_2 x + w$$

where w is a white noise process having a nonsingular covariance matrix W. This problem can be treated by the method of this section by defining the vector z as

$$z = \begin{bmatrix} z_1 \\ z_2 \end{bmatrix}$$

with

$$z_1 = \dot{y}_1 - C_1 Bu = C_1 Ax + C_1 Fv$$

$$z_2 = y_2$$

Thus

$$z = Cx + \begin{bmatrix} C_1 F & 0 \\ 0 & I \end{bmatrix} \begin{bmatrix} v \\ w \end{bmatrix} \tag{11.72}$$

where

$$C = \begin{bmatrix} C_1 A \\ C_2 \end{bmatrix}$$

The Kalman filter is governed by the equation

$$\begin{aligned}\dot{\hat{x}} &= A\hat{x} + Bu + \hat{K}(z - C\hat{x}) \\ &= A\hat{x} + Bu + \hat{K}_1(\dot{y}_1 - C_1B\hat{x}) + \hat{K}_2(y_2 - C_2\hat{x}) \qquad (11.73)\end{aligned}$$

The covariance matrix, which must be derived with proper consideration of the cross-correlation between the excitation v and the composite observation noise vector $[v', w']'$, will turn out to be singular: $C_1\hat{P} \equiv 0$. This property makes it possible to eliminate the differentiation of y_1 in the implementation of the Kalman filter, and also to reduce the order of the filter to $n - l_1$ where l_1 is the number of noise-free observations, i.e., the dimension of the vector y_1. The details of the derivation are straightforward but somewhat tedious. The final result is given in [5]. See also the discussion in Note 11.4.

The derivation of the reduced-order Kalman filter is based on differentiating the noise-free observation to produce a derived observation which contains white noise, and then showing that the differentiation of the observation is not required in the implementation. Implicit in the development is that a single differentiation of the output produces a signal that contains white noise with a nonsingular covariance matrix $CFVF'C'$. In many cases, if not most, however, the spectral density matrix $CFVF'C'$ turns out to be singular. The resulting new observation vector $z = \dot{y} - CBu$ may be thought of as having some components which are noise-free and other components which have white noise. Those components which are noise-free can be differentiated again to produce still another derived observation, which can be adjoined to the original observation so that the two together will result in an equivalent observation in which the spectral density matrix may be nonsingular. If it is still singular, another derivative is taken of those components that still do not have white noise, until finally an observation vector is constructed in which there is a nonsingular covariance matrix, or a noise-free estimate of the state can be constructed by repeated differentiation of the output. The general case, discussed in somewhat greater detail in [6], can become very complicated, so we consider a special case which is nevertheless of considerable interest. The case is that in which the matrix CF turns out to be zero. In this case

$$\dot{y} = C(Ax + Bu + Fv) = CAx + CBu$$

contains no white noise at all. In this case another time derivative is taken to give

$$\ddot{y} = CA\dot{x} + CB\dot{u} = CA(Ax + Bu + Fv) + CB\dot{u}$$

In this case we let

$$z = \ddot{y} - C(ABu + B\dot{u}) = CA^2x + CAFv \qquad (11.74)$$

This is in the same form as (11.39) and if $CAFV(CAF)'$ has an inverse, the Kalman filter is given by

$$\dot{\hat{x}} = A\hat{x} + Bu + \hat{K}(z - CA^2\hat{x}) \qquad (11.75)$$

The analysis that was used for the case in which $(CF)'VCF$ is nonsingular also applies in this case using CA instead of C. However, since z includes the second derivative of y and the transformation procedure used in obtaining the reduced-order filter only eliminates one of the derivatives, the resulting filter is likely to entail use of the first derivative of y. By assumption $\dot{y}$ does not contain any noise and hence, at least in principle, can be realized without differentiating white noise.

If differentiation of the observation is unacceptable, another approach is simply to assume the presence of additional terms in the F matrix such that $CFV(CF)'$ is a nonsingular matrix. In some cases, these additional terms may be justified as representing possible variations in the parameters of the dynamic process. (See Sec. 11.8.) In other cases, however, the only justification is that it produces an acceptable observer.

Both of these approaches are considered in the following example.

Example 11E inverted pendulum with noise-free position sensor We turn to the previous example, except in this case

$$y = \theta$$

Then

$$\dot{y} = \dot{\theta} = \omega$$

and still no noise results. The second derivative, however, is

$$\ddot{y} = \dot{\omega} = \Omega^2\theta + u + v$$

The reduced-order filter is given by

$$\begin{aligned} \dot{\hat{x}}_1 &= \hat{x}_2 + \hat{K}_1(z - CA^2\hat{x}) \\ \dot{\hat{x}}_2 &= \Omega^2\hat{x}_1 + \hat{K}_2(z - CA^2\hat{x}) + u \end{aligned} \tag{11E.1}$$

where

$$z = \ddot{y} - C(ABu + B\dot{u})$$

We find that

$$CAF = CAB = [1 \quad 0]\begin{bmatrix} 0 & 1 \\ \Omega^2 & 0 \end{bmatrix}\begin{bmatrix} 0 \\ 1 \end{bmatrix} = 1$$

$$CB = [1 \quad 0]\begin{bmatrix} 0 \\ 1 \end{bmatrix} = 0$$

Hence

$$z = \ddot{y} - u$$

and, by (11.65),

$$\hat{K} = F(CAF)^{-1} = \begin{bmatrix} 1 \\ 0 \end{bmatrix}$$

Thus (11E.1) reduces to

$$\begin{aligned} \dot{\hat{x}}_1 &= \hat{x}_2 \\ \dot{\hat{x}}_2 &= \Omega^2\hat{x}_1 + (\ddot{y} - u - \Omega^2\hat{x}_1) + u = \ddot{y} \end{aligned}$$

from which we infer that

$$\begin{aligned}\hat{x}_2 &= \dot{y}\\ \hat{x}_1 &= y\end{aligned} \tag{11E.2}$$

This result is entirely expected. If $x_1 = \theta$ can be measured without noise the optimum estimate of x_1 is the measurement itself. Since x_2 is defined as the derivative of x_1, and this derivative also contains no noise, what better estimate of x_2 can there be than $\dot{y}$?

If we are not happy with this result, we can assume that some noise gets into the first differential equation defining the process

$$\begin{aligned}\dot{\theta} &= \omega + \varepsilon v\\ \dot{\omega} &= \Omega^2\theta + u + v\end{aligned} \tag{11E.3}$$

Since $\dot{\theta} = \omega$ is a definition, it is hard to justify the εv term in (11E.3). The result, however, will be acceptable, as we shall see.

In this case

$$CF = [1 \quad 0]\begin{bmatrix}\varepsilon\\ 1\end{bmatrix} = \varepsilon$$

and, by (11.65),

$$\hat{K} = \begin{bmatrix}\varepsilon\\ 1\end{bmatrix}\frac{1}{\varepsilon} = \begin{bmatrix}1\\ 1/\varepsilon\end{bmatrix}$$

$$I - \hat{K}C = \begin{bmatrix}1 & 0\\ 0 & 1\end{bmatrix} - \begin{bmatrix}1\\ 1/\varepsilon\end{bmatrix}[1 \quad 0] = \begin{bmatrix}0 & 0\\ -1/\varepsilon & 1\end{bmatrix}$$

then

$$T = \begin{bmatrix}C\\ U_1(I - \hat{K}C)\end{bmatrix} = \begin{bmatrix}1 & 0\\ -1/\varepsilon & 1\end{bmatrix}$$

and

$$T^{-1} = \begin{bmatrix}1 & 0\\ 1/\varepsilon & 1\end{bmatrix}$$

$$TAT^{-1} = \begin{bmatrix}1 & 0\\ -1/\varepsilon & 1\end{bmatrix}\begin{bmatrix}0 & 1\\ \Omega^2 & 0\end{bmatrix}\begin{bmatrix}1 & 0\\ 1/\varepsilon & 1\end{bmatrix} = \begin{bmatrix}1/\varepsilon & 1\\ \Omega^2 - 1/\varepsilon^2 & -1/\varepsilon\end{bmatrix}$$

Hence the required submatrices of the reduced-order filter are

$$\bar{A}_{21} = \Omega^2 - 1/\varepsilon^2 \qquad \bar{A}_{22} = -1/\varepsilon \qquad L_2 = \begin{bmatrix}0\\ 1\end{bmatrix}$$

The reduced-order filter is thus given by

$$\hat{x} = \begin{bmatrix}\hat{\theta}\\ \hat{\omega}\end{bmatrix} = \begin{bmatrix}1\\ 1/\varepsilon\end{bmatrix}y + \begin{bmatrix}0\\ 1\end{bmatrix}\gamma_2$$

or

$$\hat{\theta} = y$$

$$\hat{\omega} = \frac{1}{\varepsilon}y + \gamma_2$$

with

$$\dot{\gamma}_2 = -\frac{1}{\varepsilon}\gamma_2 + \left(\Omega^2 - \frac{1}{\varepsilon^2}\right)y + u$$

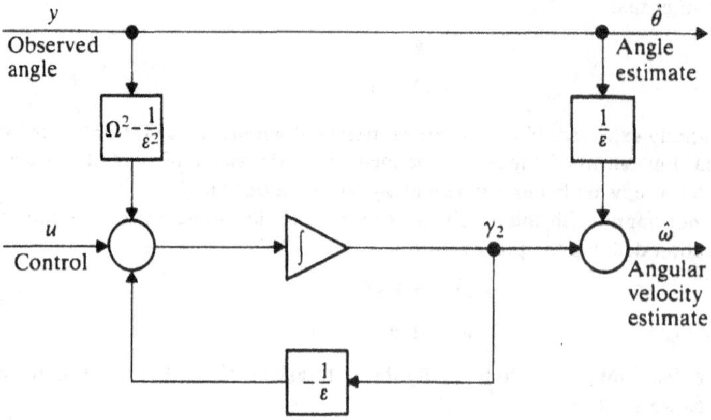

Figure 11.5 Schematic of reduced-order Kalman filter for observation of angular position. (Fictitious noise of spectral density ε added to dynamic equations.)

A block-diagram representation of this filter is shown in Fig. 11.5. It is found that the Laplace transform of $\hat{\omega}$ is given by

$$\hat{\omega} = \frac{(s + \varepsilon\Omega^2)y + \varepsilon u}{\varepsilon s + 1}$$

As the time constant $\varepsilon \to 0$, the transfer function from y to $\hat{\omega}$ tends to a pure differentiator.

The last topic we consider in this section is observations with correlated (also called "colored") noise. The observation equation for this process is

$$y = Cx + q \tag{11.76}$$

where q is not white noise but rather a Markov process

$$\dot{q} = Qq + w \tag{11.77}$$

where w is white noise with a spectral density matrix W. Combining (11.77) with the standard dynamic equation $\dot{x} = Ax + Bu + Fv$ gives the metastate and observation equations

$$\begin{aligned} \dot{\mathbf{x}} &= \mathbf{A}\mathbf{x} + \mathbf{B}u + \mathbf{F}\mathbf{v} \\ y &= \mathbf{C}\mathbf{x} \end{aligned} \tag{11.78}$$

with

$$\mathbf{x} = \begin{bmatrix} x \\ \hline q \end{bmatrix} \qquad \mathbf{v} = \begin{bmatrix} v \\ \hline w \end{bmatrix}$$

$$\mathbf{A} = \left[\begin{array}{c|c} A & 0 \\ \hline 0 & 0 \end{array}\right] \qquad \mathbf{B} = \begin{bmatrix} B \\ \hline 0 \end{bmatrix} \qquad \mathbf{F} = \left[\begin{array}{c|c} F & 0 \\ \hline 0 & I \end{array}\right] \qquad \mathbf{C} = [C \quad I]$$

The theory developed earlier applies to the metastate process: the optimum estimate is given by

$$\hat{\mathbf{x}} = \mathbf{K}y + \mathbf{L}_2\gamma_2$$

where $\hat{\mathbf{K}}$ and $\mathbf{L}_2$ are obtained as explained above. In particular, it is found that

$$\hat{\mathbf{K}} = \begin{bmatrix} \hat{K}_x \\ \hline \hat{K}_q \end{bmatrix} \qquad \mathbf{L}_2 = \begin{bmatrix} I \\ \hline L_{2q} \end{bmatrix}$$

Thus the state estimate is given by

$$\hat{x} = \hat{K}_x y + \gamma_2 \tag{11.79}$$

where γ_2 satisfies a differential equation of order $(n+m) - m = n$. After considerable algebra—see [7]—it is determined that

$$\hat{K}_x = [P_x(A'C' - C'Q) + FVF'][W + CFVF'C']^{-1}$$

where P_x satisfies the matrix quadratic equation

$$0 = \bar{A}P_x + P_x\bar{A}' - P_xZP_x + \bar{W} \tag{11.80}$$

in which

$$\begin{aligned} \bar{A} &= A - FVF'C'(W + CFVF'C')^{-1}(CA - QC) \\ Z &= (CA - QC)'(W + CFVF'C')^{-1}(CA - QC) \\ \bar{W} &= FVF'[I - C'(W + CFVF'C')^{-1}CFVF'] \end{aligned} \tag{11.81}$$

and γ_2 satisfies the differential equation

$$\dot{\gamma}_2 = \hat{A}\hat{x} - \hat{K}_x[(CA - QC)\hat{x} + Qy] + Bu \tag{11.82}$$

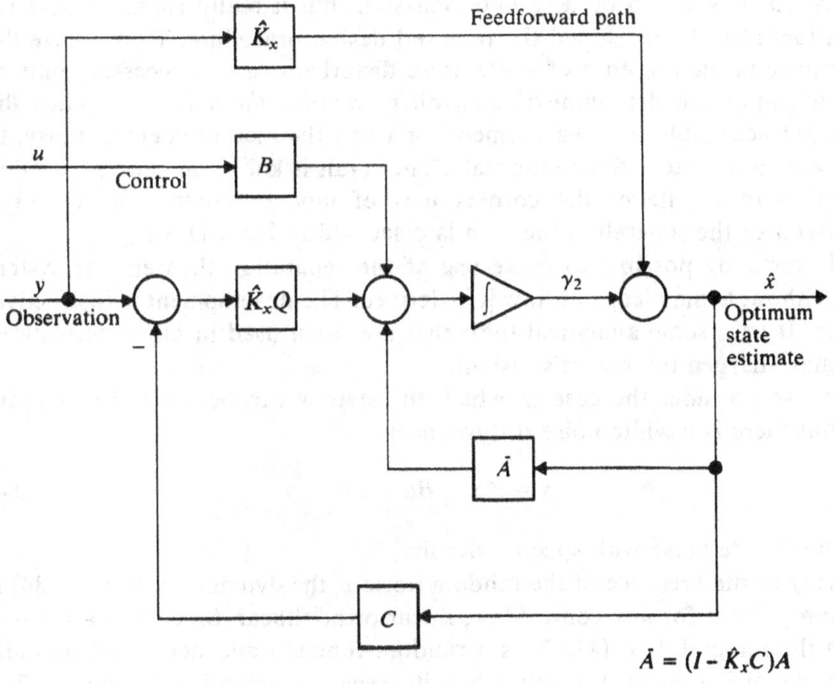

Figure 11.6 Kalman filter structure for correlated noise.

A block-diagram representation of the filter of (11.79) and (11.82) is shown in Fig. 11.6. Note that the structure is very similar to that of a filter with a white noise observation. The main difference is the feedforward path from the observed quantity y directly into the state estimate $\hat{x}$.

11.7 STOCHASTIC CONTROL: THE SEPARATION THEOREM

In Chap. 10, we studied the design of the optimum "deterministic" control law: the control law that minimizes a performance index of the form

$$J = \int_t^T (x'Qx + u'Ru)\, d\tau \tag{11.83}$$

under the assumption that the state x is accessible to observation and that there are no random disturbances acting on the process. (The symbol J is used here so that V can be used later on for the spectral density of the white noise disturbance.) We have already seen how to optimize the observer for estimating the state in the presence of white noise on the observations and white noise disturbances. But we have not yet addressed the optimization of performance in the presence of disturbances, whether or not the state is accessible to observation.

This would seem to be a serious omission, but it really isn't, because we have in fact already developed the required design procedure. To optimize the performance in the presence of white noise disturbances, it is necessary only to design the optimum deterministic controller, *ignoring the noise*, and when the state is not accessible for measurement, or when the measurement is noisy, to use a Kalman filter to estimate the state. The result is known as the "separation theorem" and is one of the cornerstones of modern control theory. (The background of the separation theorem is discussed in Note 11.5.)

It is certainly possible to make use of the separation theorem for system design without a knowledge of how it is derived. The development is instructive, however: It uses some analytical tools that are often used in other derivations and it may sharpen the reader's insight.

First we consider the case in which the state x can be measured without noise, but there is a white noise disturbance:

$$\dot{x} = Ax + Bu + Fv \tag{11.84}$$

where v is white noise with spectral density V.

Owing to the presence of the random noise v, the dynamic system (11.84) is a random process for any control law, linear or nonlinear. Since x is a random process the integral J of (11.83) is a random function and hence the optimization problem is meaningful only when its expected value is minimized. Thus the meaningful problem is to minimize

$$\bar{J} = E\{J\} = E\left\{\int_t^T (x'Qx + u'Ru)\, d\tau\right\} \tag{11.85}$$

by the choice of the control law $u = g(x)$. The optimum control law is known to be linear, but to show this is beyond the scope of this text. Thus we will proceed under the assumption that a linear control law

$$u = -Gx \tag{11.86}$$

is to be used, and optimize the selection of the gain G. With this control law, the integrand in (11.85) becomes

$$x'Qx + u'Ru = x'Qx + x'G'RGx = x'Lx$$

with

$$L = Q + G'RG \qquad \text{(Note that } L = L'\text{)} \tag{11.87}$$

Hence (11.85) becomes

$$\bar{J} = E\left\{\int_t^T x'Lx\, d\tau\right\} \tag{11.88}$$

For this control law, the closed-loop dynamics (11.84) become

$$\dot{x} = (A - BG)x + Fv$$

which has the solution

$$x(\tau) = \Phi_c(\tau, t)x(t) + z(\tau) \tag{11.89}$$

where

$$z(\tau) = \int_t^\tau \Phi_c(\tau, \lambda)Fv(\lambda)\, d\lambda \tag{11.90}$$

is the contribution to the state due to the random noise and Φ_c is the state-transition matrix corresponding to the closed-loop dynamics matrix $A - BG$. The integrand of (11.88) becomes, using (11.89) and (11.90),

$$\begin{aligned} x'(\tau)Lx(\tau) = {} & x'(t)\Phi_c'(\tau, t)L\Phi_c(\tau, t)x(t) \\ & + x'(t)\Phi_c'(\tau, t)Lz(\tau) \\ & + z'(\tau)L\Phi_c(\tau, t)x(t) \\ & + z'(\tau)Lz(\tau) \end{aligned} \tag{11.91}$$

The state $x(t)$ at the present time t is assumed known: it is not a random variable. Moreover, $z(\tau)$ is a random process resulting from excitation by the white noise process v. This process is uncorrelated with the initial state $x(t)$; hence the expected values of the second and third items in (11.91) are zero.

Thus

$$E\{x'(\tau)Lx(\tau)\} = x'(t)\Phi_c'(\tau, t)L\Phi_c(\tau, t)x(t) + E\{z'(\tau)Lz(\tau)\} \quad (11.92)$$

Note, however, that

$$z'(\tau)Lz(\tau) = \operatorname{tr}[Lz(\tau)z'(\tau)]$$

where tr N denotes the trace of the matrix N. Thus

$$E\{z'(\tau)Lz(\tau)\} = E\{\operatorname{tr}[Lz(\tau)z'(\tau)]\} = \operatorname{tr}[LP_z(\tau)] \quad (11.93)$$

where

$$P_z(\tau) = E\{z(\tau)z'(\tau)\}$$

By virtue of (11.92) and (11.93), the performance integral to be minimized is thus given by

$$\begin{aligned} \bar{J} &= \int_t^T \{x'(t)\Phi_c'(\tau, t)L\Phi_c(\tau, t)x(t) + \operatorname{tr}[LP_z(\tau)]\}\, d\tau \\ &= x'(t)M(t, T)x(t) + \int_t^T \operatorname{tr}[LP_z(\tau)]\, d\tau \end{aligned} \quad (11.94)$$

where

$$M(t, T) = \int_t^T \Phi_c'(\tau, t)L\Phi_c(\tau, t)\, d\tau \quad (11.95)$$

as already defined in Chap. 9. The first term of $\bar{J}$, given by (11.94), is thus a deterministic term $x'(t)M(t, T)x(t)$. To this term is added the integral which is the contribution due to the white noise.

The covariance matrix $P_z(\tau)$, with $z(\tau)$ given by (11.90) can be expressed as

$$P_z(\tau) = \int_t^T \Phi_c(\tau, \lambda)FVF'\Phi_c'(\tau, \lambda)\, d\lambda \quad (11.96)$$

Hence the second integral in (11.94) becomes

$$\int_t^T \operatorname{tr}[LP_z(\tau)]\, d\tau = \int_t^T \left\{\int_t^T \operatorname{tr}[L\Phi_c(\tau, \lambda)FVF'\Phi_c'(\tau, \lambda)]\, d\lambda\right\} d\tau \quad (11.97)$$

after interchanging the integration and the operation of taking the trace.

The spectral density matrix V of the noise excitation can always be factored into the product of a matrix S and its transpose

$$V = SS' \quad (11.98)$$

Moreover, it is well known that

$$\operatorname{tr}[AB] = \operatorname{tr}[BA]$$

when AB and BA are both defined. (See Appendix.) It thus follows that

$$\operatorname{tr}[L\Phi_c(\tau, \lambda)FVF'\Phi_c'(\tau, \lambda)] = \operatorname{tr}[L\Phi_c(\tau, \lambda)FSS'F'\Phi_c'(\tau, \lambda)]$$
$$= \operatorname{tr}[S'F'\Phi_c'(\tau, \lambda)L\Phi_c(\tau, \lambda)FS]$$

And, hence, again interchanging the integration and the trace operation, (11.96) becomes

$$\int_t^T \operatorname{tr}[LP_z(\tau)]\,d\tau = \int_t^T \operatorname{tr}\left\{S'F' \int_t^\tau \Phi_c'(\tau, \lambda)L\Phi_c(\tau, \lambda)\,d\lambda\, FS\right\} d\tau \quad (11.99)$$

Inverting the order of integration in (11.99)—taking careful note of how the limits of integration must be handled—we find that

$$\int_t^T \operatorname{tr}[LP_z(\tau)]\,d\tau = \int_t^T \operatorname{tr}\left\{S'F' \int_\lambda^T \Phi_c'(\tau, \lambda)L\Phi_c(\tau, \lambda)\,d\tau\, FS\right\} d\lambda \quad (11.100)$$

From (11.95), however, the inner integral is $M(\lambda, T)$. Thus (11.100) becomes

$$\int_t^T \operatorname{tr}[LP_z(\tau)]\,d\tau = \int_t^T \operatorname{tr}\{S'F'M(\lambda, T)FS\}\,d\lambda$$

Thus, finally, using (11.94), we obtain

$$\bar{J} = x'(t)M(t, T)x(t) + \int_t^T \operatorname{tr}\{S'F'M(\lambda, T)FS\}\,d\lambda \quad (11.101)$$

Note that the same matrix M that appears in the deterministic contribution $x'(t)M(t, T)x(t)$ to $\bar{J}$ due to the initial state $x(t)$ also appears under the integral sign. It follows that if the gain matrix G is chosen to minimize the performance matrix M [which is what we have done in Chap. 9 by choosing $G = \hat{G}$ as given by (9.26) and the matrix Riccati equation (9.27)] then the integral in (11.101) is also minimized. The integral is the contribution due to the noise having the spectral density matrix $V = SS'$. Thus the optimum performance is given by

$$\hat{\bar{J}} = x'(t)\hat{M}(t, T)x(t) + \int_t^T \operatorname{tr}\{S'F'\hat{M}(\lambda, T)FS\}\,d\lambda \quad (11.102)$$

This form of the minimum expected value of the performance criterion shows the symmetry of the integrand. Using $\operatorname{tr}[AB] = \operatorname{tr}[BA]$ in (11.102) restores the original spectral density matrix V:

$$\hat{\bar{J}} = x'(t)\hat{M}(t, T)x(t) + \int_t^T \operatorname{tr}\{FVF'\hat{M}(\lambda, T\}\,d\lambda \quad (11.103)$$

for the process (11.84), i.e.,

$$\dot{x} = Ax + Bu + Fv$$

with v being white noise with spectral density matrix V.

As the terminal time T becomes infinite the contribution to $\hat{\bar{J}}$ of the deterministic component remains finite at $x'(t)\bar{M}x(t)$ where $\bar{M}$ is the solution to the algebraic Riccati equation (9.29). But the contribution of the integral will in general become infinite. Thus it is more meaningful to refer to the "asymptotic cost *rate*" $\bar{J}$ which is the integrand of (11.103) and is given by

$$\bar{J} = \operatorname{tr}\{F'VF\bar{M}\} \tag{11.104}$$

Having now established that the deterministic optimum control law is identical to the stochastic optimum control law (when process noise v is present), provided a noise-free measurement can be made of the state x, we now turn to the more realistic and difficult problem of finding the optimum control law when only a noisy observation of the state:

$$y = Cx + w \tag{11.105}$$

is available. We assume, as with the Kalman filter, that w is white noise with a spectral density matrix W.

It is beyond the scope of this book to show that the optimum control law is linear, and in the form given by the separation principle of Chap. 8. (See Note 11.6.) If we are willing to accept this form of the control law, however, then we can determine the optimum controller and observer gains.

In addition to the process dynamics (11.84), we have the observer

$$\dot{\hat{x}} = A\hat{x} + Bu + K(y - C\hat{x}) \tag{11.106}$$

with the observation y given by (11.105) and the control u given by

$$u = -G\hat{x} \tag{11.107}$$

Using (11.107), we obtain from (11.88)

$$\dot{x} = Ax - BG\hat{x} + Fv \tag{11.108}$$

and, with (11.107) and (11.105), we obtain from (11.106)

$$\dot{\hat{x}} = A\hat{x} - BG\hat{x} - K[C(x - \hat{x}) + w] \tag{11.109}$$

As we did in Chap. 8, we define the estimation error by

$$e = x - \hat{x} \tag{11.110}$$

Then, (11.108) becomes

$$\dot{x} = (A - BG)x + BGe + Fv \tag{11.111}$$

And, on subtracting (11.109) from (11.111), we also get

$$\dot{e} = (A - KC)e + Fv - Kw \tag{11.112}$$

Note that (11.111) and (11.112) have the same form as deterministic equations (8.8) and (8.9) except that now the excitation noise v and the observation noise w are present.

Combine the two systems (11.111) and (11.112) into the metasystem:

$$\begin{bmatrix} \dot{x} \\ \dot{e} \end{bmatrix} = \begin{bmatrix} A - BG & BG \\ 0 & A - KC \end{bmatrix} \begin{bmatrix} x \\ e \end{bmatrix} + \begin{bmatrix} F & 0 \\ F & -K \end{bmatrix} \begin{bmatrix} v \\ w \end{bmatrix} \tag{11.113}$$

We still want to minimize

$$\bar{J} = E\left\{ \int_t^T (x'Qx + u'Ru)\, dt \right\} \tag{11.114}$$

For the sake of simplicity, we continue the development under the assumption that we are interested only in the contribution to $\bar{J}$ due to the noise and not due to the initial conditions. The effects of initial conditions can be included without any difficulty, but there will be more terms to carry around.

The integrand in (11.114) can be expressed as a quadratic form in the metastate $[x', e']'$:

$$x'Qx + u'Ru = x'Qx + \hat{x}'G'RG\hat{x} = x'Qx + (x - e)'G'RG(x - e) = \mathbf{x}'\mathbf{L}\mathbf{x} \tag{11.115}$$

where

$$\mathbf{x} = \begin{bmatrix} x \\ e \end{bmatrix} \qquad \mathbf{L} = \begin{bmatrix} Q + G'RG & -G'RG \\ -G'RG & G'RG \end{bmatrix}$$

Thus, as in the earlier development

$$\bar{J} = E\left\{ \int_t^T \mathbf{x}'\mathbf{L}\mathbf{x}\, d\tau \right\} = \int_t^T \operatorname{tr}\,[\mathbf{L}\mathbf{P}(\tau)]\, d\tau \tag{11.116}$$

for the process (11.113)

$$\dot{\mathbf{x}} = \mathbf{A}_c\mathbf{x} + \mathbf{F}\mathbf{v} \tag{11.117}$$

where

$$\mathbf{A}_c = \begin{bmatrix} A - BG & BG \\ 0 & A - KC \end{bmatrix} \qquad \mathbf{F} = \begin{bmatrix} F & 0 \\ F & -K \end{bmatrix} \tag{11.118}$$

and $\mathbf{P}(\tau)$ is the covariance matrix for the process defined by (11.117). Continuing as we did in (11.100)–(11.103) we find that

$$\bar{J} = \int_t^T \operatorname{tr}\,\{\mathbf{F}\mathbf{V}\mathbf{F}'\mathbf{M}(\tau, T)\}\, d\tau \qquad \mathbf{V} = \begin{bmatrix} V & 0 \\ 0 & W \end{bmatrix} \tag{11.119}$$

where $\mathbf{M}$ satisfies the same equation, namely (9.16), that M satisfies for the deterministic system, except that $\mathbf{A}_c$ and $\mathbf{L}$ as given by (11.118) and (11.115) are used in place of A_c and L as defined in (9.16). Thus $\mathbf{M}$ satisfies the matrix differential equation

$$-\dot{\mathbf{M}} = \mathbf{M}\mathbf{A}_c + \mathbf{A}_c'\mathbf{M} + \mathbf{L} \tag{11.120}$$

Partition **M** as follows

$$\mathbf{M} = \begin{bmatrix} M_x & M_{xe} \\ M'_{xe} & M_e \end{bmatrix} \tag{11.121}$$

In terms of the submatrices of **M** in (11.121) we find that (11.120) expands to three equations:

$$-\dot{M}_x = M_x(A - BG) + (A - BG)'M_x + Q + G'RG \tag{11.122}$$

$$-\dot{M}_{xe} = M_xBG + M_{xe}(A - KC) + (A - GB)'M_{xe} - G'RG \tag{11.123}$$

$$-\dot{M}_e = M_e(A - KC) + (A - KC)'M_e + M'_{xe}BG + G'B'M_{xe} + G'RG \tag{11.124}$$

We now have to find two matrices: the control matrix G and the observer matrix K that minimize the trace of the matrix **FVF′M**. We can proceed as we did in Sec. 9.4, by letting $\mathbf{M} = \hat{\mathbf{M}} + \mathrm{N}$, $G = \hat{G} + Z$, $K = \hat{K} + Y$, substituting these into (11.122)-(11.124), and thereby determining $\hat{G}$ and $\hat{K}$. (This derivation is left to the reader.) We can reduce the amount of calculation, however, by noting that (11.122) for M_x is exactly the same as the equation (9.20) for the deterministic gain. Thus we minimize the M_x component of **M** by computing the control gain G in exactly the same way as in the deterministic case, i.e.,

$$G = \hat{G} = R^{-1}B'\hat{M}_x \tag{11.125}$$

where

$$-\dot{\hat{M}}_x = \hat{M}_xA + A'\hat{M}_x - \hat{M}_xBR^{-1}B'\hat{M}_x + Q \tag{11.126}$$

as in (9.26) and (9.27). Having thus determined the optimum control gain we can proceed to find the optimum observer gain $\hat{K}$ using (11.123) and (11.124). In particular, (11.123) can be written

$$-\dot{M}_{xe} = (\hat{M}_xB - \hat{G}'R)G + M_{xe}(A - KC) + (A - B\hat{G})'M_{xe} \tag{11.127}$$

But, by (11.125), $\hat{M}_xB - G'R = 0$. Thus the forcing term on (11.123) vanishes and we may thus conclude that

$$\hat{M}_{xe} \equiv 0 \tag{11.128}$$

This means that the optimum matrix **M** is block diagonal. Hence

$$\mathbf{FVF'M} = \begin{bmatrix} F & 0 \\ F & -K \end{bmatrix} \begin{bmatrix} V & 0 \\ 0 & W \end{bmatrix} \begin{bmatrix} F' & F' \\ 0 & -K' \end{bmatrix} \begin{bmatrix} M_x & 0 \\ 0 & M_e \end{bmatrix}$$

$$= \begin{bmatrix} FVF'M_x & FVF'M_e \\ FVF'M_x & (FVF' + KWK')M_e \end{bmatrix} \tag{11.129}$$

The off-diagonal elements do not contribute to the trace of a matrix. Thus

$$\mathrm{tr}\,[\mathbf{FVF'M}] = \mathrm{tr}\,[FVF'M_x] + \mathrm{tr}\,[(FVF' + KWK')M_e] \tag{11.130}$$

Having already determined the control gain matrix $M_x = \hat{M}_x$ to minimize $\mathrm{tr}\,[FVF'M_x]$ we now need to determine the gain matrix K that minimizes

$\operatorname{tr}[(FVF' + KWK')M_e]$ where M_e satisfies (11.124) with $M_{xe} = 0$ and $G = \hat{G}$, i.e.,

$$-\dot{M}_e = M_e(A - KC) + (A - KC)'M_e + \hat{G}'R\hat{G} \tag{11.131}$$

With $\hat{G}$ given (11.131) is a linear equation, the solution of which can be expressed in terms of the observer transition matrix Φ_0, corresponding to $A_0 = A - KC$:

$$M_e(\tau, T) = \int_\tau^T \Phi_0'(\lambda, \tau)\hat{G}'R\hat{G}\Phi_0(\lambda, \tau)\, d\lambda$$

Then the second term in (11.130) can be written

$$\begin{aligned}
&\operatorname{tr}[(FVF' + KWK')M_e(\tau, T)] \\
&\qquad = \int_\tau^T \operatorname{tr}[\underbrace{(FVF' + KWK')\Phi_0'(\lambda, \tau)}_{A}\underbrace{\hat{G}'R\hat{G}\Phi_0(\lambda, \tau)}_{B}]\, d\lambda \\
&\qquad = \int_\tau^T \operatorname{tr}[\underbrace{(\hat{G}'R\hat{G}\Phi_0(\lambda, \tau)}_{B}\underbrace{(FVF' + KWK')\Phi_0'(\lambda, \tau)}_{A}]\, d\lambda
\end{aligned}$$

again using $\operatorname{tr}(AB) = \operatorname{tr}(BA)$. Thus we have

$$\begin{aligned}
&\operatorname{tr}[(FVF' + KWK')M_e(\tau, T)] \\
&\qquad = \operatorname{tr}\left[\hat{G}'R\hat{G} \int_\tau^T \Phi_0(\lambda, \tau)(FVF' + KWK')\Phi_0'(\lambda, \tau)\, d\lambda\right]
\end{aligned} \tag{11.132}$$

Hence, the contribution to $\bar{J}$ of (11.119) due to (11.132) is the integral thereof from t to T:

$$\begin{aligned}
J_0 &= \int_t^T \operatorname{tr}\left[\hat{G}'R\hat{G} \int_\tau^T \Phi_0(\lambda, \tau)(FVF' + KWK')\Phi_0'(\lambda, \tau)\right] d\lambda\, d\tau \\
&= \operatorname{tr} \int_t^T \hat{G}'R\hat{G}\left\{\int_t^\lambda \Phi_0(\lambda, \tau)(FVF' + KWK')\Phi_0'(\lambda, \tau)\, d\tau\right\} d\lambda
\end{aligned} \tag{11.133}$$

upon inversion of the order of integration as was done in going from (11.99) to (11.100).

The inner integral is in the form of a covariance matrix (10.57) for a system having the state transition matrix Φ_0 and an excitation noise covariance matrix $FVF' + KWK'$. Thus we can write

$$J_0 = \operatorname{tr} \int_t^T \hat{G}'R\hat{G}P(\lambda)\, d\lambda \tag{11.134}$$

where, by (10.58),

$$\dot{P} = (A - KC)P + P(A - KC)' + FVF' + KWK' \tag{11.135}$$

But (11.135) is exactly the matrix Riccati equation (11.13) for the optimum observer for the case in which u and v are uncorrelated. Thus we can conclude that J_0 is minimized by selecting the observer gain as the Kalman filter gain

$$K = \hat{K} = \hat{P}C'R^{-1} \tag{11.136}$$

where $\hat{P}$ is given by

$$\dot{\hat{P}} = A\hat{P} + \hat{P}A' - \hat{P}C'W^{-1}C\hat{P} + FVF'$$

which is the same as (11.15).

The result that we have just obtained is the celebrated "separation theorem," which can be summarized as follows:

To minimize the expected error in controlling a linear system,

$$\dot{x} = Ax + Bu + Fv$$

with observations

$$y = Cx + w$$

(*a*) Use the control law

$$u = -\hat{G}\hat{x}$$

where $\hat{x}$ is the output of a linear observer

$$\dot{\hat{x}} = A\hat{x} + Bu + \hat{K}(y - C\hat{x})$$

(*b*) Find the control gain matrix $\hat{G}$ as the solution of the corresponding deterministic optimum control problem.

(*c*) Find the observer gain matrix $\hat{K}$ as the optimum gain for the corresponding Kalman filter.

When the gains are chosen in accordance with the separation theorem, the minimum value of the expected performance is given by

$$E\left\{\int_t^T (x'Qx + u'Ru)\, d\lambda\right\} = \operatorname{tr}\left\{\int_t^T FVF'\hat{M}(\lambda, T)\, d\lambda + \int_t^T \hat{G}'R\hat{G}\hat{P}(\lambda)\, d\lambda\right\} \tag{11.137}$$

To the two integrals in (11.137) we must add the contribution of a nonzero initial condition $x'(t)\hat{M}(t, T)x(t)$. The first integral in (11.137) was already obtained in (11.102) for the case in which the entire state vector is measured without noise. The cost of measurement noise is thus given by the second integral, which is seen to involve both the optimum gain matrix $\hat{G}$ and the Kalman filter covariance matrix $\hat{P}$.

If the final time T is infinite, then, as in the case of noise-free observations, it is more meaningful to use the asymptotic cost rate

$$\bar{J} = \operatorname{tr}\{FVF'\bar{M} + \bar{G}'R\bar{G}\bar{P}\} \tag{11.138}$$

where $\bar{M}$, $\bar{G}$, and $\bar{P}$ refer to the steady state (algebraic) control and variance equations.

Example 11F Force-rebalanced accelerometer In Example 11B we considered the design of a Kalman filter for estimating the state in a "force-rebalanced" accelerometer for which the control law was obtained in Example 9B. In this example we shall study the behavior of the closed-loop system.

The complete block-diagram of the accelerometer is shown in Fig. 11.7, which is based on the dynamic model

$$\ddot{x} = a + u \tag{11F.1}$$

where a is the external acceleration and u is the control input, given by

$$u = -g_1\hat{x} + g_2\dot{\hat{x}} - \hat{a} \tag{11F.2}$$

where $\hat{x} = \hat{x}_1$, $\dot{\hat{x}} = \hat{x}_2$, and $\hat{a}$ are the estimated position, velocity, and acceleration of the proof mass. The control gains, as found in Example 9B, are

$$g_1 = c \qquad g_2 = \sqrt{2c} \tag{11F.3}$$

where c is the reciprocal of the control weighting and may be regarded as one of the design parameters.

The Kalman filter is given by

$$\begin{aligned}\dot{\hat{x}}_1 &= \hat{x}_2 + k_1(y - \hat{x}_1)\\ \dot{\hat{x}}_2 &= \hat{a} + u + k_2(y - \hat{x}_1)\\ \dot{\hat{a}} &= k_3(y - \hat{x}_1)\end{aligned} \tag{11F.4}$$

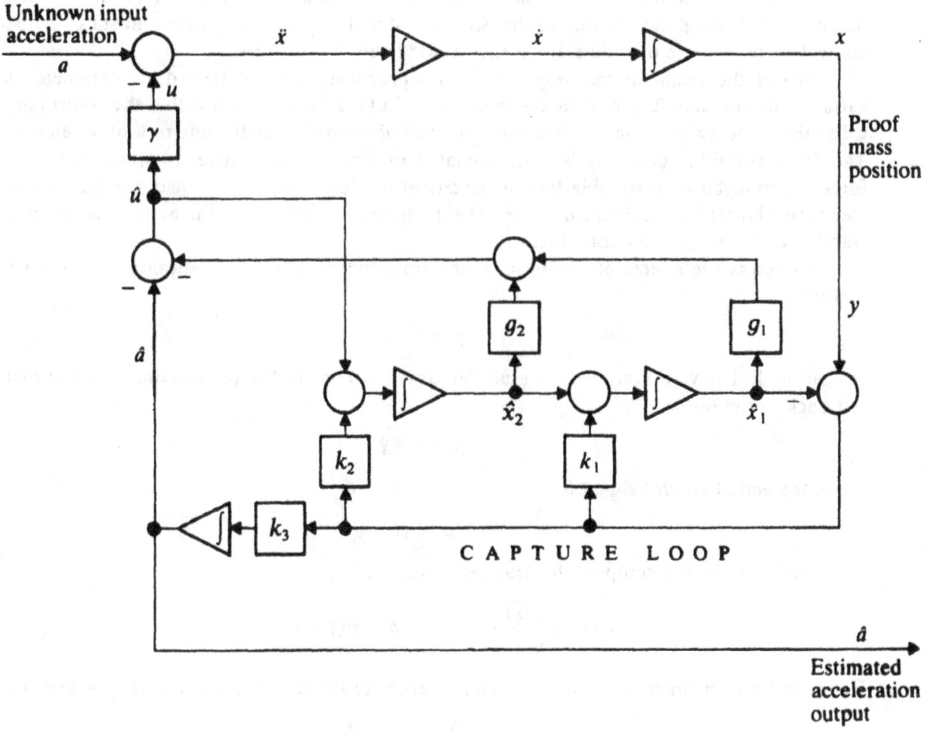

Figure 11.7 Force-rebalanced accelerometer with capture-loop dynamics.

where, as determined in Example 11B, the Kalman filter gain matrix is given by

$$K = \begin{bmatrix} k_1 \\ k_2 \\ k_3 \end{bmatrix} = \begin{bmatrix} 2\Omega \\ 2\Omega^2 \\ \Omega^3 \end{bmatrix} \qquad \Omega = \left(\frac{V}{W}\right)^{1/6} \tag{11F.5}$$

where V is the spectral density of the acceleration rate to be measured, and W is the spectral density of the noise in measuring the pick-off position. Thus, as noted in Example 11.B, Ω can be regarded as a measure of the signal-to-noise ratio.

Note that the Kalman filter (observer) in this application not only produces the feedback signal to keep the proof mass "captured," but also provides an estimate $\hat{a}$ of the unknown input acceleration a. To the extent that the input acceleration is a random walk and the noise in measuring the proof mass position is white noise, the estimate $\hat{a}$ is "optimum." Otherwise the parameters Ω and c are design parameters which can be varied to shape the dynamic characteristics of the accelerometer.

By the separation principle, the (nominal) closed-loop pole locations are the zeros of the characteristic polynomial for full-state feedback and the zeros of the characteristic polynomial of the Kalman filter. From Example 9B, the former are at

$$s = -\frac{\sqrt{2c}}{2}(1 \pm j) \tag{11F.6A}$$

and, from Example 11.B, the latter are at

$$s = -\Omega \qquad \text{and} \qquad s = -\Omega\left(\frac{1}{2} \pm j\frac{\sqrt{3}}{2}\right) \tag{11F.6B}$$

There are thus five poles in the closed-loop system having the nominal locations shown in Fig. 11.8. Three poles lie upon a circle of radius Ω: on the negative real axis and on rays at 60° angles to it. These poles are due to the Kalman filter. The poles due to the full-state feedback control lie on a circle of radius $2c$ on rays at 45° from the real axis.

One of the problems that might arise in a practical force-rebalanced accelerometer is a variation in the scale factor of the feedback loop. In (11F.1) we assumed that the control input u has the same scale as the acceleration: one unit of control force for one unit of acceleration. This ideal condition can only be approximated in a real accelerometer in which the control force is produced by a suitable type of electrical or electromagnetic transducer that converts electrical signals to mechanical force. The hallmark of quality in an accelerometer is the stability of the transducer scale factor.

To assess the effects of variation of the transducer scale factor, suppose that the scale factor is

$$\gamma = 1 + \varepsilon$$

instead of 1. The variation ε in the scale factor is not known. Hence the control signal that is fed back to the observer is

$$\hat{u} = -K\hat{x}$$

while the actual control signal is

$$u = \gamma\hat{u}$$

Let $\mathsf{H}_c(s)$ be the compensator transfer function, i.e.,

$$\mathsf{H}_c(s) = \frac{\mathsf{u}(s)}{\mathsf{y}(s)} = G(sI - A + BG + KC)^{-1}K \tag{11F.7}$$

Then the transfer function from the unknown acceleration a to the proof mass position y is

$$\mathsf{F}(s) = \frac{\mathsf{y}(s)}{\mathsf{a}(s)} = \frac{1/s^2}{1 + \gamma\mathsf{H}_c(s)/s^2} \tag{11F.8}$$

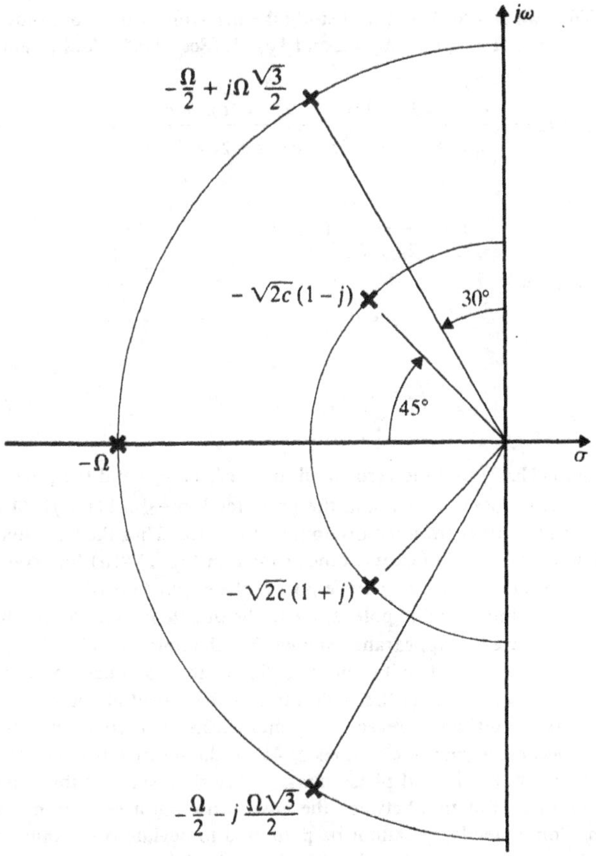

Figure 11.8 Closed-loop poles of force-rebalanced accelerometer.

Evaluating (11F.7) we find that

$$H_c(s) = \frac{(g_1k_1 + g_2k_2 + k_3)s^2 + (g_1k_2 + g_2k_3)s + g_1k_3}{s[s^2 + (k_1 + g_2)s + g_1 + k_2 + k_1g_2]} \tag{11F.9}$$

The characteristic polynomial for the system, on substituting (11F.9) into (11F.8), is

$$\begin{aligned} D(s) = s^5 &+ (k_1 + g_2)s^4 + (g_1 + k_2 + k_1g_2)s^3 \\ &+ \gamma[(g_1k_1 + g_2k_2 + k_3)s^2 + (g_1k_2 + g_2k_3)s + g_1k_3] \end{aligned} \tag{11F.10}$$

When $\gamma = 1$, (11F.10) factors into two terms:

$$D(s) = D_G(s)D_K(s) \tag{11F.11}$$

where $D_G(s) = s^2 + g_2s + g_1$ and $D_K(s) = s^3 + k_1s^2 + k_2s + k_3$. $D_G(s)$ is the characteristic polynomial of the full-state feedback system and $D_K(s)$ is the characteristic polynomial of the Kalman filter. This result is of course a consequence of the separation principle. [Verify (11F.11) by multiplying $D_G(s)$ by $D_K(s)$.]

When $\gamma \neq 1$, the closed-loop poles are given by the zeros of (11F.10). Let us study the behavior of the closed-loop poles by a root-locus analysis. From Fig. 11.8 we can see that the important parameter is the ratio of the radii of the two circles upon which the closed-loop poles lie (when $\gamma = 1$). Thus we can set $\Omega = 1$ and study the behavior of the closed-loop poles for $\gamma \neq 1$ as a function of c. For $\Omega = 1$, $k_1 = k_2 = 2$ and $k_3 = 1$. (See 11F.5.) And g_1 and g_2 are as given by (11F.3). Thus

$$\mathrm{H}_c(s) = \frac{(2c + 2\sqrt{2c} + 1)s^2 + (2c + \sqrt{2c})s + c}{s[s^2 + (2 + \sqrt{2c})s + c + 2 + 2\sqrt{2c}]} \tag{11F.12}$$

The zeros of $\mathrm{H}_c(s)$ occur at

$$s = -\frac{c}{2c + \sqrt{2c}}(1 \pm j) \tag{11F.13}$$

and the poles of $\mathrm{H}_c(s)$ occur at

$$s = 0$$

and at

$$s = -\left(\frac{2 + \sqrt{2c}}{2}\right)(1 \pm j) \tag{11F.14}$$

As $c \to 0$ we see from (11F.13) that the zeros tend to $-\sqrt{c/2}\,(1 \pm j)$ and the poles tend to $-1 \pm j$. As $c \to \infty$ the zeros tend to $-(1 \pm j)/2$ and the poles tend to $-\sqrt{c/2}\,(1 \pm j)$. Moreover, for all values of c the poles are farther from the origin than the zeros. Thus the "constellation" of open-loop poles and zeros will have the appearance shown in Fig. 11.9(a) for large c, or in Fig. 11.9(b) for small c. Note the presence of a *triple* pole at the origin. One of the poles is due to the compensator (see 11F.2) and a double pole is due to the double integration produced by the proof mass. The root loci have the appearance shown. For the nominal scale factor $\gamma = 1$, three poles lie on the unit circle (for $\Omega = 1$) and two lie on the 45° lines connecting the open-loop poles to the open-loop zeros. As the scale factor is increased above unity the loci approach the imaginary axis; further increase in γ makes the loci cross into the right half-plane and ultimately become asymptotic to lines at 45° to the positive real axis. It is thus apparent that the system has finite gain and phase margins. We shall see that the scale factor γ appears directly in the transfer function between the unknown input acceleration a and the estimated acceleration $\hat{a}$. Consequently, γ cannot be permitted to deviate from unity by more than a few percent even in an instrument of modest quality. Thus there is little danger that the closed-loop system will become unstable due to a change in γ.

A straightforward calculation produces the transfer function from the unknown input a to the estimated acceleration $\hat{a}$:

$$\frac{\hat{\mathrm{a}}(s)}{\mathrm{a}(s)} = \frac{k_3(s^2 + g_2 s + g_1)}{\mathrm{D}(s)} \tag{11F.15}$$

where $\mathrm{D}(s)$ is given by (11F.10).

When $\gamma = 1$, we found that $\mathrm{D}(s)$ factors into the product of $\mathrm{D}_G(s) = s^2 + g_2 s + g_1$ and $\mathrm{D}_K(s) = s^2 + k_1 s^2 + k_2 s + k_3$. Thus, when $\gamma = 1$, (11F.15) becomes

$$\frac{\hat{\mathrm{a}}(s)}{\mathrm{a}(s)} = \frac{k_3}{s^3 + k_1 s^2 + k_2 s + k_3} = \frac{\Omega^3}{s^3 + 2\Omega s^2 + 2\Omega^2 s + \Omega^3} \tag{11F.16}$$

which is completely independent of the control gains g_1 and g_2: The dynamics of the accelerometer depend entirely on the Kalman filter gains, which can be selected to provide whatever frequency response (bandwidth) is desired. Note also that the dc gain of the accelerometer is unity, so the accelerometer tracks an acceleration step with zero steady state error. (The control gains, however, must be chosen so that the proof mass is captured "tightly" enough; its excursion must not exceed the physical limits of motion under the largest acceleration that might be encountered during the operation of the instrument. The transfer function $\mathrm{F}(s)$ given by (11F.8) can be used to help determine these gains.)

When the scale factor $\gamma \neq 1$, then the more general transfer function (11F.15) must be used. Since, in practice $\gamma \approx 1$, the dynamic response will not be very much different from

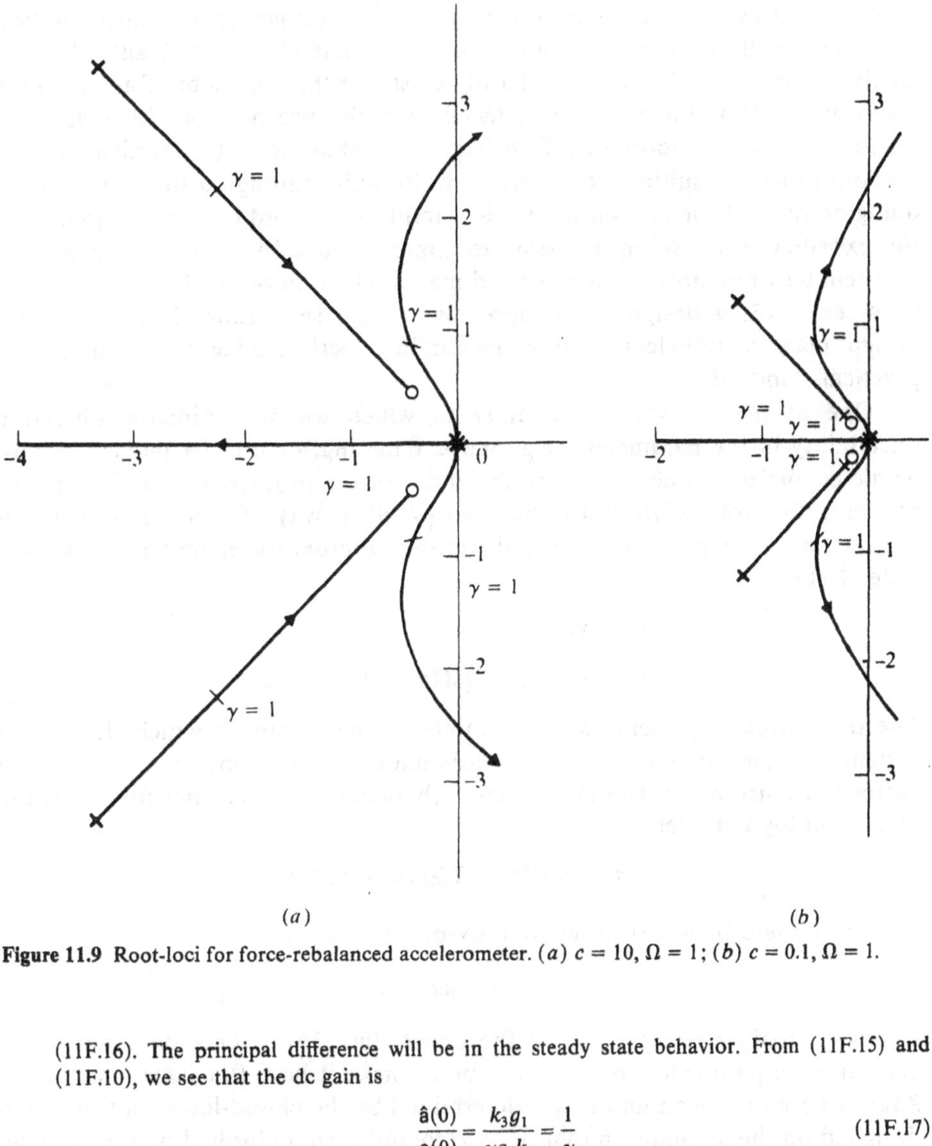

Figure 11.9 Root-loci for force-rebalanced accelerometer. (a) $c = 10, \Omega = 1$; (b) $c = 0.1, \Omega = 1$.

(11F.16). The principal difference will be in the steady state behavior. From (11F.15) and (11F.10), we see that the dc gain is

$$\frac{\hat{a}(0)}{a(0)} = \frac{k_3 g_1}{\gamma g_1 k_3} = \frac{1}{\gamma} \tag{11F.17}$$

Thus the scale factor error in producing the feedback signal is also the scale factor error of the instrument. An accelerometer of inertial navigation quality could be expected to have a scale factor error $\varepsilon = \gamma - 1$ of less than 10^{-4}. The scale factor error of a more modest quality accelerometer would be of the order of 1 percent.

11.8 CHOOSING NOISE FOR ROBUST CONTROL

White noise, for which the control law of the separation theorem is optimum, is doubtless present to some extent in every real system. But in addition to white noise, a real system has many other types of disturbances and uncertainties; compared to these, the white noise may be relatively insignificant. Unfortunately, the theoretical framework hardly exists for the treatment of anything but white noise. Thus the designer is faced with the prospect of either approximating all types of noise and disturbances as white noise (or, equivalently, as random process resulting from passing white noise through a linear system) or doing nothing. If the first option is distasteful, the second is even less palatable. For expedience it is often necessary to approximate all uncertainties present in a system by white noise. Once this is done, no claim to optimality can of course be asserted for a design based upon use of the separation theorem. Such a design may nevertheless deliver outstanding performance when judged by practical standards.

One of the types of uncertainty to which we have already alluded is uncertainty in the parameters (e.g., mass, damping, natural frequency, aerodynamic coefficients) that define the dynamics of the process. Uncertainty of this nature is far from white noise, but one practical way of dealing with it is to "cover" it with white (or correlated) noise. Suppose, for example in a second-order system

$$\dot{x}_1 = x_2$$

$$\dot{x}_2 = -\Omega^2 x_1 - 2\zeta_0 \Omega x_2 - 2\Delta\zeta x_2 + u$$

The term $2\Delta\zeta x_2$ represents an uncertainty in the dynamics which the control system designer might wish to approximate in some manner, perhaps by correlated noise of appropriate spectral characteristics and intensity. Thus one might employ a model

$$\dot{x}_2 = -\Omega^2 x_1 - 2\zeta_0 \Omega x_2 + x_3 + u$$

Where x_3 could be a first-order Markov process

$$\dot{x}_3 = -a x_3 + v$$

which approximates the effect of the uncertainty $2\Delta\zeta x_2$. One possible way of choosing the parameters of x_3 would be to make it have the same spectrum as $2\Delta\zeta x_2$ where the spectrum of x_2 is determined by the closed-loop control system designed on the assumption that $\Delta\zeta$ is zero and then multiplied by the expected value of $(\Delta\zeta)^2$. A correlated noise of this nature might "cover" the actual uncertainty $2\Delta\zeta x_2$ but can never have all the requisite statistical properties. For example $2\Delta\zeta x_2$ is completely correlated with x_2 but that correlation will be absent when the uncertainty is approximated by x_3. This method of covering uncertainty may not always be successful but no harm, and often much benefit, can come from trying it.

The use of white noise to cover process uncertainty can be theoretically justified on the basis of its improvement of the *robustness* of the resulting control system. This case is that in which the control (distribution) matrix B is uncertain.

In this case the assumption of fictitious white noise, in addition to any white noise that may actually be present, enhances the robustness and can, asymptotically, as the intensity of the fictitious noise tends to infinity, achieve the Doyle-Stein condition (8.74) described in Chap. 8. To see this, consider the open-loop process

$$\dot{x} = Ax + Bu + Fv_1 \tag{11.139}$$

Suppose that there is uncertainty as to the control distribution matrix. To cover this uncertainty, we suppose the presence of additional noise v_2 acting together with the control input. Then, (11.139) becomes

$$\dot{x} = Ax + B(u + v_2) + Fv = Ax + Bu + Fv_1 + Bv_2$$

The steady state variance equation for this observer is

$$0 = A\bar{P} + \bar{P}A' - \bar{P}C'W^{-1}C\bar{P} + FV_1F' + BV_2B' \tag{11.140}$$

where V_1 is the spectral density of the noise v_1 and V_2 is the spectral density of the noise v_2. We are shortly going to let the latter spectral density tend to infinity. For this purpose, let

$$V_2 = q^2V \tag{11.141}$$

Then (11.140) becomes

$$0 = A\bar{P} + \bar{P}A' - KWK' + FV_1F' + q^2BVB' \tag{11.142}$$

As $q \to \infty$ both $\bar{P}$ and K tend to infinity. But as shown in [8], P does not go to infinity as fast as K when $C(sI - A)^{-1}B$ has no right half-plane transmission zeros. Assuming this to be the case we conclude that

$$KWK' \to q^2BVB' \tag{11.143}$$

We can factor (11.143):

$$KW^{1/2}W^{1/2}K' \to (qBV^{1/2})(qV^{1/2}B') \tag{11.144}$$

and identify corresponding factors on the left and the right of (11.144). Thus, as the noise intensity multiplier q tends to infinity,

$$K \to qBV^{1/2}W^{-1/2} \tag{11.145}$$

This is the asymptotic gain that is used in the Doyle-Stein relation (8.74), the right-hand side of which is

$$K(I + C\Phi K)^{-1} \to qBV^{1/2}W^{-1/2}(I + C\Phi qBV^{1/2}W^{-1/2})^{-1} \to B(C\Phi B)^{-1}$$

$$\text{as} \qquad q \to \infty$$

Thus, by the use of an increasingly large noise in parallel with the input u, we can approach the robustness of the Doyle-Stein observer described in Chap. 8.

It is important to recognize that it may not be possible to use an observer gain given exactly by

$$K = qBV^{1/2}W^{-1/2}$$

because a gain matrix of this form may result in the observer closed-loop dynamics matrix $A_0 = A - KC$ not being a stability matrix. But if K is chosen as the solution of the variance equation (11.140) for any *finite* (albeit arbitrarily large) value of $V_2 = q^2 V$ *and the process is observable*, the corresponding gain matrix will necessarily result in a stable observer.

Example 11G Inverted pendulum control In the observer for control of an inverted pendulum that we considered in Example 11A, the B matrix and the F matrix were the same

$$B = F = \begin{bmatrix} 0 \\ 1 \end{bmatrix}$$

From Sec. 11.8, we know that the control law is made increasingly robust by allowing the spectral density of the excitation noise to become infinite. In this example we wish to verify this and study the closed-loop behavior.

In Example 11A we found the observer gain matrix

$$\hat{K} = \begin{bmatrix} \Omega\alpha \\ \Omega^2\alpha^2/2 \end{bmatrix} \tag{11G.1}$$

where

$$\frac{\alpha^2}{2} = 1 + \sqrt{1 + \frac{V}{\Omega^2 W}} \to \infty \quad \text{as} \quad \frac{V}{W} \to \infty \tag{11G.2}$$

Does the Doyle-Stein condition (8.74) hold? For this system

$$\Phi = (sI - A)^{-1} = \begin{bmatrix} s & -1 \\ -\Omega^2 & s \end{bmatrix}^{-1} = \frac{1}{s^2 - \Omega^2}\begin{bmatrix} s & 1 \\ \Omega^2 & s \end{bmatrix} \tag{11G.3}$$

To compute the right-hand side of (8.74) we need

$$C\Phi B = [1 \quad 0]\begin{bmatrix} s & 1 \\ \Omega^2 & s \end{bmatrix}\begin{bmatrix} 0 \\ 1 \end{bmatrix}\frac{1}{s^2 - \Omega^2} = \frac{1}{s^2 - \Omega^2} \tag{11G.4}$$

Then the right-hand side of (8.74) is

$$B(C\Phi B)^{-1} = \begin{bmatrix} 0 \\ 1 \end{bmatrix}(s^2 - \Omega^2) \tag{11G.5}$$

To compute the left-hand side of (8.74) we need:

$$I + C\Phi\hat{K} = 1 + [1 \quad 0]\begin{bmatrix} s & 1 \\ \Omega^2 & s \end{bmatrix}\begin{bmatrix} \Omega\alpha \\ \Omega^2\alpha^2/2 \end{bmatrix}\frac{1}{s^2 - \Omega^2}$$

$$= 1 + \frac{\alpha\Omega s + \alpha^2\Omega^2/2}{s^2 - \Omega^2} = \frac{s^2 - \Omega^2 + \alpha\Omega s + \alpha^2\Omega^2/2}{s^2 - \Omega^2}$$

Thus the left-hand side of (8.74) is

$$\hat{K}(I + C\Phi\hat{K})^{-1} = \begin{bmatrix} \alpha\Omega \\ \alpha^2\Omega^2/2 \end{bmatrix}\frac{s^2 - \Omega^2}{s^2 - \Omega^2 + \alpha\Omega s + \alpha^2\Omega^2/2} \tag{11G.6}$$

Thus as $\alpha \to \infty$, the left-hand side of (8.74) as given by (11G.6), approaches the right-hand side as given by (11G.5).

To study the behavior of the closed-loop system we use an observer with the gain matrix given by (11G.1) and the control gain matrix as designed in Example 9A:

$$G = [g_1, g_2] = [\beta^2\Omega^2/2, \beta\Omega] \tag{11G.7}$$

with

$$\frac{\beta^2}{2} = 1 + \sqrt{1 + \frac{c}{\Omega^2}}$$

The transfer function of the compensator is given by

$$\begin{aligned} \mathsf{H}_c(s) &= \hat{G}(sI - A + B\hat{G} + \hat{K}C)^{-1}\hat{K} \\ &= [\beta^2\Omega^2/2, \beta\Omega]\begin{bmatrix} s + \alpha\Omega & -1 \\ -\Omega^2\left(1 - \dfrac{\alpha^2 + \beta^2}{2}\right) & s + \beta\Omega \end{bmatrix}^{-1}\begin{bmatrix} \alpha\Omega \\ \alpha^2\Omega^2/2 \end{bmatrix} \\ &= \frac{\Omega^3\alpha\beta\left[\dfrac{\alpha + \beta}{2}s + \Omega\left(1 + \dfrac{\alpha\beta}{4}\right)\right]}{(s + \alpha\Omega)(s + \beta\Omega) - \Omega^2\left(1 - \dfrac{\alpha^2 + \beta^2}{2}\right)} \end{aligned} \tag{11G.8}$$

(Note that the compensator transfer function is completely symmetric with respect to α and β. Could this have been anticipated?)

The denominator of (11G.8) is

$$\mathsf{D}_c(s) = s^2 + (\alpha + \beta)\Omega s + \Omega\left(-1 + \alpha\beta + \frac{\alpha^2 + \beta^2}{2}\right)$$

with roots at

$$s = -\frac{\Omega}{2}[(\alpha + \beta) \pm \sqrt{4 - (\alpha + \beta)^2}]$$

which are two of the open-loop poles of the system. The other open-loop poles are due to the dynamics of the inverted pendulum itself and are at

$$s = \pm\Omega$$

The numerator has a zero at

$$s = \frac{2\Omega(1 + \alpha\beta/4)}{\alpha + \beta}$$

Let us investigate how the range of stability depends on the parameters α and β of the compensator. With no loss in generality we can take $\Omega = 1$. The limiting case occurs when the control "availability" parameter c and the "signal-to-noise" ratio are both zero. This in a sense is the worst possible case: The cost of control is very high and the quality of the sensor is very low so that its noise W is much larger than V. This sets the lower limit upon performance. In this case $\alpha = \beta = 2$ and the compensator transfer function is, by (11G.8),

$$\mathsf{H}_c(s) = \frac{8(s + 1)}{(s + 2)^2 + 3} \tag{11G.9}$$

Thus there is a zero at $s = -1$, on top of the stable pole of the pendulum, and poles at $s = -2 \pm j\sqrt{3}$. The loop transmission is

$$\mathsf{F}(s) = \mathsf{H}_c(s)\frac{1}{s^2 - 1} = \frac{8}{[(s + 2)^2 + 3][s - 1]}$$

The return difference in this case is given by

$$\mathsf{T}(s) = 1 + K\mathsf{H}_c(s)\frac{1}{s^2 - 1} = 1 + K\frac{8}{(s^2 + 4s + 7)(s - 1)} \tag{11G.10}$$

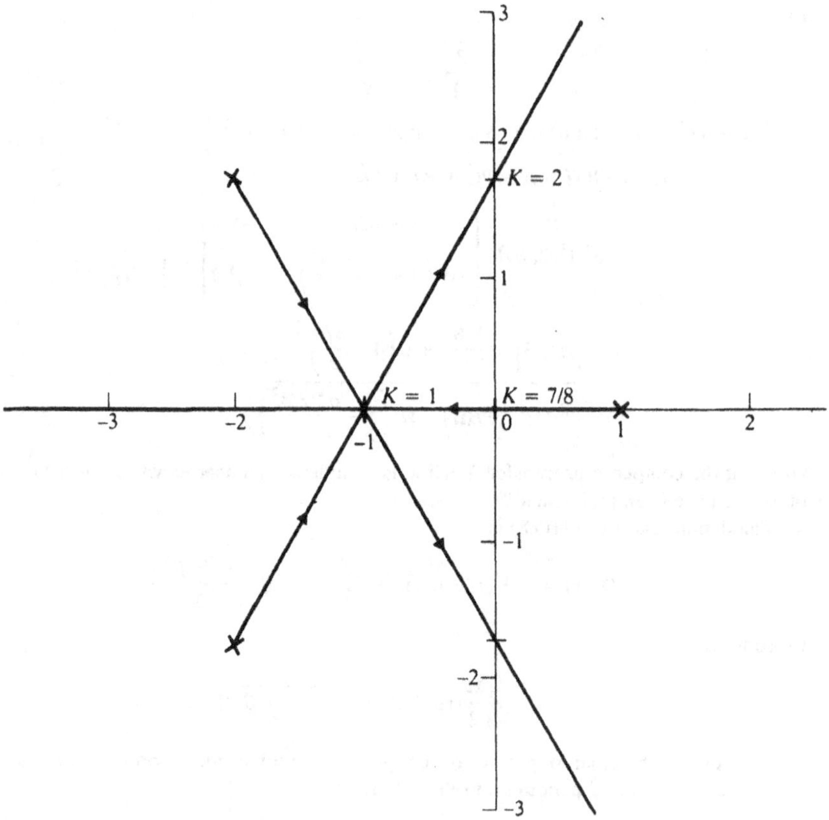

Figure 11.10 Root locus for controlled inverted pendulum ($\alpha = \beta = 2$).

The root locus, as shown in Fig. 11.10, corresponding to (11G.10), consists of three lines intersecting at $s = -1$ (when $K = 1$). Since $K = 1$ is the nominal value of the loop gain, it implies that the three closed-loop poles are all at $s = -1$, which is consistent with the results of Examples 9A and 11A. (There should also be a fourth pole at $s = -1$, but that pole is cancelled by the zero.) A Routh-Hurwitz analysis readily establishes that the range of gain for stability is

$$\tfrac{7}{8} < K < 2$$

The margin for gain reduction in this case is uncomfortably small. The margin for gain increase is better, but still not much to brag about.

More favorable margins are to be expected when the signal-to-noise ratio (or the control availability parameter c) is higher. Suppose, in particular, that β remains at its minimum value of 2 as α (and hence V/W) is increased. In this case (11G.8) gives

$$\mathsf{H}_c(s) = \frac{2\alpha(1 + \alpha/2)(s + 1)}{(s + \alpha)(s + 2) + (1 + \alpha^2/2)} \tag{11G.11}$$

There is still a zero at $s = -1$ which falls on top of the stable pendulum pole. The compensator poles occur at the zeros of the denominator of (11G.11), i.e., at

$$s = -1 - \frac{\alpha}{2} \pm j\frac{1}{2}\sqrt{\alpha^2 + 4\alpha} \tag{11G.12}$$

Note that as $\alpha \to \infty$, the compensator poles tend to $-(\alpha/2)(1 \pm j1)$. (This was to be expected in view of earlier results. See Note 9.4.) A Routh-Hurwitz stability analysis using (11G.11) gives the stability region:

$$\frac{\alpha^2/2 + 2\alpha + 1}{\alpha^2 + 2\alpha} = K_{\min} < K < K_{\max} = 1 + \frac{\alpha}{2} \tag{11G.13}$$

The lower limit $K_{\min}$ of the stability range tends to 1/2. (Since the open-loop system is unstable, the closed-loop system must of necessity be only conditionally stable.) The upper limit $K_{\max}$ of the stability range, however, becomes infinite as $\alpha \to \infty$. Thus it is possible to achieve as large a gain (increase) margin as desired by assuming a sufficiently large value of V/W.

PROBLEMS

Problem 11.1 Instrument servo

A compensator based on a Kalman filter is to be designed for the instrument servo of Prob. 9.6. Only the position error e is measured, so that

$$y = e + w$$

where w is white noise with spectral density W. The only excitation noise present occurs at the control input, so that the angular velocity is

$$\dot{\omega} = -\alpha\omega + \beta u + v$$

where v is white noise of spectral density V.

(*a*) Find and tabulate or plot the Kalman filter gains and corresponding closed-loop poles as a function of the signal-to-noise ratio V/W.

(*b*) Using the optimized gains determined in Prob. 9.6, part *a*, for several values of q_1^2, and several values of V/W, find gain margin, i.e., the range of gain variation for which the closed-loop system remains stable. Tabulate the results as functions of q_1^2 and V/W.

Problem 11.2 Inverted pendulum on cart: compensator for single output

A compensator for the inverted pendulum on the motor-driven cart of Prob. 3.6 et seq. is to be designed using the gains of Prob. 9.10 and a full-order Kalman filter as an observer.

(*a*) Assume that the only excitation noise present is coincident with the control and has the spectral density v^2, and that the only observation is the cart displacement, which is measured through white noise of spectral density w^2. Determine and plot or tabulate the Kalman filter gains and poles as a function of the ratio v^2/w^2 for $1 \leqq v^2/w^2 \leqq 10^6$.

(*b*) Using the regulator gain matrix of Prob. 9.10, part *a*, with $r^2 = 0.01$, determine the compensator transfer function $\mathsf{D}(s)$ for the range of v^2/w^2 in part *a*. Note that the compensator has *poles* in the right half-plane.

(*c*) For $r^2 = 0.01$ and $v^2/w^2 = 10^{-3}$ in part *b*, determine that the range of K for which the closed-loop system, having the return difference $\mathsf{T}(s) = 1 + K\mathsf{D}(s)\mathsf{H}(s)$ (where $\mathsf{H}(s)$ is the transfer function of the plant), is stable.

Problem 11.3 Inverted pendulum on cart: two outputs

The unstable compensator of Prob. 11.2 is undesirable for various reasons, one of which is that the compensator must not be turned on without the loop being closed. To achieve better performance it is proposed to measure the angular displacement of the pendulum as well as the linear displacement of the cart. Thus we have $y_1 = x_1$ and $y_2 = x_3 = \theta$ for outputs. Let the output spectral density matrix be $W = w^2 I$.

(*a*) For $1 \leq v^2/w^2 \leq 10^6$ find the closed-loop poles and the corresponding gain matrices.

(*b*) For the same range of v^2/w^2 as in part *a* and with the regulator gain matrix of Prob. 9.10, part *a*, with $r^2 = 10^{-2}$, find the poles of the compensator. Is it possible to achieve a stable compensator by a proper choice of v^2/w^2?

(*c*) For $v^2/w^2 = 10^5$, with $r^2 = 10^{-2}$ for the regulator, find the range of gains K for which the return difference $T(s) = 1 + KD(s)H(s)$ (where $D(s)$ and $H(s)$ are the compensator and plant transfer functions, respectively) has its zeros in the left half-plane.

Problem 11.4 Temperature control: Kalman filters

Consider the temperature control system for which the control gains were determined in Example 9C. A compensator is to be designed using a full-order Kalman filter to estimate the state using a measurement $y = x_1 + w$ where w is white noise of spectral density W.

(*a*) Assume that the exogenous temperature x_0 is a Wiener process, that is, $\dot{x}_0 = v$ where v is white noise of spectral density V. Determine the Kalman filter gains and poles as a function of the "signal-to-noise" ratio V/W.

(*b*) Using the gains determined in Example 9C for $k = 10^{-2}$ find the transfer function of the compensator. Calculate and tabulate the gain margins as functions of V/W.

Problem 11.5 Three-capacitor thermal system

A compensator for the three-capacitor thermal system of Prob. 3.7 et seq. is to be designed using the optimized gains as determined in Prob. 9.16 and a Kalman filter as the observer. To design the Kalman filter, assume that the reference temperature $\bar{v}$ and the ambient temperature are independent Wiener processes

$$E\left\{\begin{bmatrix} \bar{v}(t) \\ v_0(t) \end{bmatrix}[\bar{v}(\tau), v_0(\tau)]\right\} = V\begin{bmatrix} 1 & 0 \\ 0 & 1 \end{bmatrix}\delta(t-\tau)$$

(*a*) Assume that the noise in measurement of temperature is white noise with a spectral density of W. Find and tabulate (or plot) the Kalman filter gains and pole locations as the signal-to-noise ratio V/W varies from 10^{-4} to 1.0.

(*b*) Determine the compensator transfer function obtained by combining the control gains of Prob. 9.16 with the Kalman filter gains of part *a*. The results can be arranged in a table as follows:

Control weighting	Signal-to-noise ratio	Compensator transfer function
10^{-6}	10^{-4}	
	10^{-2}	
	1	
10^{-4}	10^{-4}	
	10^{-2}	
	1	
10^{-2}	10^{-4}	
	10^{-2}	
	1	

(*c*) Determine the gain and phase margins for several of the combinations studied in part *b*.

Problem 11.6 Aircraft stability augmentation: compensator

A compensator for stability augmentation of the aircraft considered in Example 8C et seq. is to be designed based on the separation theorem.

(*a*) Design the Kalman filter: Assume that the process excitation is white noise coincident with the control input, and has a spectral density V, and that the only observation is of the pitch rate $q = x_3$ which is corrupted by noise of spectral density W. Find and tabulate the Kalman filter gains and closed-loop poles as a function of V/W for $10^{-4} \leq V/W \leq 10^4$.

(*b*) Using the control gains determined in part *b* of Prob. 9.14, find the transfer function of the compensator and the gain and phase margins as a function of V/W in the range of part *a*.

(*c*) Discuss the similarities and differences of the compensator designed in this problem as compared to the compensator described in Example 8C.

Problem 11.7 Aircraft lateral dynamics: Kalman filter

Design a Kalman filter for the aircraft lateral dynamics (Prob. 8.9 et seq.), using the B matrix as the noise distribution matrix (i.e., assume that the noise exciting the process enters at the control points) with a noise covariance matrix $V = v^2 I$. The observations are yaw rate r and the roll rate p, with equal amounts of noise on each, i.e., $W = w^2 I$.

(*a*) Determine the gains and pole locations of the Kalman filter as a function of the "signal-to-noise" ratio v^2/w^2, for the case in which ϕ_0 is to be regarded as a determined quantity.

(*b*) Modify the Kalman filter of part *a* to include the estimation of ϕ_0.

Problem 11.8 Aircraft lateral dynamics: compensators

A compensator for the lateral motion of the aircraft studied in Prob. 11.7 is to be designed by combining the compensator designed in that problem with the optimum linear quadratic design obtained in Prob. 9.15.

(*a*) Draw the block-diagram of the compensator.

(*b*) Using the control gains found in Prob. 9.15, find the transfer function of the compensator in terms of the ratio of v^2/w^2 in the range studied in Prob. 11.7.

(*c*) For several of the values of v^2/w^2 in part *b*, plot the minimum and maximum singular values of the return difference $T_c(s)$, and use these results to estimate the gain and phase margins of the system. (See Sec. 4.9.)

Problem 11.9 Aircraft lateral dynamics: reduced-order Kalman filter

A reduced-order Kalman filter is to be designed for the aircraft lateral dynamics—See Prob. 11.7—on the basis of the assumption that the sensor noise covariance matrix W is zero.

(*a*) Design the reduced-order (i.e., second-order) filter assuming ϕ_0 is a known quantity. Compare the result with that obtained in Prob. 11.7 as $v^2/w^2 \to \infty$ in the latter.

(*b*) Repeat part *a* for the third-order filter that includes the estimation of ϕ_0.

Problem 11.10 Velocity-aided inertial navigation

Consider the velocity-aided inertial navigation system described in Example 11C.

(*a*) Assume that the accelerometer noise spectral density V_A is negligible. Find the steady state covariance submatrix

$$\begin{bmatrix} p_4 & p_5 \\ p_5 & p_6 \end{bmatrix} = E\left\{\begin{bmatrix} \Delta v^2 & \Delta v\, \Delta\phi \\ \Delta v\, \Delta\phi & \Delta\phi^2 \end{bmatrix}\right\}$$

in terms of the ratio V_G/W.

(*b*) Draw the block-diagram of the Kalman filter.

(*c*) Find the gains and closed-loop poles of the Kalman filter.

(*d*) Compare the position and velocity errors of the velocity-aided system with those of the unaided system as determined in Prob. 10.3.

Problem 11.11 Two-axis gyro: Kalman filter and compensator

Consider the two-axis gyro of Prob. 6.9 et seq. Suppose that there are only two vector noise sources present. One source of noise generates the external angular velocity:

$$\dot{\omega}_{xE} = v_x$$

$$\dot{\omega}_{yE} = v_y$$

where v_x and v_y are independent white noise processes having the same spectral density, i.e.,

$$E\left\{\begin{bmatrix} v_x(t) \\ v_y(t) \end{bmatrix}[v_x(\tau), v_y(\tau)]\right\} = v\begin{bmatrix} 1 & 0 \\ 0 & 1 \end{bmatrix}\delta(t-\tau)$$

(As discussed in Chap. 10, this makes ω_x and ω_y Wiener processes.)

The other (vector) source of noise is on the output of the position "pick-offs," the outputs of which are

$$\bar{\delta}_x = \delta_x + w_x$$

$$\bar{\delta}_y = \delta_y + w_y$$

where w_x and w_y, like v_x and v_y, are also independent white noise processes:

$$E\left\{\begin{bmatrix} w_x(t) \\ w_y(t) \end{bmatrix}[w_x(\tau), w_y(\tau)]\right\} = w\begin{bmatrix} 1 & 0 \\ 0 & 1 \end{bmatrix}\delta(t-\tau)$$

Since the noise process $[v_x, v_y]$ generates the angular velocity components that the gyro is to measure, it is really the "signal" to the gyro. Thus it is appropriate to identify the ratio v/w as the "signal-to-noise ratio" of the system.

(*a*) For the state vector $x = [\delta_x, \delta_y, \omega_{xB}, \omega_{yB}, \omega_{xE}, \omega_{yE}]'$ which the Kalman filter is to estimate, find and plot the elements of the Kalman gain matrix and the corresponding closed-loop poles, using the numerical data given in Prob. 6.9, with signal/noise ratios in the range $[0, 10^4]$.

(*b*) Draw the block-diagram of the control system showing the control outputs $u_x = \tau_x/J_D$, $u_y = \tau_y/J_D$ and the estimates $\hat{\omega}_{xE}$ and $\hat{\omega}_{yE}$ of the input angular velocity.

Problem 11.12 Distillation column: full-order Kalman filter

The compensator for the distillation column of Example 6D and Prob. 9.17 is to be designed by using a full-order Kalman filter as an observer.

(*a*) Assume that the disturbances represented by the exogenous vector are white noise with a 2×2 spectral density matrix $V = v^2 I$. The observations are of the temperature differences $y_1 = \Delta T_1$ and $y_2 = \Delta T_2$, with the observation matrix C as given by (2G.5) of Chap. 2. Assume that the 2×2 spectral density matrix of the temperature measurements is $W = w^2 I$. Find and tabulate or plot the Kalman filter gains and corresponding observer poles as the "signal-to-noise" ratio v^2/w^2 varies from 0.001 to 100.

(*b*) Using the regulator gains calculated in Prob. 9.17 for $r^2 = 1.0$, find the compensator transfer functions and estimate the gain and phase margins using (4.87) of Sec. 4.10.

(*c*) A gain margin of at least 3 is desirable for this application. If the estimates of the gain margin as determined in part *b* are less than 3, assume the presence of white noise of spectral density $q^2 I$ at the control input. Increase the value of q^2, according to the theory of Sec. 11.8, until the desired gain margin of 3 is achieved.

Problem 11.13 Double-effect evaporator: compensator design

The design which was started in Prob. 9.18 for the compensator for the double-effect evaporator will be completed now.

(*a*) Design the Kalman filter for the process using the observations x_3, x_4, and x_5, assuming that the three disturbances d_1, d_2, and d_3 are white noise with equal spectral density W. Find and tabulate the Kalman filter gains and filter poles for the following values of the signal-to-noise ratio: $V/W = 10^2, 10^3, 10^4$.

(*b*) Select a suitable value of V/W from part *a* and a gain matrix from part *a* of Prob. 9.18, and determine the transfer function (matrix) of the compensator obtained by combining these. Analyze the robustness of the design by computing the singular values of the return difference.

(*c*) A gain margin of at least 3 (9.5 dB) would be appropriate for this type of system. Does the design of part *b* appear to achieve this margin? (Use the bounds given by (4.88) of Sec. 4.10.) If the gain margin is too low, assume the presence of white noise of spectral density V_2 at each of the control inputs. For several values of V_2, recompute the Kalman filter gains and the corresponding compensator transfer functions. Then repeat the singular-value analysis to determine the value of V_2 and the corresponding compensator transfer function that achieves the desired gain margin.

Problem 11.14 Distillation column: estimation of disturbances

It is desired to achieve improved disturbance rejection in the distillation column control system by assuming that the disturbances each consist of two components: a white noise component as already considered in Prob. 11.12, and, in addition, a Wiener process component $\bar{x}_0$ (the integral of white noise, as explained in Chap. 10). Thus, for the purpose of designing the Kalman filter we use the metastate vector $\mathbf{x} = [x', \bar{x}_0']'$ which has six components. The Wiener process component $\bar{x}_0$ satisfies $\dot{\bar{x}}_0 = \bar{v}$, where $\bar{v}$ is a white noise process of spectral density $\bar{V} = \bar{v}^2 I$.

(*a*) Write the differential equations that describe the compensator and draw the block-diagram corresponding to it.

(*b*) Study the variation of the Kalman filter poles as obtained in Prob. 11.12 as the white noise $\bar{v}$ responsible for the Wiener process component of the disturbance increases in intensity.

(*c*) Find the transfer function of the compensator (which is now of sixth order) using the regulator gains of Prob. 9.17 with $r^2 = 1.0$ and the corresponding value of G^*, combined with the Kalman filter gains for $v^2/w^2 = 1.0$ and $\bar{v}^2/w^2 = 0.01$.

(*d*) Investigate the robustness of the design by finding the singular values of the return difference.

Problem 11.15 Distillation column: reduced-order Kalman filter

A compensator for the distillation column of Prob. 11.12 based on a reduced-order Kalman filter is to be designed under the assumption that the temperature measurements are noise-free, i.e., that the spectral density matrix W of the observations is zero. The exogenous disturbances are to be treated as white noise.

(*a*) Write the equations (algebraic and differential) that define the compensator that uses the reduced-order Kalman filter.

(*b*) Find the transfer function $\mathsf{D}(s)$ of the compensator from the measured output y to the control u. Compare this transfer function with that obtained in Prob. 11.12 as $w \to 0$.

(*c*) Calculate the return difference $\mathsf{T}(s) = I + \mathsf{D}(s)\mathsf{G}(s)$ and discuss the robustness of the closed-loop system.

Problem 11.16 Constant-altitude autopilot: altimeter is only sensor

A Kalman filter is to be used as the compensator for the constant-altitude autopilot considered in Prob. 6.6 et seq. The only sensor is an altimeter having an output $y = h/V + w$ where w is white noise with spectral density W.

(*a*) Assume that the only excitation noise present is coincident with the control input and has a spectral density $\bar{V}$ (not to be confused with the velocity V of the aircraft). Find and plot the Kalman filter gains and observer poles for the following range of "signal-to-noise" ratios: $1 \leqq \bar{V}/W \leqq 10^6$.

(*b*) For the $\bar{V}/W = 10^2, 10^4, 10^6$ in part *a*, and the feedback control gains determined in Prob. 9.13, with $c_1^2 = c_2^2 = 1$, and $r^2 = 0.01$, find the transfer function of the compensator, and find the range of loop gains for which the closed-loop system remains stable.

Problem 11.17 Constant-altitude autopilot: turbulence excitation

The compensator for the constant-altitude autopilot is to be designed to provide immunity to wind turbulence, which may be assumed to have a Dryden spectrum and can be modeled by a second-order system excited by white noise. The effect of the turbulence is to alter the angle of attack of the aircraft, i.e.,

$$\dot{\alpha} = q + (Z_\alpha/V)\alpha + (Z_\delta/V)\delta + (Z_\alpha/V)\alpha_W$$

where α_W is the change in angle of attack due to wind turbulence which is equal to the output of the system shown in Fig. 10.5. (The turbulence intensity is adjusted by the intensity of the white noise at the input in Fig. 10.5.)

(*a*) Using a turbulence time constant $T = 1.0$ s, find the control gains for the additional two state variables that model the turbulence using the performance criterion of Prob. 9.13 with $c_1^2 = c_2^2 = 1.0$, and $r^2 = 0.01$.

(*b*) Find the Kalman filter gains, assuming that the wind turbulence is a (correlated) random input in addition to the white noise coincident with the control input. Perform the calculation for $\bar{V}/W = 10^2, 10^4, 10^6$, and for several intensities of turbulence.

(*c*) For each of the signal-to-noise ratios $\bar{V}/W$ and turbulence intensities of part *b*, find the compensator transfer function and the range of gains for which the closed-loop system is stable. How does increasing $\bar{V}/W$ affect the gain margin of the system?

Problem 11.18 Constant altitude autopilot with rate gyro added

It is proposed to add a rate gyro to the altimeter in the constant altitude autopilot of Prob. 11.16. There are now two outputs

$$y_1 = h/V + w_1 \qquad y_2 = q + w_2$$

(*a*) Assume again that the only excitation present is white noise at the control input. Find the Kalman filter gains and observer poles for $\bar{V}/W_1 = 1, 10, 100$ and $\bar{V}/W_2 = 1, 10, 100$. Does the presence of the rate gyro significantly alter the dynamic characteristics of the observer?

(*b*) Find the permissible (for stability) range of gain variation (at the autopilot input) for each of the signal-to-noise ratios in part *a*. Does the rate gyro significantly enhance the stability of the autopilot?

Problem 11.19 Inertial navigation system with position and velocity aids

In addition to a velocity aid, it is also possible to obtain independent position measurements by which the inertial navigation system can be updated. In a realistic application, these position "fixes" would be intermittent; if they were sufficiently accurate, the role of the inertial navigation system would be to interpolate between position fixes and to aid in obtaining "smoothed" velocity estimates. Optimum filtering of intermittent or sampled data is well known (see [8], for example) but beyond the scope of this book. If the position updates are frequent (relative to the 84-minute Schuler period) they may be approximated as a continuous updating process in which the error is white noise with a spectral density $W_x = \sigma_x^2 T$, where σ_x is the one-sigma position error of each position measurement and T is the time interval between position updates.

(*a*) Draw the block-diagram of the Kalman filter.

(*b*) Assume that the velocity aid is absent. Find the steady state Kalman filter gains and closed-loop poles as a function of the ratio V_G/W_x. (The accelerometer noise is negligible.)

(*c*) A velocity aid is contemplated. How low must the spectral density of the velocity noise be in order that the steady state position error be significantly reduced (i.e., by a factor of 2)?

Problem 11.20 Hydraulically actuated gun turret

A Kalman filter is to be used as an observer in realizing the compensator for the azimuth channel of the hydraulically actuated gun turret of Prob. 9.11.

(*a*) For purpose of design the exogenous vector x_0 is to be represented as white noise with a diagonal spectral density matrix with elements made proportional to the square of the expected reference or disturbance amplitudes. On the basis of the data given in Ref. [1] of Chap. 2 it would thus be appropriate to use a spectral density matrix

$$V = \left[\begin{array}{c|ccc} 1 & 0 & 0 & 0 \\ \hline 0 & 10^{-4} & 0 & 0 \\ 0 & 0 & 10^{-4} & 0 \\ 0 & 0 & 0 & v^2 \end{array}\right]$$

where v^2 is a design parameter which can be varied for purposes of achieving robustness. (Figure 2.9 shows that d_q, the fourth component of the exogenous vector, occurs at the same point of the control input.) The observed quantity is the gun azimuth error ($y = \theta - \theta_0 = x_1$) with spectral density matrix $W = w^2$. Find and tabulate the Kalman filter gains and poles as the ratio v^2/w^2 is varied from 1 to very large values.

(*b*) For the range of filter gains studied in part *a* find the transfer function of the compensator using the control gains determined in Prob. 9.11 with $q^2 = 1000$.

(*c*) Find the return difference and corresponding stability (gain and phase) margins for three of the values of v^2/w^2 studied in part *a*. Verify that the margins increase as v^2/w^2 is increased.

(*d*) Assume that the reference input θ_0 and d_τ are Wiener processes:

$$\dot{\theta}_0 = \bar{v}_1 \qquad \dot{d}_\tau = \bar{v}_3$$

where $\bar{v}_1$ and $\bar{v}_3$ are white noise processes each with spectral density of unity. Adjoin these two dynamic variables to the original four to obtain a six-component metastate vector. For the range of v^2/w^2 studied in part *a*, find the Kalman filter gains, and determine and tabulate or plot the corresponding closed-loop poles.

(*e*) Draw the block-diagram of the compensator and discuss the results for the filter model of part *d*.

Problem 11.21 Kalman filter gains depend on "signal-to-noise ratio"

Consider the matrix Riccati equation for the error covariance matrix in the Kalman filter

$$\dot{\hat{P}} = A\hat{P} + \hat{P}A' - \hat{P}C'W^{-1}C\hat{P} + FVF'$$

Suppose the observation noise covariance matrix V and excitation noise covariance matrix W both are multiplied by the *same* positive constant, say α.

Show that although $\hat{P}$ changes, the Kalman filter gain $\hat{K} = \hat{P}C'V^{-1}$ does not change. (This justifies the claim that the Kalman filter gains depend only on the "ratio" of W to V.)

Is this result reasonable? (Consider the dual problem of optimum deterministic control.)

Problem 11.22 Missile autopilot: Kalman filter

A compensator is to be designed for the missile autopilot considered in Example 9D. The sensors consist of an accelerometer which measures $e = a_{NC} - a_N$ and a rate gyro that measures the pitch rate q. The only excitation noise present is coincident with the control u and has a spectral density V. The spectral density of the accelerometer noise is W_A and the spectral density of the gyro noise is W_G.

(*a*) Assume that the commanded acceleration a_{NC}, an exogenous variable, can be measured. Draw the block-diagram of the Kalman filter. Find and plot or tabulate the filter gains and poles for $1 \leq V/W_A \leq 10^6$, $1 \leq V/W_G \leq 10^6$.

(*b*) How must the Kalman filter design of part *a* be modified if a_{NC}, assumed to be constant, must be estimated?

Problem 11.23 Missile autopilot: compensator

A compensator for the missile of Prob. 11.22 is to be designed by combining the full-state feedback controller of Example 9D with the Kalman filter of Prob. 11.22.

(*a*) Using the controller gains of Example 9D with $R = 10^7$, find the compensator transfer functions for the range of V/W_A and V/W_G considered in Prob. 11.22.

(*b*) A gain margin of at least 10 is needed. What values of V/W_A and V/W_G are required to achieve this margin?

NOTES

Note 11.1 Background of Kalman filtering

The basic idea of least-squares filtering is traceable to Gauss,[12] the fountainhead of many concepts in pure and applied mathematics. But the use of statistical methods in control and communications systems is largely due to the work of Norbert Wiener[13]. Wiener was concerned with the separation of a signal from a background of noise in which it is imbedded. In his analysis, Wiener assumed that the signal and the background noise had statistical properties that were characterized by their correlation functions. He expressed the *impulse response* of the optimum filter in the form of an integral equation—the famous Wiener-Hopf equation—involving these correlation functions. A frequency-domain interpretation of Wiener's results was later given by Bode and Shannon[14] who provided a way of obtaining the *transfer function* of the optimum filter as the tandem combination of two other transfer functions, the first of which transforms the observed data into white noise and the second of which is the optimum filter for a white noise input.

Statistical optimization of control systems by frequency-domain methods was a popular research topic of the 1950s, but it held little interest for Kalman who was more concerned with the state-space representation of deterministic systems and their properties (See Note 1.1). In 1958, however, he turned his attention to random processes. His familiarity with state-space methods (which he was instrumental in developing) made it natural for him to suggest that stochastic processes should be characterized neither by their correlation functions nor by their spectral densities, but rather as the responses of linear systems to white noise. Pursuing this idea, he discovered that the optimum estimation problem was "dual" (Kalman's expression[15]) to the deterministic optimum control problem. His fundamental papers[1, 2] on the state-space approach to filtering and prediction followed directly.

Kalman was concerned with the optimum filter in the probabilistic sense. But it was subsequently shown that his results (in the discrete-time case, at least) could be interpreted as the solution to the deterministic least-squares problem that had occupied the interests of applied mathematicians and scientists since the time of Gauss. (See [12].)

Note 11.2 $\tilde{V}$ is positive semidefinite

A well-known criterion for a matrix to be positive semidefinite is that its determinant and the determinants of all its principal minors be nonnegative. Thus

$$Z = \begin{bmatrix} V & X \\ X' & W \end{bmatrix}$$

is positive semidefinite if and only if

$$|V| \geqq 0 \qquad |W| \geqq 0 \qquad \begin{vmatrix} V & X \\ X' & W \end{vmatrix} \geqq 0$$

Now, by writing $\tilde{V} = V - XW^{-1}X'$, we assume that W^{-1} exists. Hence $|W| \geqq 0$. There is a well-known determinantal formula[10] which asserts that

$$\begin{vmatrix} A & B \\ C & D \end{vmatrix} = |D||A - BD^{-1}C|$$

Using this formula with $V = A$, $B = X$, $C = X'$, and $D = W$ gives

$$\begin{vmatrix} V & X \\ X' & W \end{vmatrix} = |W||V - XW^{-1}X'| \geqq 0 \qquad (*)$$

Since $|W| \geqq 0$ we can divide both sides of (*) by W to obtain

$$|\tilde{V}| = |V - XW^{-1}X'| \geqq 0$$

which implies that $\tilde{V}$ is positive definite.

Note 11.3 "Innovations" terminology

The introduction of the terms "innovations" or "innovations process" in the context of Kalman filtering is due to Kailath,[11] although its use in mathematical statistics goes back farther. The idea underlying the term is that the new information used to "update" the estimate of the state is contained in the innovations process $r = y - \hat{y}$. The term "residual" is also appropriate as suggesting that something is left over when $\hat{y}$ does not match y and the Kalman filter operates as a feedback system to reduce the leftover to zero.

Note 11.4 Singular spectral density matrix of observation noise

The case in which the observation noise is present but has a singular spectral density matrix has been considered by Hutton.[5] The basic idea is to transform the observation vector into the form $\bar{y}_1 = \bar{C}_1 x$ and $\bar{y}_2 = \bar{C}_2 x + w_2$ where w_2 has a nonsingular spectral density matrix. The dimension of $\bar{y}_2$ is the rank of the starting spectral density matrix. From that point, the derivation of the optimal filter follows the outlines of Sec. 11.6. The order of the resulting filter is $k - l_1$, where l_1 is the dimension of $\bar{y}_1$, that is, the "nullity" of the starting spectral density matrix.

Note 11.5 The separation theorem

The earliest statement of the separation theorem in the literature of control theory was given by Joseph and Tou[16] for discrete-time systems in 1961, only a year after Kalman's seminal paper on discrete-time optimal filtering.[1] Two years later, Gunckel and Franklin[17] published a slightly more general statement of the theorem. The earliest statement of the separation theorem, although not in the control literature, was given by the econometrician H. A. Simon in 1956.[18] Simon considered a very simple linear system with a quadratic performance criterion. His result, expressed in control terminology, is that a control law that uses the present *expected* value of the state is equivalent to a control law based on certain knowledge of the future. Simon called this the "certainty equivalence" method and pointed out that linear dynamics and a quadratic performance criterion were apparent requirements for the validity of the certainty equivalence method.

The separation theorem for discrete-time systems was so reasonable and convenient that it was immediately adopted, without rigorous proof, for continuous-time systems. A rigorous proof was given by Wonham[19] in 1968. Wonham showed that the separation theorem is valid under more general conditions than were previously thought required. In particular, the requirement of a quadratic performance criterion (and hence a linear feedback law) was shown to be unnecessary.

REFERENCES

1. Kalman, R. E., "A New Approach to Linear Filtering and Prediction Problems," *Trans. ASME (J. Basic Engineering),* vol. 82D, no. 1, March 1960, pp. 35–45.

2. Kalman, R. E., and Bucy, R. S., "New Results in Linear Filtering and Prediction Theory," *Trans. ASME (J. Basic Engineering)*, vol. 83D, no. 1, March 1961, pp. 95–108.
3. Richman, J., and Friedland, B., "Design of Optimum Mixer-Filter for Aircraft Navigation Systems," *Proc. NAECON, Dayton, OH*, May 1967, pp. 429–438.
4. Gantmacher, F. R., *The Theory of Matrices*, Chelsea Publishing Company, New York, vol. 1, 1959.
5. Hutton, M. F., "Solutions of the Singular Stochastic Regulator Problem," *Trans. ASME (J. Dynamic Systems, Measurement & Control)*, vol. 95G, no. 4, December 1973, pp. 414–417.
6. Bryson, A. E., Jr., and Johansen, D. E., "Linear Filtering for Time-Varying Systems Using Measurements Containing Colored Noise," *IEEE Trans. on Automatic Control*, vol. AC-10, no. 1, January 1965, pp. 4–10.
7. Friedland, B., "Limiting Forms of Optimum Stochastic Linear Regulators," *Trans. ASME (J. Dynamic Systems, Measurement & Control)*, vol. 93G, no. 3, September 1971, pp. 134–141.
8. Kwakernaak, H., and Sivan, R., *Linear Optimal Control Systems*, Wiley-Interscience, New York, 1972.
9. Kalman, R. E., "When is a Linear Control System Optimal?," *Trans. ASME (J. Basic Engineering)*, vol. 86D, no. 1, March 1964, pp. 51–60.
10. Gantmacher, F. R., *The Theory of Matrices*, Chelsea Publishing Company, New York, vol. 2, 1959.
11. Kailath, T., "An Innovations Approach to Least-Squares Estimation, Part I: Linear Filtering in Additive White Noise," *IEEE Trans. on Automatic Control*, vol. AC-13, no. 6, December 1968, pp. 646–655.
12. Sorenson, H. W., "Least-Squares Filtering: From Gauss to Kalman," *IEEE Spectrum*, vol. 7, no. 7, July 1970, pp. 63–68.
13. Wiener, N., *The Extrapolation, Interpolation, and Smoothing of Stationary Time Series*, John Wiley & Sons, New York, 1949.
14. Bode, H. W., and Shannon, C. E., "A Simplified Derivation of Least-Squares Smoothing and Prediction Theory," *Proceedings of the IRE*, vol. 38, no. 4, April 1950, pp. 481–492.
15. Kalman, R. E., "On the General Theory of Control Systems," *Proc. First International Congress, IFAC, Moscow, USSR*, June 1960, pp. 481–492.
16. Joseph, P. D., and Tou, J. T., "On Linear Control Theory," *Trans. AIEE, Pt. II*, vol. 80, no. 3, September 1961, pp. 193–196.
17. Gunckel, T. L., III, and Franklin, G. F., "A General Solution for Linear Sampled Data Control," *Trans. ASME (J. Basic Engineering)*, vol. 85D, 1963, pp. 197–201.
18. Simon, H. A., "Dynamic Programming Under Uncertainty With a Quadratic Criterion Function," *Econometrica*, vol. 24, 1956, pp. 74–81.
19. Wonham, W. M., "On The Separation Theorem of Stochastic Control," *SIAM J. on Control*, vol. 6, no. 2, 1968, pp. 312–326.

APPENDIX

MATRIX ALGEBRA AND ANALYSIS

A.1 INTRODUCTION

Matrices are widely applied in many branches of contemporary engineering, ranging from stress analysis, to electrical circuits, to engineering economics. An engineering curriculum without an introduction to matrix theory is difficult to conceive.

This appendix is not intended as a substitute for a text on matrices and their applications in engineering; there are many excellent texts available for this purpose. Rather it is intended as a concise summary of the facts about matrices that the reader will need to know in reading this book. Having them all at hand will minimize the need to consult a standard reference book. It also serves to define the notation and terminology which are, regrettably, not entirely standard.

While some derivations and proofs are given to provide motivation and insight, no attempt has been made at proving or even justifying every statement that is made. The interested reader is urged to consult a suitable textbook for details of proof, for other properties of matrices, and for many additional applications.

A.2 LINEAR EQUATIONS AND MATRIX NOTATION

Consider a system of linear equations

$$\begin{aligned} y_1 &= a_{11}x_1 + \cdots + a_{1m}x_m \\ y_2 &= a_{21}x_1 + \cdots + a_{2m}x_m \\ &\cdots\cdots\cdots\cdots\cdots\cdots \\ y_n &= a_{n1}x_1 + \cdots + a_{nm}x_m \end{aligned} \tag{A.1}$$

To reduce the amount of writing, (A.1) can be expressed as

$$\begin{bmatrix} y_1 \\ \vdots \\ y_n \end{bmatrix} = \begin{bmatrix} a_{11} & \cdots & a_{1m} \\ \cdots & \cdots & \cdots \\ a_{n1} & \cdots & a_{nm} \end{bmatrix} \begin{bmatrix} x_1 \\ \vdots \\ x_m \end{bmatrix} \tag{A.2}$$

The arrays of numbers enclosed by the square brackets are known as *matrices.* In general a rectangular array having m columns and n rows, of the form

$$A = \begin{bmatrix} a_{11} & \cdots & a_{1m} \\ \cdots & \cdots & \cdots \\ a_{n1} & \cdots & a_{nm} \end{bmatrix} \tag{A.3}$$

is called an $n \times m$ matrix. A single letter A is used to designate the entire $n \times m$ matrix. The matrices in (A.2) having only one column, namely

$$y = \begin{bmatrix} y_1 \\ \vdots \\ y_m \end{bmatrix} \quad \text{and} \quad x = \begin{bmatrix} x_1 \\ \vdots \\ x_n \end{bmatrix} \tag{A.4}$$

are generally called *vectors.* (The three-dimensional vectors of classical physics are special cases of the general mathematical vectors used in this book. When, as in Chap. 3, we want to refer specifically to a physical vector such as force or velocity, we use an arrow over the letter, viz., $\vec{f}$, $\vec{v}$.)

Since a vector is a $1 \times n$ matrix, any result that applies to all $n \times m$ matrices also applies to vectors. As is customary in texts on systems, vectors are generally denoted by lowercase letters and other matrices are generally denoted by capital letters.

In terms of the notation introduced above, the system of linear equations (A.1) can be written simply as

$$y = Ax \tag{A.5}$$

Equation (A.5) can be read as "(the vector) y is the result of multiplying (the vector) x by (the matrix) A." Thus we have defined the operation of multiplying a vector by a matrix, as the equivalent of the system of linear equations (A.1). Multiplying a vector by a matrix is a special case of multiplying one matrix by another. We shall consider this operation and others in the next section.

A 1×1 matrix (i.e., a single component vector) which is just a single number, is known as a *scalar.*

To save writing, the matrix A is sometimes exhibited as

$$A = [a_{ij}]$$

where a_{ij} is a *typical* element of the matrix A.

A.3 MATRIX OPERATIONS

Addition and subtraction Matrices can be combined by use of the operations of addition, subtraction, and multiplication in much the same manner as scalars. These operations may be defined in terms of systems of simultaneous equations. Thus, suppose

$$\begin{aligned} y_1 &= a_{11}x_1 + \cdots + a_{1n}x_n \\ y_2 &= a_{21}x_1 + \cdots + a_{2n}x_n \\ &\cdots\cdots\cdots\cdots \\ y_m &= a_{m1}x_1 + \cdots + a_{mn}x_n \end{aligned} \tag{A.6}$$

and

$$\begin{aligned} z_1 &= b_{11}x_1 + \cdots + b_{1n}x_n \\ z_2 &= b_{21}x_1 + \cdots + b_{2n}x_n \\ &\cdots\cdots\cdots\cdots \\ z_m &= b_{m1}x_1 + \cdots + b_{mn}x_n \end{aligned} \tag{A.7}$$

then, adding each equation in (A.6) to the corresponding equation in (A.7) gives

$$\begin{aligned} y_1 + z_1 &= (a_{11} + b_{11})x_1 + \cdots + (a_{1n} + b_{1n})x_n \\ y_2 + z_2 &= (a_{21} + b_{21})x_1 + \cdots + (a_{2n} + b_{2n})x_n \\ &\cdots\cdots\cdots\cdots\cdots\cdots \\ y_m + z_m &= (a_{m1} + b_{m1})x_1 + \cdots + (a_{mn} + b_{mn})x_n \end{aligned}$$

or, in the matrix shorthand,

$$\begin{bmatrix} y_1 + z_1 \\ y_2 + z_2 \\ \vdots \\ y_m + z_m \end{bmatrix} = \begin{bmatrix} a_{11} + b_{11} & \cdots & a_{1n} + b_{1n} \\ a_{21} + b_{21} & \cdots & a_{2n} + b_{2n} \\ \cdots & \cdots & \cdots \\ a_{m1} + b_{m1} & \cdots & a_{mn} + b_{mn} \end{bmatrix} \begin{bmatrix} x_1 \\ x_2 \\ \vdots \\ x_n \end{bmatrix}$$

or

$$y + z = (A + B)x = Cx$$

where

$$C = A + B = \begin{bmatrix} a_{11} + b_{11} & \cdots & a_{1n} + b_{1n} \\ \cdots & \cdots & \cdots \\ a_{m1} + b_{m1} & \cdots & a_{mn} + b_{mn} \end{bmatrix} \tag{A.8}$$

Thus each term in the sum of two matrices is the sum of the corresponding elements of the summands. It is clear from the definition that two matrices can be added only when they are both of the same dimensions: $m \times n$, and, when $A + B$ is defined,

$$A + B = B + A \tag{A.9}$$

Subtraction of two matrices is defined by

$$C = A - B = \begin{bmatrix} a_{11} - b_{11} & \cdots & a_{1n} - b_{1n} \\ \cdots & \cdots & \cdots \\ a_{m1} - b_{m1} & \cdots & a_{mn} - b_{mn} \end{bmatrix} \tag{A.10}$$

Multiplication Multiplication of matrices is defined in terms of substitution of one linear system of equations into another. Consider

$$\left.\begin{aligned} y_1 &= a_{11}x_1 + \cdots + a_{1n}x_n \\ y_2 &= a_{21}x_1 + \cdots + a_{2n}x_n \\ &\cdots\cdots\cdots\cdots \\ y_m &= a_{m1}x_1 + \cdots + a_{mn}x_n \end{aligned}\right\} \text{ i.e., } y = Ax \tag{A.11}$$

and

$$\left.\begin{aligned} x_1 &= b_{11}w_1 + \cdots + b_{1l}w_l \\ x_2 &= b_{21}w_1 + \cdots + b_{2l}w_l \\ &\cdots\cdots\cdots\cdots \\ x_n &= b_{n1}w_1 + \cdots + b_{nl}w_l \end{aligned}\right\} \text{ i.e., } x = Bw \tag{A.12}$$

Substitute (A.11) into (A.12) to obtain

$$\begin{aligned} y_1 &= a_{11}(b_{11}w_1 + \cdots + b_{1l}w_l) + \cdots + a_{1n}(b_{n1}w_1 + \cdots + b_{nl}w_l) \\ y_2 &= a_{21}(b_{11}w_1 + \cdots + b_{1l}w_l) + \cdots + a_{2n}(b_{n1}w_1 + \cdots + b_{nl}w_l) \\ &\cdots\cdots\cdots\cdots\cdots\cdots \\ y_m &= a_{m1}(b_{11}w_1 + \cdots + b_{1l}w_l) + \cdots + a_{mn}(b_{n1}w_1 + \cdots + b_{nl}w_l) \end{aligned} \tag{A.13}$$

Collecting coefficients of the w_i in (A.13) gives

$$\begin{aligned} y_1 &= (a_{11}b_{11} + \cdots + a_{1n}b_{n1})w_1 + \cdots + (a_{11}b_{1l} + \cdots + a_{1n}b_{nl})w_l \\ y_2 &= (a_{21}b_{11} + \cdots + a_{2n}b_{n1})w_1 + \cdots + (a_{21}b_{1l} + \cdots + a_{2n}b_{nl})w_l \\ &\cdots\cdots\cdots\cdots\cdots\cdots \\ y_m &= (a_{m1}b_{11} + \cdots + a_{mn}b_{n1})w_1 + \cdots + (a_{m1}b_{1l} + \cdots + a_{mn}b_{nl})w_l \end{aligned} \tag{A.14}$$

or

$$y = \begin{bmatrix} y_1 \\ \vdots \\ y_m \end{bmatrix} = \begin{bmatrix} c_{11} & \cdots & c_{1l} \\ \cdots & \cdots & \cdots \\ c_{m1} & \cdots & c_{ml} \end{bmatrix} \begin{bmatrix} w_1 \\ \vdots \\ w_l \end{bmatrix} = Cw$$

where

$$C = \begin{bmatrix} (a_{11}b_{11} + \cdots + a_{1n}b_{n1}) & \cdots & (a_{11}b_{1l} + \cdots + a_{1n}b_{nl}) \\ \cdots\cdots & \cdots & \cdots\cdots \\ (a_{m1}b_{11} + \cdots + a_{mn}b_{n1}) & \cdots & (a_{m1}b_{1l} + \cdots + a_{mn}b_{nl}) \end{bmatrix} \tag{A.15}$$

Now, since in "shorthand" notation (A.11) and (A.12) are

$$y = Ax \qquad \text{and} \qquad x = Bw$$

we can write

$$y = A(Bw) = ABw = Cw$$

where C is the *product* matrix defined by (A.15). In words, the (i, j)th element c_{ij} of the product matrix is computed by multiplying and summing the elements of the ith row of the first matrix A by the jth column of the second matrix B:

$$c_{ij} = \underbrace{\overrightarrow{\boxed{a_{i1}, a_{i2}, \ldots, a_{in}}}}_{i\text{th row}} \overset{j\text{th column}}{\boxed{\begin{matrix} b_{1j} \\ \vdots \\ b_{nj} \end{matrix}}}\Big\downarrow = \sum_{k=1}^{n} b_{ik}b_{kj}$$

Note that the product $C = AB$ of two matrices A and B is meaningful when the number of rows of B is equal to the number of columns of A. Matrices having this property are said to be *conformable*. It is possible for the product AB to be defined, and the product BA to be undefined, because the number of columns of B may not equal the number of rows of A. Even when BA and AB *both exist*, it is not necessary that $AB = BA$. For example, suppose A is a "row vector"

$$A = [a_1, \ldots, a_n]$$

and B is a "column vector"

$$B = \begin{bmatrix} b_1 \\ \vdots \\ b_n \end{bmatrix}$$

Then

$$AB = [a_1, \ldots, a_n]\begin{bmatrix} b_1 \\ \vdots \\ b_n \end{bmatrix} = (a_1b_1 + \cdots + a_nb_n) \quad \text{a scalar} \tag{A.16}$$

and

$$BA = \begin{bmatrix} b_1 \\ \vdots \\ b_n \end{bmatrix}[a_1, \ldots, a_n] = \begin{bmatrix} b_1a_1 & \cdots & b_1a_n \\ \cdots & \cdots & \cdots \\ b_na_1 & \cdots & b_na_n \end{bmatrix} \quad \text{an } n \times n \text{ matrix} \tag{A.17}$$

Clearly $AB \neq BA$.

But even when A and B are both square matrices (of dimension $n \times n$) and hence both products AB and BA exist, it is not always true that $AB = BA$. When $AB = BA$ the matrices are said to *commute.*

A 1×1 (or scalar) matrix c is comformable and commutes with any matrix.

$$cA = Ac = c\begin{bmatrix} a_{11} & \cdots & a_{1n} \\ \cdots & \cdots & \cdots \\ a_{m1} & \cdots & a_{mn} \end{bmatrix} = \begin{bmatrix} ca_{11} & \cdots & ca_{1n} \\ \cdots & \cdots & \cdots \\ ca_{m1} & \cdots & ca_{mn} \end{bmatrix}$$

With scalars, the operation of division by a quantity b, defined as multiplication by the reciprocal of b, that is,

$$c = \frac{a}{b} = a \cdot b^{-1} = b^{-1} \cdot a$$

where $b^{-1} = 1/b$. There are several complications in trying to extend the concept of division to matrices. First, we need to define the reciprocal (called an *inverse*) of a matrix. Next, we need to determine whether the inverse exists. (We know that the reciprocal of every *scalar* except zero does exist.) And finally, even if the inverse B^{-1} does exist, it is possible that $AB^{-1} \neq B^{-1}A$ since there is no assurance that B^{-1} and A commute. Division of one matrix by another is thus not a very useful concept.

A.4 DETERMINANTS AND MATRIX INVERSION

A system of n linear equations in n unknowns $x_1, \ldots, x_n$

$$\begin{aligned} y_1 &= a_{11}x_1 + \cdots + a_{1n}x_n \\ &\cdots\cdots\cdots\cdots \\ y_n &= a_{n1}x_1 + \cdots + a_{nn}x_n \end{aligned} \tag{A.18}$$

may or may not have a unique solution. It is well known that a unique solution exists if and only if the *determinant* of the matrix A is nonzero:

$$|A| = \begin{vmatrix} a_{11} & \cdots & a_{1n} \\ \cdots & \cdots & \cdots \\ a_{n1} & \cdots & a_{nn} \end{vmatrix} \neq 0 \tag{A.19}$$

The reader probably knows various definitions of the determinant of a square array or matrix. The basic definition of the determinant is the sum of all possible products of n elements, each taken from a *different* column, and with the sign of the product taken as (+) if the columns from which the elements are taken are in "lexographic order" and as (−) if they are not. ("In lexographic order" means that the columns are taken from right to left without jumping backward.)

A recursive definition of the determinant of a matrix A is

$$|A| = a_{11}D_{11} - a_{12}D_{12} + a_{13}D_{13} - \cdots + (-1)^n a_{1n}D_n \tag{A.20}$$

where D_{ij} is the determinant of the $(n-1) \times (n-1)$ subarray formed from A by deleting the ith row and jth column from the original matrix. There are many other algorithms for the calculation of the determinant which can be used as the basic definition.

If $|A| \neq 0$ and hence (A.18) has a unique solution, that solution can be written

$$\begin{aligned} x_1 &= b_{11}y_1 + \cdots + b_{1n}y_n \\ &\cdots\cdots\cdots\cdots\cdots\cdots \\ x_n &= b_{n1}y_1 + \cdots + b_{nn}y_n \end{aligned} \tag{A.21}$$

In matrix notation (A.18) and (A.21) are

$$y = Ax \tag{A.22}$$

and

$$x = By \tag{A.23}$$

Substitute (A.23) into (A.22) to obtain

$$y = ABy = Iy \tag{A.24}$$

which, written out in detail, is

$$\begin{bmatrix} y_1 \\ y_2 \\ \vdots \\ y_n \end{bmatrix} = \begin{bmatrix} 1 & 0 & \cdots & 0 \\ 0 & 1 & \cdots & 0 \\ \cdots & \cdots & \cdots & \cdots \\ 0 & 0 & \cdots & 1 \end{bmatrix} \begin{bmatrix} y_1 \\ y_2 \\ \vdots \\ y_n \end{bmatrix} \tag{A.25}$$

The square matrix in (A.25) having 1's along the main diagonal and 0's everywhere else, is called the $(n \times n)$ *identity matrix*

$$I = \begin{bmatrix} 1 & 0 & \cdots & 0 \\ 0 & 1 & \cdots & 0 \\ \cdots & \cdots & \cdots & \cdots \\ 0 & 0 & \cdots & 1 \end{bmatrix} \tag{A.26}$$

Thus we conclude that the product of A and B in (A.24) is the identity matrix

$$AB = I \tag{A.27}$$

If (A.27) holds then B and A are said to be inverses and written

$$B = A^{-1} \qquad A = B^{-1} \tag{A.28}$$

(It is readily shown that if $AB = I$, then $BA = I$; substitute (A.22) into (A.23), for example.)

If the determinant of a matrix is zero, the matrix has no inverse and is said to be *singular*; if the determinant of a matrix is nonzero and its inverse does exist, the matrix is said to be *nonsingular.*

Devising efficient algorithms for computing the determinant and the inverse of large matrices has been an important branch of numerical analysis. The obvious algorithms are often unreliable when the dimension of the matrix exceeds 100 or so.

For theoretical developments, the inverse of a matrix can be expressed by the formula usually called *Cramer's rule*

$$A^{-1} = \frac{\text{adj}\,(A)}{|A|} \tag{A.29}$$

where the matrix written "adj (A)" is called the *adjoint* (or *adjugate*) matrix. The (i, j)th element of this matrix is the *cofactor* of a_{ji}. The cofactor, or "signed minor" is the determinant of the submatrix obtained by deleting the row and column containing a_{ij} and having a (+) sign if $i+j$ is even and a (−) sign if $i+j$ is odd. For example, the determinants and adjoint matrices for 2×2 and 3×3 matrices are as follows

$n = 2$:

$$A = \begin{bmatrix} a_{11} & a_{12} \\ a_{21} & a_{22} \end{bmatrix} \qquad \text{adj}\, A = \begin{bmatrix} a_{22} & -a_{12} \\ -a_{21} & a_{11} \end{bmatrix}$$
$$|A| = a_{11}a_{22} - a_{21}a_{12} \tag{A.30}$$

$n = 3$:

$$A = \begin{bmatrix} a_{11} & a_{12} & a_{13} \\ a_{21} & a_{22} & a_{23} \\ a_{31} & a_{32} & a_{33} \end{bmatrix} \qquad \text{adj}\, A = \begin{bmatrix} \begin{vmatrix} a_{22} & a_{23} \\ a_{32} & a_{33} \end{vmatrix} & -\begin{vmatrix} a_{12} & a_{13} \\ a_{32} & a_{33} \end{vmatrix} & \begin{vmatrix} a_{12} & a_{13} \\ a_{22} & a_{23} \end{vmatrix} \\ -\begin{vmatrix} a_{21} & a_{23} \\ a_{31} & a_{33} \end{vmatrix} & \begin{vmatrix} a_{11} & a_{13} \\ a_{31} & a_{33} \end{vmatrix} & -\begin{vmatrix} a_{11} & a_{13} \\ a_{21} & a_{23} \end{vmatrix} \\ \begin{vmatrix} a_{21} & a_{22} \\ a_{31} & a_{32} \end{vmatrix} & -\begin{vmatrix} a_{11} & a_{12} \\ a_{31} & a_{32} \end{vmatrix} & \begin{vmatrix} a_{11} & a_{12} \\ a_{21} & a_{22} \end{vmatrix} \end{bmatrix}$$

$$|A| = a_{11}\begin{vmatrix} a_{22} & a_{23} \\ a_{32} & a_{33} \end{vmatrix} - a_{12}\begin{vmatrix} a_{21} & a_{23} \\ a_{31} & a_{33} \end{vmatrix} + a_{13}\begin{vmatrix} a_{21} & a_{22} \\ a_{31} & a_{32} \end{vmatrix} \tag{A.31}$$

Cramer's rule is one of the least efficient algorithms for computing the inverse of a matrix and is recommended only for the proverbial person stranded on a desert island with no references to more efficient algorithms and no desire to invent any.

It can be shown that the determinant of the product of two $n \times n$ matrices is the product of the determinants:

$$|AB| = |A| \cdot |B|$$

Moreover, it is readily seen that

$$|I| = 1$$

Thus

$$1 = |AA^{-1}| = |A| \cdot |A^{-1}|$$

or

$$|A^{-1}| = \frac{1}{|A|}$$

A.5 PARTITIONED MATRICES

We often deal with linear systems of vector-matrix equations

$$\begin{aligned} y_1 &= A_{11}x_1 + A_{12}x_2 + \cdots + A_{1m}x_m \\ y_2 &= A_{21}x_1 + A_{22}x_2 + \cdots + A_{2m}x_m \\ &\cdots\cdots\cdots\cdots\cdots\cdots\cdots\cdots \\ y_n &= A_{n1}x_1 + A_{n2}x_2 + \cdots + A_{nm}x_m \end{aligned} \tag{A.32}$$

in which the x_i's are themselves vectors and the A_{ij}'s are matrices. We can write (A.32) as

$$\left[\begin{array}{c} y_1 \\ \hline y_2 \\ \hline \vdots \\ \hline y_n \end{array}\right] = \left[\begin{array}{c|c|c|c} A_{11} & A_{12} & \cdots & A_{1m} \\ \hline A_{21} & A_{22} & \cdots & A_{2m} \\ \hline \cdots & \cdots & \cdots & \cdots \\ \hline A_{n1} & A_{n2} & \cdots & A_{nm} \end{array}\right] \left[\begin{array}{c} x_1 \\ \hline x_2 \\ \hline \vdots \\ \hline x_m \end{array}\right] \tag{A.33}$$

The vectors $y_1, \ldots, y_n, x_1, \ldots, x_m$ are *subvectors* of vectors of higher dimension

$$y = \left[\begin{array}{c} y_1 \\ \hline \vdots \\ \hline y_n \end{array}\right] \qquad x = \left[\begin{array}{c} x_1 \\ \hline \vdots \\ \hline x_m \end{array}\right] \tag{A.34}$$

and the matrices A_{ij} are submatrices of a larger matrix

$$A = \left[\begin{array}{c|c|c} A_{11} & \cdots & A_{1m} \\ \hline \cdots & \cdots & \cdots \\ \hline A_{n1} & \cdots & A_{nm} \end{array}\right] \tag{A.35}$$

The vectors x and y in (A.34) and the matrix A in (A.35) are known as *partitioned matrices*. The broken lines are used to show the partitioning and are particularly useful when the elements of the submatrices are explicitly shown, for example

$$\left[\begin{array}{c} y_1 \\ y_2 \\ \hline y_3 \\ y_4 \end{array}\right] = \left[\begin{array}{cc|c} a_{11} & a_{12} & a_{13} \\ a_{21} & a_{22} & a_{23} \\ \hline a_{31} & a_{32} & a_{33} \\ a_{41} & a_{42} & a_{43} \end{array}\right] \left[\begin{array}{c} x_1 \\ x_2 \\ \hline x_3 \end{array}\right] \tag{A.36}$$

In this case y and x are both partitioned into two subvectors and A is partitioned into $2 \times 2 = 4$ submatrices of different dimension.

The dashed lines indicating the partitioning are sometimes omitted for the sake of appearance when the context makes it clear that partitioned matrices are being considered. For example (A.36) might be written

$$\begin{bmatrix} \mathbf{y}_1 \\ \mathbf{y}_2 \end{bmatrix} = \begin{bmatrix} A_{11} & A_{12} \\ A_{21} & A_{22} \end{bmatrix} \begin{bmatrix} \mathbf{x}_1 \\ \mathbf{x}_2 \end{bmatrix} \tag{A.37}$$

where

$$\mathbf{y}_1 = \begin{bmatrix} y_1 \\ y_2 \end{bmatrix} \qquad \mathbf{y}_2 = \begin{bmatrix} y_3 \\ y_4 \end{bmatrix} \tag{A.38}$$

The boldface notation for the subvectors x_1, x_2, y_1, y_2 are used in this case to distinguish *subvectors* $\mathbf{y}_1$ and $\mathbf{y}_2$ from the components y_1, y_2 (both of which appear in subvector $\mathbf{y}_1$).

Vectors and matrices may be partitioned into as many subdivisions as are convenient for a given application. The only rule that must be observed is that the submatrices must be of the proper dimension: The submatrix A_{ij} in (A.33) must have as many columns as the corresponding subvector x_j and as many rows as the subvector y_i.

Matrices that are appropriately partitioned may be added, subtracted, and multiplied. For example

$$\left[\begin{array}{c:c} A_{11} & A_{12} \\ \hdashline A_{21} & A_{22} \end{array}\right] \left[\begin{array}{c:c} B_{11} & B_{12} \\ \hdashline B_{21} & B_{22} \end{array}\right] = \left[\begin{array}{c:c} A_{11}B_{11} + A_{12}B_{21} & A_{11}B_{12} + A_{12}B_{22} \\ \hdashline A_{21}B_{11} + A_{22}B_{21} & A_{21}B_{12} + A_{22}B_{22} \end{array}\right] \tag{A.39}$$

is valid provided the partitioning of A and B is such that all the matrix sums and products on the right-hand side of (A.39) are of matrices of conformable dimensions.

Partitioning can sometimes be useful for obtaining inverses. Consider the system

$$y_1 = Ax_1 + Bx_2 \tag{A.40}$$

$$y_2 = Cx_1 + Dx_2 \tag{A.41}$$

i.e.,

$$\begin{bmatrix} y_1 \\ y_2 \end{bmatrix} = \begin{bmatrix} A & B \\ C & D \end{bmatrix} \begin{bmatrix} x_1 \\ x_2 \end{bmatrix}$$

The inverse of

$$M = \begin{bmatrix} A & B \\ C & D \end{bmatrix}$$

can be expressed in terms of the inverses of the submatrices. Suppose submatrix A has an inverse. Then we can solve (A.40) for x_1

$$x_1 = A^{-1}y_1 - A^{-1}Bx_2 \tag{A.42}$$

Substitute this into (A.41) to obtain

$$y_2 = CA^{-1}y_1 - (CA^{-1}B - D)x_2$$

Whence

$$x_2 = (CA^{-1}B - D)^{-1}(CA^{-1}y_1 - y_2) \tag{A.43}$$

which is substituted into (A.42) to yield

$$x_1 = [A^{-1} - A^{-1}B(CA^{-1}B - D)^{-1}CA^{-1}]y_1 + A^{-1}B(CA^{-1}B - D)^{-1}y_2 \tag{A.44}$$

The solution to (A.40) and (A.41) has thus been obtained, and hence we have the inverse of M:

$$\begin{aligned} M^{-1} &= \left[\begin{array}{c|c} A & B \\ \hline C & D \end{array}\right]^{-1} \\ &= \left[\begin{array}{c|c} A^{-1} - A^{-1}B(CA^{-1}B - D)^{-1}CA^{-1} & A^{-1}B(CA^{-1}B - D)^{-1} \\ \hline (CA^{-1}B - D)^{-1}CA^{-1} & -(CA^{-1}B - D)^{-1} \end{array}\right] \end{aligned} \tag{A.45}$$

Thus we have managed to express the inverse of a big matrix M in terms of the inverses of smaller matrices A and $CA^{-1}B - D$.

If D has an inverse, then the same procedure can be used to obtain another expression for M^{-1}

$$\begin{aligned} M^{-1} &= \left[\begin{array}{c|c} A & B \\ \hline C & D \end{array}\right]^{-1} \\ &= \left[\begin{array}{c|c} -(BD^{-1}C - A)^{-1} & (BD^{-1}C - A)^{-1}BD^{-1} \\ \hline D^{-1}C(BD^{-1}C - A)^{-1} & D^{-1} - D^{-1}C(BD^{-1}C - A)^{-1}BD^{-1} \end{array}\right] \end{aligned} \tag{A.46}$$

Since (A.45) and (A.46) give different expressions for the same matrix, each submatrix on the right-hand side of (A.46) must equal the corresponding submatrix of (A.45). In particular, we must have

$$-(BD^{-1}C - A)^{-1} = A^{-1} - A^{-1}B(CA^{-1}B - D)^{-1}CA^{-1}$$

This is a version of a matrix-inversion lemma often attributed to Schur. The more familiar form is obtained by replacing D by $-D$

$$(A + BD^{-1}C)^{-1} = A^{-1} - A^{-1}B(CA^{-1}B + D)^{-1}CA^{-1} \tag{A.47}$$

This lemma is the basis of many important results in systems and control theory.

A.6 TRANSPOSITION AND RELATED OPERATIONS: SPECIAL MATRIX TYPES

If the rows of one matrix are the columns of another, the matrices are called the *transposes* of each other:

$$A = \begin{bmatrix} a_{11} & a_{12} & \cdots & a_{1m} \\ a_{21} & a_{22} & \cdots & a_{2m} \\ \cdots & \cdots & \cdots & \cdots \\ a_{n1} & a_{n2} & \cdots & a_{nm} \end{bmatrix} \qquad A' = \begin{bmatrix} a_{11} & a_{21} & \cdots & a_{n1} \\ a_{12} & a_{22} & \cdots & a_{n2} \\ \cdots & \cdots & \cdots & \cdots \\ a_{1m} & a_{2m} & \cdots & a_{nm} \end{bmatrix} \tag{A.48}$$

In (A.48) A and A' are transposes of each other. The "prime" (′) symbol is used in this book to denote transposition. In other books, transposition is sometimes denoted by a superscript T, that is, A^T.

The following properties are easily proved:

$$\begin{aligned} (A')' &= A \\ (AB)' &= B'A' \\ |A| &= |A'| \qquad \text{(if } A \text{ is a square matrix)} \end{aligned} \tag{A.49}$$

A (square) matrix whose columns and rows are the same

$$A = \begin{bmatrix} a_{11} & a_{12} & \cdots & a_{1n} \\ a_{12} & a_{22} & \cdots & a_{2n} \\ \cdots & \cdots & \cdots & \cdots \\ a_{1n} & a_{2n} & \cdots & a_{nn} \end{bmatrix} = A' \tag{A.50}$$

is said to be *symmetric.* Symmetric matrices play important roles in systems and control.

A matrix whose columns are the *negatives* of its rows is called *skew-symmetric.* A skew-symmetric matrix has the general form

$$S = \begin{bmatrix} 0 & a_{12} & \cdots & a_{1n} \\ -a_{12} & 0 & \cdots & a_{2n} \\ \cdots & \cdots & \cdots & \cdots \\ -a_{1n} & -a_{2n} & \cdots & 0 \end{bmatrix} = -S' \tag{A.51}$$

(Note that the diagonal elements of a skew-symmetric matrix are always zero.) Skew-symmetric matrices occur in studying the dynamics of certain types of conservative systems but are not encountered as often as symmetric matrices.

A matrix whose transpose is equal to its inverse is said to be *orthogonal*

$$A' = A^{-1} \tag{A.52}$$

From (A.52)

$$A'A = AA' = I \tag{A.53}$$

This means that the sum of products of the elements

$$\sum_{i=1}^{n} a_{ij}a_{ik} = \delta_{jk} = \begin{cases} 1 & \text{for} \quad j = k \\ 0 & \text{for} \quad j \neq k \end{cases} \tag{A.54}$$

i.e., the sum of products of a row (or column) by the corresponding elements of the *same* row (or column) is 1, but the sum of products of a row (or column) by the corresponding elements of a *different* row (or column) is zero.

When the elements of a matrix are complex numbers, it is more usual to work with the transpose of the complex conjugate of the matrix, which is written A^H

$$A = \begin{bmatrix} a_{11} & a_{12} & \cdots & a_{1m} \\ \cdots & \cdots & \cdots & \cdots \\ a_{n1} & a_{n2} & \cdots & a_{nm} \end{bmatrix}, \qquad A^H = \begin{bmatrix} a_{11}^* & \cdots & a_{n1}^* \\ a_{12}^* & \cdots & a_{n2}^* \\ \cdots & \cdots & \cdots \\ a_{1m}^* & \cdots & a_{nm}^* \end{bmatrix} \tag{A.55}$$

The complex conjugate transpose of a matrix A is also, regrettably, called the *adjoint* matrix of A. There is no connection between the adjoint of this section and the adjoint as defined in Sec. A.4; fortunately the two different adjoints rarely occur in the same context.

A matrix that equals its adjoint ($A^H = A$) is said to be *hermitian*, and one which is the negative of its adjoint ($A^H = -A$) is said to be *skew-hermitian*.

A matrix whose adjoint equals its inverse

$$A^H = A^{-1} \tag{A.56}$$

or

$$AA^H = A^H A = I \tag{A.57}$$

is said to be *unitary*.

Some other special matrix types are frequently encountered in control theory. A square matrix whose elements are zero in all positions above the principal diagonal

$$A = \begin{bmatrix} a_{11} & 0 & \cdots & 0 \\ a_{12} & a_{22} & \cdots & 0 \\ \cdots & \cdots & \cdots & \cdots \\ a_{1n} & a_{2n} & \cdots & a_{nn} \end{bmatrix} \tag{A.58}$$

is called *lower triangular*; its transpose

$$A' = \begin{bmatrix} a_{11} & a_{12} & \cdots & a_{1n} \\ 0 & a_{22} & \cdots & a_{2n} \\ \cdots & \cdots & \cdots & \cdots \\ 0 & 0 & \cdots & a_{nn} \end{bmatrix} \tag{A.59}$$

is called *upper triangular*. The determinant of a (lower or upper) triangular matrix is the product of the diagonal elements. Hence a triangular matrix is nonsingular if and only if its diagonal elements are all nonzero.

A matrix whose only nonzero elements lie on the main diagonal

$$\Lambda = \begin{bmatrix} \lambda_1 & 0 & \cdots & 0 \\ 0 & \lambda_2 & \cdots & 0 \\ \cdots & \cdots & \cdots & \cdots \\ 0 & 0 & \cdots & \lambda_n \end{bmatrix} = \text{diag}\,[\lambda_1, \lambda_2, \ldots, \lambda_n] \tag{A.60}$$

is called a *diagonal* matrix and it is customarily written with a capital Greek lambda (Λ).

A.7 LINEAR INDEPENDENCE AND RANK

A set of mathematical objects $a_1, a_2, \ldots, a_r$ (specifically, in our case vectors or columns of a matrix) is said to be *linearly dependent*, if and only if there exists a set of constants $c_1, c_2, \ldots, c_r$, not *all* zero, such that

$$c_1 a_1 + c_2 a_2 + \cdots + c_r a_r = 0$$

If no such set of constants exists, the set of objects is said to be linearly independent.

Suppose A is a matrix (not necessarily square) with $a_1, a_2, \ldots, a_n$ as its columns

$$A = [a_1 \mid a_2 \mid \cdots \mid a_n]$$

The rank of A, sometimes written rank(A) or $r(A)$ is the largest number of independent columns of A. The rank of A cannot be greater than n of course, but it can be smaller than n.

A fundamental theorem regarding the rank of a matrix can be stated as follows:

> The rank of A is the dimension of the largest nonzero determinant formed by deleting rows and columns from A. (A.61)

Thus we can say that the rank of a matrix is the maximum number of linearly independent columns of the matrix, the test for which is the largest (in dimension) nonsingular determinant found "embedded" in the matrix.

The rank of a matrix cannot exceed the number of columns of a matrix. Also, because determinants correspond to square arrays, the rank cannot exceed the number of rows of a matrix. Thus the rank of a rectangular matrix cannot exceed the lesser of the number of rows or the number of columns. A matrix whose rank is equal to the lesser of the number of rows and the number of columns is said to be *of full rank*.

Numerical determination of the rank of a matrix is not a trivial problem: If the brute-force method of testing is used, a goodly number of determinants must be evaluated. Moreover, some criterion is needed to establish how close to zero a numerically computed determinant must be in order to be declared zero. The basic numerical problem is that rank is not a continuous function of the elements of a matrix: a small change in one of the elements of a matrix can result in a discontinuous change of its rank.

The rank of a product of two matrices cannot exceed the rank of either factor

$$\text{rank}\,(AB) \leqq \min\,[\text{rank}\,(A), \text{rank}\,(B)] \tag{A.62}$$

But if either factor is a nonsingular (square) matrix the rank of the product is the rank of the remaining factor:

$$\begin{aligned} \text{rank}\,(AB) &= \text{rank}\,(A) \text{ if } B^{-1} \text{ exists} \\ \text{rank}\,(AB) &= \text{rank}\,(B) \text{ if } A^{-1} \text{ exists} \end{aligned} \tag{A.63}$$

A.8 EIGENVECTORS, EIGENVALUES, CHARACTERISTIC EQUATIONS, SIMILAR MATRICES

A vector v is called an *eigenvector* of the matrix A if

$$sv = Av \tag{A.64}$$

i.e., if v is parallel to Av. In order for (A.64) to hold we need

$$(sI - A)v = 0 \tag{A.65}$$

Except for the trivial case $v = 0$, (A.65) can hold only when the matrix $\Phi(s) = (sI - A)^{-1}$ (called the *resolvent* matrix) is singular. In order for $\Phi(s)$ to be singular we must have

$$\mathsf{D}(s) = |sI - A| = \begin{vmatrix} s - a_{11} & -a_{12} & \cdots & -a_{1n} \\ -a_{21} & s - a_{22} & \cdots & -a_{2n} \\ \cdots & \cdots & \cdots & \cdots \\ -a_{n1} & -a_{n2} & \cdots & s - a_{nn} \end{vmatrix} = 0 \tag{A.66}$$

On expanding the determinant, we find that $\mathsf{D}(s)$, which is called the *characteristic polynomial*, is a polynomial of degree n:

$$|sI - A| = \mathsf{D}(s) = s^n + a_1 s^{n-1} + \cdots + a_{n-1}s + a_n \tag{A.67}$$

To verify that $\mathsf{D}(s)$ must be of degree n, note that one term in $\mathsf{D}(s)$ is the product of the diagonal elements $(s - a_{11})(s - a_{22}) \cdots (s - a_{nn})$, which is clearly a polynomial of degree n with the coefficient of s^n being unity. Every other term in the determinant will have, at most, $n - 1$ diagonal elements and will thus be a polynomial of degree $n - 1$ or less. Hence the coefficient of s^n in (A.67) must be unity.

The equation

$$\mathsf{D}(s) = s^n + a_1 s^{n-1} + \cdots + a_{n-1}s + a_n = 0 \tag{A.68}$$

is called the *characteristic equation* of the matrix A and its n roots are called the *characteristic roots*, or *characteristic values*, or *eigenvalues* of the system. When A is the dynamics matrix of a linear system, the eigenvalues determine the dynamic response of the system and also turn out to be the *poles* of the system. (See Chap. 3.)

The resolvent $\Phi(s)$ can be written as follows:

$$\Phi(s) = (sI - A)^{-1} = \frac{\operatorname{adj}(sI - A)}{|sI - A|} \tag{A.69}$$

The adjoint matrix adj $(sI - A)$ is a matrix polynomial of the form

$$\operatorname{adj}(sI - A) = E_1 s^{n-1} + E_2 s^{n-2} + \cdots + E_n \tag{A.70}$$

where $E_1, E_2, \ldots, E_n$ are $n \times n$ matrices. (This form can be inferred via Cramer's rule (A.29).) Thus we can write

$$(sI - A)^{-1}|sI - A| = E_1 s^{n-1} + \cdots + E_n \tag{A.71}$$

and, on multiplying both sides of (A.71) by $sI - A$,

$$|sI - A|I = (sI - A)(E_1 s^{n-1} + \cdots + E_{n-1}s + E_n) \tag{A.72}$$

From (A.67), the left-hand side of (A.72) is

$$s^n I + a_1 s^{n-1} I + \cdots + a_n I \tag{A.73}$$

and, upon performing the indicated multiplication, the right-hand side of (A.72) is

$$E_1 s^n + (E_2 - AE_1)s^{n-1} + \cdots + (E_n - AE_{n-1})s - AE_n \tag{A.74}$$

Equating coefficients of (A.73) and (A.74) results in the following set of equations for the coefficient matrices of adj$(sI - A)$:

$$\begin{aligned} E_1 &= I \\ E_2 - AE_1 &= a_1 I \\ E_3 - AE_2 &= a_2 I \\ &\cdots\cdots \\ E_n - AE_{n-1} &= a_{n-1} I \\ -AE_n &= a_n I \end{aligned} \tag{A.75}$$

This set of equations provides an algorithm by which it is possible to compute $E_2, E_3, \ldots, E_n$ recursively given the coefficients a_i of the characteristic polynomial. It is possible to incorporate the calculation of the a_i into the algorithm as discussed in Chap. 3. Our purpose here, however, is not to explore this algorithm but to develop a fundamental property of the characteristic equation.

To this end we write

$$E_2 = AE_1 + a_1I = AI + a_1I = A + a_1I$$
$$E_3 = AE_2 + a_2I = A^2 + a_1A + a_2I$$
$$E_4 = AE_3 + a_3I = A^3 + a_1A^2 + a_2A + a_3I$$
$$\vdots$$

and thus

$$E_n = A^{n-1} + a_1A^{n-2} + \cdots + a_{n-1}I \tag{A.76}$$

Thus, multiplying both sides of (A.76) by A, we get

$$AE_n = A^n + a_1A^{n-1} + \cdots + a_{n-1}A \tag{A.77}$$

But, from the last equation in (A.75)

$$AE_n + a_nI = 0 \tag{A.78}$$

Thus, adding a_nI to (A.77) gives the final equation

$$A^n + a_1A^{n-1} + \cdots + a_{n-1}A + a_nI = 0 \tag{A.79}$$

This is the famous relation known as the *Cayley-Hamilton theorem* which some regard as the fundamental theorem of matrix algebra. Note that this equation is the same as the characteristic equation (A.68) with the scalar s^i in the latter replaced by the matrix A^i $(i = 1, 2, \ldots n)$ in (A.78). Thus, another way of stating the Cayley-Hamilton theorem is:

"Every matrix satisfies its own characteristic equation."

If the eigenvalues of A, that is, the zeros of $\mathsf{D}(s)$, are *distinct*, then A does not satisfy any equation of lower than nth degree. If one or more eigenvalues are repeated, however, A may also satisfy an equation of lower than nth degree

$$A^k + \alpha_1A^{k-1} + \cdots + \alpha_kI = 0 \qquad k < n \tag{A.80}$$

If A satisfies an equation of lower degree than n, the matrix is called *derogatory*; the corresponding scalar polynomial

$$M(s) = s^k + \alpha_1s^{k-1} + \cdots + \alpha_k \tag{A.81}$$

is called the *minimum polynomial.* It can be shown that the zeros of the minimum polynomial coincide with the zeros of the characteristic polynomial, that is, $M(s)$ is a factor of the characteristic polynomial. Thus, if A has distinct eigenvalues, the minimum and the characteristic polynomials are identical.

The theory for matrices with repeated eigenvalues is much more involved and luxuriant than the theory for matrices with distinct eigenvalues. The assumption of distinct eigenvalues usually makes proofs and derivations much simpler than they would have to be in the general (i.e., repeated eigenvalues) case.

For each eigenvalue, say s_i, we can find an eigenvector v_i satisfying (A.64)

$$s_i v_i = A v_i$$

If the eigenvalues are distinct, then we can find n eigenvectors

$$s_1 v_1 = A v_1, \quad s_2 v_2 = A v_2, \quad \ldots, \quad s_n v_n = A v_n \tag{A.82}$$

It should be understood that the *directions* of the eigenvectors, but not their lengths, are determined by (A.82), since any vector v_i in (A.82) can be multiplied ("scaled") by any constant without affecting the validity of the equations.

The eigenvector equations (A.82) can be arranged in matrix form

$$[v_1 \mid v_2 \mid \cdots \mid v_n] \begin{bmatrix} s_1 & 0 & \cdots & 0 \\ 0 & s_2 & \cdots & 0 \\ \cdots & \cdots & \cdots & \cdots \\ 0 & 0 & \cdots & s_n \end{bmatrix} = A[v_1 \mid v_2 \mid \cdots \mid v_n]$$

or, more simply,

$$V\Lambda = AV \tag{A.83}$$

where V is the matrix whose columns are the eigenvectors

$$V = [v_1 \mid v_2 \mid \cdots \mid v_n]$$

and Λ is a diagonal matrix

$$\Lambda = \begin{bmatrix} s_1 & 0 & \cdots & 0 \\ 0 & s_2 & \cdots & 0 \\ \cdots & \cdots & \cdots & \cdots \\ 0 & 0 & \cdots & s_n \end{bmatrix} = \operatorname{diag}(s_1, \ldots, s_n) \tag{A.84}$$

It can be shown that the eigenvectors are linearly independent and hence the matrix V is nonsingular. Thus we can express (A.83) as

$$A = V\Lambda V^{-1} \tag{A.85}$$

This equation is known as a "similarity transformation" and we can say that the matrices A and V are "similar." Also note that

$$\Lambda = V^{-1}AV \tag{A.86}$$

In the case of distinct eigenvalues, we can thus say that A is "similar to a diagonal matrix" by a similarity transformation. Another way of saying this is that "A can be transformed to a diagonal matrix" by a similarity transformation.

When the eigenvalues of A are not distinct it is sometimes, but not always, possible to transform A to a diagonal matrix by a similarity transformation. It is always possible, however, to transform A into one of the *canonical* (i.e., standard) forms, such as the *Jordan normal form* or one of the *companion forms* described in Chap. 3. Any pair of matrices A and B related by the transformation

$$B = TAT^{-1} \tag{A.87}$$

where T is any (nonsingular) matrix are said to be "similar" to each other.

Similar matrices have the same characteristic polynomial. This is readily seen with the aid of the Cayley-Hamilton theorem. Let the characteristic equation for A be

$$s^n + a_1 s^{n-1} + \cdots + a_{n-1}s + a_n = 0$$

Then, by the Cayley-Hamilton theorem,

$$A^n + a_1 A^{n-1} + \cdots + a_{n-1}A + a_n I = 0 \tag{A.88}$$

If B is similar to A

$$A = TBT^{-1}$$
$$A^2 = (TBT^{-1})(TBT^{-1}) = TB^2T^{-1}$$
$$\cdots\cdots\cdots\cdots\cdots\cdots\cdots$$
$$A^n = TB^nT^{-1}$$

Then (A.88) becomes

$$T(B^n + a_1 B^{n-1} + \cdots + a_{n-1}B + a_n I)T^{-1} = 0$$

or

$$B^n + a_1 B^{n-1} + \cdots + a_{n-1}B + a_n I = 0 \tag{A.89}$$

Comparing (A.88) and (A.89) we see that A and B have the same characteristic equation.

The coefficients $a_1, a_2, \ldots, a_n$ of the characteristic polynomial are thus invariant under a similarity transformation and are sometimes referred to as the *invariants* of A. By setting s to zero in $D(s)$, as given by (A.67), we see that

$$|-A| = (-1)^n|A| = a_n \tag{A.90}$$

the other coefficients of the characteristic polynomial can be expressed by determinants of submatrices of A. In particular

$$a_1 = -\text{tr}\,(A) = \sum_{i=1}^{n} a_{ii} \tag{A.91}$$

(The symbol tr (A) is read "trace of A" and, as indicated in (A.91), is the sum of the diagonal elements of A.) The other coefficients can be written as

$$a_i = (-1)^i\,\text{tr}_i\,(A)$$

where $\text{tr}_i\,(A)$ signifies the sum of the "principal minors" or A of order i, that is, the determinant of the array obtained by deleting the rows and columns containing $n-i$ diagonal elements; $\text{tr}\,(A) = \text{tr}_i\,(A)$ and $|A| = \text{tr}_n\,(A)$.

The trace of a matrix is used often in matrix analysis. The algorithm given in Chap. 3 for recursive calculation of the resolvent is expressed in terms of traces of matrices.

One frequently useful relation is

$$\operatorname{tr}(AB) = \operatorname{tr}(BA) \tag{A.92}$$

provided that AB and BA are both defined. It is not necessary for A and B to be square matrices, in which case AB and BA will be matrices of different dimensions. For example, if A is a row vector and B is a column vector, i.e.,

$$A = [a_1, \ldots, a_n] \qquad B = \begin{bmatrix} b_1 \\ \vdots \\ b_n \end{bmatrix}$$

Then, from (A.16) and (A.17) we see that

$$\operatorname{tr}(AB) = \operatorname{tr}(BA) = a_1 b_1 + \cdots + a_n b_n$$

This expression is called the *inner-* or *dot-product* of the vectors A and B.

To verify (A.92) note that the (i, j)th element of AB is

$$[AB]_{ij} = \sum_{k=1}^{m} a_{ik} b_{kj}$$

where m is the number of columns of A and rows of B. Thus

$$\operatorname{tr}(AB) = \sum_{i=1}^{n} \sum_{k=1}^{m} a_{ik} b_{ki}$$

where n is the number of rows of A and columns of B. In a similar manner we find

$$\operatorname{tr}(BA) = \sum_{k=1}^{m} \sum_{i=1}^{n} b_{ki} a_{ik}$$

On interchanging the order of summation in these expressions we see that they are identical.

A.9 MATRIX FUNCTIONS AND ANALYSIS

The resolvent $\Phi(s) = (sI - A)^{-1}$ is an example of a matrix function of the (scalar) variable s; in this case it is expressible as the ratio of a matrix polynomial to a scalar polynomial—the characteristic polynomial. The reciprocal of the characteristic polynomial could be expressed as a power series in s, which when multiplied by $\operatorname{adj}(sI - A)$ would result in a power series in A

$$\Phi(s) = C_0 + C_1 s + C_2 s^2 + \cdots$$

More generally, a matrix function of a scalar variable may be conceived of as a power series in that variable with matrix coefficients

$$F(x) = F_0 + F_1 x + F_2 x^2 + \cdots \tag{A.93}$$

The rules of calculus that apply to scalar functions can generally be applied to matrix functions. Thus the derivative of a matrix function is given by

$$\frac{dF(x)}{dx} = F_1 + 2F_2 x + 3F_3 x^2 + \cdots \tag{A.94}$$

and the indefinite integral is

$$\int F(x)\,dx = F_0 x + F_1 \frac{x^2}{2} + F_2 \frac{x^3}{3} + \cdots \tag{A.95}$$

Another way of looking at a matrix function of a scalar is as an array in which each of the elements is a scalar function

$$F(x) = \begin{bmatrix} f_{11}(x) & \cdots & f_{1m}(x) \\ \cdots & \cdots & \cdots \\ f_{n1}(x) & \cdots & f_{nm}(x) \end{bmatrix}$$

From this point of view

$$\frac{dF(x)}{dx} = \begin{bmatrix} \dfrac{df_{11}(x)}{dx} & \cdots & \dfrac{df_{1m}(x)}{dx} \\ \cdots & \cdots & \cdots \\ \dfrac{df_{n1}(x)}{dx} & \cdots & \dfrac{df_{nm}(x)}{dx} \end{bmatrix} \tag{A.96}$$

and

$$\int F(x)\,dx = \begin{bmatrix} \int f_{11}(x)\,dx & \cdots & \int f_{1m}(x)\,dx \\ \cdots & \cdots & \cdots \\ \int f_{n1}(x)\,dx & \cdots & \int f_{nm}(x)\,dx \end{bmatrix} \tag{A.97}$$

Both ways of looking at matrix functions of a scalar variable are consistent. Thus, to obtain the derivative of a function of a matrix, one can either take the derivative of each term or one can express the matrix function as a power series and then use the formula (A.94) for the derivative of a function defined by a power series.

The rules of calculus of scalar functions generally apply to matrix functions, for example,

$$\frac{d}{dx}(A(x)B(x)) = A(x)\frac{dB(x)}{dx} + \frac{dA(x)}{dx}B(x) \tag{A.98}$$

Note that the order in which the matrices appear in (A.90) is important. It would not be correct to interchange $A(x)$ and $dB(x)/dx$ unless these two matrices commute.

To find the derivative of the inverse of a matrix requires some care. Suppose

$$A(x)B(x) = I \tag{A.99}$$

and hence $A(x)$ and $B(x)$ are inverses of each other. Since the derivative of a constant matrix (I, in this case) is zero, we have

$$A(x)\frac{dB(x)}{dx} + \frac{dA(x)}{dx}B(x) = 0$$

or

$$A(x)\frac{dB(x)}{dx} = -\frac{dA(x)}{dx}B(x)$$

On use of (A.99) this becomes

$$\frac{d(A^{-1}(x))}{dx} = -A^{-1}(x)\frac{dA(x)}{dx}A^{-1}(x) \tag{A.100}$$

This is the matrix version of the familiar scalar formula

$$\frac{df^{-1}(x)}{dx} = -\frac{df(x)/dx}{f^2(x)}$$

The matrix functions of greatest importance in the study of linear dynamic systems are the resolvent $(sI - A)^{-1}$ already defined, and the exponential function

$$e^{At} = I + At + A^2\frac{t^2}{2!} + A^3\frac{t^3}{3!} + \cdots \tag{A.101}$$

(The variable t instead of x is used, because this variable is usually *time*, for which t is a more suitable symbol.)

The derivative of the exponential is

$$\begin{aligned}\frac{d}{dt}(e^{At}) &= A + A^2t + A^3\frac{t^2}{2!} + \cdots \\ &= A\left(I + At + A^2\frac{t^2}{2!} + \cdots\right) = A\,e^{At}\end{aligned} \tag{A.102}$$

which is the same as it is in the scalar case. The definite integral

$$\int_0^t e^{A\tau}\,d\tau = It + A\frac{t^2}{2} + A^2\frac{t^3}{3!} + \cdots \tag{A.103}$$

This expression is valid whether or not A is singular. But if A is nonsingular,

(A.103) can be written

$$\int_0^t e^{A\tau}\,d\tau = A^{-1}\left(At + A^2\frac{t^2}{2!} + A^3\frac{t^3}{3!} + \cdots\right)$$

$$= A^{-1}(e^{At} - I) \qquad \text{(A.104)}$$

which is a generalization of the scalar formula

$$\int_0^t e^{a\tau}\,d\tau = \frac{e^{at} - 1}{a}$$

A.10 QUADRATIC FORMS, NORMS, AND SINGULAR VALUES

An expression of the form

$$q = \sum_{j=1}^{n} a_{ij} x_i x_j \qquad \text{(A.105)}$$

is known as a *quadratic form*. In matrix notation, a quadratic form can be written as

$$q = x'Ax = [x_1, \ldots, x_n]\begin{bmatrix} a_{11} & \cdots & a_{1n} \\ \cdots & \cdots & \cdots \\ a_{n1} & \cdots & a_{nn} \end{bmatrix}\begin{bmatrix} x_1 \\ \vdots \\ x_n \end{bmatrix} \qquad \text{(A.106)}$$

Note that the matrix A, in terms of which the quadratic form is expressed, is not unique. The coefficient of the product $x_i x_j$ is $(a_{ij} + a_{ji})$, and another matrix $\bar{A}$ for which $(\bar{a}_{ij} + \bar{a}_{ji}) = (a_{ij} + a_{ji})$ will produce the same quadratic form. To associate a unique matrix with a quadratic form, the matrix A is generally taken to be symmetric ($A' = A$).

A quadratic form q is said to be *positive definite* if and only if

$$q > 0 \qquad \text{for all } x \neq 0 \qquad \text{(A.107)}$$

(Every quadratic form is zero at $x = 0$.)

A quadratic form is said to be *positive semidefinite* if and only if

$$q \geqq 0 \qquad \text{for all } x \qquad \text{(A.108)}$$

A positive semidefinite form can never be negative, but it may go to zero for values of x other than zero. (The regions for which q goes to zero are lines or hyperplanes in the n-dimensional "hyperspace.")

A quadratic form q is said to be *negative definite* if $-q$ is positive definite, and is *negative semidefinite* if $-q$ is positive semidefinite.

A quadratic form that is neither positive nor negative definite or semidefinite is said to be *indefinite*. An indefinite form is positive for some values of x (i.e., for some region of the n-dimensional hyperspace) and negative for other values of x (i.e., for some other region in the space).

When x is a two-dimensional vector, the curves

$$q = a_{11}x_1^2 + 2a_{12}x_1x_2 + a_{22}x_2^2 = c$$

define (nested) ellipses when q is positive definite; parabolas, when q is positive semidefinite; and hyperbolas when q is indefinite, as shown in Fig. A.1. The geometric interpretation generalizes to more than two dimensions. Positive definite quadratic forms correspond to hyperellipsoids; semidefinite forms correspond to paraboloids. The situation with indefinite forms, however, is more complex. The indefinite form

$$q = (x_1 - x_2)^2 - x_3^2$$

generates parabolas in the subspace defined by $\{x_1, x_2, 0\}$ but hyperbolas in the subspace $\{0, x_2, x_3\}$ and hence is a very complex "quadric surface" in the general three-dimensional space.

Consider the transformation of variables

$$x = Ty$$

The quadratic form $q = x'Ax$ can be expressed in terms of y as

$$q = y'T'ATy = y'By$$

The matrix B of the quadratic form $y'By$ is related to the original matrix by the relationship

$$B = T'AT \tag{A.109}$$

Matrices related to each other by a transformation such as given by (A.109) are said to be *congruent.* If A is real then it is always possible to find an orthogonal matrix T, that is, $T' = T^{-1}$ as defined by (A.52), for the transformation, and moreover, that B is a diagonal matrix. Thus, if A is symmetric, we can also say it is *orthogonally similar* to a diagonal matrix

$$\Lambda = T^{-1}AT = T'AT \tag{A.110}$$

The quadratic form $q = x'Ax$ is transformed by T to the quadratic form

$$q = y'\Lambda y = [y_1 \cdots y_n] \begin{bmatrix} s_1 & 0 & \cdots & 0 \\ 0 & s_2 & \cdots & 0 \\ \cdots & \cdots & \cdots & \cdots \\ 0 & 0 & \cdots & s_n \end{bmatrix} \begin{bmatrix} y_1 \\ y_2 \\ \vdots \\ y_n \end{bmatrix}$$

$$= s_1y_1^2 + s_2y_2^2 + \cdots + s_ny_n^2 \tag{A.111}$$

We see immediately that if A is real (and, as already assumed, symmetric) it must have real eigenvalues, and the nature of these eigenvalues determine the nature of the quadratic form, as summarized in Table A1.

Quadratic forms are used extensively as the performance integrands in optimum control problems. (See Chap. 9.) In addition, they are useful in defining the norm of a matrix, which is a measure of the "size" of the matrix

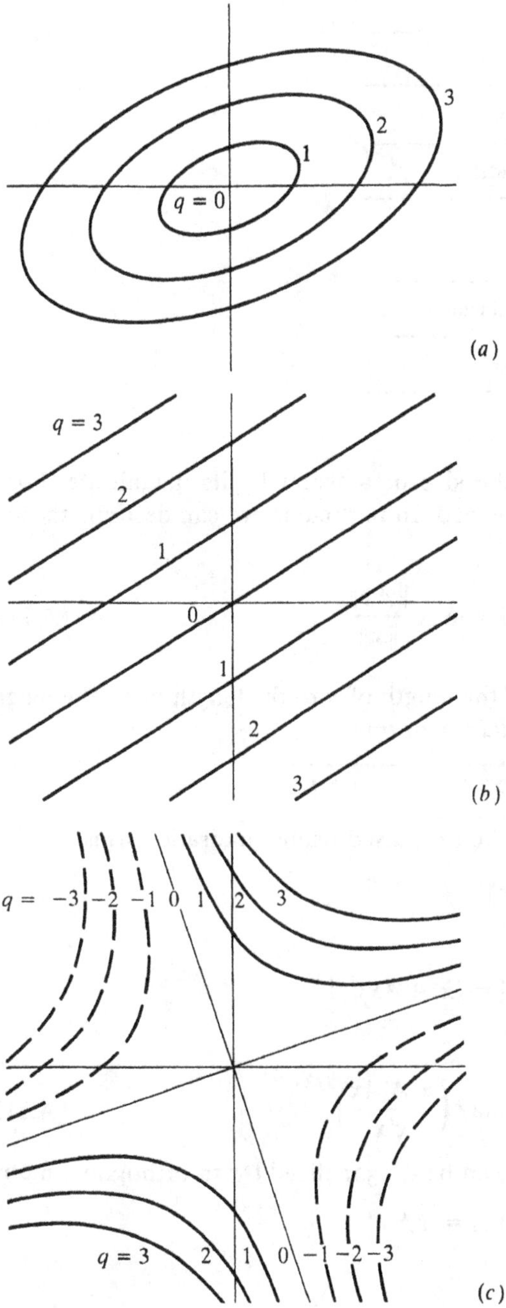

Figure A1 Three types of quadratic forms. (*a*) Positive definite; (*b*) Positive semidefinite; (*c*) Indefinite.

Table A1 Eigenvalues of A determine nature of quadratic form

Eigenvalues	Quadratic form
$s_i > 0$, all i	Positive definite
$s_i \geqq 0$, all i	Positive semidefinite
$s_i \geqq 0$, some i $s_i \leqq 0$, some i	Indefinite
$s_i \leqq 0$ for all i	Negative semidefinite
$s_i < 0$ for all i	Negative definite

(not its *dimension*). We describe the size of a scalar by its magnitude. For a matrix, a more general concept is needed. In particular, we can describe the size A by its *norm*

$$\|A\| = \max_x \frac{\|Ax\|}{\|x\|} \tag{A.112}$$

i.e., the largest value of the ratio of the length of y to the length of x. The length $\|x\|$ of a vector x is defined as *euclidean* norm

$$\|x\| = \sqrt{x_1^2 + x_2^2 + \cdots + x_n^2}$$

as expected. The ratio (A.112) can be expressed using quadratic forms

$$\|x\| = (x'x)^{1/2}$$

and

$$\|Ax\| = (x'A'Ax)^{1/2}$$

Thus

$$\|A\| = \max_x \left(\frac{x'A'Ax}{x'x}\right)^{1/2} \tag{A.113}$$

Since $A'A$ is a symmetric matrix it can be diagonalized by an orthogonal matrix

$$A'A = T\Lambda T'$$

Let

$$x = Ty$$

then

$$x'A'Ax = y'T'A'ATy = y'\Lambda y$$

and

$$x'x = y'T'Ty = y'y$$

because $T' = T^{-1}$. Thus (A.113) becomes

$$\|A\| = \max_y \left(\frac{y'\Lambda y}{y'y}\right)^{1/2} = \max_y \left(\frac{s_1 y_1^2 + s_2 y_2^2 + \cdots + s_n y_n^2}{y_1^2 + y_2^2 + \cdots + y_n^2}\right)^{1/2} \tag{A.114}$$

where $s_1, s_2, \ldots, s_n$ are the eigenvalues of $A'A$. It is an easy matter to show that the maximum value of the ratio in (A.114) is s_n. Thus we conclude that

$$\|A\| = |\text{maximum eigenvalue of } A'A|^{1/2} \tag{A.115}$$

The square roots of the eigenvalues of $A'A$ (or for $A^H A$ when A is complex) are called the *singular values* of A and are useful in numerical analysis. The norm of A is the largest singular value. The ratio of the largest to the smallest singular value of A is a measure of how close the matrix A comes to being singular. This is used in robustness analysis. (See Chap. 4.)

BIBLIOGRAPHY

Airy, G. B., "On the Regulator of the Clockwork for Effecting Uniform Movement of the Equatoreals," *Memoirs of the Royal Astronomical Society*, vol. 11, 1840, pp. 249-267.

Åström, K. J., *Introduction to Schostic Control Theory*, Academic Press, New York, 1970.

Athans, M., and Falb, P. L., *Optimal Control: An Introduction to the Theory and its Applications*, McGraw-Hill Book Co., New York, 1966.

Bass, R. W., and Gura, I., "High-Order System Design Via State-Space Considerations," *Proc. Joint Automatic Control Conf., Troy, NY*, June 1965, pp. 311-318.

Bass, R. W., "Robustified LQG Synthesis to Specifications," *Proc. 5th Meeting of Coord. Group On Modern Control Theory, Part II, Dover, NJ*, October 1983, pp. 11-93.

Bergen, A. R., and Ragazzini, J. R., "Sampled-Data Processing Techniques for Feedback Control Systems," *Trans. AIEE, Pt. II*, vol. 73, no. 6, November 1954, pp. 236-244.

Black, H. S., "Inventing the Negative Feedback Amplifier," *IEEE Spectrum*, vol. 14, no. 1, January 1977, pp. 54-60.

Bode, H. W., *Network Analysis and Feedback Amplifier Design*, D. Van Nostrand, New York, 1945.

Bode, H. W., and Shannon, C. E., "A Simplified Derivation of Least Squares Smoothing and Prediction Theory," *Proceedings of the IRE*, vol. 38, no. 4, April 1950, pp. 481-492.

Brockett, R. W., *Finite Dimensional Linear Systems*, John Wiley & Sons, New York, 1965.

Bryson, A. E., Jr., and Johansen, D. E., "Linear Filtering for Time-Varying Systems Using Measurements Containing Colored Noise," *IEEE Trans. on Automatic Control*, vol. AC-10, no. 1, January 1965, pp. 4-10.

Bryson, A. E., Jr., and Ho, Y.-C., *Applied Optimal Control*, Blaisdell Publishing Co., Waltham, MA, 1969.

Chang, S. S. L., *Synthesis of Optimum Control Systems*, McGraw-Hill Book Co., New York, 1961.

Churchill, R. V., *Introduction to Complex Variables and Applications*, McGraw-Hill Book Co., New York, 1948.

Coddington, E. A., and Levinson, N., *Theory of Ordinary Differential Equations*, McGraw-Hill Book Co., New York, 1955.

D'Azzo, J. J., and Houpis, C. H., *Linear Control System Analysis and Design: Conventional and Modern*, McGraw-Hill Book Co., New York, 1981.

Doob, J. L., *Stochastic Processes*, John Wiley & Sons, New York, 1953.

Doyle, J. C., and Stein, G., "Robustness with Observers," *IEEE Trans. on Automatic Control*, vol. AC-24, no. 4, August 1979, pp. 607-611.

Dynkin, E. B., *Markov Processes*, Springer-Verlag, New York, 1965.

Etkin, B., *Dynamics of Flight*, John Wiley & Sons, New York, 1959.

Evans, W. R., "Graphical Analysis of Control Systems," *Trans. AIEE, Pt. II*, vol. 67, 1948, pp. 547-551.

Farrell, J., *Integrated Aircraft Navigation*, Academic Press, New York, 1976.

Fisher, D. G., and Seborg, D. E., "Model Development, Reduction, and Experimental Evaluation for an Evaporator," *Ind. Eng. Chem. Process Design and Development*, vol. 11, no. 2, February 1972, pp. 213-221.

Fisher, D. G., and Seborg, D. E., "Advanced Computer Control Improves Process Performance," *Instrumentation Technology*, vol. 20, no. 9, September 1973, pp. 71-77.

Fisher, D. G., and Seborg, D. E. (ed.), *Multivariable Computer Control: A Case Study*, North-Holland Publishing Co., Amsterdam, 1976.

Friedland, B., "Controllability Index Based On Conditioning Number," *Trans. ASME (J. Dynamic Systems, Measurement & Control)*, vol. 97G, no. 4, December 1975, pp. 444-445.

Friedland, B., "On The Calibration Problem," *IEEE Trans. on Automatic Control*, vol. AC-22, no. 6, December 1977, pp. 899-905.

Friedland, B., "Limiting Forms of Optimum Stochastic Linear Regulators," *Trans. ASME (J. Dynamic Systems, Measurement & Control)*, vol. 93G, no. 3, September 1971, pp. 134-141.

Fuller, A. T., "The Early Development of Control Theory," *Trans. ASME (J. Dynamic Systems, Measurement & Control)*, vol. 96G, no. 2, June 1976, pp. 109-118.

Fuller, A. T., "The Early Development of Control Theory, II," *Trans. ASME (J. Dynamic Systems, Measurement & Control)*, vol. 98G, no. 3, September 1976, pp. 224-235.

Gantmacher, F. R., *The Theory of Matrices*, 2 vol., Chelsea Publishing Company, New York, 1959.

Gilles, E. D., and Retzbach B., "Reduced Models and Control of Distillation Columns with Sharp Temperature Profiles," *IEEE Trans. on Automatic Control*, vol. AC-28, no. 5, May 1960, pp. 628-630.

Gilles, E. D., Retzbach, B., and Silberberger, F., "Modeling, Simulation, and Control of an Extractive Distillation Column," *Computer Applications to Chem. Eng.* (ACS Symp. ser. no. 124), 1980, pp. 481-492.

Goldstein, H., *Classical Mechanics*, Addison-Wesley Publishing Co., Reading MA, 1953.

Gunckel, T. L., III, and Franklin, G. F., "A General Solution for Linear Sampled Data Control," *Trans. ASME (J. Basic Engineering)*, vol. 85D, January 1963, pp. 197-201.

Hooke, R., "Lampas, or Description of Some Mechanical Improvements of Lamps and Waterpones Together with Some Other Physical and Mechanical Discoveries," *Proc. Royal Society (London)*, vol. 8, 1831.

Householder, A. S., *The Theory of Matrices in Numerical Analysis*, Blaisdell Publishing Co., Waltham, MA, 1964.

Hurwitz, A., "Über die Bedingungen, unter welchen einer Gleichung nur Wurzeln mit negativen reelen Teilen besitzt," *Math. Ann.*, vol. 146, 1895, pp. 273-284.

Hutton, M. F., "Solutions of the Singular Stochastic Regulator Problem," *Trans. ASME (J. Dynamic Systems, Measurement & Control)*, vol. 95G, no. 4, December 1973, pp. 414-417.

Huygens, C., *Horologium Oscillatorium* (1673) in Oeuvres complètes, Nijhoff, Amsterdam, vol. 17-18, December 1932.

James, H. M., Nichols, N. B., and Phillips, R. S., *Theory of Servomechanisms* (MIT Radiation Laboratory Series, Vol. 25), McGraw-Hill Book Co., New York, 1947.

Jazwinski, A. H., *Stochastic Processes and Filtering Theory*, Academic Press, New York, 1970.

Joseph, P. D., and Tou, J. T., "On Linear Control Theory," *Trans. AIEE, Pt. II*, vol. 80, no. 3, September 1960, pp. 193-196.

Kailath, T., "An Innovations Approach to Least-Squares Estimation, Part I: Linear Filtering in Additive White Noise," *IEEE Trans. on Automatic Control*, vol. AC-13, no. 6, December 1968, pp. 646-655.

Kailath, T., *Linear Systems*, Prentice-Hall Inc., Englewood Cliffs, NJ, 1980.

Kalman, R. E., "A New Approach to Linear Filtering and Prediction Problems," *Trans. ASME (J. Basic Engineering)*, vol. 82D, no. 1, March 1960, pp. 35-45.

Kalman, R. E., "On the General Theory of Control Systems," *Proc. First International Congress, IFAC, Moscow, USSR*, 1960, pp. 481-492.

Kalman, R. E., "Contributions to the Theory of Optimal Control," *Proc. 1959 Mexico City Conf. on Differential Equations, Mexico City*, 1960, pp. 102-199

Kalman, R. E., "Mathematical Description of Linear Dynamic Systems," *SIAM J. on Control*, ser. A., vol. 1, no. 2, 1963, pp. 152-192.

Kalman, R. E., "When Is a Linear Control System Optimal?," *Trans. ASME (J. Basic Engineering)*, vol. 86D, no. 1, March 1964, pp. 51-60.

Kalman, R. E., and Bucy, R. S., "New Results in Linear Filtering and Prediction Theory," *Trans. ASME (J. Basic Engineering)*, vol. 83D, no. 1, March 1961, pp. 95-108.

Kalman, R. E., Ho, Y. C., and Narendra, K. S., "Controllability of Linear Dynamic Systems," *Contributions to Differential Equations*, vol. I, no. 2, 1963, pp. 189-213.

Klein, F., and Sommerfeld, A., *Über die Theorie des Kreisels* (2 vol.), Teubner, Leipzig, 1897.

Kleinman, D. L., "On an Iterative Technique for Riccati Equation Computations," *IEEE Trans. on Automatic Control*, vol. AC-13, no. 1, February, 1968, pp. 114-115.

Kushner, H. J., *Introduction to Stochastic Control Theory*, Holt, Rinehart & Winston, New York, 1971.

Kwakernaak, H., and Sivan, R., *Linear Optimal Control Systems*, Wiley-Interscience, New York, 1972.

Lehtomaki, N. A., Sandell, N. R., Jr., and Athans, M., "Robustness Results on LQG Based Multivariable Control System Designs," *IEEE Trans. on Automatic Control*, vol. AC-26, no. 1, February 1981, pp. 75-93.

Locke, A. S. (ed.), *Guidance*, D. Van Nostrand, New York, 1955.

Loh, N. K., Cheok, K. C., and Beck, R. R., "Modern Control Design for a Gun-Turret Control," *Southcon/83 Convention Record, Atlanta GA*, paper no. 1/5, 1983.

Luenberger, D. G., "Observing the State of a Linear System," *IEEE Trans. on Military Electronics*, vol. MIL-8, April 1964, pp. 74-80.

Luenberger, D. G., "Observers for Multivariable Systems," *IEEE Trans. on Automatic Control*, vol. AC-11, no. 2, April 1966, pp. 190-197.

Luenberger, D. G., "An Introduction to Observers," *IEEE Trans. on Automatic Control*, vol. AC-16, no. 6, December 1971, pp. 596-602.

Lyapunov, M. A., "Le problème général de la stabilité du mouvement," *Ann. Fac. Sci. Toulouse*, vol. 9, 1907, pp. 203-474.

MacFarlane, A. J. G. (ed.), *Frequency Response Methods in Control Systems*, IEEE Press, New York, 1979.

Madiwale, A. N., and Williams, D. E., "Some Extensions of Loop Transfer Recovery," *Proc. American Control Conference, Boston, MA*, June 1985, pp. 790-795.

Mason, S. J., "Feedback Theory: Some Properties of Signal Flow Graphs," *Proceedings of the IRE*, vol. 41, no. 9, September 1953.

Maxwell, J. C., "On Governors," *Philosophical Magazine*, vol. 35, 1868, pp. 385-398.

Middleton, D., *An Introduction to Statistical Communication Theory*, McGraw-Hill Book Co., New York, 1960.

Minorsky, N., *Nonlinear Oscillations*, D. Van Nostrand, New York, 1962.

Moore, B. C., "Principal Component Analysis in Linear Systems: Controllability, Observability, and Model Reduction," *IEEE Trans. On Automatic Control*, vol. AC-26, no. 1, February 1981, pp. 17-32.

Nyquist, H., "Regeneration Theory," *Bell System Technical Journal*, vol. 11, 1932, pp. 126-147.

Ogata, K., *Modern Control Engineering*, Prentice-Hall Inc., Englewood Cliffs, NJ, 1970.

Oppelt, W., "A Historical Review of Autopilot Development, Research, and Theory in Germany," *Trans. ASME (J. Dynamic Systems, Measurement & Control)*, vol. 98G, no. 3, September 1976, pp. 215-223.

Papoulis, A., *Probability, Random Variables, and Stochastic Processes*, John Wiley & Sons, New York, 1965.

Parks, P. C., "A New Proof of the Routh-Hurwitz Stability Criterion Using the Second Method of Lyapunov," *Proc. Cambridge Philosophical Society*, vol. 58, pt. 4, 1962, pp. 694–702.

Retzbach, B., "Einsatz von systemtechnische Methoden am Beispeil einer Mehrstoffdestillation," *Chem.-Ing.-Techn.*, vol. 55, no. 3, 1983, p. 235.

Richman, J., and Friedland, B., "Deisgn of Optimum Mixer-Filter for Aircraft Navigation Systems," *Proc. NAECON, Dayton, OH*, May 1967, pp. 429–438.

Rosenbrock, H. H., *Computer Aided Control System Design*, Academic Press, New York, 1974.

Routh, E. J., *A Treatise on the Stability of a Given State of Motion*, Macmillan & Co., London, 1877.

Rynaski, E. G., "Flight Control Synthesis Using Robust Output Observers," *Proc. AIAA Guidance and Control Conference, San Diego, CA*, September 1982, pp. 825–831.

Safonov, M. G., *Stability and Robustness of Multivariable Feedback Systems*, MIT Press, Cambridge, MA, 1980.

Schuler, M., "Die Störung von Pendel- und Kreiselapparaten durch die Beschleuningung des Fahrzeuges," *Physicalische Zeitschrift*, vol. 24, 1923, pp. 344–350.

Schwarz, R. J., and Friedland, B., *Linear Systems*, McGraw-Hill Book Co., New York, 1965.

Seckel, E., *Stability and Control of Airplanes and Helicopters*, Academic Press, New York, 1964.

Simon, H. A., "Dynamic Programming Under Uncertainty With a Quadratic Criterion Function," *Econometrica*, vol. 24, 1956, pp. 74–81.

Sorenson, H. W., "Least-Squares Filtering: From Gauss to Kalman," *IEEE Spectrum*, vol. 7, no. 7, July 1970, pp. 63–68.

Special Issue on Linear Multivariable Control Systems, *IEEE Trans. on Automatic Control*, vol. AC-26, no. 1, February 1981.

Special Issue: Three-Mile Island and the Future of Nuclear Power, *IEEE Spectrum*, vol. 16, no. 11, November 1979.

Synge, J. L., and Griffith, B. A., *Principles of Mechanics*, McGraw-Hill Book Co., New York, 1949.

Wax, N. (ed.), *Selected Papers on Noise and Stochastic Processes*, Dover Publications, New York, 1954.

Wiener, N., *The Extrapolation, Interpolation, and Smoothing of Stationary Time Series*, John Wiley & Sons, New York, 1949.

Wonham, W. M., "On The Separation Theorem of Stochastic Control," *SIAM J. on Control*, vol. 6, no. 2, 1968, pp. 312–326.

Wonham, W. M., *Linear Multivariable Control: a Geometric Approach*, 2d ed., Springer-Verlag, New York, 1979.

INDEX OF APPLICATIONS

INDEX